Electroceramic Materials and Applications

Electroceramic Materials and Applications

Ceramic Transactions Series, Volume 196

Proceedings of the 6th Pacific Rim Conference on Ceramic and Glass Technology (PacRim6); September 11–16, 2005; Maui, Hawaii

Edited by
R. W. Schwartz

The American Ceramic Society

WILEY-INTERSCIENCE

A JOHN WILEY & SONS, INC., PUBLICATION

Published by John Wiley & Sons, Inc., Hoboken, New Jersey
Published simultaneously in Canada.

Library of Congress Cataloging-in-Publication Data is available.

ISBN-13 978-0-470-08295-9
ISBN-10 0-470-08295-X

10 9 8 7 6 5 4 3 2 1

Contents

Preface

This volume is a collection of papers presented at the 6th PAC RIM meeting of the American Ceramic Society in Kapalua, Maui, Hawaii, September 11–16, 2005. The program for the Electronics Division at this meeting consisted of six topical symposia, including Emerging Advances in Electronic, Optical, Magnetic, and Elastic Ferroics, Scanning Probe Microscopy of Electronic Ceramics, Thermoelectric and Energy Harvesting Materials for Solid State Power Conversion, Chemical Methods of Preparation of Electroceramics, The International Symposium on Transparent Conducting Oxides, and General Topics in Electroceramics. In addition to these symposia, the Electronics Division also co-sponsored symposia on Advanced Ceramics for Clean Energy Applications, Novel Processing of Ceramics, Thermochemistry and Metrology of Interfacial Interactions, and Processing of High Temperature Superconductors. In all, more than 225 papers affiliated with the Electronics Division were presented during this international meeting, which was attended by more than 750 scientists from around the world. All papers included in this volume were reviewed by experts in the field.

The editor gratefully acknowledges the important contributions of the following individuals who assisted with the organization of the symposia and review of the manuscripts: David Cann, Paul Clem, Fatih Dogan, Ryoji Funahashi, Venkatram Gopalan, Amit Goyal, Tetsuya Hasegawa, Hideo Hosono, Kazumi Kato, Ramamoorthy Ramesh, Jim Smay, Ichiro Taguchi, Terry Tritt, Bruce Tuttle, Dwight Viehland, Winnie Wong-Ng, and Jihui Yang. The editor would also like to acknowledge the staff of The American Ceramic Society for their assistance in the preparation of this proceedings volume.

R. W. Schwartz
Department of Materials Science and Engineering
University of Missouri—Rolla

Electroceramic Materials and Applications
Edited by R. W. Schwartz

High Temperature Superconductors: Processing and Properties

Electroceramic Materials and Applications
Edited by R. W. Schwartz

EFFECT OF INITIAL PHASES IN PRECURSOR ON THE FORMATION AND PROPERTIES OF Bi2223 SHEATHED WITH Ag-Cu ALLOY

Yuichi Nakamura[1], Sohei Nakashima[1,2], Naoyuki Torii[1], Manabu Saigo[1], Ryoji Inada[1] and Akio Oota[1]
[1] Toyohashi University of Technology
1-1 Tempaku-cho
Toyohashi, Aichi 441-8580, Japan

[2] Present address : NTT FACILITIES
Atsuta Media Wing, 7-30 Gohonmatsu-cho, Atsuta-ku
Nagoya, Aichi 456-0016, Japan

ABSTRACT

Ag sheathed Bi2223 tapes are required to have sufficient mechanical strength for withstanding the mechanical forces. The use of alloy sheath is effective to improve the mechanical strength. However, when the alloy sheath directly contacts with filaments, superconducting properties are usually degraded due to undesirable reactions between alloy element(s) and filaments such as contamination and formation of impurity phases. The Bi2223 tapes sheathed with Ag-Cu alloys were fabricated using several types of precursor. It was found that the J_c property could be recovered by reducing the Cu contents of precursor powder while the formation and growth of large $(Sr,Ca)_{14}Cu_{24}O_x$ (14:24) particles limited the recovery of J_c. This formation and growth of large 14:24 particles was found to occur in early stage of sintering. This means that the reaction to form the 14:24 phase is faster than the formation reaction of Bi2223 from Bi2212, Ca_2CuO_3 (2:1) and CuO phases consisting the precursor. Although the precursor containing fine 14:24 particles as an initial phase instead of the 2:1 phase is used to suppress the enlargement of 14:24 particles, the formation of large 14:24 particles could not suppressed. This suggests that the formation of the 14:24 phase may attribute to the decomposition of Bi2212 phase under this average composition of precursor and the sintering condition. The selection of proper composition to stabilize the Bi2212 phase under the high Cu potential in Ag-Cu alloy sheaths may suppress the formation and enlargement of 14:24 particles, and will improve the J_c values.

INTRODUCTION

The Ag-sheath $(Bi,Pb)_2Sr_2Ca_2Cu_3O_x$ (Bi2223) superconducting tapes by the powder-in-tube (PIT) method are the most applicable high-T_c wires for 77K usage in the present stage while the 123 coated conductors are under developments. Fabrication of the Bi2223 tapes consists of first calcination of the precursor powder that is packed in an Ag or Ag-alloy tube, and repeated deformation and sintering process to form the *c*-axis oriented Bi2223 phase. Applying these Bi2223 tapes to actual applications, the Ag sheathed Bi2223 tapes are required to have sufficient mechanical strength for withstanding mechanical forces in various power applications. The use of alloy sheath is effective to improve the mechanical strength. Many Ag-alloy sheath materials such as Ag-Cu [1-5], Ag-Mg [6] and Ag-Al [5] have been examined to achieve high mechanical strength. However, when the Ag-alloys are used for sheath material directly contacting superconducting filaments, the superconducting properties of the tapes are degraded. Such the degradation is caused by the undesired reaction between the sheath with alloy element(s) and Bi2223 filaments, which induces the contamination and/or the formation of impurity phases in the filaments. The formation of impurity particles decreases the volume fraction of the Bi2223 phase and blocks the transport current pass in the filaments. Ishizuka et al. [2,3] reported the effects of the starting composition and the additives for Ag-Cu alloy sheath on the superconducting properties. Aloysius et al. [4] also reported the effect of initial Cu composition with respect to the Cu stoichiometry in the Ag-Cu alloy sheath tape. They reported the combination of Cu compositions in superconducting core and alloy sheath is important to improve the J_c value of the alloy sheathed tapes. This means the control of chemical potential between the core and alloy sheath is important to suppress the reaction and to keep the core stoichiometry. In our previous work [7], we fabricated the Bi2223 tapes sheathed with Ag-Cu alloys using Cu deficient precursors. The microstructural observation of the tapes revealed that large $(Sr,Ca)_{14}Cu_{24}O_x$ (14:24) particles were formed preferably near the interface between Ag-Cu alloy sheath and oxide filaments. These particles disturbed the crystal orientation of the Bi2223 phase and decreased the J_c values. This reduction of J_c property could be recovered by controlling the initial Cu contents of precursor powder and the thickness of the filaments, while the formation and growth of large 14:24 particles still limited the recovery of J_c. Hence, the suppression of the formation and growth of large 14:24 particles is required to achieve the high J_c values for Ag-Cu alloy sheathed tape. In this study, the effects of initial composition and the consisting phases of precursor power on the phase development of the Bi2223 multifilamentary tapes sheathed with Ag-Cu alloy are investigated to reduce the size and volume of the 14:24 phases and to achieve the high J_c.

EXPERIMENTAL

2223 precursors were prepared with two-powder process [8], in which 2212 and Ca-Cu-O precursors were mixed. In this study, several types of the precursor powder were used. The average compositions of the 2223 precursor were selected to be the cation ratios of Bi:Pb:Sr:Ca:Cu = 1.7:0.3:1.9:2.0:3.0-x. To see the effect of Cu content of the precursor, nominal composition of $x = 0$ and several Cu-poor compositions of $x = 0.2$, 0.4, 0.6 and 0.8 were used for Ag-Cu alloy sheathed tapes. The standard 2223 precursors A are the mixture of the 2212 powder with the composition of Bi:Pb:Sr:Ca:Cu = 1.7:0.3:1.9:1.0:2.0 calcined at 825 °C and the Ca-Cu-O powder with the composition of Ca:Cu = 1:1-x (x = 0~0.8) calcined at 950 °C. The Ca-Cu-O powder consists of Ca_2CuO_3 (2:1 phase) and CuO phases. On the other hand, the fine 14:24 phase based precursor instead of the standard Ca-Cu-O powder was also used to see the effect of the initial constituent phases of precursor. In this precursor B, the 2212 powder with the composition of Bi:Pb:Sr:Ca:Cu = 1.7:0.3:1.6:1.7:1.6 and the 14:24 phase powder with the average size of 1 μm having the composition of Sr:Ca:Cu = 7:7:24 were mixed to have the average composition same as that of the standard precursor A of x=0.4.

All Bi2223 tapes were fabricated by standard PIT method. The Ag-4.2%Cu alloy, that was a commercially used alloy in the Ag-Cu system, was selected as the sheath material. Each precursor powder was packed into the Ag-4.2wt%Cu alloy tube and then drawn down as a wire. These wires were stacked into the Ag-4.2wt%Cu alloy tube and then drawn down again to form multifilamentary wires. In this experiment, we rolled these wires and fabricated 61 filaments Bi2223 green tapes sheathed with the Ag-4.2wt%Cu alloy. For comparison, pure Ag sheathed tape was also fabricated by the same method using the standard precursor A with the normal composition (x=0). These tapes were sintered at 840 °C for 50h in air and rolled again to the final thickness of about 0.17~0.29mm. After the second rolling step, every tape was finally sintered at 840 °C for 150h in air. To see the phase development during first sintering process, some of the tapes were quenched in air after predetermined sintering time.

Transport critical current density (J_c) properties were evaluated at 77K self field by the standard DC 4-probe method with the criterion of 1 μV/cm. X-ray diffraction (XRD) analysis was performed for phase identification after etching out the sheath material using a mixture of H_2O_2 and NH_3OH. The microstructure of samples was observed by a scanning electron microscopy (SEM), and the compositional analysis was performed with an energy dispersive X-ray (EDX).

RESULTS AND DISCUSSION

Figure 1 shows the result of XRD measurements of the Ag-Cu alloy sheathed tapes using the nominal composition (x=0) and the Cu deficient compositions (x=0.4 and 0.8) after the final sintering. The peak intensities of unreacted impurity phases such as 14:24 and Bi2212 phases decreased in the tapes using the Cu deficient precursors comparing to the tape using the nominal composition. This means that the formation of impurity 14:24 phases could be suppressed by using Cu deficient precursors to balance the Cu supply form the Ag-Cu alloy sheath. However, the intensities of unreacted impurity phases were not decreased by further reduction of the Cu content in precursor from x=0.4 to 0.8. The microstructures of these tapes were observed with SEM to confirm the difference in microstructure by the change in initial Cu contents. Figure 2 shows the microstructure of longitudinal cross sections of the tapes sheathed with the Ag-Cu

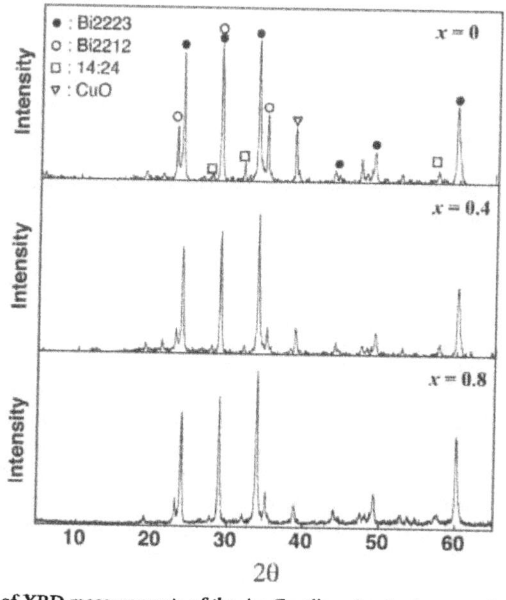

Fig. 1 The results of XRD measurements of the Ag-Cu alloy sheathed tapes using the standard precursor A with (a) the nominal composition (x=0) and the Cu deficient compositions of (b) x=0.4 and (c) x=0.8 after the final sintering.

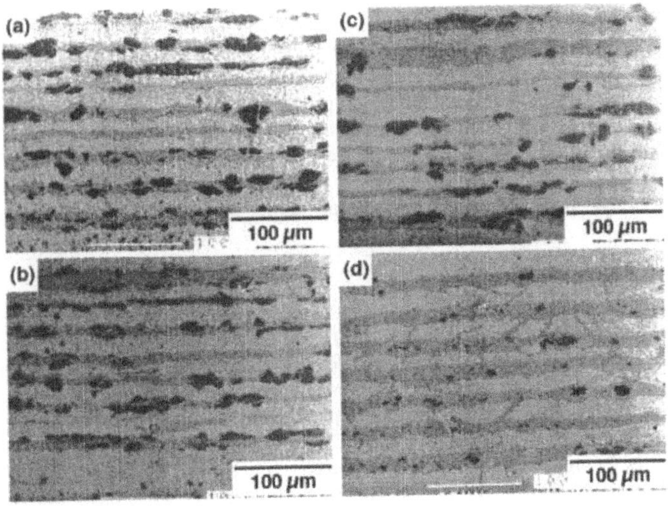

Fig. 2 the microstructure of longitudinal cross sections of the tapes sheathed with Ag-Cu alloy using the standard precursor A with (a) the nominal composition (x=0) and the Cu deficient compositions of (b) x=0.4 and (c) x=0.8 after the final sintering, and (d) sheathed with pure Ag.

alloy, where light matrices are the Ag-alloy sheath, light gray regions are the filaments, and dark gray particles in filaments are the 14:24 phase, respectively. For comparison, the microstructure of the tape sheathed with pure Ag is also shown in Fig. 2 (d). As shown in this figure, there are many 14:24 particles, which have cation ratios close to Sr:Ca:Cu = 7:7:24 from EDX analysis, in the filaments of alloy sheathed tapes. The amount of 14:24 particles in the alloy sheathed tapes is very large comparing to that in the tape sheathed with pure Ag. In addition, the size of those particles is about the same as the filament thickness of about 10 μm and sufficient to interrupt the superconducting current path. Similar to the result of XRD measurements shown in Fig. 1, the amount of 14:24 phase in the tape slightly decreased by using the Cu deficient compositions of x=0.4 and 0.8 in comparison with the sample using the nominal composition of x=0. These results mean that the reaction to form the 14:24 phase is enhanced by the Cu element of the sheath material. This enlargement of the 14:24 particles makes the formation reaction of the Bi2223 phase difficult by decreasing the area of interface between 14:24 phase and another phases, and many unreacted 14:24 particles will remain after the final sintering.

Figure 3 shows the effect of the Cu contents in precursor on the J_c property of the alloy sheathed tapes. As shown in Fig. 3, the J_c value of the tapes increased as the Cu content of precursor decreased from x=0 to x=0.4, but further improvement of J_c was not attained by the reduction of Cu more than x=0.4. This behavior of J_c values corresponds to the microstructural changes as shown in Figs. 1 and 2. These results suggest that the J_c value of the Ag-Cu alloy sheathed tapes could be partially improved by the control of the Cu content in precursor through the suppression of the formation of 14:24 particles in the filaments. This reduction of Cu content in precursor is, however, not sufficient for the further improvement of microstructure and for

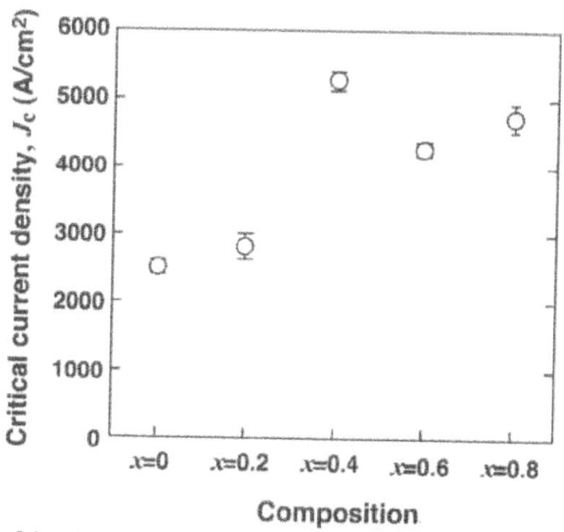

Fig. 3 J_c property of the alloy sheathed tapes using the different Cu contents in precursor.

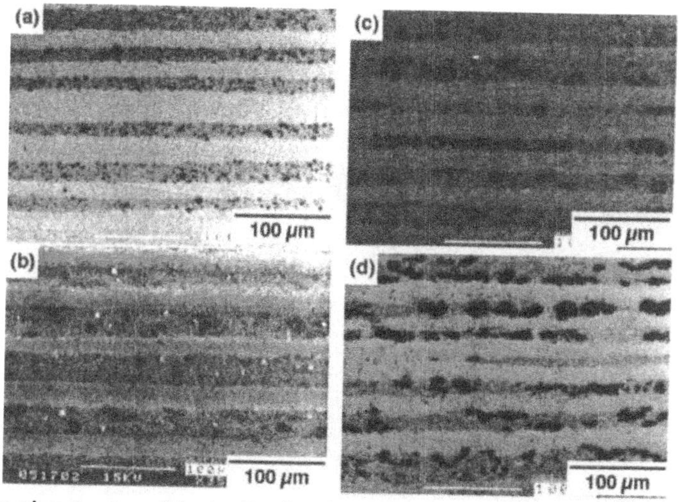

Fig. 4 the microstructure of the Ag-Cu alloy sheathed tapes with the precursor of x=0.4. (a) before sintering, after (b) 9 h, (c) 11 h, and (d) 50 h sintering. Small black particles in (a) are the 2:1 and CuO phases. Dark large particles in (b), (c) and (d) are the 14:24 phase.

attaining enough high J_c value of the Ag-Cu alloy sheathed tapes; that is, the formation of large 14:24 particles must be suppressed for the further improvement.

To clarify the formation and growth process of the 14:24 particles, the development of microstructure during the first sintering was observed. Figure 4 shows the microstructure development of the Ag-Cu alloy sheathed tapes with the precursor of x=0.4. Fig. 4 (a) shows the longitudinal cross section of the green tape, where black particles in filaments are the 2:1 and CuO phases. As shown in Fig. 4 (a), the 2:1 and CuO particles with a few μm in size are finely dispersed in the precursor filaments of the green tape. Figs. 4 (b), (c) and (d) show the microstructures of the tapes sintered for 9, 11 and 50 h, respectively. In these photos, dark particles with about 10 μm are the 14:24 particles. As shown in these photos, the 14:24 phase particles began to form after 9 h of sintering, and many 14:24 particles were grown to a size comparable to filament thickness after 11h sintering. This formation and growth of the 14:24 particles is thought to be an intermediate product of alkaline-earth cuprate (AEC) from the 2:1 phase to form the Bi2223 phase. The 14:24 phase should be, therefore, more stable than the 2:1 phase in this average composition and sintering condition since there is no 14:24 particle in the precursor. Under this condition, once the 14:24 phase particle nucleates in the filament, the growth of the nucleated particle is easier than the nucleation of another 14:24 phase particles due to the difference in kinetics factors. Hence, the nucleated 14:24 particles may easily grow larger and larger. Based on this consideration, if the enlargement of the 14:24 particles arises by the decomposition of the 2:1 phase, the enlargement of the 14:24 phase may be suppressed by losing the 2:1 phase instead using fine 14:24 particles in precursor.

Figure 5 shows the phase development measured by the XRD analysis of the alloy sheathed tapes using the precursor B containing the fine 14:24 particles. This precursor B includes phases such as 2201 and Ca_2PbO_4 (CP) phases in addition to Bi2212, 14:24 and CuO phases to compensate the stoichiometry as shown in Fig. 5 (a). After the first sintering of 50 h,

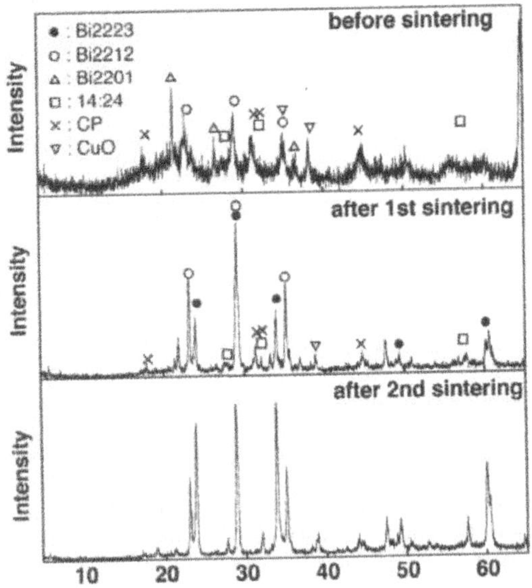

Fig. 5 The results of XRD measurements of the alloy sheathed tapes using the precursor B containing the fine 14:24 particles. (a) before sintering, (b) after first sintering of 50h, and (c) after final sintering.

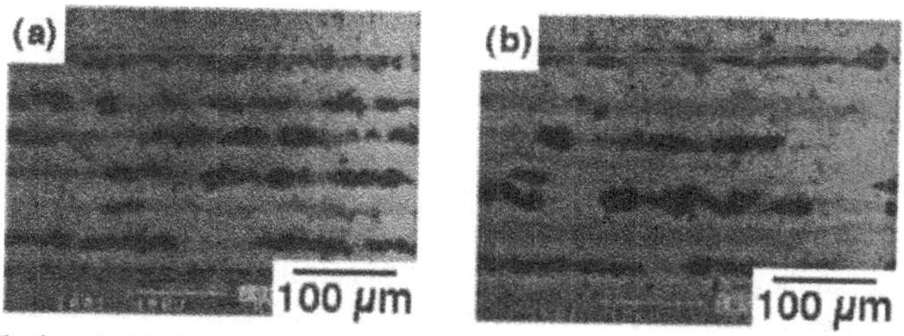

Fig. 6 The longitudinal cross sections of the tapes using the precursor B after (a) the first and (b) the second sintering.

the detected phases became similar to those using the standard precursor A, and the peak intensities after final sintering were almost the same as those using the precursor A shown in Fig. 1. Figure 6 shows the longitudinal cross sections of the tapes using the precursor B after the first and the second sintering. As shown in this figure, even though the fine 14:24 particles were contained initially in the precursor, many large 14:24 particles existed in the filaments after the first sintering similar to the case of standard precursor A consisting of the 2:1 and CuO phases. The J_c value of this tape was only as small as 2×10^3 A/cm^2. This suggests that the formation of large 14:24 phase is not simply prevented by the use of fine 14:24 phase particles. Remembering that the 14:24 phase in the tape contains Sr element, the formation of the 14:24 phase must be accompanied by the decomposition of some amount of the Bi2212 phase, which is the main phase containing Sr element in the precursor. That is, the formation of 14:24 phase is thought to attribute to the instability of Bi2212 phase in this average composition and sintering temperature under the high Cu potential of the Ag-Cu alloy.

The reported phase diagrams[9-14] indicate that the Bi2212 and Bi2223 phases have the region in equilibrium with the 2:1 and CuO phases without the14:24 phase. The equilibrium region including the Bi2212, 14:24 and CuO phases is reported to exist in the Sr and Cu rich side from the stoichiometric 2223 composition although the composition used in this experiment is slightly Sr poor. Based on the quasi-quaternary phase diagram[12] assuming the tie-line connecting the stoichiometric Bi2212 and CuO compositions, the 14:24 phase will not be formed using the average composition used in this experiment even though the Cu is supplied from the alloy sheath. This inconsistency could be explained by thinking of the solubility range of Bi2212 and Bi2223 phases. The Bi2212 and Bi2223 phases are actually solid solutions while the reported phase diagrams usually neglect the solubility of those phases for simplicity. The solubility ranges of these superconducting phases are reported to depend on the sintering temperature as well as the oxygen partial pressure. Although many researches have been performed to find out the solubility limits and primary phase fields of these phases[9,12,15-18], the solidus compositions precise enough have not confirmed yet because it is difficult to achieve the equilibrium among related phases in which too many constituents of 5 cations exist except for oxygen. In spite of this difficulty, if we can select a proper composition to stabilize the Bi2212 phase under the high Cu potential in Ag-Cu alloy sheaths, the formation and enlargement of 14:24 particles must be suppressed. When the undesired reaction such as the formation of large 14:24 particles through the decomposition of Bi2212 phase could be suppressed by the proper selection of initial phase compositions, we can focus on the conversion reaction and control of the Bi2223 phase, and the J_c values as high as those of the Bi2223 tapes sheathed with pure Ag could be obtained using the Ag-Cu alloy sheath.

CONCLUSION

The Bi2223 tapes sheathed with Ag-Cu alloys were fabricated, and the effects of precursor composition and constituent phases on the microstructure and the J_c property were investigated. The use of Cu deficient precursor up to $x = 0.4$ can slightly suppress the formation of large 14:24 particles and improve the J_c values of the alloy sheathed tapes while the further reduction of Cu content could not improve the microstructure and J_c values. This formation of large 14:24 particles from the 2:1 and Bi2212 phases was found to occur in the early stage of sintering. Although the precursor containing fine 14:24 particles as an initial phase instead of the 2:1 phase was used to suppress this formation of large 14:24 particles, the formation of large 14:24 particles could not suppressed. These results suggest that the reaction to form the 14:24 phase may attribute to the decomposition of Bi2212 phase under this average composition of precursor and the sintering condition. The selection of proper composition to stabilize the Bi2212 phase under the high Cu potential in Ag-Cu alloy sheaths may suppress the formation and enlargement of 14:24 particles, and the J_c values as high as high as those of the Bi2223 tapes sheathed with pure Ag will be achieved using the Ag-Cu alloy sheath.

ACKNOWLEDGEMENTS

This work was partially supported in The 21st Century COE Program "Intelligent Human Sensing", from the ministry of Education, Culture, Sports, Science and Technology, and supported in part by a Grant-in-Aid for Scientific Research (No. 15360149) from Japan Society for the Promotion of Science.

REFERENCES

[1] Yoshiaki Tanaka, Toshihisa Asano, Tomoyuki Yanagiya, Masao Fukutomi, Kazunori Komori, Hiroshi Maeda, "Fabrication and superconducting properties of AgCu alloy-sheathed BiSrCaCuO oxide tapes" *Jpn. J. Appl. Phys.* **31**, L235-38 (1992).

[2] Masayuki Ishizuka, Yoshiaki Tanaka, Hiroshi Maeda, "Superconducting properties and microstructures of Bi-2223 Ag-Cu alloy sheathed tapes doped with Ti, Zr or Hf" *Physica C* **252**, 339-47 (1995).

[3] Masayuki Ishizuka, Yoshiaki Tanaka, Takehiro Hashimoto, Hiroshi Maeda, "Influences of Cu composition and sintering condition in Bi-2223 tapes using Ag-Cu alloy sheath doped with Ti, Zr or Hf" *Physica C* **290**, 265-74 (1997).

[4] R.Zeng, Y.C. Guo, Y. Tanaka, J. Horvat, M. Ionescu, T.P. Beales, M. Appley, H.K. Liu, S.X. Dou, "Effect on the phase formation of Bi-2223 in some Ag-alloy sheathed PIT tapes" *Physica C* **307**, 229-36 (1998).

[5] R.P. Aloysius, A. Sobha, P. Guruswamy, K.G.K. Warrier, U. Syamaprasad, "Fabrication of (Bi,Pb)-2223/Ag-Cu alloy sheathed tapes: optimization of Cu composition in the sheath with respect to the Cu stoichiometry in the system" *Physica C* **328**, 221-29 (1999).

[6] W. Goldacker, E. Mossang, M. Quilitz, M. Rikel, "On the phase formation in Ag and AgMg sheathed BSCCO (2223) tapes" *IEEE Trans. Appl. Supercond.* **7**, 1407-10 (1997).

[7] Y. Nakamura, S. Nakashima, R. Inada, A. Oota, "Effect of composition on the fabrication and properties of Ag–Cu alloy sheathed (Bi,Pb)2223 tapes" *Physica C* **412-414**, 1085-90 (2004).

[8] Y. Nakamura, R. Inada, A. Oota, "Phase formation mechanism and properties of Ag-sheathed (Bi,Pb)-2223 tapes prepared by two-powder method" *Physica C* **392-396**, 1015-19 (2003).

[9] Byungsun Hong, Thomas O. Mason, "Solid-solution Ranges of the n=2 and n=3 superconducting phases in $Bi_2(Sr_xCa_{1-x})_{n+1}Cu_nO_y$ and the effect on Tc", *J. Am. Ceram. Soc.*, **74**, 1045-52 (1991).

[10] P. Strobel, J.C. Tolédano, D. Morin, J. Schneck, G. Vacquier, O. Monnereau, J. Primot, T. Fournier, "Phase diagram of the system $Bi_{1.6}Pb_{0.4}Sr_2CuO_6$-$CaCuO_6$ between 825 °C and 1100 °C" *Physica C* **201**, 27-42 (1992).

[11] Stefanie Kaesche, Peter Majewski, Fritz Aldinger, "Phase relations and homogeneity region of the high temperature superconducting phase $(Bi,Pb)_2Sr_2Ca_2Cu_3O_{10+d}$" *J. Electron. Mater.* 24, 1829-34 (1995).

[12] Stefanie Kaesche, Peter Majewski, Fritz Aldinger, "Phase equilibria in the system Bi_2O_3-PbO-SrO-CaO-CuO with special regard to the $(Bi,Pb)_{2+x}Sr_2Ca_2Cu_3O_{10+d}$ phase", *Z. Metallk.*. **87** 587-93 (1996).

[13] Peter J. Majewski, Stefanie Kaesche, Fritz Aldinger, "Fundamental material aspects underlying the preparation of high-temperature superconducting $(Bi,Pb)_{2+x}Sr_2Ca_2Cu_3O_{10+d}$ ceramics" , *J. Am. Ceram. Soc.* **80**, 1174-80 (1997).

[14] P. Majewski, "Phase diagram studies in the system Bi-Pb-Sr-Ca-Cu-O-Ag", *Supercond. Sci. Technol.* **10**, 453-67 (1997).

[15] Peter Majewski, Stefanie Kaesche, Huang-Lung Su, Fritz Aldinger, "The Pb solubility of the Bi-based high-Tc superconductors "Bi2Sr2CaCu2O8" and "Bi2Sr2Ca2Cu3O10" as a function of temperature", " *Physica C* **221**, 295-98 (1994).

[16] P. Majewski, Huang-Lung Su, M. Quilitz, "Relationships between the chemical composition and properties of the high-temperature superconductor $Bi_{2+x}Sr_{2-y}Ca_{1+y}Cu_2O_{8+\delta}$", *J. Mater. Sci.* **32**, 5137-41 (1997).

[17] W. Wong-Ng, L.P. Cook, F. Jiang, W. Greenwood, U. Balachandran, M. Lanagan, "Subsolidus phase equilibria of coexisting high-Tc Pb-2223 and 2212 superconductors in the (Bi,Pb)-Sr-Ca-Cu-O system under 7.5% O_2", *J. Mater. Res.* **12**, 2855-65 (1997).

[18] W. Wong-Ng, L.P. Cook, A. Kearsley, W. Greenwood, "Primary phase field of the Pb-doped 2223 high-Tc superconductor in the (Bi,Pb)-Sr-Ca-Cu-O system", *J. Res. Natl. Inst. Stand. Technol.* **104**, 277-88 (1999).

Electroceramic Materials and Applications
Edited by R. W. Schwartz

THICKNESS DEPENDENCE OF CRITICAL CURRENT DENSITY IN IBAD/PLD YBCO COATED CONDUCTOR

T. Matsushita*, M. Kiuchi, K. Kimura
Faculty of Computer Science and Systems Engineering, Kyushu Institute of Technology
680–4 Kawazu, Iizuka, Fukuoka, 820–8502, Japan

S. Miyata, A. Ibi, T. Muroga, Y. Yamada
Superconductivity Research Laboratory
2–4–1 Mutsuno, Atsuta-ku, Nagoya, Aichi 456–8587, Japan

Y. Shiohara
Superconductivity Research Laboratory,
1–10–13 Shinonome, Koto-ku, Tokyo 135–0062, Japan

ABSTRACT

The dependence of the critical current properties on superconducting layer thickness was investigated in the range of 0.5 to 1.5 μm for YBCO coated tapes made by the PLD processing on IBAD substrates. The critical current density at low fields in the direction of the *c*-axis was found to decrease with the increasing thickness d in proportion to $d^{-1/2}$. This seems to agree with the prediction of the two-dimensional collective pinning of random point pins. However, this dependence did not change over a wide temperature range of 5 to 77 K and the critical thickness for the two-dimensional pinning was found to be much smaller than 0.5 μm. This result suggests that the observed thickness dependence does not come from the pinning mechanism but simply from the change in the superconducting layer structure with increasing thickness. The effect of flux creep on the critical current properties at high temperatures is also investigated, and the observed thickness dependences of irreversibility field and n-value are compared with the theoretical predictions of the flux creep-flow model. It is clarified that the effect of the transverse flux bundle size on the pinning potential should also be considered for a detailed analysis of the thickness dependence of the irreversibility field.

INTRODUCTION

It has been clarified that the critical current properties of YBCO coated tapes depend on the thickness of superconducting layer. That is, the critical current density in PLD(Pulsed Laser Deposition)-processed tapes at low magnetic fields decreases with increasing thickness, d[1]. This is similar to the results on thin films deposited on single crystal substrates[2]. In these cases the critical current density J_c approximately obeys the relationship:

$$J_c \propto d^{-1/2} \tag{1}$$

predicted by the collective pinning theory[3] for the case of two-dimensional pinning by random point pins. The two-dimensional pinning occurs for thin films of thickness d thinner than the pinning correlation length approximately given by

$$l_{44} \simeq \left(\frac{Ba_f}{2\pi\mu_0 J_c}\right)^{1/2}, \tag{2}$$

*Correspondence author. E-mail: matusita@cse.kyutech.ac.jp

where $a_f = (\sqrt{3}\phi_0/B)^{1/2}$ is the flux line spacing with ϕ_0 denoting the flux quantum.

However, recent investigation on a series of coated tapes fabricated by the PLD process on IBAD(Ion Beam Assisted Deposition)-processed substrates has clarified that J_c depends on the thickness obeying Eq. (1) in the thickness region much below the correlation length given by Eq. (2) and that this dependence does not change with temperature[4]. This throws doubt on the explanation of the thickness dependence in coated tapes by the two-dimensional collective pinning mechanism, indicating other causes such as the growth of a-axis aligned grains in thick PLD processed tapes. This speculation seems to be compatible with the fact that J_c does not depend appreciably on d for the TFA-MOD(TriFluoroAcetate-Metal Organic Deposition)-processed tapes[5]. Thus, it is necessary to clarify the correct mechanism which causes the thickness dendent J_c in the PLD-processed tapes.

On the other hand, J_c shows the opposite dependence at high fields; it is higher for a thicker superconducting tape[4,6]. This is connected to the thickness dependence of the irreversibility field. This is ascribed to the larger pinning potential U_0. That is, U_0 takes a small value for a thin tape, resulting in a significant reduction of J_c due to the flux creep, since the flux bundle volume is limited by the thickness. In the previous measurements, however, the observed thickness dependence of the irreversibility field was much weaker than the theoretical prediction of the flux creep-flow model[4,6]. For the improvement of the theoretical analysis the effect of the transverse flux bundle size should also be taken into account.

For these reasons the critical current density of YBCO coated tapes depends on the superconducting layer thickness in a complicated manner. Hence, it is necessary to clarify all the mechanisms which influence the critical current properties and to find out the method of estimation of J_c for the optimum design of superconducting layer thickness for each application.

In this paper, the thickness dependence of J_c is investigated for a new series of PLD-processed tapes in the thickness range of 0.5 to 1.5 μm. In order to clarify the dimensionality of pinning the thickness dependence is measured over a wide temperature range to change l_{44} which determines the dimensionality of the pinning to compare the result with the prediction of the collective pinning theory. The effect of flux creep which is significant at high temperatures is also investigated in detail.

EXPERIMENTS

The specimens measured were YBCO coated tapes fabricated by the PLD method on IBAD substrates with $Gd_2Zr_2O_7$ inner layer and CeO_2 cap layer, and the thickness of superconducting layer was changed from 0.5 to 1.5 μm. The critical temperature of these specimens are listed in Table 1.

Table 1: Specification of specimens.

Specimen	thickness d (μm)	T_c (K)
#1	0.5	88.4
#2	1.0	88.2
#3	1.5	88.4

Relaxation of DC magnetization was measured with a SQUID magnetometer in a magnetic field parallel to the c-axis of the specimens. Measurements were done up to 7 T in the temperature range of 5 to 85 K. E-J characteristics were estimated from the results[6] and J_c was determined using the electric field criterion of $E_c = 1.0 \times 10^{-8}$ V/m. The n value, the index expressing the E-J characteristics as $E \propto J^n$, was estimated in the field range $E = 10^{-9}$–10^{-8} V/m. The irreversibility field B_i was defined by the magnetic field at which J_c reduced to 1×10^8 A/m^2.

RESULTS AND DISCUSSION

Figure 1(a) and (b) show the critical current density of the three specimens at 5 and 77.3 K. It is clearly seen that J_c decreases with increasing d at 5 K and this dependence does not change with the magnetic field. At 77.3 K, on the other hand, although the thickness dependence of the same trend is seen at low magnetic fields, it is fairly weak and J_c of the thinnest specimen rapidly falls off with increasing magnetic field. This degradation of J_c of thin specimen at high temperature and high magnetic field is caused by the flux creep.

The thickness dependence of J_c at 0.1 T at various temperatures is shown in Fig. 2. It is found that J_c obeys Eq. (1) as shown by the straight lines over a wide temperature range. The present result is compared with the previous result[4] on another series of specimens in Fig. 3. The thickness dependence is the same between the two series of specimens, although the value of critical current density is different. Hence, the same conclusion as the previous parer[4] is obtained: the degradation of J_c in thick tapes does not originate from the two-dimensional collective pinning but probably from a change in the structure of superconducting layers. Figure 4 shows the theoretically estimated l_{44} at 0.1 T using J_c of the thickest specimen 3. It is about 0.085 μm at 5 K and hence the three-dimensional pinning with no thickness

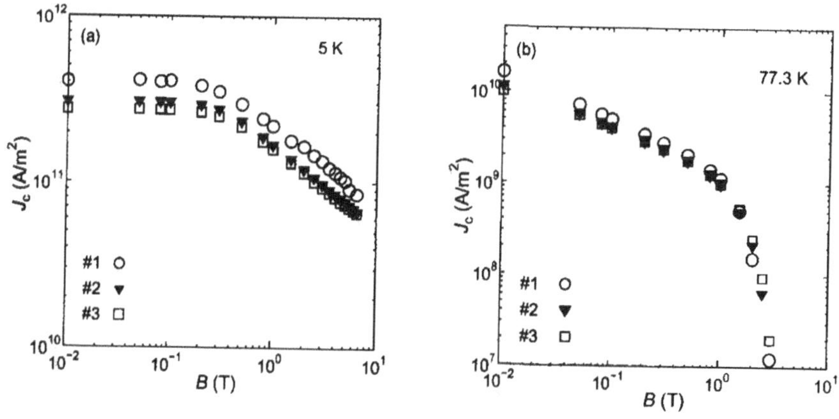

Figure 1: Critical current density of the three specimens at (a) 5 K and (b) 77.3 K.

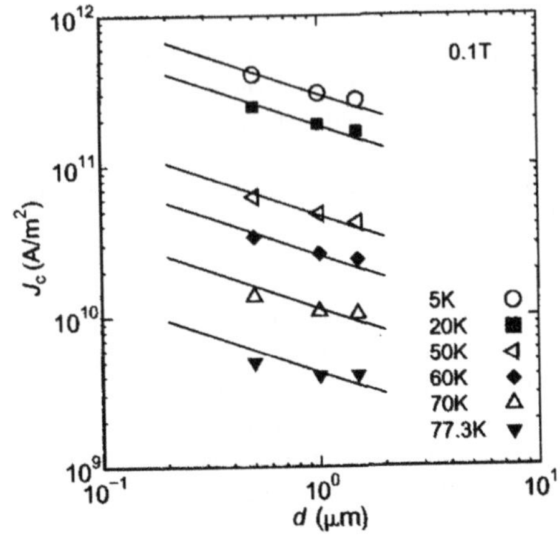

Figure 2: Thickness dependence of critical current density at 0.1 T at various temperatures. Straight lines show the dependence given by Eq. (1).

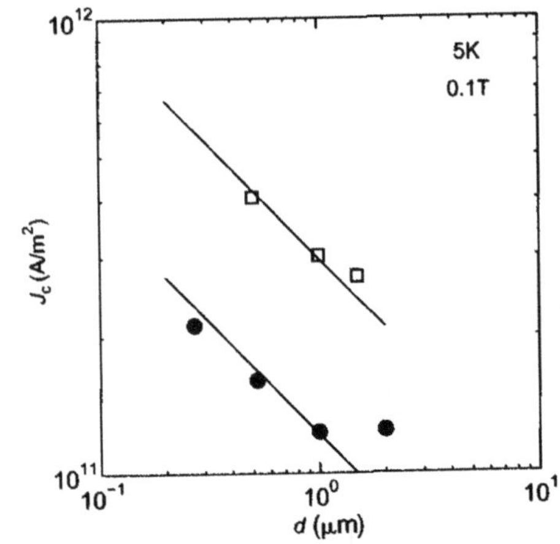

Figure 3: Comparison of thickness dependence between the present (open squares) and previous (solid circles) series of YBCO coated tapes at 5 K and 0.1 T.

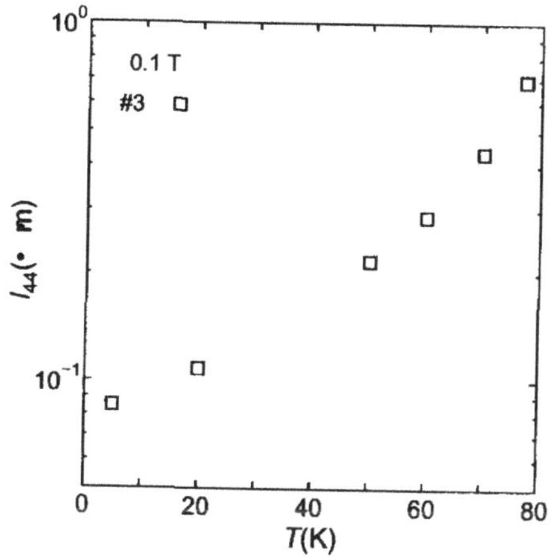

Figure 4: Temperature dependence of l_{44} estimated at 0.1 T using J_c of specimen 3.

dependence is expected from the collective pinning mechanism within the thickness range of the present investigation. The situation does not change with temperature. This also confirms the above conclusion.

The difference in J_c by a factor of about 2.5 at 5 K between the previous and present specimens is considered to be caused by the difference in the quality in the superconducting layer. Thus, there seems to be room for further improvement of J_c by optimizing conditions of the PLD process.

Figure 5 shows the temperature dependence of the irreversibility field. It is found that the irreversibility field is the highest for the thickest specimen in spite of the lowest J_c at high temperatures. The thickness dependence of the irreversibility field at 77.3 K is shown in Fig. 6. This clearly demonstrates the advantage of thicker tapes for applications at high magnetic fields.

The critical current properties were analysed using the flux creep-flow model[6]. In the analysis the scaling law of the virtual critical current density J_{c0} in the ideal creep-free case is assumed as

$$J_{c0} = A\left(1 - \frac{T}{T_c}\right)^m B^{\gamma-1}\left(1 - \frac{B}{B_{c2}}\right)^\delta, \tag{3}$$

where A, m, γ and δ are the pinning parameters. A which represents the pinning strength is assumed to be statistically distributed as

$$f(A) = K\exp\left[-\frac{(\log A - \log A_m)^2}{2\sigma^2}\right], \tag{4}$$

where A_m is the most probable value of A and σ^2 is a parameter representing a distribution width. Another important parameter which influences the flux creep is g^2 representing the

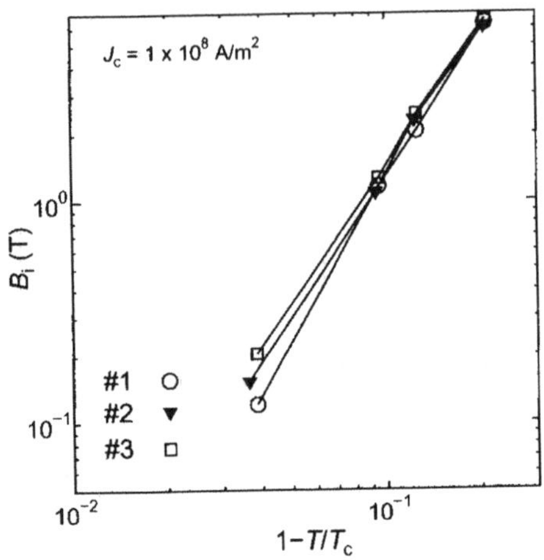

Figure 5: Temperature dependence of irreversibility field.

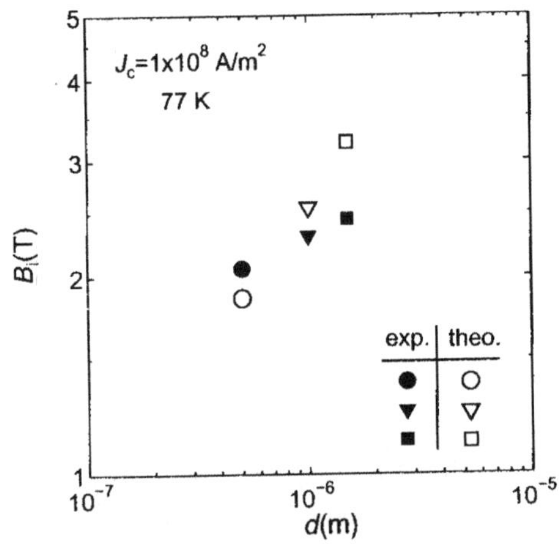

Figure 6: Thickness dependence of irreversibility field at 77.3 K. The solid symbols are experimental results and open symbols are theoretical results.

Table 2: Pinning parameters of specimens in the low and medium temperature regions.

specimen	A_m	m	σ^2(5 K/20 K)	γ	g^2
#1	6.0×10^{11}	2.5	0.0042/0.0048	0.63	3
#2	4.8×10^{11}	2.7	0.0047/0.0051	0.63	3
#3	4.3×10^{11}	2.7	0.0042/0.0046	0.63	3

Table 3: Pinning parameters of specimens in the high temperature region.

specimen	A_m	m	σ^2(70 K/77.3 K)	γ	g^2
#1	7.7×10^{11}	1.9	0.023/0.027	0.59	3
#2	5.7×10^{11}	1.9	0.020/0.021	0.59	3
#3	5.0×10^{11}	1.8	0.021/0.023	0.58	3

number of flux lines in the flux bundle[7]. This value is determined so as to maximize the critical current density under the flux creep[8] and takes a larger value in a more three-dimensional superconductor such as YBCO. In the present analysis we assumed $g^2 = 3$,

$$B_{c2}(T) = B_{c2}(0)\left(1 - \frac{T}{T_c}\right) \tag{5}$$

with $B_{c2}(0) = 80.5$ T and $\delta = 2$ in Eq. (3), and other parameters A_m, σ^2, m and γ are adjusted so that the calculated E-J curves agree with the experiments. The parameters obtained are listed in Tables 2 and 3.

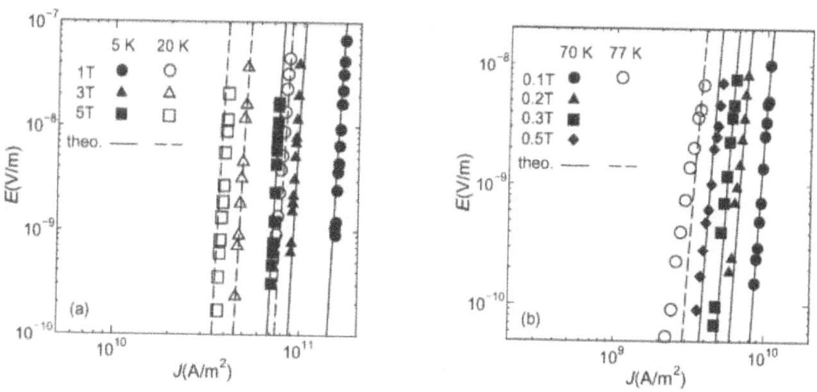

Figure 7: E-J characteristics of specimen at (a) low temperatures and (b) high temperatures. Solid lines are theoretical results.

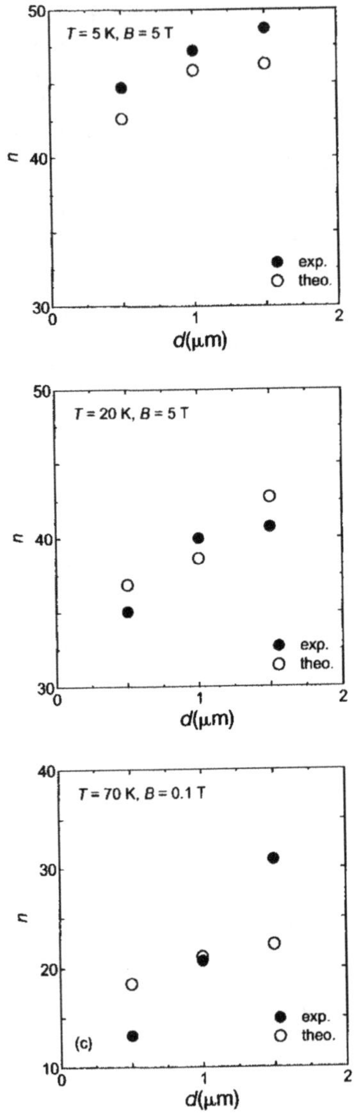

Figure 8: Thickness dependence of n value at (a) low, (b) medium and (c) high temperatures. Solid and open symbols represent the experimental and theoretical results, respectively.

In Fig. 7(a) and (b) the observed E-J characteristics of specimen 2 are compared with the theoretical results at low and high temperatures, respectively. It is found that the agreement is good except the case at 77.3 K. Experimental and theoretical n-values are compared in Fig. 8. Although the n value does not appreciably depend on the thickness at low temperatures, it becomes appreciable with a smaller value for a thinner specimen at medium temperatures. This dependence is emphasized at high temperatures. Thus, the thicker tapes are desirable for applications to superconducting coils operated in a persistent current mode in high magnetic fields such as NMR magnets.

The theoretically estimated irreversibility field in each specimen is compared with the experimental result in Fig. 6. The theoretically predicted thickness dependence is stronger than observed, although the trend of higher irreversibility field in thicker tape agrees with experiments. This is the same as in the previous investigation[4,6]. This seems to be caused by the assumption of the constant g^2 value. Recently it was found that g^2 takes a larger value in a specimen thinner than the pinning correlation length for Bi-2212 superconductor[7]. This means that the transverse flux bundle size tends to become large so as to reduce the significant effect of flux creep. If the thickness dependence of g^2 is neglected, a simple argument derives the pinning potential as

$$U_0 \propto J_{c0} d g^2 \propto d^{1/2}, \qquad (6)$$

where Eq. (1) is used for the thickness dependence of J_{c0}. Since $B_i^{(3-2\gamma)/2}$ is proportional to U_0, we have $B_i \propto d^{1/2}$ in case $\gamma = 1/2$ as shown by open symbols in Fig. 6.

The exact expression including the thickness dependence of g^2 leads to

$$U_0 \propto (J_{c0} d g_e^2)^{-1/3} \propto d^{1/6}, \qquad (7)$$

where g_e^2 is a bulk g^2 value independent of the film thickness. Hence, the same argument leads to $B_i \propto d^{1/6}$ which explains the experimental result shown by solid symbols in Fig. 6. This means that the transverse flux bundle size is enlarged to reduce the significant effect of flux creep, although this brings about a slight reduction in J_{c0}. Such a behavior can be understood that the flux bundle size is determined so as to maximize the critical current density under the flux creep[8].

SUMMARY

The dependence of superconducting layer thickness on the critical current properties was investigated for the YBCO coated tapes made by PLD processing on IBAD substrates, and the following results were obtained:

(1) It was found that the critical current density at low fields decreased with the increasing thickness d in proportion to $d^{-1/2}$ as suggested by the two-dimensional collective pinning mechanism. However, the estimated pinning correlation length was much thinner than the specimen thickness. Hence, it was concluded that the observed thickness dependence does not come from the two-dimensional collective pinning mechanism.

(2) The above result suggests that the observed thickness dependence originates from the change in the structure of the superconducting layer in PLD processed tapes with increasing thickness. This is supported by the fact that the observed thickness dependence did not change over a wide range of temperature in which the pinning correlation length changed remarkably.

(3) At high temperatures, the irreversibility field and the n-value increased with increasing thickness. These behaviors can be explained by the flux creep-flow model. Thus, for the better performance at high temperatures the thicker tapes are more suitable.

ACKNOWLEDGEMENT

This work is partly supported by the New Energy and Industrial Technology Development Organization (NEDO) as Collaborative Research and Development of Fundamental Technologies for Superconductivity Applications.

REFERENCES

[1]L. Civale, B. Maiorov, A. Serquis, J. O. Willis, J. Y. Coulter, H. Wang, Q. X. Jia, P. N. Arendt, M. Jaime, J. L. MacMaus-Driscoll, M. P. Mayley and S. R. Foltyn,"Understanding high critical currents in $YBa_2Cu_3O_7$ thin films and coated conductors", *J. Low Temp. Phys.* **135** 87 (2004).

[2]R. Wördenweber, "Growth of high-T_c thin films", *Supercond. Sci. Technol.* **12** R86 (1999).

[3]A. I. Larkin and Yu N. Ovchinnikov, "Pinning in type II superconductors", *J. Low Temp. Phys.* **34** 409 (1979).

[4] T. Matsushita, M. Kiuchi, K. Kimura, S. Miyata, A. Ibi, T. Muroga, Y. Yamada and Y. Shiohara, "Dependence of critical current properties on thickness of superconducting layer in YBCO coated tapes", *3rd Workshop on Machano-Electromagnetic Property of Composite Superconductors*, 17–20 July 2005, Kyoto.

[5]T. Izumi, Y. Tokunaga, H. Fuji, R. Teranishi, J. Matsuda, S. Ssada, T. Honjo, Y. Shiohara, T. Muroga, S. Miyata, T. Watanabe, Y. Yamada, Y. Iijima, T. Saitoh, T. Goto, A. Yoshinaka and A. Yajima, "Progress in development of coated conductors by TFA-MOD processing", *Physica C* **412–414** 885 (2004).

[6]T. Matsushita, T. Watanabe, Y. Fukumoto, K. Yamauchi, M. Kiuchi, E. S. Otabe, T. Kiss, T. Watanabe, S. Miyata, A. Ibi, T. Muroga, Y. Yamada and Y. Shiohara, "Flux pinning characteristics of YBCO coated conductor", *Physica C* **426–431** 1096 (2005).

[7]T. Matsushita, M. Kiuchi, T. Yasuda, H. Wada, T. Uchiyama and I. Iguchi, "Thickness dependence of pinning properties in Bi-2212 superconductor", *Supercond. Sci. Technol.* **18** 1348 (2005).

[8]T. Matsushita, "On the flux bundle size in weakly pinned superconductors", *Physica C* **217** 461 (1993).

Electroceramic Materials and Applications
Edited by R. W. Schwartz

DRASTICALLY INNOVATIVE BSCCO LONG LENGTH WIRE BY CONTROLLED OVER PRESSURE SINTERING

K. Yamazaki, S. Kobayashi, T. Kato, K. Ohkura, K. Fujino, J, Fujikami, E. Ueno, N. Ayai, M. Kikuchi, K. Hayashi, K. Sato and R. Hata

Sumitomo Electric Industries, Ltd
1-1-3, Shimaya, Konohana-ku, Osaka, 554-0024
Japan

ABSTRACT

Since ConTrolled Over Pressure (CT-OP) sintering process increased the relative density in filaments of Bi2223 long length wire to almost 100%, the performance of wire has improved drastically: (a) increasing critical current, (b) higher phase purity, (c) healing cracks and porosities, (d) increasing mechanical property and others. The uniformity of longitudinal direction of Bi2223 long length wires is required for applications such as cable, magnet, motor and so on. Highly dense state of filaments in wires fabricated by CT-OP have achieved the uniform Bi2223 long length wires (> 1000 m) which the average I_C was 152 A and n-value was 21 end to end. The magnetic property of Bi2223 wires was improved by CT-OP process at high and low temperature and CT-OP process achieved enhancement of the irreversibility field of Bi2223 wires.Bi2223 wires are expected to apply to products cooling by liquid nitrogen and there was the problem which ballooning occurred in normal atmosphere sintering wire. The ballooning will result in significant degradation in critical current because filament shape is deformed. CT-OP wires with the filaments of almost 100% relative density are not ballooned at liquid nitrogen tight state. We can manufacture Bi2223 long length wire which is over 1,000m, homogeneous, high performance and anti-ballooned, and have begun commercial production. At present, projects of power cable, magnets, motors and many others have carried out using CT-OP long length wires.

INTRODUCTION

Since the discovery of high-T_C superconductor (HTS) in 1986, we have proceeded with the research of HTS wires and magnet[1], power transformer[2] and power cable[3-6] have been developed using $(Bi,Pb)_2Sr_2Ca_2Cu_3O_X$ (Bi2223) wires. At present Bi2223 wire is close to practical application and progressed additionally in improvement of performance and development of application such as cable, magnet, motor and others. It is still necessary for practical application to improve performance of Bi2223 wires. Jc value is confirmed 300 kA/cm^2 at local regions in grains of Bi2223 monofilamentary tape by magneto optical imaging method[7]. So we could achieve improvement of performance of Bi2223 wires if high J_C local regions could increase and improve connectivity between Bi2223 grains. And then the uniformity of longitudinal direction of Bi2223 long length wires also is required for practical application. Bi2223 long length wires will gain the trust of HTS users from the high performance and the high uniformity.

Recently we have reported effects of controlled over pressure (CT-OP) which is the method to apply gas pressure to wires during sintering process[8-10]. These effects are that 100 % relative density and improvement of Bi2223 phase homogeneity, grain alignment, grain connectivity and mechanical property compared to normal sintering (atmosphere sintering), and so Ic value of Bi2223 wire is improved drastically. The theoretical mass density of Bi2223 used in this study is 6.3 g/cm^3 [11].We have developed a large furnace (called as CT-OP furnace) where temperature, pressure, and oxygen partial pressure that are related to the performance of Bi2223 wires can be controlled independently and kilometer lengths of Bi2223 wires can be sintered.

Actually Bi2223 wire sintered in CT-OP furnace is more than 1500 m and good performance[8] and Sumitomo Electric manufacture two types Bi2223 CT-OP wires of High I_C Type and High Strength Type. Recently many applications using Bi2223 CT-OP wires have been carried out, for example: (1) HTS motor which has been developed by IHI, SEI, Fukui Univ., Taiyo Nippon Sanso, Nakashima Propeller, Niigata Power System, Hitachi and Fuji, and will be miniaturized, be efficient, be reduced the effects on the environment and others compared to normal motor. (2) transformer which has been developed by Railway Technical Research Institute and SEI, and will be miniaturized and be trimmed weight compared to normal transformer. To be closer to practical application, we have expected that the performance of Bi2223 wires increase by optimization of CT-OP process. In this paper, the performance of Bi2223 CT-OP wires which progressed in optimization of CT-OP additionally is reported.

EXPERIMENTAL

Ag-sheathed Bi2223 superconducting wires were fabricated using powder in tube (PIT) method. The Ag-alloy was prepared to increase the mechanical properties. We applied CT-OP after first heat treatment and second rolling process. The gas pressure is up to 30 MPa using a booster. The width was 4.1-4.5 mm and the thickness was 0.20-0.24 mm. The Ag ratio was 1.5-2.2. The length of Bi2223 CT-OP wires were 500-1500m.

The microstructure was observed with SEM using wires polished longitudinal section. Critical current (I_C) was determined from I-V curve using a four probe method at 77 K and self-field and I_C criterion was 1 μV/cm. The average n-value was defined in the region of 10^{-6} to 10^{-5} V/m. The ballooning test in liquid nitrogen under 1 MPa for 24 hours was carried out. Tensile stress tests were carried out at room temperature and liquid nitrogen temperature (77 K).

RESULTS AND DISCUSSION

Electrical property

High performance and uniformity in longitudinal direction of Bi2223 wires are strongly required for Bi2223 application. CT-OP furnace can manufacture long length wires in the kilometer class. Figure 1 shows the I_C and n-value distribution in the longitudinal direction of (a) High I_C Type and (b) High Strength Type wires at 77K and self-field. I_C of long length wires was measured every 4 m. Since Ag ratio of High I_C Type wires are smaller than that of High Strength Type wires, I_C of High I_C Type are higher than that of High Strength Type. The average I_C values are (a) 152 A for High I_C Type and (b) 110A for High Strength Type and length of both type wires are more than 1,000 m. The average n-value was 21 for both type wires. As can be seen in fig. 1, we can manufacture Bi2223 long length CT-OP wires which are uniformity in longitudinal direction. For High I_C Type wires, I_C of normal atmosphere sintering was 90 A at 77 K and self-field, and so I_C becomes 160% or higher by CT-OP. We once reported that I_C becomes 130% or higher by CT-OP[6-8] and we suggest that the performance of Bi2223 CT-OP wires which progressed in optimization of CT-OP additionally improved.

Figure 2 shows an I_C performance histogram of recent commercial production (year: 2004, 2005, present) which wire length was more than 900 m and number of lots was more than 100 lots. As can be seen in fig. 2, I_C value has shifted higher in the process of time, because of optimized CT-OP process. The average I_C of present commercial production was 150 A, which we can manufacture Bi2223 long length wires in a performance reproducibility, and the maximum I_C was 163 A for long length wires. The maximum I_C was 168 A for a 4 m range of the average 163 A long wires. From results of Fig.1 and Fig.2, we suggest that CT-OP is a quiet effective process for high performance, uniformity and a high mass productivity of Bi2223 long length wires.

Microstructure

Figure 3 shows the SEM images of Bi2223 wires after (a) normal sintering and (b) CT-OP process. Fig. 3(a) shows that non-superconducting phase such as alkaline earth cuprates (Ca-Sr-Cu-O) and lead-rich phase (Ca-Pb-O) and some porosity which cause decrease of Bi2223 phase homogeneity and poor connectivity between Bi2223 grains respectively, exist in filaments of wires and so the performance of wires is poor. But Fig.3 (b) shows large gray regions and no porosities (high density) in filaments. We suggest that connectivity improves by CT-OP and this showed previously that CT-OP is able to suppress defects in filaments by magneto optical image method[6,7]. But non-superconducting phases exist still in filaments after CT-OP and we have tried to decrease these phases, and then I_C value will increase more than a present average (152 A). Bi2212-intergrowh phase existed in Bi2223 grains is also one of important factors of performance of Bi2223 wires, but it is difficult to observe and evaluate intergrowth with SEM images. And so we consider evaluation of intergrowth using TEM which is able to observe at sub-micron level.

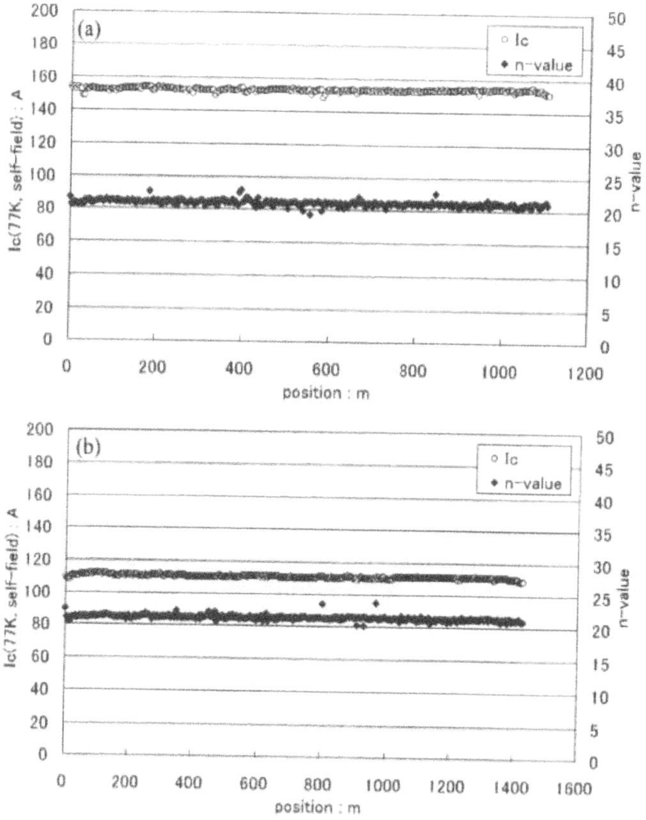

Figure 1. the I_c value and n-value distribution in the longitudinal direction of (a) High I_C Type and (b) High Strength Type at 77 K and self-field. I_C of long length wires was measured every 4 m.

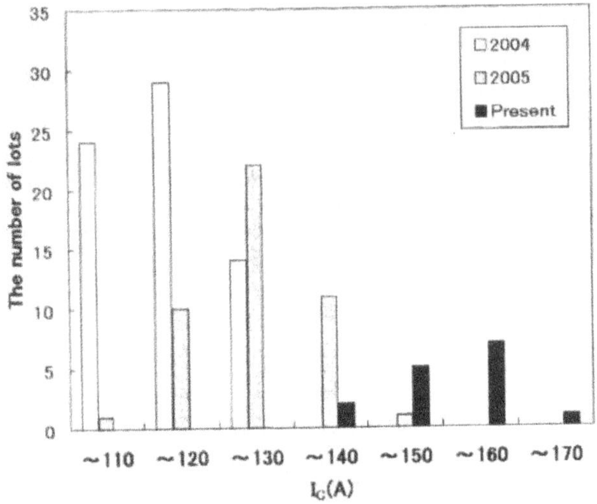

Figure 2. I_C performance histogram of recent commercial production. I_c was measured at 77 K and self-field. The length and number of lots for CT-OP wires were more than 900 m and more than 100 lots, respectively.

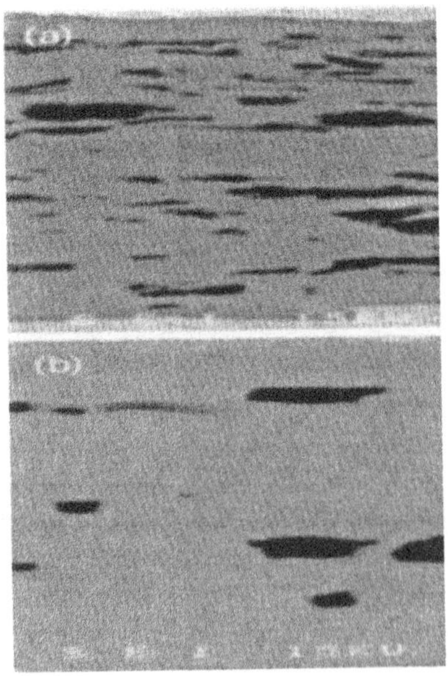

Figure 3. SEM images of (a) normal atmosphere sintering and (b) CT-OP sintering. The gray regions, black regions, white regions and tiny irregular black regions are Bi2223, alkaline earth cuprates (AEC), Ca-Pb-O and porosities, respectively.

I_C-B property

Using vortex creep model, Matsushita et. al., showed that irreversibility field of CT-OP wires enhanced compared to that of normal atmosphere sintering wires and suggested that weak parts of pin force distribution were improved by high density state of filaments in CT-OP wires[12]. Figure 4 shows I_C-*B* properties at 77 K of Bi2223 wires in present CT-OP production, 2004 year CT-OP production and normal atmosphere sintering production which I_C values at 77 K and self-field were 152 A, 120 A and 90 A, respectively. Magnetic field was applied perpendicular to the wire surface. Compared 2004 year to normal atmosphere sintering, I_C of 2004 year was higher than that of normal atmosphere sintering at a region of magnetic field (0-0.5 T). This result is relative to enhance irreversibility field by CT-OP. At next, compared present production to 2004 year, I_C of present production was certainly higher than that of 2004 year at a range of magnetic field (0-0.5 T). The connectivity between Bi2223 grains in filaments was improved additionally by progressing optimization of CT-OP and the irreversibility field also was improved. From result of fig. 4, we suggest that uniformity in filaments was higher than 2004 year and optimization of CT-OP is effective for improvement of I_C-*B* property.

I_C at 20 K under 3 Tesla perpendicular to the wire surface was 284 A for High I_C CT-OP wires. 20K is a typical temperature of operating magnet application. CT-OP wires represented high performance at a low temperature.

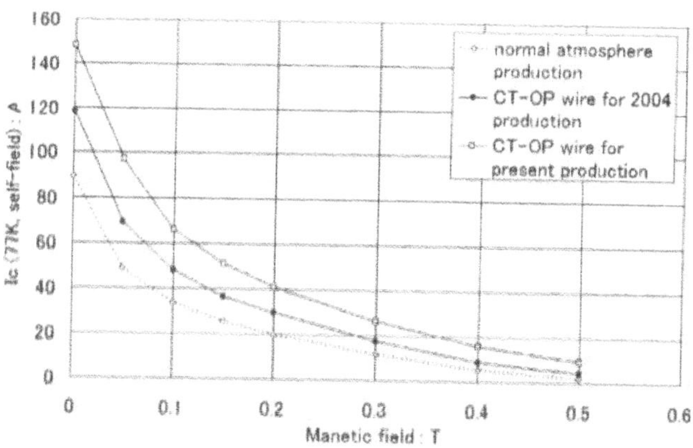

Figure 4. I_C-*B* property at 77 K of Bi2223 wires in present CT-OP, 2004 year CT-OP and normal sintering. I_C criterion was 1μV/cm and magnetic field applied perpendicular to the wire surface.

Anti-ballooning test

Liquid nitrogen cooling system is essential for applications such as power cables and transformer using Bi2223 wires. When cooled in liquid nitrogen and warmed up to room temperature, the wires may balloon up by vaporization of liquid nitrogen that has entered into the filaments through the defects of the sheath. The ballooning will result in significant degradation in Ic because filament shape is deformed[13]. To stop liquid nitrogen entering into filaments, the defect of the sheath of Bi2223 wires were covered with additional hermetic material. But wire surface was covered over and wire edge was not covered over perfectly. And so liquid nitrogen entered into filaments from wire edge and at last the ballooning occurred using Bi2223 wires with additional hermetic material. Seeing ballooning from filaments in Bi2223 wires, if no porosities exist in filaments, the wires would not be ballooned up. Because CT-OP process removes porosities and heals cracks in the filaments (relative density ~ 100%)[6,7], it is expected that CT-OP wires are not ballooned up even without additional hermetic materials.

Figure 5 shows I_C distribution of longitudinal direction of High Strength Type CT-OP wire before and after anti-ballooning test in liquid nitrogen under 1 MPa for 24 hours. After anti-ballooning test, I_C degradation was not seen in all length and so the uniformity in filaments which is deformed by porosities and cracks wires improved by CT-OP. It is so far confirmed that several wires were not ballooned after cooled in liquid nitrogen under 1 MPa for 24 hours.

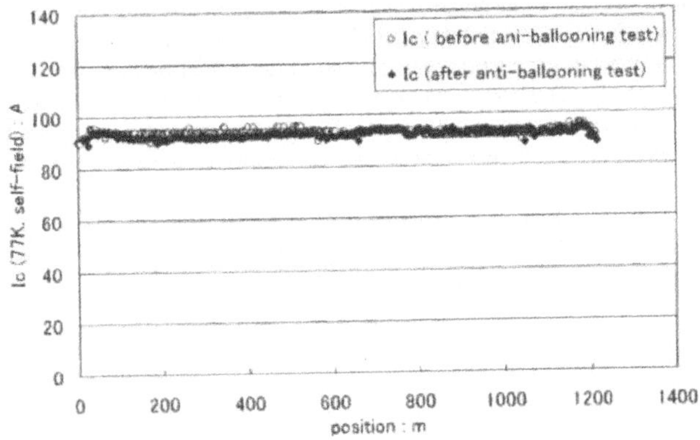

Figure 5. I_C distribution of longitudinal direction of High Strength Type CT-OP wire after anti-ballooning test. Anti-ballooning test was carried out in liquid nitrogen under 1 MPa for 24 hours.

Mechanical Property

The mechanical property is one of important factors for practical application of Bi2223 superconducting wires. This property is successfully increased when using silver-alloy sheath[14]. As the amount of additive in alloy increases, mechanical property of Bi2223 wires is improved, but workability of wires decreases. Furthermore as the Ag ratio of wires is increased, mechanical property is also improved. But decrease in I_C is a disadvantage, since the volume of superconducting filaments decreases as well. Considering Bi2223 wires consist of filaments and silver or silver alloy, filaments would also affect the mechanical property of wires. Since phases in filaments after sintering process are mainly Bi2223 grains, we consider that it is important for high mechanical property to improve the connectivity between grains. If no porosities and cracks exit in filaments, the connectivity would be improved and mechanical property would be also improved. Using CT-OP, we can manufacture Bi2223 wires with highly dense filaments which the relative density is almost 100 %. Because of investigations of mechanical property of CT-OP wires, we carried out a tensile stress tolerance test. In a tensile stress test, I_C was measured at 77 K and self-field using stressed CT-OP wires at room temperature and liquid nitrogen temperature (77 K). In this test, normal atmosphere sintering wires were measured for comparison.

Figure 6 shows I_C/I_{C0} degradation dependence on tensile stress for stressed High I_C Type wires at (a) room temperature and (b) 77 K, respectively. I_{C0} of vertical axis represents I_C of no stressed samples. The critical tensile stress values which I_C/I_{C0} degradation took place at room temperature and 77 K were increased to124 MPa from 88 MPa and 181 MPa from 111 MPa by CT-OP process, respectively. The improvement of tensile stress for CT-OP wires showed that the connectivity between grains increased compared to normal atmosphere sintering wires and mechanical property of filaments in Bi2223 wires is one of important factors to decide mechanical property of wires.

Figure 7 shows I_C/I_{C0} degradation dependence on tensile stress for stressed High Strength Type wires at (c) room temperature and (d) 77 K, respectively. The critical tensile stress values which I_C/I_{C0} degradation took place at room temperature and 77 K were increased to 203 MP from 123 MPa and 268 MPa from 147 MPa by CT-OP process, respectively. Since the Ag ratio of High Strength Type wires is higher than that of High I_C Type wires, mechanical property of High Strength Type wires is better. From results of fig.6 and fig.7, CT-OP wires which are dense in filaments have excellent mechanical properties without additional sheath.

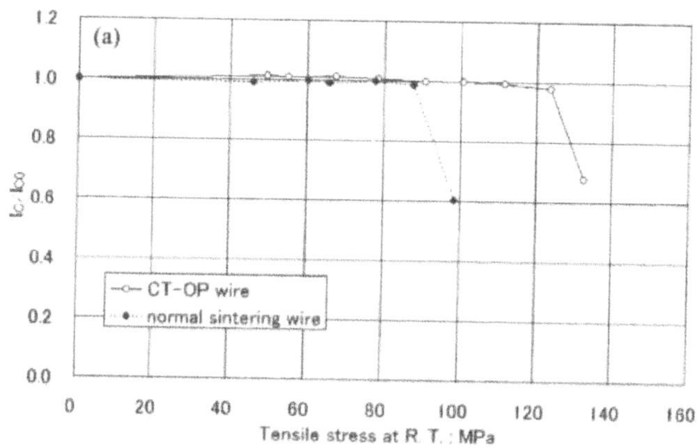

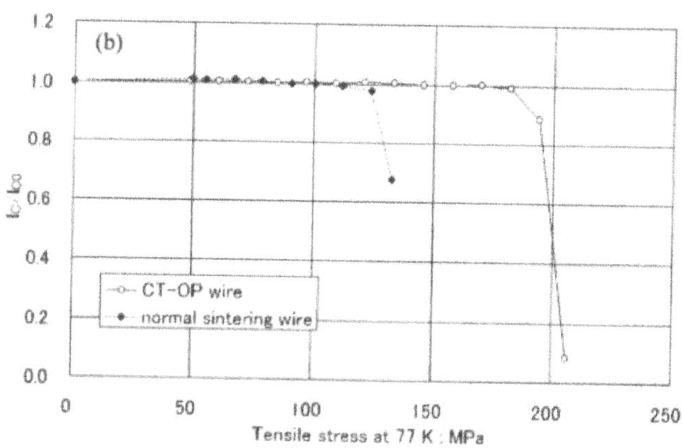

Figure 6. I_C/I_{C0} degradation dependence on tensile stress for stressed High I_C Type wires at (a) room temperature and (b) 77 K, respectively. I_{C0} of vertical axis represents I_C of no stressed samples and I_C was measured at 77 K and self-field.

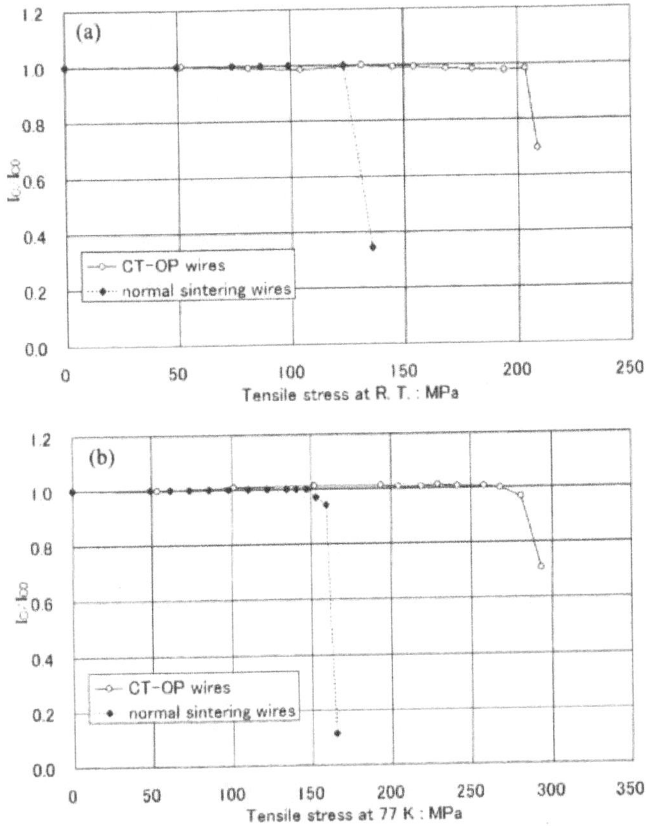

Figure 7. I_C/I_{C0} degradation dependence on tensile stress for stressed High Strength Type wires at (a) room temperature and (b) 77 K, respectively. I_{C0} of vertical axis represents I_C of no stressed samples and I_C was measured at 77 K and self-field.

CONCLUTION

We have progressed in optimization of CT-OP which is a method for sintering process of Bi2223 superconducting wires. We manufacture two type wires of High I_C Type and High Strength Type. At present, the average I_C value and n-value of Bi2223 CT-OP wires was 152A (High IC Type), 110 A (High Strength Type) and 21 end to end for over 1,000 m length, respectively. CT-OP wires were the uniformity in longitudinal direction. The maximum I_C value of CT-OP long length wires was 163A. I_C of Bi2223 wires fabricated by CT-OP process increased by more than 60 % compared to those fabricated by the normal sintering. According to

I_C performance histogram of recent commercial production, I_C has shifted higher in the process of time. The performance of B2223 wires has been improved by progressed optimization of CT-OP process. CT-OP process achieved improvement of some properties. The Bi2223 phase becomes homogenous in microstructure with SEM. The I_C-B property of wires at a high and low temperature which progressed optimization of CT-OP process was improved and irreversibility field have enhanced additionally. For applications using liquid nitrogen, there was a problem which ballooning occurred, but CT-OP wires with highly dense state were not ballooned at liquid nitrogen tight state. And CT-OP wires which are dense in filaments have excellent mechanical properties without additional sheath. We suggest that CT-OP is a quiet effective process for high performance, uniformity and a high mass productivity of Bi2223 long length wires (> 1000 m), and Bi2223 wires fabricated by CT-OP process have to be clearly distinct from normal atmosphere sintering wires. And so we call CT-OP wires as Drastically Innovative BSCCO (DI-BSCCO) wires.

ACKNOWLDEMENT

The authors would like to thank Dr. Hitoshi Kitaguchi of National Institute for Materials Science for his cooperation and evaluations of I_C-B data presented in this paper.

REFERENCES

[1]T. Kato, K. Ohkura, M. Ueyama, K. Ohmatsu, K. Hayashi, and K. Sato, "Development of High-T_C Superconducting Magnet Using Ag-Sheathed Bi2223 Tapes," *in: Proceedings of Fifteenth International Conference on Magnet Technology, Science Press,* 793-95 (1998).

[2]K. Funaki, M. Iwakuma, K. Kajikawa, M. Hara, J. Suehiro, T. Ito, Y. Takata, T. Bohno, S. Nose, M. Konno, Y. Yagi, H. Maruyama, T. Ogata, S. Yoshida, K. Ohashi, H. Kimura, and K. Tsutsumi, "Development of a 22 kV/6.9 kV Single-phase Model for a 3 MVA HTS Power Transformer," *IEEE. Trans. Appl. Supercond.*, **11**, 1578-81 (2001).

[3]T. Masuda, T. Kato, H. Yumura, M. Watanabe, Y. Ashibe, K Ohkura, C. Suzawa, M. Hirose, S. Isojima, K. Matsuo, S. Honjo, T. Mimura, T. Kuramochi, Y. Takahashi, H. Suzuki, and T. Okamoto, "Development of a 100 m, 3-core 114 MVA HTSC Cable System," *Physica C*, **372-376**, 1580-84 (2002).

[4]T. Masuda, T. Kato, H. Yumura, M. Watanabe, Y. Ashibe, K. Ohkura, C. Suzawa, M. Hirose, S. Isojima, K. Matsuo, S. Honjo, T. Mimura, T. Kuramochi, Y. Takahashi, H. Suzuki, and T. Okamoto, "Verification Tests of a 66 kV HTSC Cable System for Practical Use (first cooling tests)," *Physica C*, 378-381, 1174-80 (2002).

[5]*Superconductivity Communications*, **11**, No.6, Dec. 2002.

[6]T. Masuda, H. Yumura, M. Watanabe, H. Takigawa, Y. Ashibe, C. Suzawa, T. Kato, K. Okura, Y. Yamada, M. Hirose, K. Yatsuka, K. Sato and S. Isojima, "High-temperature Superconducting Cable Technology and Development Trends," *SEI Technical Review*, 59, 8-13 (2005).

[7]K. Patnaik, D. M. Feldmann, Y. Polyanskii, X. Y. Cai, E. E. Hellstrom, D. C. Larbalestier and Y. Hung, "Local Measurement of Current Density by Magneto-optical Current Reconstruction in Normally and Overpressure Processed Bi-2223 Tapes," *IEEE. Tran. Appl. Supercond.*, **13**, 2930-33 (2003).

[8]K. Yamazaki, T. Kato, K. Ohkura, M. Ueyama, N. Ayai, H. Takigawa, E. Ueno, K. Hayashi and K. Sato, "Development of Ag-sheathed Bi2223 Superconducting Wires," *Trans. Mater. Res. Soc.* Jpn., **29**, 1229-33 (2004).

[9]T. Kato, S. Kobayashi, K. Yamazaki, K. Ohkura, M. Ueyama, N. Ayai, J. Fujikami, E. Ueno, M. Kikuchi, K. Hayashi and K. Sato, "Development of High Performance Ag Sheathed Bi2223 Wire," Physica C, **412-414**, 1066-1072 (2004).

[10]S. Kobayashi, T. Kato, K. Yamazaki, K. Ohkura, K. Fujino, J. Fujikami, E. Ueno, N. Ayai, M. Kikuchi, K. Hayashi, K. Sato and R. Hata, "Controlled Over Pressure Processing of Bi2223 long length wires," *IEEE Trans. Appl. Supercond.* **15**, 2534-37 (2005).

[11]Q. Y. Hu, H. K. Liu and S. X. Dou, "Effect of Mechanical Deformation on The Mass Density of Ag-clad (Bi,Pb)2Sr2Ca2Cu3O10 Wire and Tape," *Appl. Supercond.*, **4**, 17-24 (1996).

[12]T. Matsushita, Y. Himeda, M. Kikuchi, J. Fujikami, K. Hayashi, "Characterization of Critical Current Density in Silver-sheathed Bi-2223 Tape," *IEEE Trans. Appl. Supercond.* **15**, 2518-21 (2005).

[13]L. Masur, D. Parker, M. Tanner, E. Podtburg, D. Buczek, J. Scudiere, P. Caracino, P. Corsaro, and M. Nassi, "Long Length Manufacturing of High Performance BSCCO-2223 Tape for the Detroit Edison Power Cable Project,"*IEEE Trans. Appl. Supercond.*, 11, 3256-3260, (2001).

[14]K. Hayashi, N. Saga, K. Ohkura and K. Sato, "Ag Alloy Sheathed BPSCCO Superconducting Wires," *Advances in Superconductivity*, **8**, 811-814 (1996).

Electroceramic Materials and Applications
Edited by R. W. Schwartz

SINGLE GRAIN YBa2Cu3Oy POROUS CERAMIC SUPERCONDUCTORS

J. G. Noudem and E. S. Reddy

CRISMAT, CNRS-UMR 6508, ENSICAEN, 6, Bd Maréchal Juin,
14050 CAEN Cedex 4, FRANCE

ABSTRACT

Using a combination of standard ceramic processing and an infiltration technique, single grain superconductor foams of $YBa_2Cu_3O_y$ (Y123) have been produced. The process, involving negligible shrinkage, results in an improved microstructure with a uniform distribution of submicron $Y_2Ba_1Cu_1O_5$ (Y211) inclusions in the Y123 matrix. The foam structure could be highly favourable for good oxygenation, better cooling and consequently for an efficient heat extraction from the superconducting element in view to overcome the typical hot-spot phenomena commonly observed in superconducting fault current limiters. In this work, the processing conditions and the magneto-transport measurements of these foams are reported. The investigations reveal superconducting properties being similar to those of bulk melt processed materials. The magnetic hystereses vs. field measurements show a high anisotropy of the critical current density in good agreement with the microstructural observations. The advantages of the infiltration growth process and of the novel porous superconductor (Y123) ceramic are discussed.

INTRODUCTION

Since the discovery of the high temperature superconductivity in BaLaCuO[1], the YBCO oxide ceramics have been considered as one of the promising candidates for superconducting applications. Huge effort has been engaged to the development of the processing techniques to prepare the material with properties suitable for practical applications. The applications of superconductors for different systems require various geometries and forms as pellets, rods, thin and thick films, and complex bulk shapes.

The development of different melt processing methods have led to Y123 ceramics with properties acceptable for many applications[2-5]. However, despite these properties some problems exist for e.g. (i) oxygenating the whole bulk material without inducing any micro-cracks during the tetragonal-to-orthorhombic phase transition[6]. (ii) suppression of hot spots, commonly observed during the transport current application. To overcome these problems, the superconducting material processed as a foam structure or with the artificial drilled holes have been developed[7-11]. Using a combination of standard superconducting ceramic processing and an infiltration technique[12-14], single grain superconductor foams of $YBa_2Cu_3O_y$ (Y123) have been processed. An important feature of this process is that it offers the flexibility to produce large samples of near-net shape without distortions and cracks. In this process the liquid source (barium rich liquid phase and copper oxide) infiltrates into porous preformed Y_2BaCuO_5 (Y211) pellet, above the peritectic temperature (T_p) and reacts to form Y123. This can be expressed by the chemical reaction:

$$Y_2BaCuO_5 + (3BaCuO_2 + 5CuO) \longrightarrow 2\ YBa_2Cu_3O_x$$

The novel morphology of superconducting foam material processed seems to be a good candidate for increasing interfacial flux pinning if the pores can be made sufficiently small. Many other prospects are related to the foam structure like e.g.

- more efficient heat transfer,
- faster oxygenation and less related micro-cracking, possibility of reinforcement and of interlocking connections etc..

Objectives of the present work are to give an overview of:

- single grain Y123 foam/porous bulk fabrication
- the properties of the superconducting foam material.

EXPERIMENTAL

The Y211 precursor foams of different dimensions were fabricated as replica of commercial polyurethane foam (figure 1) with a porosity of 10-100 ppi. Details of $YBa_2Cu_3O_y$ (Y123) foam preparation have been reported elsewhere[7, 8]. Figure 2 shows the schematic diagram of different steps involved in the preparation of Y123 foam by an infiltration-growth process.

The liquid phase ($3BaCuO_2 + 5CuO$) or $Ba_3Cu_5O_8$ (Y035) was prepared by solid state milling of stoichiometric amounts of high purity $BaCO_3$ and CuO in the agate bowl of a planetary mill for 2 hours. The mixed powder was then heat-treated at 870°C, 24 hours with intermediate grindings. The preparation of Y211 foams involves the coating/impregnation of polyurethane foam of desired porosity with optimised water based Y211 slurry. This slurry was prepared by mixing commercially available Y211 fine (~2µm) powder in water with 5 wt% polyvinyl-alcohol (PVA) as binder. The process of coating/impregnation of polyurethane foams involves dipping the foams into the Y211 slurry and drying at room temperature. Repeating the step several times results in a thick Y211 coating. The organic components PVA and polyurethane were burned by slow heating at a rate of 50°C/h to 600°C for 6 hours in air. The final Y211 foam was sintered in air at 1100°C, 20 hours for densification. The final green Y211 pre-form was transformed into single-grain Y123 superconducting foam by infiltration growth process reported elsewhere[7, 12-14] in using the configuration shown in figure 2. The sample was rapidly heated to 1050°C for 30 min. At this stage the liquid phase infiltrate into the interstitial spaces between Y211 particles due to the capillary action. The peritectic reaction occurring between the green Y211 reservoir and the liquid phase source Y035 during slow cooling stage (1010°C to 980°C at 0.3°C/h) from the peritectic temperature results in the growth of Y123 grain. The samples, after grain growth, from 980°C were cooled to room temperature in 6 hours. The nucleation and orientation of Y123 phase is controlled by placing a *c*-axis oriented seed-either $NdBa_2Cu_3O_y$ (Nd123)/$SmBa_2Cu_3O_y$ (Sm123) or MgO-on top of the foam with a Y_2O_3 cloth[15] as support for the seed and to facilitate a homogeneous grain growth into the foam. In addition, the as-processed samples were oxygenated at 450° C for 150 hours in flowing oxygen. For microstructural examination, the samples were mounted using cold resin and polished on a rotating machine (Tegrapol-31, Struers Inc.) using varying grades of diamond particle solution until ¼ µm as a final. The polished surface were examined using an optical microscope (Olympus, BH2-UMA) and scanning electron microscope, SEM (Philips XL 30 FEG) with EDX. The texture of the sample was performed using x-ray diffraction and pole figure measurements of the as-processed foam using X'Pert (Philips) device.

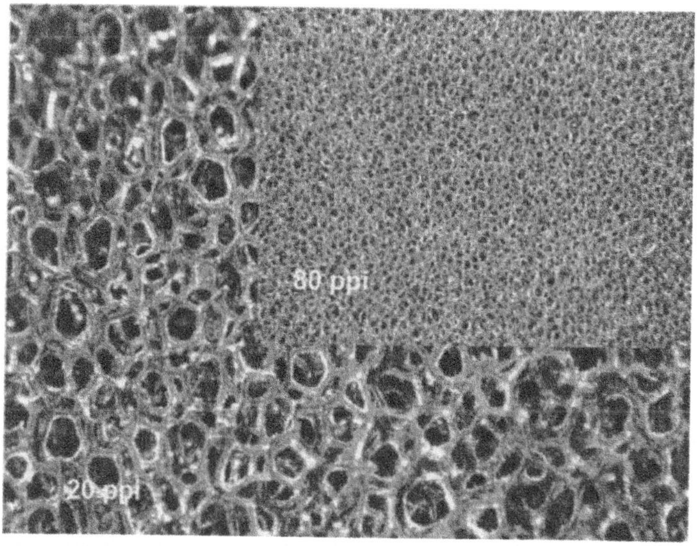

Figure1: commercial polyurethane foams of various pore sizes

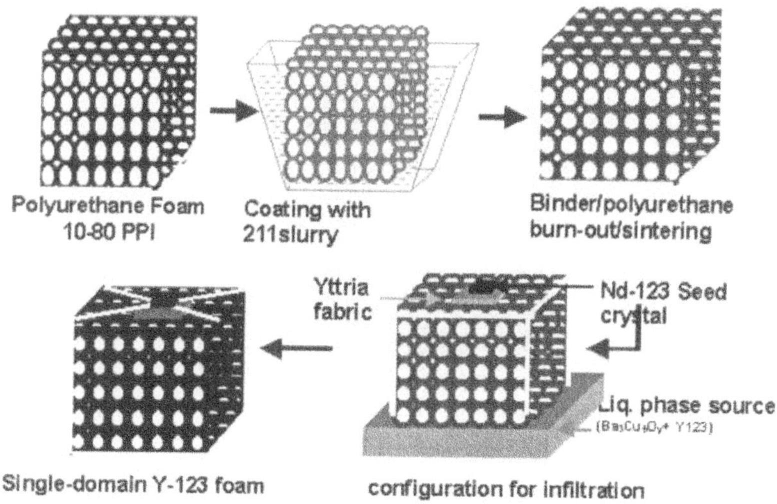

Figure 2: Schematic diagram illustrating the processing of single grainY123 foam

Resistivity measurements were performed using the standard four-point contact method. Contacts between the sample and the current leads have been fabricated by painting silver paste (4929 Dupont Inc.) and subsequent annealing at 900°C for half an hour to ensure a good silver diffusion followed by oxygenation at low temperatures. A commercial PPMS (Magnet Power Supply) apparatus and a SQUID (Quantum Design) magnetometer were used for resistivity and magnetization measurements.

RESULTS AND DISCUSSION

Microstructure

The macroscopic top of the Y123 foams view are presented in figure 3a. We can observe that, the resulting Y123 foam sample is single domain. The sample has an open porosity of 20 ppi (pores per inch) with a strut thickness of about 300 μm (figure 3b). Various number of single-grain Y123 foams of pore size between 10 to 100 ppi were reported elsewhere[8]. Polished surface (figure 3c) of the struts reveal the typical microstructural features known from melt processed bulk materials, with parallel platelet gaps and trapped Y211 particles in a Y123 matrix. From the microstructures the volume content of Y211 particles is estimated to be around 40 volume percent with an average particle size around 2-5 μm. No specific efforts like doping[16-17] or irradiation[18] have been made to optimize the Y211 particle size and defect density with respect to high critical current densities.

The homogeneous distribution of Y211 inclusions in the infiltration-process can be explained as due to the small Y211 particles in starting Y211 pre-form. In this process, the Y211 particles are further dissolved in the liquid phase to form Y123 phase resulting in finer spherical 211 particles, in contrast to larger acicular size Y211 and their inhomogeneous distribution in the sample processed conventionally without any dopants[14].

Figure 3a: Photograph of a single domain Y123 foam with a porosity of 20 ppi (pores per inch)

Textural analysis

The Y123 foam was analyzed using XRD diffraction to check the texture and the final composition. The diffraction spectrum of the polished surface of the sample is shown in figure 4. The single-grain signature of the material, presence of only $\{00\ell\}$ peaks, can be evidenced from

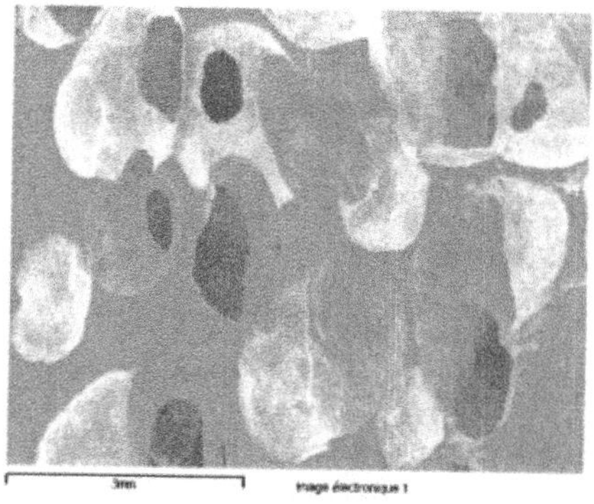

b)

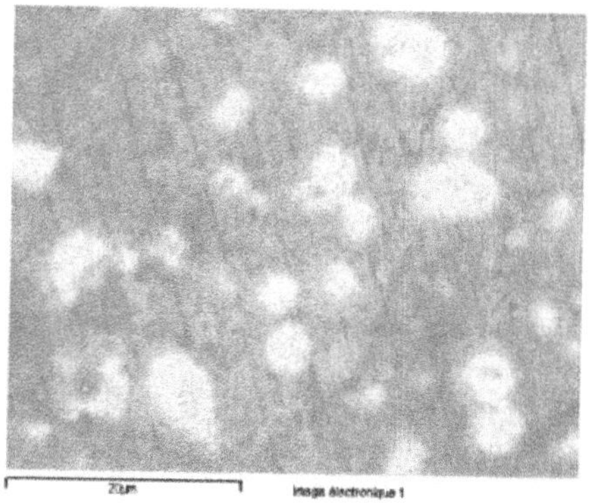

c)

Figure 3: SEM picture, b) of a fractured surface of a superconducting Y123 foam. c) of the polished surface of the struts reveals residual Y211 particle (white) inclusions in the 123 matrix.

X-ray diffraction pattern. The XRD pattern also shows that the sample is a single phase with no observable impurities e.g extra liquid phase. Probably during infiltration process, according to the chemical reaction give in the introduction part and after the liquid saturation, the extra liquid is rejected like commonly observed on the sample supports after melt processing.

Resistivity

The temperature dependence of the resistivities of various bars were measured. The bars were cut parallel and perpendicular to the surface of the seeded plane in agreement with the

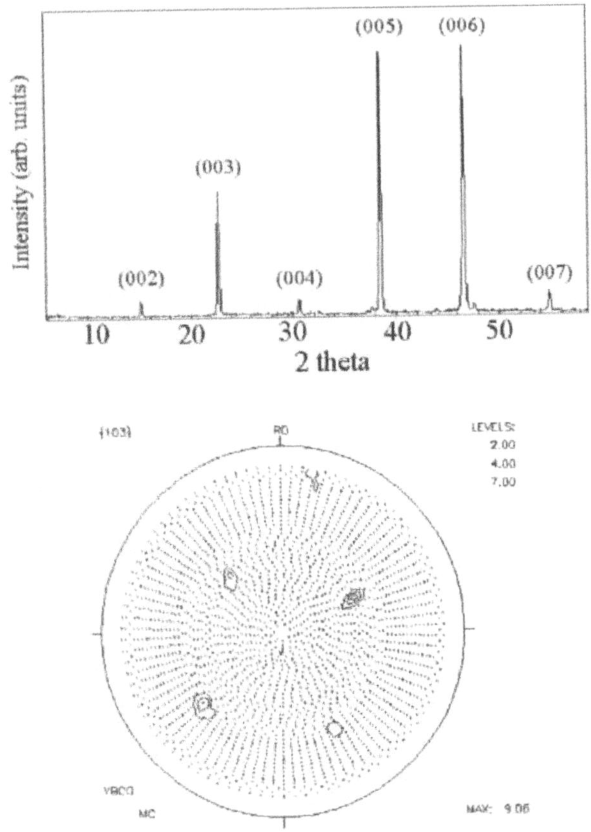

Figure 4: (*a*) X-ray diffraction pattern and (*b*) 103 pole figure recording of the single domain YBa2Cu3O7−*x* foam achieved from foams embedded in epoxy resin. A polished surface parallel to the seed crystal is used for x-ray characterization. The intense {00ℓ} peaks indicate the c-axis texture and the presence of four-fold symmetric poles confirms the single domain nature of the foam.

microstructure. Suitable contacts allowed the current to be passed either in the (*ab*) planes or parallel to the c-axis. Figure 5a shows the resistive transition of a typical Y123 foam sample. Both configurations exhibited a narrow superconducting transition ($\Delta T \approx 1.5$ K) with T_c (onset) ≈ 91.5 K and T_c ($\rho = 0$) ≈ 90 K. The room temperature resistivity along the ab-planes and along the c-axis are 2 and 16 mΩ.cm respectively. The resistivity ratio of about 8 for the parallel and perpendicular configuration confirms the texture/single grain nature of the foam superconductor.

Magnetization Jc

The zero field cooled magnetization curve, figure 5b, shows a very narrow transition with an onset at 92 K, comparable to values taken from above resistivity R (T) curves. The steep transition characterizes an essentially pure Y123 phase and indicates no substantial contaminations originating from the initial polyurethane foam and organic compounds being burnt to form Y211 precursor foam.

In order to correlate sample texture and magnetic anisotropy, hysteresis cycles with magnetic fields applied parallel and perpendicular to the direction of the single domain surface

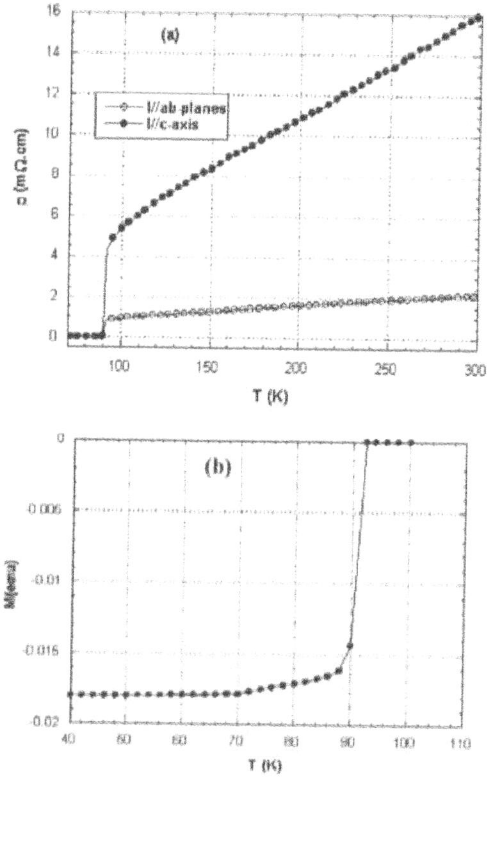

Figure 5: (a) Electrical resistivity versus temperature according to two configurations, the current, I, being injected parallel to the *ab*-planes or parallel to the *c*-axis into bar-shaped foam samples. (b) Magnetization vs. temperature M (T).

have been performed at 77 K on a 2x2x2 mm^3 sample, figure 6a. The magnetic J_c values are estimated for both directions from M-H cycles on the basis of the Bean model taking $J_c = 30\Delta M/d$, where, d, is the sample size in cm and ΔM in emu/cm^3 is the hysteresis of the magnetization. Figure 6b presents the J_c (H) determined from the width of hysteresis loops at two configurations. J_c^{ab} corresponds to the average critical current density in the ab-planes under magnetic field and J_c^c, the critical current density along the c-axis.

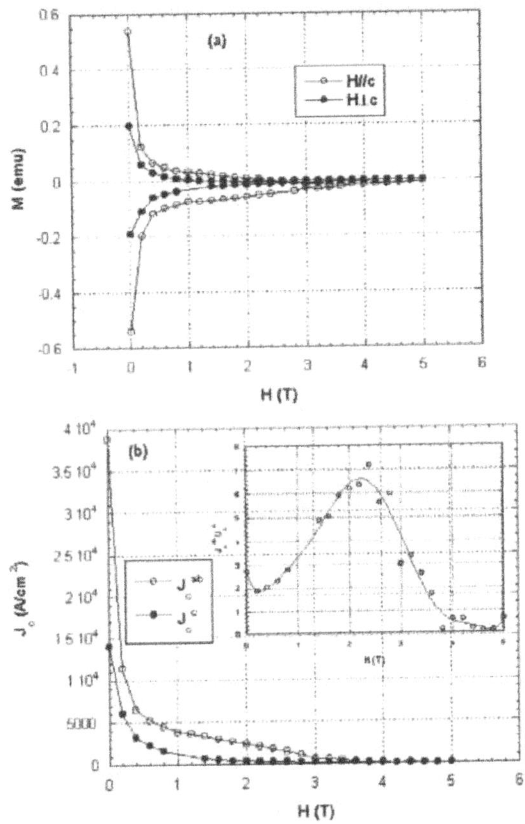

Figure 6: (a) Magnetic hysteresis cycle M(H) at 77 K following two directions of the measuring magnetic field parallel and perpendicular to the c-axis, (b) field dependence of the critical current density J_c deduced from M (H). The inset shows the anisotropy of J_c as a function of the external magnetic field.

A critical current density of $J_c^{ab} \approx 40$ kA/cm^2 at 0 T can be deduced for current parallel to the ab-planes. A $J_c^c \approx 15$ kA/cm^2 is obtained for currents parallel to the c-axis, inset figure 6b, leading to an anisotropy factor J_c^{ab}/J_c^c of about 7 at 2 T.

These values are similar to those of single domain bulks with a non-optimized Y211 content and size distribution. There is a lot of scope for further improvement of J_c in the Y123-foams by refining the Y211 distribution in their microstructure using e.g. doping methods known from bulk materials processing[16, 17]. The transport current measurements have been performed using pulse current[7]. The current across textured Y123 foam exceeded 1000 A during 150 ms. This corresponds to the current carry through 0.1 cm^2 cross section and critical transport current density exceeding 10^4 A/cm^2 at 77 K in self field.

Outlook

Foams of conventional ceramic materials or porous structures, such as alumina and zirconia, are established components in a number of industrial applications such as filters, structures for catalysts, elements for thermal insulation and flame barriers. The combination of high surface area and low density of YBCO superconducting foams makes them interesting candidates for both a variety of novel applications and for fundamental studies. YBCO foams of strut thickness of a few hundred μm or artificial perforated Y123 bulk into desired structure (figure 7), for example, are good candidates for resistive elements in superconducting fault current limiters. In this application, the low thickness of the struts or wall between the holes allows more efficient heat transfer between foam or perforated superconductor and cryogenic coolant during an over-current fault compared to conventional bulk material. Superconducting bulks with artificial array of holes can be reinforced continuously with alloys[5] or high strength resins to improve their mechanical and thermal properties to overcome the forces encountered in levitation and quasi-permanent magnet applications. The high surface area of the foams, which may be adjusted by varying the pore size, makes them interesting candidates for studying fundamental aspects of flux pinning, since the extent of surface pinning, and hence J_c, are expected to differ significantly from bulk YBCO grains of similar microstructure.

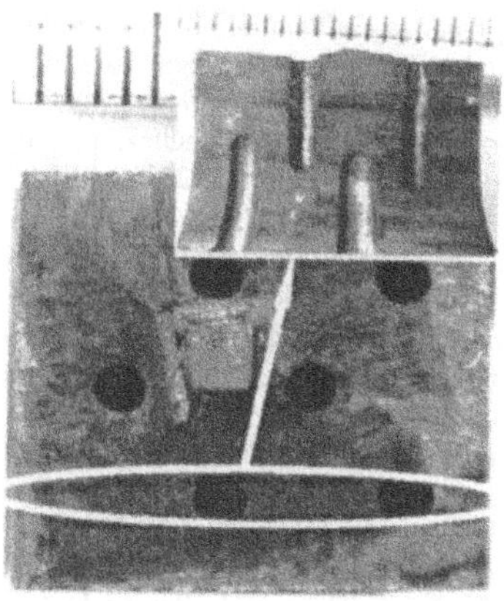

Figure7: Feasibility of c-axis meander shape (inset) for fault current limiter issue from single domain monolith sample with 2 mm holes.

CONCLUSION

The textured YBCO (Y123) foams have been successfully prepared from commercially available polyurethane using infiltration-growth process. The single grain nature and microstructure resulted in properties similar to those of bulk materials.

The properties of the novel superconducting Y123 foams have been investigated. From R (T) and M (T) measurements, a T_c of 92 K with a (narrow) transition width of 1.5 K has been deduced indicating an uncritical contamination by residuals of the organic raw materials used to manufacture the foams. Magnetic hysteresis curves reveal a critical current density of $J_c^{ab} \approx 40$ kA/cm^2 and a high anisotropy $J_c^{ab}/J_c^c \sim$ of up to 7 in an external field of 2 T. First transport current measurements with a pulsed current source indicate a transport J_c exceeding 10 kA/cm^2. Currently this value is mainly limited by insufficient quality of the contacts. Further investigations on contact manufacture, transport J_c, optimization of Y211 particle size and distribution as well as on the thermal and mechanical properties of the superconducting foams are necessary. The possibility of fabricating high J_c 123 foam in various structures allows a host of applications using this new form of material, one of these applications being the superconducting element of a resistive fault current limiter. The important property required for a FCL element is the rapid dissipation of thermal energy during fault conditions without burning the superconductor. The use of 123 foam with a large surface area allows us to effectively extract the heat generated in the superconductor.

REFERENCES

[1]J.G. bernorz and K.A. Müller, "possible high Tc superconductivity in the Ba-La-Cu-O system" Z. Phys. B-Condensed Matter 64, 189-193 (1986).

[2]W. Paul, M. Lakner, J. Rhyner, P. Unternahner, Th. Baumann, M. Chen, L. Widenhorn and A. Guérig, "Test of 1.2 MVA high- Tc superconducting fault current limiter", Supercond. Sci. Technol., 10, 914-918 (1997) 914.

[3]X. Obradors, T. Puig, E. Mendoza, J. Plain, J. Figueras, X. Granados, A.E. Carillo, E. Vares, F. Sandiumengue and P. Tixador, "Tuning the critical currents in bulk MTG YBCO for current limiting devices", Supercond. Sci. Technol., 13879-885(2000).

[4]J. Noudem, J. M. Barbut, O. Belmont, J. Sanchez, P. Tixador and R. Tournier, "Current limitation at 1080 A under 1100 V with bulk Bi-2223", IEEE Trans. Appl. Supercond., 9, 664-667 (1998).

[5]M. Tomita and M. Murakami, "High-temperature superconductor bulk magnets that can trap magnetic fields of over 17 tesla at 29 K", Nature 421, 517-520 (2003).

[6]D. Isfort, X. Chaud, R. Tournier and G. Kapelski, "Cracking and oxygenation of YBaCuO bulk superconductors: application to *c*-axis elements for current limitation ", Physica C 390, 341-355 (2003).

[7]E.S.Reddy and G. J. Schmitz, "Superconducting foams", Supercond, Sci. Technol 15, L21-L24 (2002).

[8] E. S. Reddy, M. Herweg and G. J. Schmitz, "Processing of Y_2BaCuO_5 foams", Supercond. Sci. Technol. 16, 608-612 (2003).

[9]X. Chaud, S. Meslin, J. Noudem, C. Harnois, L. Porcar, D. Chateigner, R. Tournier, "Isothermal growth of large YBaCuO single domains through an artificial array of holes", Journal of Crystal Growth., 275, 855-860 (2005).

[10]J.G. Noudem, S. Meslin, C. Harnois, D. Chateigner and X. Chaud , "Melt textured $YBa_2Cu_3O_{7-x}$ bulk with artificially patterned holes: a new way of processing c-axis fault current limiter meander", Supercond. Sci. Technol. 17, 931- 936(2004).
[11]E Sudhakar Reddy, NHari Babu, Y Shi, D A Cardwell and G J Schmitz, "Processing of large grain Y-123 superconductors with pre-defined porous structures", Supercond. Sci. Technol. 18, 15-18 (2005).
[12]E. S. Reddy E and T. Rajasekharan, "Fabrication of textured $REBa_2Cu_3O_7/RE_2BaCuO_5$ (RE = Y, Gd) composites by infiltration and growth of RE_2BaCuO_5 preforms by liquid phases", Supercond. Sci.Technol. 11, 523-534(1998).
[13]N. H. Babu and T. Rajasekharan, "Infiltration-growth Processing of NdBa2Cu3O7-d Superconductor", J. Am. Ceram. Soc, 82, 2978-84 (1999).
[14] S. Meslin, and J.G. Noudem, "Infiltration and top seeded growth mono-domain $YBa_2Cu_3O_{7-x}$ bulk superconductor", Supercond. Sci. Technol. 17, 1324 -1328(2004).
[15]Zircar Products Inc., www. Zircarzirconia.com, Florida NY, USA.
[16]C. Harnois, "Attempts and success in texturing large single domains of cerium and silver doped YBaCu3Oy", Supercond. Sci. and Technol., 15, 864-870 (2002).
[17]F. Giovannelli and I. Monot-Laffez,"YBCO bulk material processed in an oxygen controlled melt-growth process", Supercond. Sci. and Technol., 15, 1193-1198 (2002).
[18]P. Prozorov, A. Tsameret, Y. Yeshurun, G. Koren, M. Konczykowski and S. Bouffard, "Frequency dependent irreversibility line and undirectional magnetic anistrophy in thin $YBa_2Cu_3O_{7-x}$ films irradiated with heavy ions", Physica C 235, 3063-64 (1994).

Electroceramic Materials and Applications
Edited by R. W. Schwartz

BENEFITS OF USING A NANOSTRUCTURED TARGET IN LASER ABLATION OF THIN YBCO FILMS

P. Paturi, H. Huhtinen, M. Peurla[1], Yu. P. Stepanov[2], J. Raittila and R. Laiho
Wihuri Physical Laboratory,
Department of Physics,
University of Turku, FIN-20014 Turku, Finland
[1]also at Graduate School of Materials Research, Turku, Finland
[2]also at A.F. Ioffe Institute, St. Petersburg, Russia

ABSTRACT

Nanostructured targets with grain size of about 100 nm have been used for pulsed laser deposition of $YBa_2Cu_3O_{6+x}$ (YBCO) thin films. The targets were prepared by sintering from nanosized powder (1-3 nm high and 50-100 nm wide particles). Using nanostructured targets in PLD leads to smaller growth islands on the substrates than using micron size grained targets. The smaller growth island size leads to smaller twin domain size and faster relaxation of the twin lattice. The denser defect structure in the nanofilms leads to better pinning and hence higher J_c, even up to 8 MA/cm^2 at 77 K. The pinning can be further enhanced by doping the target with $BaZrO_3$ (BZO), which improves the high field critical current density of the films with 350 % at 5 K and 5 T. The BZO is found to orient itself with the YBCO lattice and affect the twin lattice relaxation and therefore we believe that the BZO is situated in the growth island boundaries.

INTRODUCTION

Superconducting $YBa_2Cu_3O_{6+x}$ (YBCO) thin films are of great interest for applications such as electronics[1] and coated conductors.[2–4] Specially for coated conductors the dependence of critical current density, J_c, on magnetic field and temperature is important. J_c is largely defined by the density and nature of the magnetic flux pinning sites present in the film[5] and the way of increasing the J_c is to increase the number of strong pinning sites in the film. This can be achieved e.g. by irradiating the samples by high energy particles,[6,7] by adding secondary phases[8–10] or by increasing the number of intrinsic pinning sites by changing the deposition parameters.[11–13] Irradiating leads to columnar pinning sites, which enhance zero field J_c with small irradiation doses but decreases J_c with higher doses. Small doses of irradiation also increase the characteristic or accommodation field B^*, where the vortices have accommodated all the pinning sites and their interactions become more important. At B^* J_c also begins to decrease with increasing field. Adding small amounts of secondary phases such as $BaZrO_3$, Y_2O_3 or Y_2BaCuO_5 increases B^*, but does not affect zero field J_c or even lowers it slightly. Both J_c and B^* increase when the deposition parameters are changed to increase the amount of intrinsic pinning sites. In practice for applications it is much easier to change the deposition parameters to reach higher J_c and B^* than to irradiate all the products with high energy particles.

Pulsed laser deposition (PLD) is the method that produces the best quality films, although it might not be the preferred choice in large scale applications.[2] PLD and the films produced by PLD, however, give a standard and easy way of producing high quality films, where the physics behind

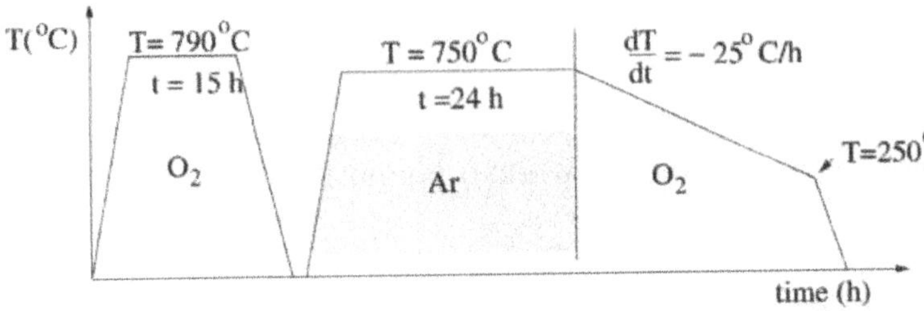

Figure 1: The annealing process for the YBCO nanopowder.[15]

the high J_c can be easily studied.

In this paper we review the work done so far on YBCO films deposited from a target which has nanometer sized grains. The preparation of the original nanosized powder, the targets and the films are first briefly described, then the differences in the deposition processes of different kind of targets are described. The properties of the films are compared to films deposited from a micron size grained target and finally the recent results on films deposited from targets doped with $BaZrO_3$ are presented.

PREPARATION OF THE TARGET AND FILMS

The nanostructured target for the laser ablation was pressed and sintered from nanosized YBCO powder, which was produced by the citrate-gel method.[14,15] In the citrate-gel method the nitrates of the metals Y, Ba and Cu are mixed with the correct 1:2:3 ratio and citric acid is added to act as a chelating agent. Then ethylenediamine is added dropwise until the pH is 6. The solution is slowly turned to a gel in an oven at 80°C, after which it is fired at 500°C in air to remove the organic compounds. The precursor powder is then annealed in oxygen and argon according to the process shown in Fig 1. This results in YBCO powder with $T_c = 92$ K and average particle height of 3 nm and radius of 50 nm.[16] The particle size distribution as measured by AFM is shown in Fig. 2.

The target was pressed from the nanopowder using 150 MPa pressure and subsequent 20 h sintering at 900°C. The density of the target is 4.7 g/cm^3. According to Rietveld refinement the target is pure YBCO with grain size of 100 nm. In the case of the $BaZrO_3$ (BZO) doped targets, BZO was added already in the solution phase in form of metal salts.[17] The actual BZO content of the target was determined from the Rietveld analysis of the XRD data.

The YBCO films were usually deposited on single crystal $SrTiO_3$ (100) substrates using the nanopowder targets (n-targets) and micrograined target (μ-target). A XeCl excimer laser ($\lambda = 308$ nm) was used to excite the plume inside a deposition chamber made from stainless steel.[18] The oxygen pressure in the chamber was 0.4 Torr and the temperature of the substrate was 725°C during the deposition of the best films.[11] The films ablated from the n- and μ-target are hereafter called n- and μ-films, respectively.

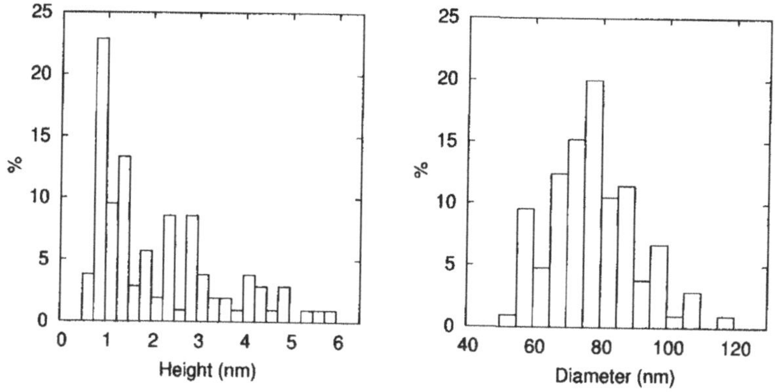

Figure 2: The size distribution of the nanopowder.[15]

THE DEPOSITION PROCESS

The deposition process has been investigated by measuring the emission spectra of the plume and particle sizes and shapes in the plume.[19] The emission spectra did not show any significant differences between the plumes from the n-target and the μ-target, although the intensities of the lines in the previous case were slightly stronger, indicating a higher temperature of the plume emitted from the n-target.

The size and the shape of the particles in the plume were obtained from AFM images of particles captured on pieces of a silicon wafer placed into the plume at different distances from the surface of the target (Fig. 3). At the deposition distance (25–40 mm) the average diameters of the particles in the plume from the n-target (35 – 65 nm) were found to be smaller than those in the plume from the μ-target (45 – 110 nm).[19,20] Slightly surprisingly, it was found that near the target (10 mm) the particles from the n-target are wider than those from the μ-target. Though, the largest particles do not reach the substrate. The heights of the particles in the n-target plume are always smaller than those of the μ-target. The sizes of the particles in the n-target plume correlate well with the particle size in the target, whereas in the μ-target plume the particles are much smaller than in the target. Therefore it is plausible that whole grains are detached from the n-target by the laser pulse and the smallest of them migrate to the substrate to act as growth islands. In the μ-target this process is very rare and therefore the growth islands have to form from the atoms and molecules in the plume by clusterization,[21] which is also seen as the growth of the particle size at longer distances from the target (Fig. 3). This difference in the target evaporation processes might explain the differences seen between the films.

The kinetics of sparse ablated particles has also been studied[19] by using samples ablated with smaller laser energy (4 mJ) and only a few pulses. After ablation the samples were studied by AFM and annealed at different temperatures for 1 h in flowing oxygen. The particles deposited from the μ-target started to grow larger at lower temperatures than the particles from the n-target. The

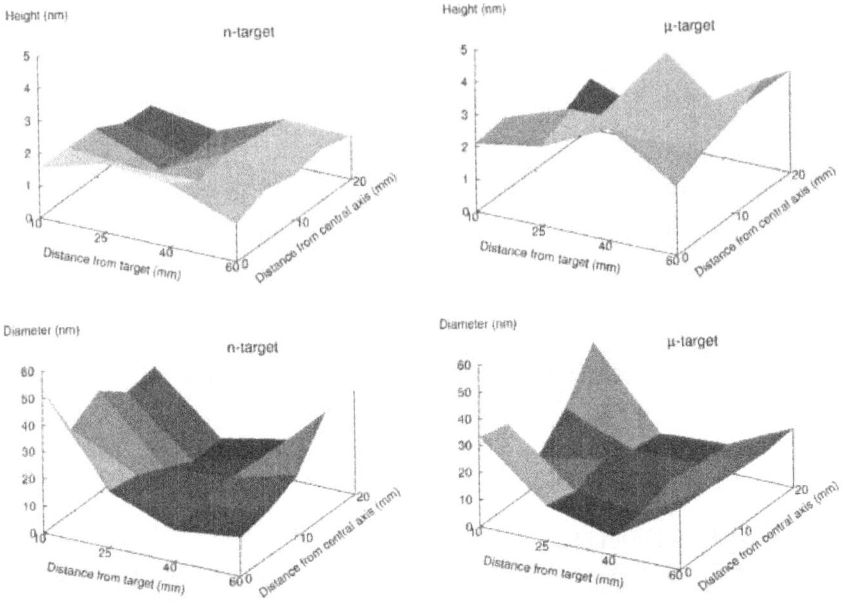

Figure 3: The heights and diameters of the particles captured at different distances from the plume for the n- and μ-target.

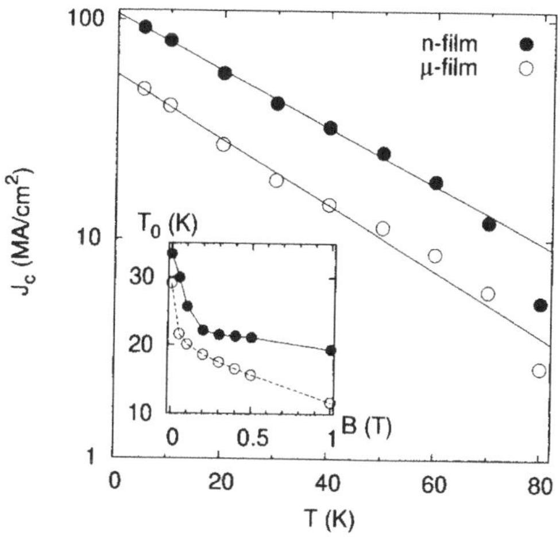

Figure 4: The temperature dependence of J_c for μ- and n-films in zero applied field. The lines are fits to Eq. (1). The inset shows the field dependence of the characteristic temperature T_0 obtained from the fits.[22]

growth temperature of the particles from the μ-target was about 400°C after which their height and diameter increased quite rapidly. A clear increase in the height and the diameter of the particles from the n-target was not observed until at 800°C. The density of the particles deposited from the μ-target and the n-target on $SrTiO_3$ substrates decreased rapidly above the same temperatures of 400°C and 800°C, respectively. It is noteworthy that the growth temperature of the particles from the n-target is higher than the usual deposition temperature of 725°C.

PROPERTIES OF THE FILMS

The best pure films deposited from the n-target have $T_c = 92$ K and $J_c = 8{\cdot}10^6$ A/cm^2 at 77 K in zero field.[11] The temperature dependences of zero field J_c for n- and μ-films are shown in Fig. 4. When fitted with a function of the form

$$J_c(T) = J_0 e^{-T/T_0}, \tag{1}$$

the characteristic temperature, T_0, (inset in Fig.4) describes the dependences of J_c on temperature. It is found to be higher for n-films than μ-films in the whole field range. Thus J_c decreases faster with temperature in the μ-films than in the n-films. Also the shapes of the $T_0(B)$-curves are somewhat different: after the steep fall in the low fields, T_0 of the μ-film decreases much faster than the one of the n-film.[22]

The surface RMS roughness of the n-films has been found very low, down to 1.5 nm.[11] It is worth noting that the RMS roughnesses of the films correlate quite well with the average height of the particles observed in the plume (Fig. 3), 1.5 nm for the n-films and 4 nm for the μ-film.

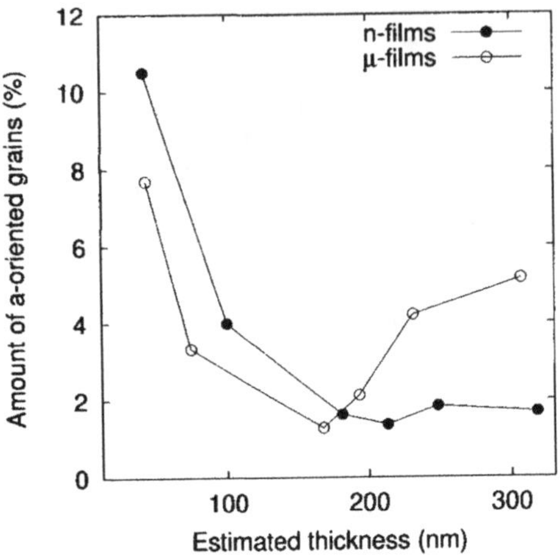

Figure 5: The percentage of the material with its a-axis perpendicular to the plane of the substrate in the μ- and n-films.[24]

Detailed analysis of the magnetic pinning and relaxation in the films[22] indicates that the higher J_c in the n-films is due to higher density of pinning sites rather than stronger pinning sites. By using a theory of strong sparse pinning sites[23] we can estimate the physical size and density of the pinning sites. It turns out that the density of the pinning sites is a decade larger for the n-films than the μ-films, whereas the sizes of the pinning sites were found to be about the same size, a few nanometers in a/b-direction, but extended to over 100 nm in the c-direction. This is the natural result if the pinning sites are considered to be e.g. dislocations, growth island boundaries and twin planes.

Structural XRD analysis of the films of different thicknesses[24] has shown that split in the XRD peak, due to the relaxation of the strains in the lattice by turning one twin colony, reaches its fully relaxed value of 0.9° faster in the n-films than in the μ-films. This is thought to be due to the smaller size of the domains, which is due to the smaller growth island size observed by AFM measurements.[19] Smaller growth island size leads to larger amount of defects in the film, which are situated where the growth islands grow toghether, also the twin boundaries prefer these places.[25] The twin domain size in the n-films has been found to be about 10 nm by high-resolution TEM, whereas it is about 40 nm in the μ-films on MgO.[26] Since the substrate induced strain is smaller in films on STO the twin spacing is expected to be even larger in μ-films on STO.[26] In cross section images the twins are also observed in the first 100 nm of the thickness of the n-film, whereas in the μ-films they are observed only higher up in the film. This is in perfect agreement with the faster relaxation of the lattice in n-films than in μ-films.

Thus the evidence from both structural and magnetic data shows that the use of a nanostructured target leads to higher density of the extended pinning sites. X-ray photoemission spectroscopy

(XPS), on the other hand has shown that there are no differences in the chemical composition of the films.[27] This agrees well with the evidence obtained from the deposition process.

Furthermore, structural XRD data shows[24] that as the n-films get thicker (above 200 nm) there is much less formation of grains with their a/b-axis perpendicular to the substrate than in the μ-films (Fig. 5). The good quality of the n-films all the way to the surface is probably one of the reasons of the high J_c and can prove to be crucial in electronic applications along with the very smooth surface of the films.

Recent magneto-optical microscopy data[28] also shows that when the critical current is measured as a function of the direction in the ab-plane, no dependency is observed for the n-films whereas for the μ-films the J_c changes with the angle. This can be explained by the smaller twin domains in the n-films, which leads to "dead-end" twins, which block the movement of the vortices.[29] Whereas in the μ-films the twin boundaries channel the movement of the vortices in the films. Magneto-optical and magnetometer data also shows that the accommodation field B^* is much higher for the n-films ($\approx$ 150 mT) than the μ-films ($\approx$ 40 mT). The B^* has been found to depend on the dislocation density of the film,[12] which also agrees with the previous conclusion of the n-films having a higher density of pinning sites.

PROPERTIES OF THE BZO DOPED FILMS

By doping the n-films with varying amounts of BZO the pinning properties can still be greatly enhanced. According to the systematic study[30] of different amounts of doping there is an optimum value for BZO in the YBCO lattice, which is about 4 wt-%. With doping the T_c decreases slightly but the zero field J_c increases and also the B^* increases radically (Fig. 6). The accommodation field B^* of 4 wt-% BZO doped film is 500 mT, which indicates a very large amount of strong extended pinning sites in the films.

The shape of the $J_c(B)$ curves also changes radically with BZO doping. Usually the decrease of J_c above B^* can be described with the exponential law $J_c \propto B^{\alpha}$, where the exponent α is predicted to be -0.5 – -3 by different pinning theories.[22] The pure n-films obey the $\alpha \approx -0.5$ dependency predicted by the strong pinning theories, but in the BZO doped films $\alpha \approx -0.2$ which is more than predicted by any theory. It also means that even above B^* the decrease of J_c is much slower than in any other YBCO films studied so far. This causes the enhancement of J_c caused by BZO doping to be even larger in high fields, as much as 350% in 5 T at 5 K (Fig. 7).

The XRD studies of BZO doped films[30] show that the BZO interferes with the twin structure of the films. The peak split in ϕ, which is 0.9° for totally relaxed film, decreases with increasing BZO content (Fig. 8), indicating that the strain caused by the orthorhombicity of YBCO is relaxed by the BZO instead of turning the twin domains. The pole figure (Fig. 9) of the (110) peak of BZO shows that the BZO is perfectly aligned with the surrounding YBCO matrix. This and the large width of the x-ray peak of BZO indicating small particle size, gives us reason to believe that the BZO is situated in the twin/growth island boundaries in the form of nanoparticles. TEM confirms this and reveals that the BZO particles are aligned in "chains" spaced about 30–40 nm apart. According to TEM the average size of the BZO particles is 5–10 nm.

CONCLUSIONS

We have reviewed the properties of the thin films laser ablated from nanograined targets. It

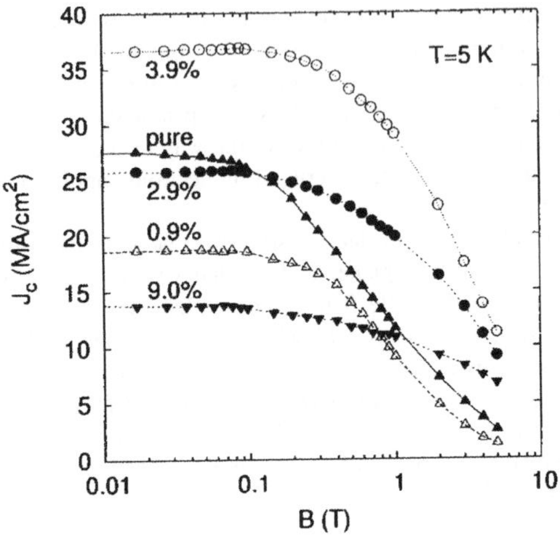

Figure 6: The dependency of J_c on magnetic field for the differently doped films.[30]

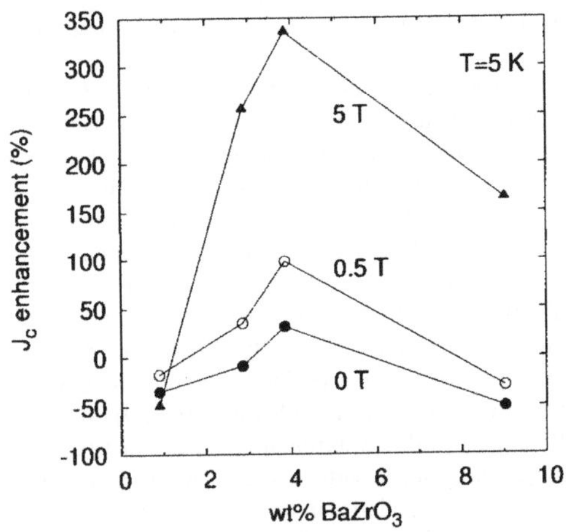

Figure 7: The enhancement of J_c at 5 K in different external magnetic fields for the differently BZO doped films.[30]

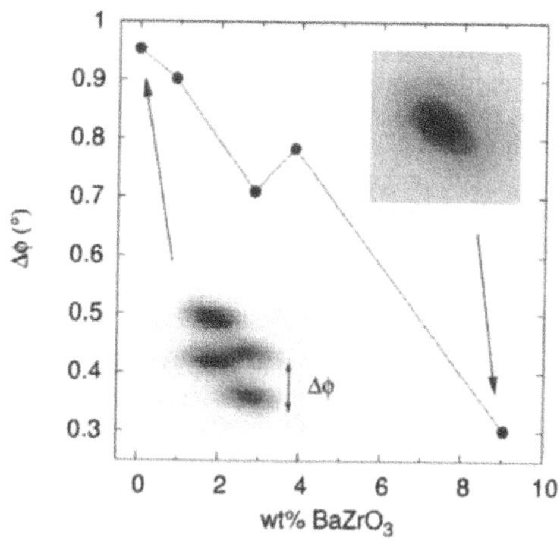

Figure 8: Dependency of the twin produced split in the (212)-peak of BZO-doped films as the function of the dopant concentration.[30]

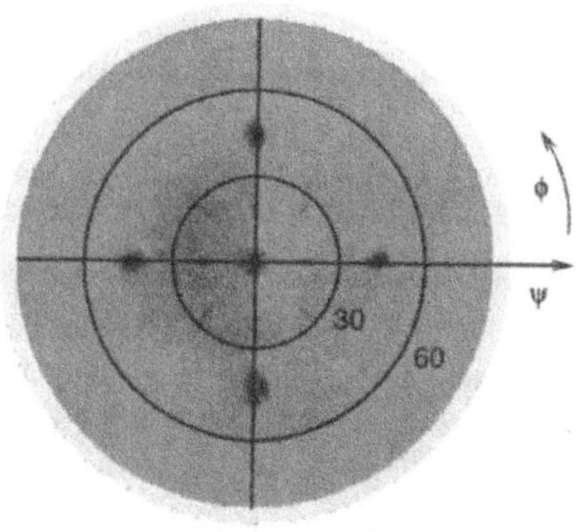

Figure 9: The pole figure of the (110) BZO peak for the 9 wt-% doped film.[17] The large peaks at $\psi = 45°$ indicate that the BZO is perfectly aligned with the surrounding YBCO lattice. The little peaks at $\psi = 28°$ arise from the MgO substrate.

can be concluded that by using a nanograined target the J_c of the films can be enhanced, which is due to additional strong pinning sites caused by the smaller growth island size on the substrate. The films from the nanograined target also have smoother surface as compared with films from a micrograined target. This is probably due to the smaller particle size in the plume during deposition.

By doping the target with $BaZrO_3$ the J_c can be further increased at the same time as the accommodation field B^* increases from 150 mT to 500 mT in 5 K. The J_c enhancement is even more pronounced in high fields and can be as high as 350 % in 5 T at 5 K. Thus the doped films retain their properties much better in high fields than the undoped films.

ACKNOWLEDGMENTS

We would like to thank Mr J. Järvinen for finding the plume particle size data from his old files. This work has been continuously supported by the Academy of Finland and the Wihuri Foundation.

REFERENCES

[1] N. Klein, High-frequency applications of high-temperature superconductor thin films, Rep. Prog. Phys. 65 (2002) 1387.

[2] T. Watanabe, Y. Shiohara, T. Izumi, Progress and future prospects of research and development on coated conductors in Japan, IEEE T. Appl. Supercon. 13 (2003) 2445.

[3] A. Usoskin, H. C. Freyhardt, A. Issaev, J. Dzick, J. Knoke, M. P. Oomen, M. Leghissa, H.-W. Neumuller, Large area YBCO-coated stainless steel tapes with high critical currents, IEEE T. Appl. Supercon. 13 (2003) 2452.

[4] Y. Iijima, K. Kakimoto, T. Saitoh, Fabrication and transport characteristics of long length YBCO coated conductors processed by IBAD and PLD, IEEE T. Appl. Supercon. 13 (2003) 2466.

[5] G. Blatter, M. V. Feigel'man, V. B. Geshkenbein, A. I. Larkin, V. M. Vinokur, Vortices in high-temperature superconductors, Reviews of Modern Physics 66 (1994) 1125.

[6] E. Crescio, R. Gerbaldo, G. Ghigo, L. Gozzelino, C. Camerlingo, A. Monaco, C. Nappi, G. Cuttone, A. Rovelli, Interplay between as grown defects and heavy ion induced defects in YBCO films, Intl. J. Mod. Phys. B 13 (1999) 1177.

[7] L. Gozzelino, D. Botta, R. Cherubini, A. Chiodoni, R. Gerbaldo, G. Ghigo, F. Laviano, B. Minetti, E. Mezzetti, Magneto-optical analysis of the critical current density dependence on temperature in proton irradiated YBCO films, Supercond. Sci. Technol. 17 (2004) S500.

[8] J. L. MacManus-Driscoll, S. R. Foltyn, Q. X. Jia, H. Wang, A. Serquis, L. Civale, B. Maiorov, M. E. Hawley, M. P. Maley, D. E. Peterson, Strongly enhanced current densities in superconducting coated conductors of $YBa_2Cu_3O_{7-x}$+$BaZrO_3$, Nat. Mater. 3 (2004) 439.

[9] T. Haugan, P. N. Barnes, R. Wheeler, F. Meisenkothen, M. Sumption, Addition of nanoparticle dispersions to enhance flux pinning of the $YBa_2Cu_3O_{7-x}$ superconductor, Nature 430 (2004) 867.

[10] T. A. Campbell, T. J. Haugan, I. Maartense, J. Murphy, L. Brunke, P. N. Barnes, Flux pinning effects of Y_2O_3 nanoparticulate dispersions in multilayered YBCO thin films, Physica C 423 (2005) 1.

[11] H. Huhtinen, P. Paturi, E. Lähderanta, R. Laiho, Laser deposition of thin superconducting films from a nanocrystalline YBCO target, Supercond. Sci. Technol. 12 (1999) 81.

[12] B. Dam, J. M. Huijbregtse, F. C. Klaassen, R. C. F. van der Geest, G. Doornbos, J. H. Rector, A. M. Testa, S. Freisem, J. C. Martinez, B. Stuble-Pumpin, R. Griessen, Origin of high critical currents in YBCO superconducting thin films, Nature 399 (1999) 439.

[13] F. C. Klaassen, G. Doornbos, J. M. Huijbregtse, R. C. F. van der Geest, B. Dam, R. Griessen, Vortex pinning by natural linear defects in thin films of YBCO, Phys. Rev. B 64 (2001) 184523.

[14] E. Blinov, V. G. Fleisher, H. Huhtinen, R. Laiho, E. Lähderanta, P. Paturi, Y. P. Stepanov, L. Vlasenko, Preparation of one to three unit cell thick $YBa_2Cu_3O_{7-\delta}$ powders and investigation of their magnetic and microwave properties, Supercond. Sci. Technol. 10 (1997) 818.

[15] J. Raittila, H. Huhtinen, P. Paturi, Y. P. Stepanov, Preparation of superconducting $YBa_2Cu_3O_{7-\delta}$ nanopowder by deoxydation in Ar before final oxygenation, Physica C 371 (2002) 90.

[16] P. Paturi, J. Raittila, H. Huhtinen, V.-P. Huhtala, R. Laiho, Size dependent properties of YBCO nanopowder, J. Phys. Cond. Mat. 15 (2003) 2103.

[17] M. Peurla, H. Huhtinen, P. Paturi, Y. P. Stepanov, J. Raittila, R. Laiho, YBCO films prepared by PLD using nanocrystalline targets doped with $BaZrO_3$ or Y211, IEEE T. Appl. Supercon. 15 (2005) 3050.

[18] P. Paturi, H.Huhtinen, R. Laiho, Inexpensive substrate heater for oxidizing environments, Rev. Sci. Instrum. 69 (1998) 3945.

[19] H. Huhtinen, J. Järvinen, R. Laiho, P. Paturi, J. Raittila, Laser deposition from a nanostructured YBaCuO target: Analysis of the plume and growth kinetics of particles on $SrTiO_3$, J. Appl. Phys. 90 (2001) 1521.

[20] P. Paturi, H. Huhtinen, R. Laiho, J. Raittila, Uniform size Y-Ba-Cu-O particles obtained by laser ablation and post-annealing, IEEE T. Appl. Supercon. 11 (2001) 3449.

[21] D. B. Geohegan, A. A. Puretzky, D. J. Rader, Gas-phase nanoparticle formation and transport during pulsed laser deposition of YBCO, Appl. Phys. Lett. 74 (1999) 3788.

[22] M. Peurla, H. Huhtinen, P. Paturi, Magnetic relaxation and flux pinning in YBCO films prepared by PLD from a nanocrystalline target, Supercond. Sci. Technol. 18 (2005) 628.

[23] C. J. van der Beek, M. Konczykowski, A. Abal'oshev, I. Abal'osheva, P. Gierlowski, S. J. Lewandowski, M. V. Indenbom, S. Barbanera, Strong pinning in high-temperature superconducting films, Phys. Rev. B 66 (2002) 24523.

[24] P. Paturi, M. Peurla, K. Nilsson, J. Raittila, Crystalline orientation and twin formation in YBCO thin films laser ablated from a nanocrystalline target, Supercond. Sci. Technol. 17 (2004) 564.

[25] G. Kästner, D. Hesse, R. Scholtz, H. Koch, F. Ludwig, M. Lorenz, H. Kittel, Microstructure defects in YBCO thin films - A TEM study to discuss their influence in device properties, Physica C 243 (1995) 281.

[26] S. K. Streiffer, E. M. Zielinski, B. M. Lairson, J. C. Bravman, Thickness dependence of the twin density in YBCO thin films sputtered onto MgO substrates, Appl. Phys. Lett. 58 (1991) 2171.

[27] P. Paturi, H. Huhtinen, K. Laajalehto, R. Laiho, Reason for high critical current in YBCO thin films ablated from nanostructured target, Supercond. Sci. Technol. 13 (2000) 622.

[28] P. Paturi, M. Peurla, J. Raittila, N. H. Andersen, Dependence of critical current density on crystalline direction in thin YBCO films, Physica C (submitted).

[29] A. Gurevich, L. D. Cooley, Anisotropic flux pinning in a network of planar defects, Phys. Rev. B 50 (1994) 13563.

[30] M. Peurla, P. Paturi, Y. P. Stepanov, A. C. Bodi, J. Raittila, R. Laiho, Effect of $BaZrO_3$ doping on superconducting and structural properties of YBCO films prepared by PLD from a nanograined target, Supercond. Sci. Technol. (2005) (to be submitted).

Electroceramic Materials and Applications
Edited by R. W. Schwartz

IMPURITY-DOPING EFFECTS ON CRITICAL CURRENT PROPERTIES IN $ErBa_2Cu_3O_y$ FILMS

S. Horii
Department of Applied Chemistry, University of Tokyo, and Japan Science and Technology Agency (JST)
7-3-1, Hongo, Bunkyo-ku
Tokyo, Tokyo, 113-8656, Japan

Y. Ichino and Y. Yoshida
Nagoya University, and JST
Furo-cho, Chikusa-ku
Nagoya, Aichi, 464-8603, Japan

K. Matsumoto and T. Horide
Kyoto University, and JST
Yoshida-Honmachi, Sakyo-ku
Kyoto, Kyoto, 606-8501, Japan

M. Mukaida
Kyushu University, and JST
Motooka, Nishi-ku
Fukuoka, Fukuoka, 819-0395, Japan

A. Ichinose
Central Research Institute of Electric Power Industry
2-6-1, Nagasaka
Yokosuka, Kanagawa, 240-0196, Japan

R. Kita
Shizuoka University
3-5-1, Johoku
Hamamatsu, Shizuoka, 432-8561, Japan

J. Shimoyama and K. Kishio
Department of Applied Chemistry, University of Tokyo
7-3-1, H ongo, Bunkyo-ku
Tokyo, Tokyo, 113-8656, Japan

ABSTRACT

We report on the critical current properties of the $BaZrO_3$(BZO)- and Co-doped $ErBa_2Cu_3O_y$ thin films grown by the pulse laser deposition method. The BZO-doped Er123 films processed through the oxygen post-annealing showed clear enhancement of critical current density (J_c) in comparison with normal Er123 thin films. Even in the BZO doping level of 0.5 wt.% for Er123, TEM observation revealed that nano-rods of BZO were embedded in Er123 matrix, which was qualitatively consistent with a result of angular dependence of J_c at 77 K. The T_c values of the slightly Co-doped films depend on the substrate temperature, while irreversibility lines for the temperature normalized by T_c were almost coincident with one another, which were overlapped on that of the normal Er123 thin film. The positive Co-doping effect on J_c emerged at low temperatures. We believe that local regions with lower T_c, in which the Co doping levels in the CuO_2 planes were relatively higher than that of matrix, act as vortex pinning centers.

INTRODUCTION

It is generally recognized that for the cuprates with high critical temperature (T_c) that superconducting properties can be controlled by the control of carrier doping levels of the CuO_2 planes. In the case of $REBa_2Cu_3O_y$ (RE123, RE: rare earth element), superconducting properties (carrier doping levels) are dominated by the oxygen content (y) in the chain site because of their large oxygen non-stoichiometries[1,2]. Namely, T_c, irreversibility field (H_{irr}), critical current density (J_c) and pinning potential of a vortex depend on y in RE123 phases largely, and an oxygen post-annealing process is indispensable to improve superconducting properties. Furthermore, the annealing temperature to optimize a carrier-doping level depends on a type of RE in RE123[2]. $ErBa_2Cu_3O_y$ (Er123) exhibited the highest annealing temperature among all RE123 compounds and the temperature in flowing oxygen was ~500°C [2,3]. In practice, for the melt-solidified bulks and single crystals of Er123, the annealing temperatures at 450-500°C brought their high T_c and J_c[4].

Research and development of second-generation coated conductors of RE123 is desirable for accomplishment of higher J_c and H_{irr}. Pulsed laser deposition (PLD) has proven useful for the production of RE123 films as has metal-organic deposition[5] using trifluoroacetates (TFA-MOD)[6]. For the former method, in which films were generally grown through a non-thermal-equilibrium process using a plasma state, optimization of superconducting properties on the as-grown films has been controlled mainly by deposition conditions. Recently, our group reported the effects of oxygen post-annealing on superconducting properties in Er123/$SrTiO_3$ (STO) thin films grown by PLD, and found that the oxygen annealing was effective for improving J_c in $ErBa_2Cu_3O_y$ films fabricated through a non-thermal-equilibrium process even though the as-deposited films showed a T_c value of ~90K[3,7].

In addition to the choices of production methods and appropriate RE123 compounds[6,7,8], introduction of artificial pinning centers (APC) into matrix of RE123 film as novel approach for improving J_c at 77K has been demonstrated and discussed. The Y123 films grown on STO substrates dispersed Y_2O_3 nano-particles[9] and $BaZrO_3$(BZO)-doped RE123 films (RE=Y[10], Er[11]) have been reported as the systems containing the c-axis-correlated APC. Moreover, critical current properties under high magnetic fields have been improved also in the Sm123 films containing Sm-rich 123 nano-particles[12] and the Y123 films containing Y_2BaCuO_5 nano-particles[13]. It is recognized from these investigations that lower-T_c and/or T_c-suppressed regions which are finely dispersed in orders of nanometers are effective as APC.

In this study, the Er123 thin films containing dilute concentration levels of Co ions or BZO nano-particles have been fabricated by the PLD method. In the case of the slightly Co-doped system, it is expected that local T_c-suppressed CuO_2 planes contaminated by Co play a role of APC if the average Co-Co distance in a CuO_2 plane exceeds the coherence length. Using thus impurity-doped Er123 films, we evaluated the changes of superconducting properties including T_c, H_{irr} and J_c for the oxygen post-annealing conditions, and discuss the effects of the BZO and Co doping in PLD-Er123 films as novel APC.

EXPERIMENTAL

The BZO and Co doped Er123 films were grown on (100) $SrTiO_3$ substrates by a conventional PLD method using an ArF excimer laser. Compositions of the target bulks for growth of the BZO-doped Er123 films were 0.5, 1.0 and 1.5 wt% of BZO added to $ErBa_2Cu_3O_y$, and that for the Co-doped films was $ErBa_2(Cu_{0.995}Co_{0.005})_3O_y$. Ex situ post-annealing was performed at 325-500°C in flowing oxygen. Incidentally, the annealing process at 325°C includes a slow-cooling process down to room temperature after a waiting period at 325°C for 15 min to obtain fully oxidized films for the 325°C-annealed specimens, and a rapid cooling process after a waiting period for 15 min was adopted for the specimens annealed above 400°C to preserve y of the films at the annealed temperatures.

J_c and irreversibility temperature (T_{irr}) of the as-deposited and oxygen-annealed samples with roughly 100 μm in width were characterized by the four-probe method under various magnetic fields parallel to the c-axis. T_c^{zero} under self-field and T_{irr} were defined as the temperature at which electrical resistivity (ρ) shows $\rho / \rho(95K) = 10^{-3}$ under each magnetic field. Critical current was defined as a value of current at which electric field of 5 μV/cm at 77K or 1 μV/cm at 50 and 30 K was generated under each magnetic field.

RESULTS AND DISCUSSION

Critical Current Properties of The $BaZrO_3$-Doped Er123 films

Figure 1 shows the temperature dependence of the normalized inplane resistivity [$\rho(T) / \rho(95K)$] under various external magnetic fields from 0 to 9 T for $H//c$ in the 0.5 wt.% BZO – Er123 thin film annealed at 475°C in flowing oxygen. The T_c^{zero} value of the film was 89.7 K, which is almost coincident with that of the pure Er123 film[14]. Those of 1.0wt.% and 1.5wt.%

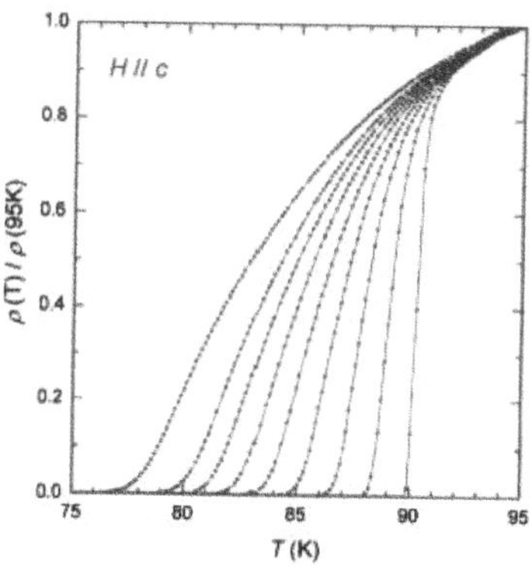

Fig.1 Temperature dependence of the normalized resistivity for the 0.5 wt.% of BZO-doped Er123 film

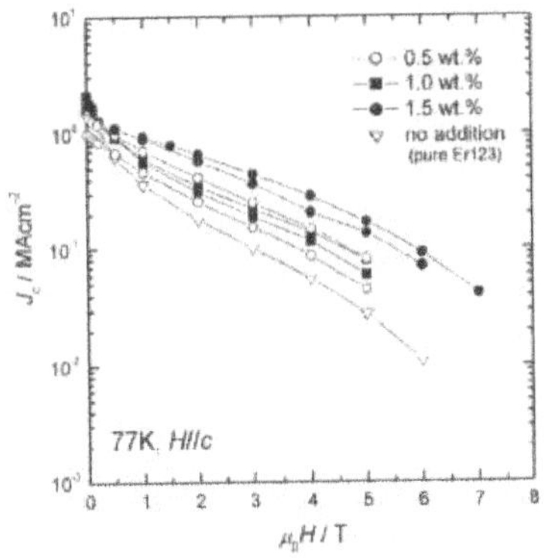

Fig. 2 External magnetic field dependence of J_c at 77 K for the BZO-doped Er123 films annealed at 450°C.

BZO-doped Er123 films also exhibited ~ 90K. That is, the superconductivity of the matrix was not affected by the BZO-doping, suggesting that the BZO phase was precipitated separately from the matrix of Er123. The T_{irr} values determined from Fig. 1 were in good agreement with those of the pure Er123 film[14]. The irreversibility lines were apparently insensitive to the doping levels of BZO.

Figure 2 shows the external field dependence of J_c at 77 K for $H//c$ in the oxygen-annealed BZO-doped Er123 films, also with the J_c-μ_0H curve for a standard pure Er123 film without BZO-doping. Incidentally, all the as-deposited films had self-field J_c values in the range 0.3 - 0.6 MA/cm^2, while post-annealed films had enhanced J_c values, at over 1 MA/cm^2, as

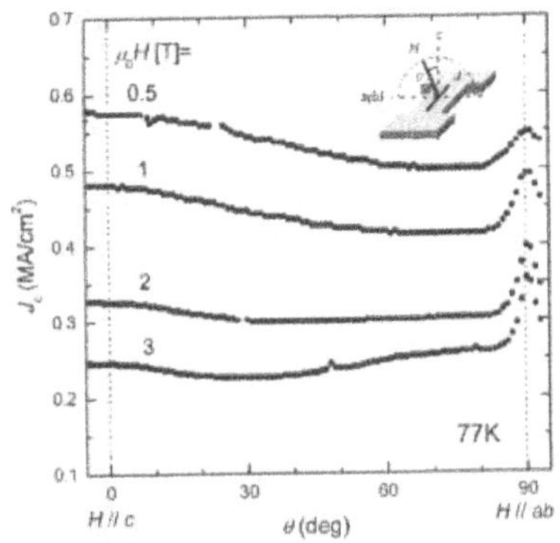

Fig. 3 Angular dependence of J_c at 77K for the 0.5 wt.% of BZO-doped Er123 film.

shown in Fig. 2. Obviously, the J_c values of the BZO-doped Er123 exceeded that of the pure Er123 film in a whole external field region above 0.5 T, and furthermore were systematically increased with the increase of the BZO-doping levels. For the 1.5 wt.% BZO-doped film, the J_c

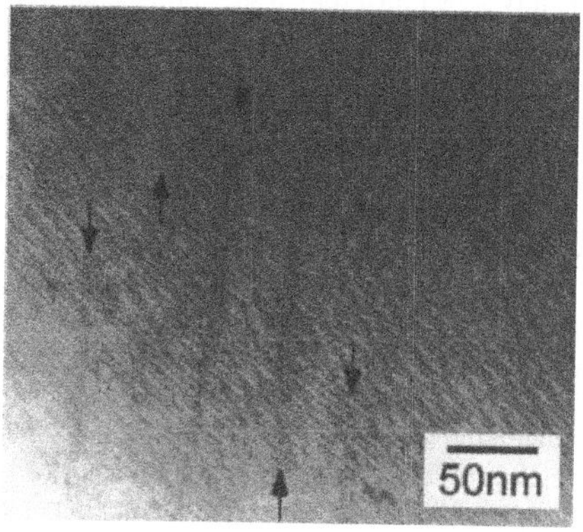

Fig. 4 A TEM image of cross-section for the 0.5 wt.% of BZO-doped Er123 film.

value showed 0.17 $MAcm^{-2}$ in 5 T for $H//c$ at 77 K.

In order to clarify the mechanism of J_c enhancement in the BZO-doped Er123 films, the angular dependence of the inplane J_c was clarified at 77 K under various magnetic fields for the 0.5 wt.% BZO-doped film, as shown in Fig.3. Incidentally, this measured film was annealed at 450°C in flowing oxygen, and was rotated maintaining $J \perp H$ in order to obtain anisotropic flux pinning properties systematically, as shown in inset of Fig.3. One can clearly see that angular peaks for $H//c$ were observed under all the measured external fields up to 3 T. Interestingly, for the J_c - θ curve under $\mu_0 H$ = 0.5 T, the J_c value for $H//c$ exceeded that for $H//ab$, which is strongly suggesting that the doped BZO plays a role of nano-rod type of artificial pinning center, which is recognized as the c-axis-correlated pinning center, even in a BZO-doping level of 0.5 wt.%. Here, addition of 0.5 wt.% of BZO into Er123 corresponds approximately to 1.3 mol%, while Driscoll et al. [10] have reported the J_c enhancement in the Y123 films containing 5 mol% of BZO. As shown in Fig. 4, embedded nano-rods in Er123 film were obviously confirmed from cross-sectional transmission electron microscopy. Therefore, the doped BZO was embedded as a nano-rod type of defect and this defect acts as the c-axis-correlated pinning center properly even in approximately four times lower BZO-doping level compared with the doping level performed by Driscoll et al. [10].

Critical Current Properties of Slightly Co-Doped Er123 films

Using the target with a composition of $ErBa_2(Cu_{0.995}Co_{0.005})_3O_y$, the Co-doped Er123 films were fabricated at several substrate temperatures. For thus obtained films, the oxygen post-annealing was performed and a T_c value of each film was determined from temperature dependence of resistivity. Figure 5 shows annealing temperature dependence of T_c for the Co-doped Er123 films which were grown at four different substrate temperatures: 770, 800, 825 and 850°C. Interestingly, the T_c values were systematically increased with the growth temperature when we compare their T_c values at a certain annealing temperature. Renevier et al. [15] reported that Co ions were dominantly substituted for Cu sites of the chain. However, taking into account that carrier-doping conditions of these films were homogeneous and sufficient, our result indicates that some of the doped Co ions can be substituted for Cu sites of the CuO_2 planes and the doping levels into the CuO_2 planes can be controlled by control of the growth temperature. Here, from a report regarding the relationship between T_c and concentration levels of Zn in Y123[16], Co-substitution levels in the CuO_2 plane were estimated for these Co-doped films. Supposing that the T_c-suppression effect in the CuO_2 planes of RE123 by the Zn-doping corresponds with that by the Co-doping, the estimated substitution levels of Co ions into the CuO_2 planes were 0.15, 0.25, 0.5 and 1.0 at.% for the growth temperatures of 850, 825, 800 and 770°C, respectively. Apparently, for the growth temperature at 800°C, the substitution level of Co ions was coincident with the nominal composition of the target. When we focus on the change of T_c as a function of annealing temperature, one can see that the T_c values were systematically decreased with the annealing temperature for the 800°C -, 825°C – and 850°C-depositon films. This indicates that the overdoping conditions of carriers for these films oxygen-annealed below 450°C are preserved though the superconductivity deteriorated by the Co-doping and carrier doping states seem apparently shifted toward underdoping states. Shiraki et al. [17] reported that the carrier-doping states were changed toward under-doping conditions with increase of Co incorporation levels into the chain sites for the Y123 polycrystalline bulks. However, in the dilute doping levels, carrier-doping states were not largely affected even for the change of T_c by the Co-doping. On the other hand, the appropriate annealing temperature was lowered only for the 770°C-deposition film. Above this Co-doping level, the shift of the appropriate temperature can be understood in terms of the change of carrier-doping levels toward the underdoping states, as reported in the polycrystalline bulks[17].

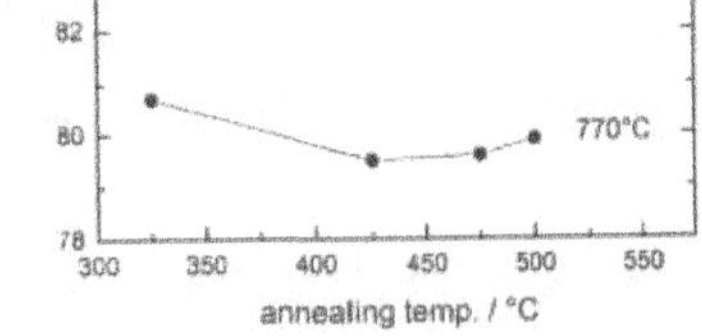

Fig. 5 Annealing temperature dependence of T_c for the Co-doped Er123 films grown at four different substrate temperatures.

Figure 6 shows the changes of H_{irr} for the Co-doped Er123 films grown at these four different substrate temperatures as a function of temperature normalized by T_c (T/T_c), also with data of the pure Er123 film and a Zn-doped Er123 film ($T_c \sim 80K$) grown using a target with a composition of $ErBa_2(Cu_{0.997}Zn_{0.003})_3O_y$. Clearly, the irreversibility lines of all the Co-doped films almost overlap. Moreover, these trends were very close to the line of the pure Er123, which indicates that the anisotropy factors (γ^2) are preserved at a certain normalized temperature even though the T_c values are reduced by the Co-doping. This behavior is on a quite contrast with the result of the Zn-doped film, that is, the irreversibility line was shifted toward lower normalized temperature region by the Zn-doping.

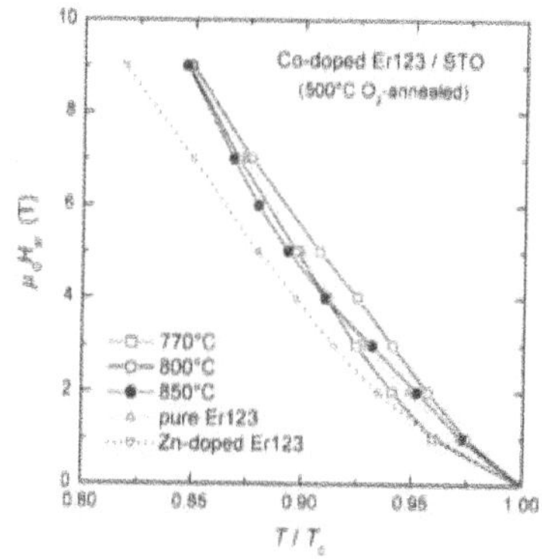

Fig. 6 The normalized temperature dependence of H_{irr} for the Co-doped Er123 films. For comparison, the data of the pure Er123 and Zn-doped Er123 films were also shown.

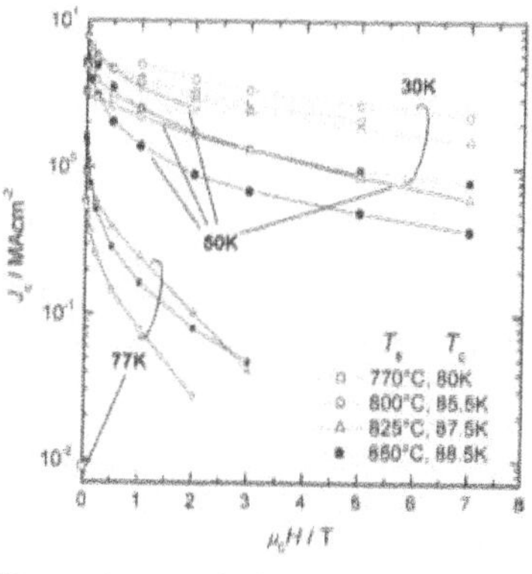

Fig. 7 External magnetic field dependence of J_c at 77, 50 and 30 K for the Co-doped Er123 films.

For the purpose of the evaluation of critical current properties of the Co-doped Er123 films, external magnetic field dependence of J_c at 77, 50 and 30K was measured for the films grown at these four different substrate temperatures, as shown in Fig. 7. The J_c values at 77 K depend on the T_c values and were increased with T_c (or substrate temperature). However, the behaviors of J_c at low temperatures of 30 and 50 K were on the contrary to the result at 77 K. That is, the J_c values of the films grown at lower substrate temperatures (or the film which showed lower T_c) were always higher than those of the films with higher T_c. Although the mechanism on the J_c enhancement at low temperatures is still an open problem, we deduce that lower T_c regions containing relatively more Co ions in the CuO_2 planes play a role of vortex pinning centers. In order to clarify the above effect, further investigations, such as angular dependence of J_c, cross-sectional TEM observation and the estimation and comparison of number of defects are required.

CONCLUSION

The BZO- and Co-doped Er123 films were fabricated using the PLD method and the critical current properties of the oxygen post-annealed films were revealed. The T_c values were insensitive to the change of BZO doping levels for the BZO-doped films, while they largely depended on the substrate temperature for the Co-doped films. The J_c values of the BZO-doped films at high magnetic fields were clearly enhanced with increase of the BZO-doping level. Even in 0.5 wt.% (1.3 mol%) of the doping level, the BZO nano-rods were observed clearly using the TEM, and broad peaks for *H*//*c* emerged in the J_c-θ curves. For the Co-doped Er123 films, in spite of decrease of T_c with the decrease of substrate temperature, irreversibility lines of all the measured films were almost overlapped on that of pure Er123 film in the H_{irr} - T/T_c diagram. The J_c enhancement effect by the Co-doping emerged at low temperatures, which is presumably because local regions showing lower T_c compared with T_c of the matrix act as additional vortex pinning centers.

REFERENCES

[1] K. Kishio, J. Shimoyama, T. Hasegawa, K. Kitazawa, and K. Fueki, "Determination of oxygen nonstoichiometry in a high- T_c superconductor $Ba_2YCu_3O_{7-\delta}$," *Jpn. J. Appl. Phys.*, **26**, L1228-L1230 (1987).

[2] J. Shimoyama, S. Horii, K. Otzschi and K. Kishio, " How to optimize critical current performance of RE123 materials by controlling oxygen content," *Mat. Res. Soc. Symp. Proc.*, **689**, 265-270 (2002).

[3] S. Horii, A. Ichinose, M. Mukaida, K. Matsumoto, T. Ohazama, Y. Yoshida, J. Shimoyama, and K. Kishio, " Enhancement of critical current density on $ErBa_2Cu_3O_y$ thin films by a post-annealing process," *Jpn. J. Appl. Phys.*, **43**, L1223-L1225 (2004).

[4] T. Nakashima, T. Maruyama, M. Honzumi, S. Horii, J. Shimoyama, and K. Kishio, "Control of microstructure in Er123 melt-solidified bulks for enhanced critical current properties," *IEEE Transactions on Applied Superconductivity*, **15**, 3176-3179 (2005).

[5] P.C. McIntyre, M.J. Cima, J.A. Smith Jr., R.B. Hallock, M.P. Siegal, and J.M. Phillips, "Effect of growth conditions on the properties and morphology of chemically derived epitaxial thin films $Ba_2YCu_3O_{7-x}$ on (001) $LaAlO_3$," *J. Appl. Phys.*, **71**, 1868 (1992).

[6] T. Iguchi, T. Araki, Y. Yamada, H. Ikuta, I. Hirabayashi, Y. Shiohara, and U. Mizutani, "Metal-organic deposition of RE-Ba-Cu-O (RE=Dy, Ho, Er, and Tm) films using trifluoroacetates," *Physica*, **C392-396**, 900-904 (2003).

[7] S. Horii, M. Mukaida, Y. Ichino, K. Matsumoto, T. Ohazama, A. Ichinose, Y. Yoshida, J. Shimoyama, and K. Kishio, "Oxygen post-annealing effects on critical current properties of PLD-$ErBa_2Cu_3O_y$ films grown at several substrate temperatures," *Physica*, **C426-431**, 1015-1020 (2005).

[8] K. Sudoh, Y. Ichino, M. Itoh, Y. Yoshida, Y. Takai and I. Hirabayashi, " Correlation between film thickness and critical current density of $Sm_{1+x}Ba_{2-x}Cu_3O_{6+\delta}$ films deposited by pulsed laser deposition," *Jpn. J. Appl. Phys.*, **41**, L983-L985 (2002).

[9] K. Matsumoto, T. Horide, A. Ichinose, S. Horii, Y. Yoshida and M. Mukaida, " Critical Current Control in $YBa_2Cu_3O_{7-\delta}$ Films Using Artificial Pinning Centers," *Jpn. J. Appl. Phys.*, **44**, L246-L248 (2005).

[10] J.L. Macmanus-Driscoll, S.R. Foltyn, Q.X. Jia, H. Wang, A. Serquis, B. Maiorov, L. Civale, M.E. Hawley, M.P. Maley, and D.E. Peterson, " Strongly enhanced current densities in superconducting coated conductors of $YBa_2Cu_3O_{7-x}$ + $BaZrO_3$," *Nature Materials*, **3**, 439 (2004).

[11] M. Mukaida, T. Horide, R. Kita, S. Horii, A. Ichinose, Y. Yoshida, O. Miura, K. Matsumoto, K. Yamada, and N. Mori, " Critical current density enhancement around a matching field in $ErBa_2Cu_3O_{7-x}$ with $BaZrO_3$ nano-rods," *Jpn. J. Appl. Phys.*, **44**, L952-L954 (2005).

[12] Y. Yoshida, K. Matsumoto, Y. Ichino, M. Ito, A. Ichinose, S. Horii, M. Mukaida, and Y. Takai, " High-critical-current-density epitaxial films of $SmBa_2Cu_3O_{7-x}$ in high fields," *Jpn. J. Appl. Phys.*, **44**, L129-L132 (2005).

[13] T. Haugan, P.N. Barnes, R. Wheeler, F. Meisenkothen, and M. Sumption, " Addition of nanoparticle dispersions to enhance flux pinning of the $YBa_2Cu_3O_{7-x}$ superconductor," *Nature*, **430**, 867-870 (2004).

[14] S. Horii, M. Mukaida, K. Matsumoto, T. Ohazama, A. Ichinose, Y. Yoshida, J. Shimoyama, and K. Kishio, " Post-annealing effects of superconducting properties on $ErBa_2Cu_3O_y$ films fabricated by pulsed laser deposition," *Physica*, **C412-414**, 1306-1309 (2004).

[15] H. Renevier, J.L. Hodeau, M. Marezio, and A. Santoro, " Electron- and powder neutron-diffraction studies of $YBa_2(Cu_{1-y}Co_y)_3O_{6+x}$ with 0.05<y<0.33," *Physica*, **C220**, 143-159 (1994).

[16] J.M. Tarascon, P. Barboux, P.F. Miceli, L.H. Greene, G.W. Hull, M. Eibschutz, and S.S. Sunshine, " Structural and physical properties of the metal (M) substituted $YBa_2Cu_{3-x}M_x)_3O_{7-y}$ perovskite," *Phys. Rev.* , **B37**, 7458-7469 (1988).

[17] M. Shiraki, J. Shimoyama, S. Horii, and K. Kishio, "Bulk superconductivity observed in $(Co,Cu)(Sr,Ba)_2(Y,Ca)Cu_2O_y$," *Physica*, **C426-431**, 487-491 (2005).

Electroceramic Materials and Applications
Edited by R. W. Schwartz

SYNTHESIS OF A NEW COPPER OXIDE "$In_{0.3}Cu_{0.7}Ba_2LaCu_2O_y$" AND CHARACTERIZATION OF ITS PHYSICAL PROPERTIES

Masaaki Fujihara, Fumiaki Sato, Tatsuaki Nakano, Shiro Kambe, and Osamu Ishii.

Graduate School of Science and Engineering, Yamagata University.
4-3-16, Jonan,Japan
Yonezawa, Yamagata, 992-8510

Masahiko Nishijima

Institute for Materials Research, Tohoku University
Katahira 2-1-1, Aoba-ku,Japan
Sendai, Miyagi, 980-8577

ABSTRACT

Though we previously reported a copper oxide, $InBa_2CuOy$ with an a-axis length of 4.18Å, its electric property was insulating. Based on the assumption that the insulating property comes from the existence of the oxygen vacancy correlated with the extremely long a-axis length, shortening the a-axis length below 4.0Å was tried to diminish the oxygen vacancy. The compound that we expected is represented by $In_xCu_{1-x}Ba_2LnCu_2O_y$ (In1212). To prepare the $In_xCu_{1-x}Ba_2LnCu_2O_y$, various kind of lanthanoides revealing that the smallest amount of impurity phases was formed when Ln = La. With changing the In content, x in $In_xCu_{1-x}BaLa_2Cu_2O_y$ from 0.2 to 1.0, the least impurity was prepared when x was 0.3. The XRD pattern of $In_{0.3}Cu_{0.7}Ba_2LaCu_2O_y$ was refined by the Rietveld analysis, suggesting that the structure is tetragonal with a-axis length of3.91±0.03 Å and a c-axis length of 11.81±0.09 Å, and is similar in structure to $LaBa_2Cu_3O_y$ except the partial substitution of In for Cu(2) Its resistivity was semiconducting. After annealing in O_2, the resistivity was drastically decreased, but superconductivity was not observed above the temperature of 10K. According to the Hall effect measurement, the hole density of the oxygen-annealed sample was 7.9×10^{19} cm^{-3} at room temperature. While the ratio of itinerant holes *v/p* in the $YBa_2Cu_3O_{6.76}$ superconductor was as large as 100-200%, the ratio for $In_{0.3}Cu_{0.7}Ba_2LaCu_2O_y$ was 0.001% to 10%, suggesting that the majority of holes are maybe localized in the In,Cu(2) plane.

INTRODUCTION

We previously reported a new blocking layer with an a-axis length of 4.18Å, $InBa_2CuO_{4.44}$[1]. The blocking layer supplies on apical oxygen to the CuO_2 layer and produces an oxygen vacancy in the plane of the CuO_2 layer. By matching the blocking layer with the CuO_2 layer, a new layered cuprate, $InBa_2CuO_{4.44}$, was prepared. Its space group was Pmmm with lattice parameters of a=4.1820(6)Å, b=4.1820(5)Å and c=8.0889(9)Å. The fact that the a- and b-axis lengths of $InBa_2CuO_{4.44}$ compound are much longer than of the other cuprate compounds is explained by the existence of an oxygen vacancy in the CuO_2 layer. Its resistivity at room temperature was as large as 28.8MΩcm. The insulating property probably comes from the existence of the oxygen vacancy in the CuO_2 layer.

The objective of this research is to make the In-based cuprate conducting. For that, it is required to shorten the a-axis length below 4.0Å and to diminish oxygen vacancy in the CuO_2 layer. We tried to prepare a compound that had double CuO_2 layers. The compound we expected is represented by $In_xCu_{1-x}Ba_2LnCu_2O_y$ (In-1212), where Ln means a lanthanoide.

EXPERIMENTAL

The $In_xCu_{1-x}Ba_2LnCu_2O_y$ ($0 \leq x \leq 1$) was prepared by the conventional solid state reaction method. Samples were prepared by mixing In_2O_3, BaO, La_2O_3 and CuO in an agate mortar. The powder was put into an aluminum crucible and calcined at 800 - 900°C in air for 10 h, ground in the mortar, and pressed into a pellet at a pressure of 600 kgf/cm^2. The pellet was put into an aluminum crucible again, and sintered at 860 – 1000°C in air for 20h. After sintering, the pellets were quenched to room temperature and ground to powder. Sintering was repeated until a single phase was prepared. To decrease the oxygen vacancies in the CuO_2 layer, $In_xCu_{1-x}Ba_2LnCu_2O_y$ was annealed under pressures of 0.1 – 35MPa oxygen at temperatures of 400°C for 10 hours. The phase of $In_xCu_{1-x}Ba_2LnCu_2O_y$ was ascertained by powder X-ray diffraction using Cu-Kα radiation. The crystal structure was refined using the Rietveld refinement program RIETAN [2]. Oxygen contents and average Cu valences were determined by a newly developed coulometric method [3]. Hole concentration was determined using the Hall effect. We made cross-shaped samples using a computer aided modeling machine (Roland CAMM-2). Currents of 1μA – 10mA were used in the measurements in a magnetic field of 1.5 T at room temperature. To compare the hole concentration with the Cu valence determined by coulometry, hole concentration was expressed not by hole density, n_H [cm^{-3}] but by hole number per Cu site, p. The correlation between n and p is expressed by

$$p = \frac{n}{x}.$$

Here, x is given by

$$x = \frac{s}{V},$$

where s and V represent the number of Cu atoms per unit cell and the volume of a unit cell, respectively. The ratio of itinerant holes was calculated by $100 \times p/v$, where v = (Cu valence –2). It should be noted that the hole concentration (p) is measured in a polycrystalline sample, whose Hall coefficient mainly reflects the hole concentration in the planar CuO_2 layer.

RESULTS AND DISCUSSION

To prepare a layered cuprate with double CuO_2 layers, we tried to prepare $In_xCu_{1-x}Ba_2LnCu_2O_y$, where Ln= La, Ce, Pr, Nd, Sm, Eu, Gd, Tb, Dy, Ho, Er, Tm, Yb and Lu. Among the lanthanoides, $In_xCu_{x-1}Ba_2LaCu_2O_y$ had the smallest amount of impurity phases. We therefore focused our attention on this system. Fig.1 shows the XRD pattern of $In_xCu_{1-x}Ba_2LaCu_2O_y$ ($0.2 \leq x \leq 1$). The smallest size of the impurity peaks was observed when x=0.3. After fixing the composition at $In_{0.3}Cu_{0.7}Ba_2LaCu_2O_y$, the sintering temperature was changed from 860°C to 1000°C. The XRD patterns of the specimens annealed at 860°C-1000°C are shown in Fig.2. We observed impurity peaks originating from $InBa_2CuO_y$ for the specimen annealed at 1000°C and peaks from $BaCuO_2$ and $InBa_2CuO_y$ for that sintered at 860°C. On the other hand, in the specimen sintered at 960°C, minimum impurity peaks were observed. From the coulometric measurement, the oxygen content was determined to be 6.79.

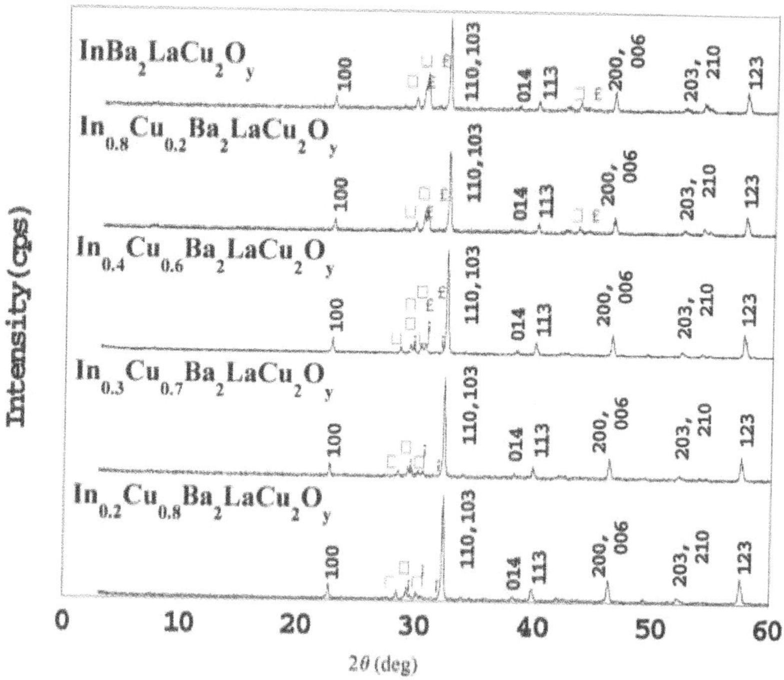

Fig. 1 XRD patterns of $In_xCu_{1-x}Ba_2LaCu_2O_y$ ($1 \leq x \leq 0.2$) samples sintered at 860°C for 20 h. The symbols ▲ and ■ represent impurity peaks from $InBaCuO_y$ and $BaCuO_2$, respectively.

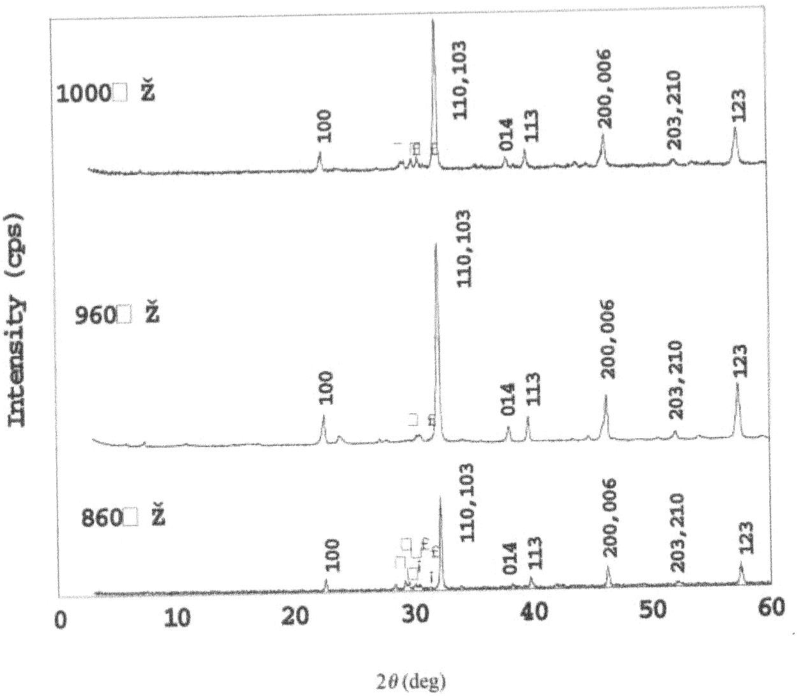

Fig. 2 XRD patterns of $In_{0.3}Cu_{0.7}Ba_2LaCu_2O_y$ samples sintered at 860 °C, 960 °C, and 1000 °C for 20h. The symbols of ▲ and ■ represent impurity peaks from $InBaCuO_y$ and $BaCuO_2$, respectively. For the specimen annealed at 960 °C, the smallest peaks from impurity phases were observed.

Based on the results of XRD and coulometry of the $In_{0.3}Cu_{0.7}Ba_2LaCu_2O_y$, we refined its crystal structure using Rietveld analysis. The impurity peaks were deleted before the Rietveld calculation. As shown in Fig.3, the observed XRD pattern (dotted line) coincided with the one (solid line) calculated by Rietveld analysis. The refined crystallographic data of $In_{0.3}Cu_{0.7}Ba_2LaCu_2O_{6.79}$ are listed in Table 1. The reliability factors of $In_{0.3}Cu_{0.7}Ba_2LaCu_2O_{6.79}$ were R_{wp}=5.85%, Rp=4.21%, Re=2.13%, Ri=2.741%, respectively. By Rietveld analysis, it was found that In was substituted for Cu in the CuO_2 layer. The structure was compared with the TEM image of $In_{0.3}Cu_{0.7}Ba_2LaCu_2O_{6.79}$ shown in Fig. 4. While the distance A - B is 4.1Å, that of B - C is 3.6Å. By comparing the distances with the results from Rietveld analysis, it is found that the A site corresponds to the Cu site and the B and C sites correspond to the (In.Cu) sites. From the positional relations of the A, B and C sites, the larger shadows D and E in Fig. 4 are considered to be Ba and La, respectively. From the structure deduced from Rietveld analysis and TEM image, it is confirmed that the structure of $In_{0.3}Cu_{0.7}Ba_2LaCu_2O_{6.79}$ is tetragonal with a = 3.91±0.03Å, and c=11.81±0.09Å. As shown in Table 1, the occupation of oxygen sites (2) and (3), which are located in the CuO_2 layer, are found to be 1.0, revealing that the oxygen vacancy in the CuO_2 layer of $InBa_2CuO_{4.44}$ is diminished in $In_{0.3}Cu_{0.7}Ba_2LaCu_2O_{6.79}$.

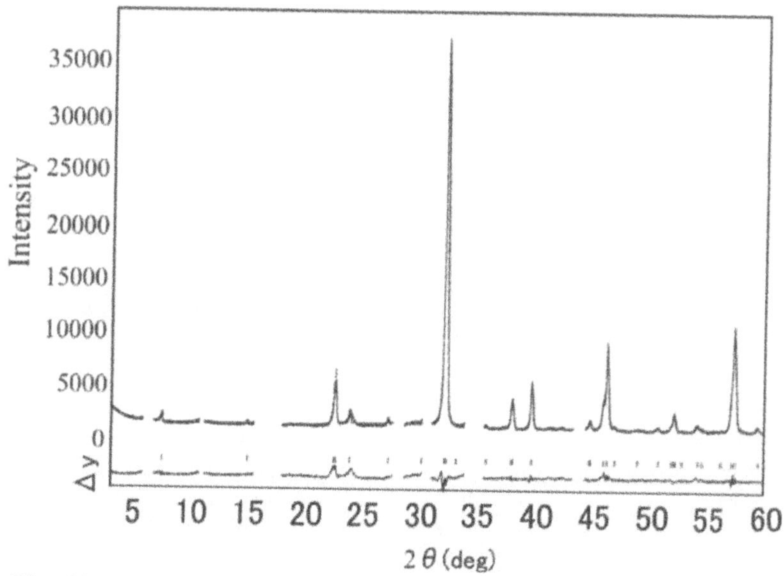

Fig. 3 Rietveld refinement pattern of $In_{0.3}Cu_{0.7}Ba_2LaCu_2O_y$. The dotted and solid lines represent observed and calculated patterns, respectively. The markers below the diffraction pattern indicate positions of Bragg reflections, and Δy is the difference between observed and calculated intensities.

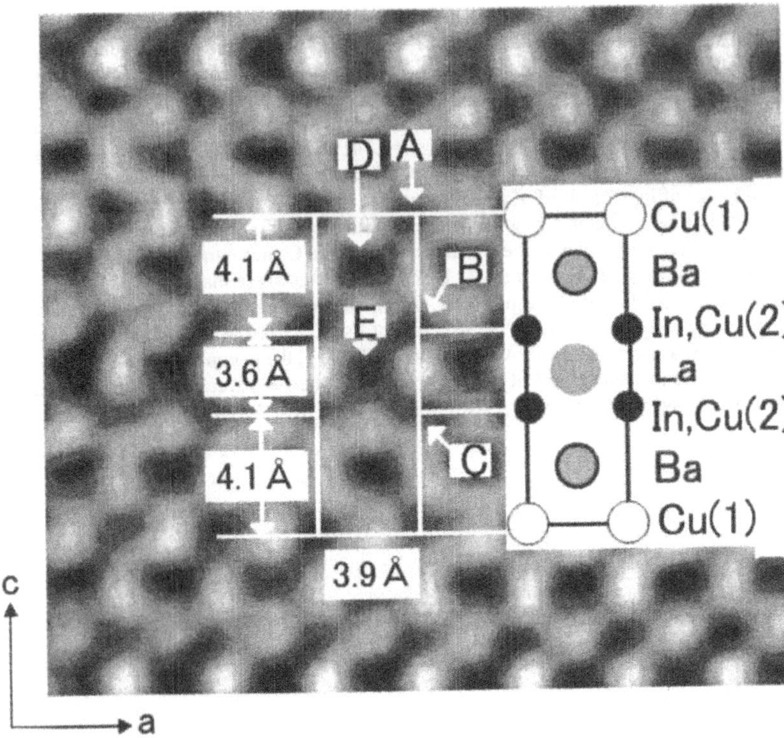

Fig. 4 High resolution TEM of $In_{0.3}Cu_{0.7}Ba_2LaCu_2O_{6.79}$ taken with [010] incidence.

Table I The refined crystallographic data of the $In_{0.3}Cu_{0.7}Ba_2LaCu_2O_{6.79}$.

$In_{0.3}Cu_{0.7}Ba_2LaCu_2O_{6.79}$

Atom	Site	x	y	z	B	Occupancy
Cu(1)	a	0	0	0	0.718(9)	1
Ba	h	0.5	0.5	0.1896(2)	0.376(4)	1
La	d	0.5	0.5	0.5	0.173(2)	1
In	g	0	0	0.3446(6)	2.641(3)	0.15
Cu(2)	g	0	0	0.3446(6)	2.641(3)	0.85
O(1)	i	0	0.5	0.374(2)	0.500(6)	0.7
O(2)	g	0	0	0.170(2)	0.500(6)	1
O(3)	f	0	0.5	0	0.500(6)	1

R_{WP}=5.85% R_P=4.21% R_I=6.55% R_e=2.13% S=2.741

Space group, tetragonal Pmmm(Vol. A.No.123) a=3.91(3)Å c=11.81(9)Å

As described in the introduction section, copper oxides having CuO_2 layers whose a-axis length is between 3.914Å and 3.944Å, between which there should exist the boundary of hole and electron doping, have not been reported in detail. Therefore, it is interesting to determine the electric properties of $In_{0.3}Cu_{0.7}Ba_2LaCu_2O_{6.79}$. In Fig. 5, the dependence of resistivity on temperature for $In_{0.3}Cu_{0.7}Ba_2LaCu_2O_{6.79}$ is shown. Its resistivity at room temperature was 60Ωcm, and it was semiconducting at lower temperatures.

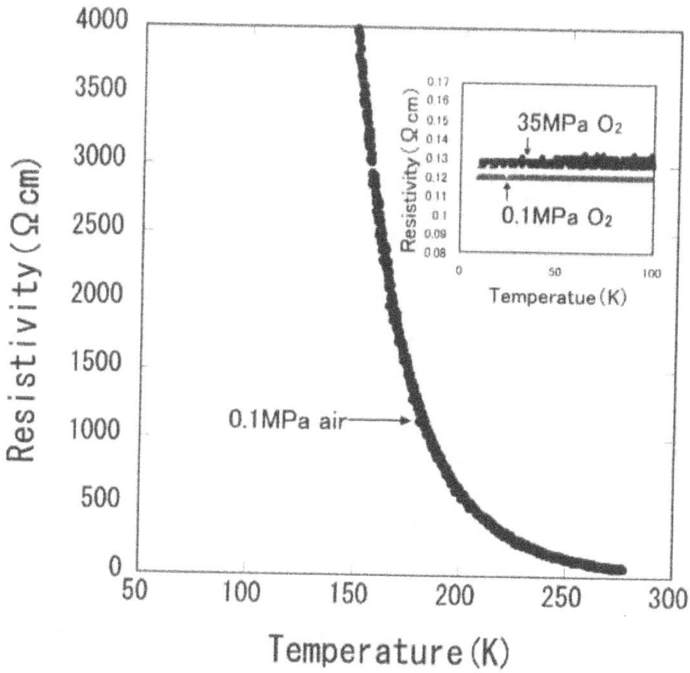

Fig. 5 Dependence of resistivity on temperature for $In_{0.3}Cu_{0.7}Ba_2LaCu_2O_y$ specimens annealed in 0.1MPa air, 0.1MPa O_2, and 35MPa O_2.

To increase the conductivity, the $In_{0.3}Cu_{0.7}Ba_2LaCu_2O_{6.79}$ was annealed in oxygen atmospheres of 0.1 and 35MPa. Annealing at O_2 pressures higher than 35MPa led to formation of impurity phases. From the coulometric experiment, it was revealed that the oxygen content increased from 6.79 to 6.88, which corresponded to the increase in Cu valence from 2.10 to 2.17. Considering the Cu valence higher than 2.0 and the increase in conductivity (inset of Fig. 5) by O_2 annealing, $In_{0.3}Cu_{0.7}Ba_2LaCu_2O_{6.79}$ is a hole-doped conductor. From the fact that the Cu valence 2.17 is larger than that of the 60 K-class $YBa_2Cu_3O_{6.76}$, the Cu valence 2.17 is large enough to induce superconductivity, and optimization of the Cu valence for the occurrence of superconductivity should be achieved for the $In_{0.3}Cu_{0.7}Ba_2LaCu_2O_{6.88}$. By annealing in 0.1MPa O_2, the resistivity at room temperature decreased drastically from 60 Ωcm to ~0.1 Ωcm (Fig. 5). Annealing in 35MPa O_2 also decreased the resistivity to ~0.1 Ωcm. These drastic decreases in resistivity can be explained by the increase in oxygen content and the Cu valence. However, the $In_{0.3}Cu_{0.7}Ba_2LaCu_2O_{6.88}$, whose Cu valence was optimized to be 2.13, was not superconducting above 10K. Unfortunately the resistivity below 10K was not able to be measured. According to the Hall effect measurement, the hole density of the oxygen-annealed sample was 7.9×10^{19} cm^{-3} at room temperature. This value is much smaller than 3.1×10^{21} cm^{-3} for a $YBa_2Cu_3O_{6.76}$ sample. In Fig. 6, the hole density n_H and the ratio of itinerant holes p / v are shown. With an increase in the Cu valence from 2.11 to 2.18, the hole density was increased from $1.0 \times 10^{16} cm^{-3}$ to $6.9 \times 10^{19} cm^{-3}$. These values of hole density measured by the Hall effect measurement are much smaller than those calculated by the Mott-Hubbard model assuming that an increase in the Cu valence from 2.0 to3.0 dopes one hole to the Cu ion represented by the dashed line in Fig. 6(a), with the result that the hole density determined in the Hall voltage measurement did not reach the hatched optimum hole density region, necessary for the occurrence of superconductivity. In spite of the optimization of the Cu valence for $In_{0.3}Cu_{0.7}Ba_2LaCu_2O_{6.88}$, its hole density was extremely small in comparison with the $YBa_2Cu_3O_{6.76}$ sample superconductor. Therefore, it is concluded that holes induced by the excess oxygen are not doped to the CuO_2 sufficiently. While the ratio of itinerant holes v/p in the $YBa_2Cu_3O_{6.76}$ superconductor was as large as 100-200%, the ratio for $In_{0.3}Cu_{0.7}Ba_2LaCu_2O_y$ was 0.001% to 10%, suggesting that the majority of holes are maybe localized in the In,Cu(2) plane.

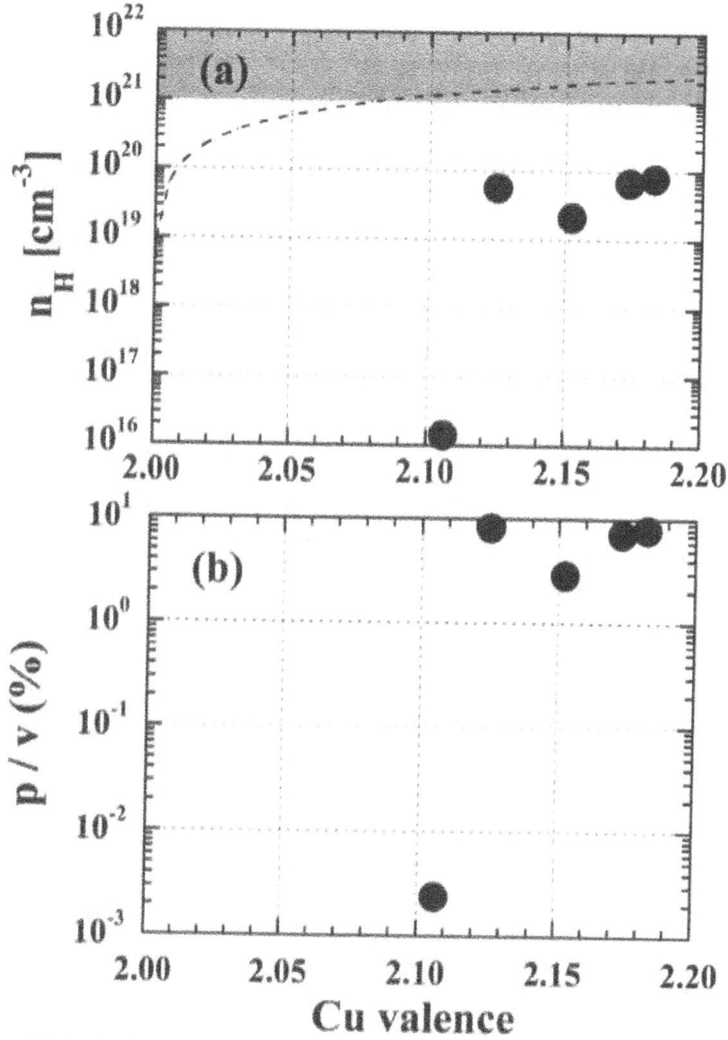

Fig. 6 Hole density (n_h) and ratio of itinerant holes (p/v) against Cu valence

CONCLUSION

To prepare the In-based copper oxides with having double CuO_2 layers, formation of $In_xCu_{1-x}BaLa_2Cu_2O_y$ was examined. Less impurity was found when x was 0.3. By varying the sintering temperature from 860°C to 1000°C, it was found that the optimum sintering temperature for forming less impurity was around 960°C. The XRD pattern of $In_{0.3}Cu_{0.7}Ba_2LaCu_2O_y$ was refined by the Rietveld analysis, suggesting that the structure is tetragonal with a-axis length of 3.91±0.03 Å and a c-axis length of 11.81±0.09Å, and is similar in structure to $LaBa_2Cu_3O_y$ except the substitution of In for Cu(2). The structure was confirmed by the high-resolution TEM image. Its resistivity showed semiconducting. After annealing in O_2, the resistivity was drascitally decreased, but superconductivity was not observed above the temperature of 10K. According to the Hall effect measurement, the hole density of the oxygen-annealed sample was $7.9 \times 10^{19} cm^{-3}$ at room temperature, which was much smaller than $3.1 \times 10^{21} cm^{-3}$ for $YBa_2Cu_3O_{6.76}$. While the ratio of itinerant holes *v/p* in the $YBa_2Cu_3O_{6.76}$ superconductor was as large as 100-200%, the ratio for $In_{0.3}Cu_{0.7}Ba_2LaCu_2O_y$ was 0.001% to 10%, suggesting that the majority of holes are maybe localized in the In,Cu(2) plane.

ACKNOWLEDGMENTS

The authors thank Jeffrey Willis of Los Alamos National Laboratory for his helpful discussions. This work was partly supported by the fund of Yamagata Prefecture.

A part of the research was supported by the joint research program with the Institute for Solid State Physics, Tokyo Univ.

REFERENCES

[1]S. Kambe, I. Shime, S. Ohshima, and K. Okyama, "Structure of Ba_2InCuO_y: A new layered cuprate with a blocking layer of $BaInO_y$ perovskite" Physica C **220** 119-126 (1994).

[2]F. Izumi, and T. Ikeda, Mater. Sci. Forum. **198** 321-324 (2000).

[3]F. Sato, N. Komiyama, S. Kambe, and O. Ishii, "Development of micro-coulometry for measuring oxygen content in copper oxides" Phys. Lett. submitted (2003).

Electroceramic Materials and Applications
Edited by R. W. Schwartz

MICRO-COULOMETRY FOR MEASURING OXYGEN CONTENT IN COPPER OXIDES

Fumiaki Sato, Masaaki Fujihara, Shiro Kambe and Osamu Ishii
Graduate School of Science and Engineering, Yamagata University
4-3-16, Jonan Yonezawa 992-8510, Japan.

ABSTRACT

A micro-coulometry system was newly developed and employed for measuring the oxygen content in copper oxides using a reduced amount of material. To achieve this reduction in sample size, Ar flow rate, and coulometric current were optimized. When using 5.0 mg of $YBa_2Cu_3O_{6.71}$, which is about 1 / 10 the amount used in conventional coulometry, the oxygen content was successfully determined to be 6.68 ± 0.03 by correcting the side reaction. It was found to be necessary to continue to pass Ar through the solution before coulometry more than ten minutes and during the coulometric measurement. This method will be useful for measuring the oxygen content of oxide samples smaller than 10 mg, i.e. oxide materials produced in small quantities for electronic applications.

INTRODUCTION

Many kinds of oxides, such as superconductive cuprates, magnetoresistive materials, conducting films, and tin oxide sensors, have been studied for electronic applications. The electronic properties of such oxide materials are strongly correlated with their oxygen content. For example, the critical temperature of copper oxide superconductors is so sensitive to the oxygen content that it is necessary to control and to optimize this quantity. In response to the necessity to control the oxygen content, various methods for measuring this quantity have been developed.[1]

Thermogravimetry (TG) is a conventional way to measure the weight loss of an oxide during controlled heating. However, even in the case that the material does not contain unstable elements, such as Hg or Tl, the accuracy of the oxygen content observed by TG is less than that achievable by chemical analysis.[2] Among the chemical analysis methods for measuring the oxygen content, coulometry is a commonly used method. The analytical device is compact and allows high accuracy. Iodometry is also an easy way of measuring the oxygen content, but the end point of titration is usually determined by the change in color, leading to the problem that it is rather difficult to determine the end point precisely. One merit of coulometry is the ease of determining the end point of the redox reaction, which minimizes measurement errors. With conventional coulometric methods, typically about 50-100 mg of a specimen is dissolved in hydrochloric acid, and the change of the redox potential is measured. Furthermore, to improve the reliability of the measurement, several experiments are required, with the result that several hundred mg of a specimen is required in total. An improved coulometry method was found to yield the oxygen content of an $YBa_2Cu_3O_{7-\delta}$ reference sample with a reproducibility within ±0.01 for samples with a mass of ~10 mg.[3, 4]

As functional oxide materials are now being used in the application of nanotechnology to electronics, the amount of functional oxide material available for analysis is further being decreased. For the important synthesis techniques for functional oxide films, such as physical and chemical vapor deposition, the mass of the samples is as small as 1–5 mg. It has, however, been difficult to measure the oxygen content of such a small quantity of oxide material.

The object of this study is to establish the oxygen content of copper oxide samples of less than 5-10 mg. For that purpose, we newly fabricated a smaller coulometry measurement system, which consists of developed miniaturized electrodes suitable for a smaller cell.

EXPERIMENTAL

Two coulometry systems, one with a conventional 300 ml cell (a) and a newly developed 30 ml cell (b) were used (Fig. 1). The 300 ml cell system (a) is made up of a reaction cell and anode/cathode/redox (vs. Ag/AgCl) electrodes, and the 30 ml cell system (b) is made up of the newly developed cell (30ml) and miniaturized anode/cathode/ORP (Oxidation-redaction potential) electrodes. We call the systems (a) "conventional coulometry" and (b) "micro-coulometry."

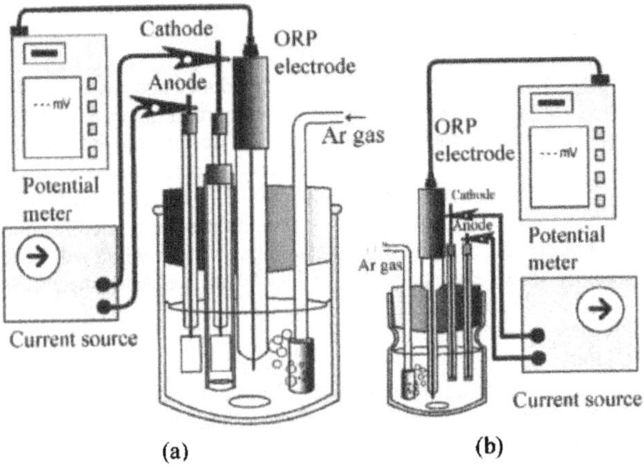

Fig. 1. Illustrations of conventional (a) and micro (b) coulometry system

The redox potential is expressed by Nernst's equation. For the Cu^{+}/Cu^{2+} redox system, the redox potential $\boldsymbol{E}$ vs. Ag/AgCl ($\boldsymbol{E}_{Ag/AgCl}$) is given by

$$E_{Ag/AgCl} = -0.047 - 0.0591 log \frac{[Cu^{+}]}{[Cu^{2+}]}. \qquad (1)$$

To prevent oxidation of Cu^{+} ions by oxygen dissolved in the Cu^{+}/Cu^{2+} solution, coulometry was carried out in a cell using a flow of 0~0.1 *l*/min Ar to remove dissolved oxygen. An amount of 2-15 mg of copper (I) chloride (High Purity Chemicals Co.) was dissolved in aq. 2.0 mol/*l* HCl. A d.c. current of 0.15-5 mA was passed through the anode and cathode.

To decrease the mass of the sample used for coulometry, it is required to decrease the volume of reaction chemicals. Since the amount of oxygen is usually measured by the reaction of Cu^{+} in CuCl and Cu^{3+} in copper oxide, the actual amount of Cu^{+} in CuCl used has to be estimated. This was checked by coulometric titration, and the result was taken into account through a correction parameter, *f*, calculated as

$$f = \frac{ItM_{\mathrm{CuCl}}}{Fm_{\mathrm{CuCl}}}, \tag{2}$$

where I, t, M_{CuCl}, F, and m_{CuCl} are dc current applied, coulometric time, molecular weight of CuCl, Faraday's constant, and mass of CuCl used for the experiment, respectively.

Pre-measurement was performed by dissolving 2-5 mg of CuCl in the solution and by applying current using a constant-current source (Hokuto Denko: HA-301) until the redox potential against the Ag/AgCl electrode reached 930 mV. Actual analysis of the sample was carried out after a pre-measurement in the same cell, and the coulometric time was determined when the redox potential reached 930 mV vs. Ag/AgCl again reached.

In the Cu^{+}/Cu^{2+} redox system, high valent metal compounds that come from the oxide sample are reduced with Cu^{+} in the solution. The high valent compound $YBa_2Cu_3O_{6.71}$, whose average Cu valence was 2.14, was chosen to test the new system. $YBa_2Cu_3O_{6.71}$ superconductor was prepared by solid-state reaction from a mixture of Y_2O_3, BaO, and CuO. The materials were sintered at 900-950°C for 10 h in air. To increase the oxygen content in the Y123 phase, they were annealed at 400-500°C for 10h in O_2.

The crystal structure was determined by the powder X-ray diffraction method using CuKα (α=1.54056 Å) radiation, indicating that the samples were composed of single phase $YBa_2Cu_3O_{6.71}$. The onset superconducting critical temperature (Tc^{on} = 82 K) was determined by the conventional four probe method.

The $YBa_2Cu_3O_y$ specimen was dissolved in the hydrochloric solution, where Cu^{3+} in $YBa_2Cu_3O_y$ reacts with Cu^{+} in the solution and makes two Cu^{2+} ions:

$$Cu^{3+} + Cu^{+} \rightarrow 2Cu^{2+}. \tag{3}$$

The rest of Cu^{+} in the solution is oxidized on the cathode as follows:

$$Cu^{+} \rightarrow Cu^{2+} + e^{-}. \tag{4}$$

The excess amount of Cu^{+} (Cu^{+}_{excess}) is calculated from the coulometric time, t, according to Faraday's law.

$$(Cu^{+}_{\mathrm{excess}}) = \frac{It}{F}, \tag{5}$$

The amount of Cu^{3+} in $YBa_2Cu_3O_y$ is calculated by

$$(Cu^{3+}) = \frac{fm_{\mathrm{CuCl}}}{M_{\mathrm{CuCl}}} - n(Cu^{+}_{\mathrm{excess}}). \tag{6}$$

The average Cu valence, z_{Cu}, therefore, is obtained by

$$z_{Cu} = \frac{3(Cu^{3+}) + 2[3M_{YBa_2Cu_3O_{78}} - (Cu^{3+})]}{3M_{YBa_2Cu_3O_{78}}} \tag{7}$$

where $M_{YBa_2Cu_3O_{7-\delta}}$ is the molar quantity of $YBa_2Cu_3O_y$.

The oxygen content, y, was calculated from charge neutrality as follows,

$$y = \frac{7 + 3z_{Cu}}{2}. \qquad (8)$$

RESULTS & DISCUSSION

As shown in Table I and Fig. 2, the Cu valence was measured with varying Ar flow rates from 0.02 to 0.1 *l*/min. The mass ratio, therefore, was fixed at 0.2, that is, 2 mg of CuCl and 10 mg of $YBa_2Cu_3O_{6.71}$ (Conditions A, B, and C). For comparison, conventional coulometry with the 300 ml cell was also done under condition D. Without Ar flow, the measured redox potential was too unstable to perform coulometry. Also, Ar flows greater than 0.1 *l*/min splashed the solution, which made it impossible to perform the coulometric measurement.

When the Ar flow rate is 0.02, the measured Cu valence values are located within 2.16 ± 0.09. Although the average Cu valence of 2.16 is an appropriate result compared with that of 2.14 measured by conventional coulometry, the error range ± 0.09 is rather larger than that of ± 0.03 of the conventional one. As the Ar flow rate is increased from 0.02 to 0.1, the error range becomes smaller, decreasing from ± 0.09 to ± 0.01, revealing that the increase in Ar flow leads to a decrease in the measurement error to within ± 0.01 and that an Ar flow rate of 0.1 *l*/min is the most suitable for micro-coulometry. It is to be noted that the error of ± 0.01 of micro-coulometry is much smaller than that of ± 0.03 of the conventional one. The accuracy possibly is a result of the miniaturized electrodes, and the experiments are now carried out this way for more precise measurements.

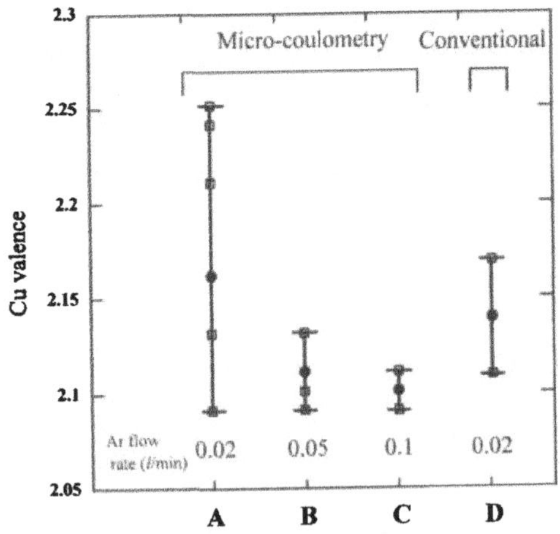

Fig. 2. Change in Cu valence measured by micro- and conventional coulometry with various Ar flow rates.

Table. I Cu valances of Y123 standard (Cu valence = 2.14) measured with various Ar flow rate

Condition	A	B	C	D
Coulometry	Micro			Conventional
Specimen	$YBa_2Cu_3O_{6.71}$			
Ar gas flow rate (*l*/min)	0.02	0.05	0.1	0.02
Current (mA)	2.5			5.0
Weight of CuCl (mg)	2.0			10
Weight of $YBa_2Cu_3O_{6.71}$ (mg)	10			50
Cu valence	2.16±0.09	2.11±0.02	2.10±0.01	2.14±0.03

To clarify the change in the dissolved oxygen content by the Ar flow, the content of dissolved oxygen in the acid solution was measured by dissolved oxygen (DO) electrode (OE-270AA DKK·TOA Co. Ltd). In Fig.3, the change in the dissolved oxygen content of aq. 2.0 mol/*l* HCl solution against Ar flow time is shown. Before starting the measurement, the solution was left stirring in the atmosphere for ten minutes. Just after flowing Ar at a rate of 0.1 *l*/min, the oxygen content decreased rapidly from 10 to 0.2 mg/*l* for 200 sec. After that, it continued to decrease slowly from 0.2 to under 0.01 mg/*l* till 600 sec. Further Ar flowing kept the dissolved oxygen under 0.01 mg/*l*, revealing that Ar flowing over 600 sec before coulometry is required to ignore the effect of the dissolved oxygen on the redox potential.

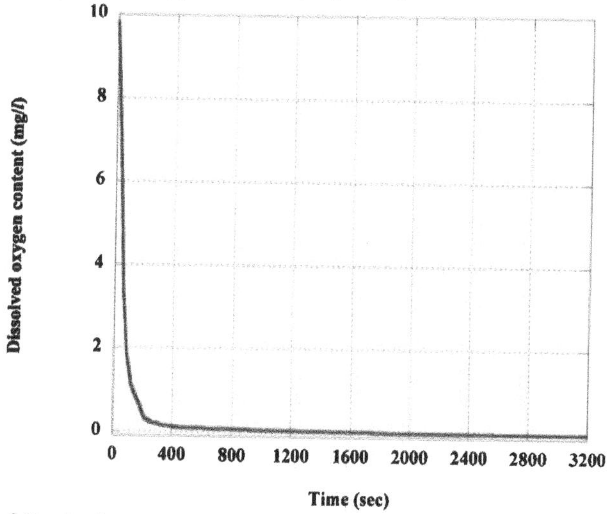

Fig. 3. Variation of dissolved oxygen content against Ar flowing time. For 200 sec, the content decreases rapidly, and after that it reaches under 0.01 mg/*l* in 600 sec.

The first step of coulometry is to introduce 10 mg of CuCl to the solution and to hold the redox potential between 240 and 250 mV. If the dissolved oxygen gas exists there, the redox potential will change by the redox reaction between Cu^+ and O_2. To examine the effect of dissolved oxygen, the redox potential and dissolved oxygen content were measured against the time after introduction of CuCl as shown in Fig. 4. Before introducing the CuCl, Ar was bubbled through the solution at a rate of 0.1 *l*/min for 600 sec to keep the oxygen content under 0.01 mg/*l*. After introduction of CuCl without Ar flow, the dissolved oxygen content increased from ~0 to 7.5 mg/*l* for 6000 sec with increasing the redox potential from 230 to 550 mV. On the other hand, continuous Ar flow after the CuCl introduction resulted in holding of the oxygen content around zero and the redox potential between 230 and 250 mV, indicating that Ar flow after the introduction of CuCl is also necessary to keep the redox potential between 230 and 250 mV.

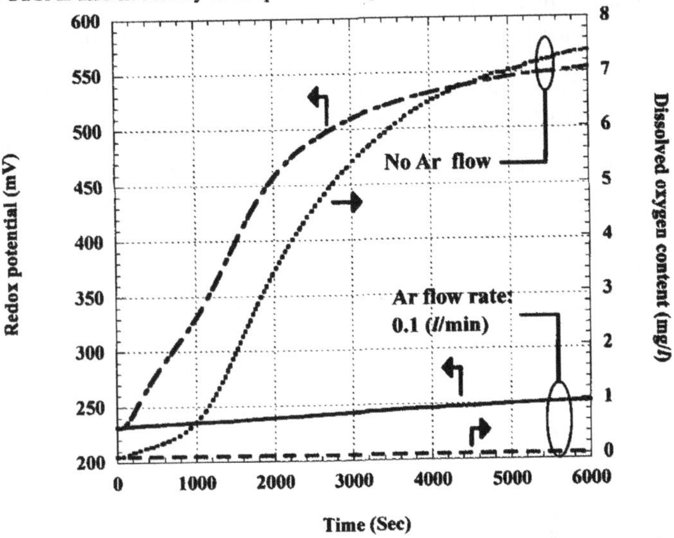

Fig. 4. The changes of redox potential and dissolved oxygen content against the time after CuCl introduction. Ar flowing keeps the dissolved oxygen content under 0.01 mg/*l* and drastically improves the increase in the redox potential by the dissolved oxygen.

In series A of Fig.5, Cu valances measured by coulometry with 10 mg of $YBa_2Cu_3O_{6.71}$ are listed and plotted against the coulometric current. When the coulometric current is set between 1.0 and 5.0 mA, the mass of CuCl and $YBa_2Cu_3O_{6.71}$ used is 2.0 and 10 mg, respectively. Otherwise, when the coulometric current is set smaller than 1.0 mA, the smaller masses of 1 mg CuCl and 5 mg $YBa_2Cu_3O_{6.71}$ are used. This is because smaller coulometric currents drastically lengthen the coulometric time. For example, the expected coulometric time when using 2.0 mg CuCl and 10 mg $YBa_2Cu_3O_{6.71}$ and 0.15 mA current is as long as 6-8 hours, which is too long for practical use. By using half the mass of CuCl and $YBa_2Cu_3O_{6.71}$, the coulometric time was shortened.

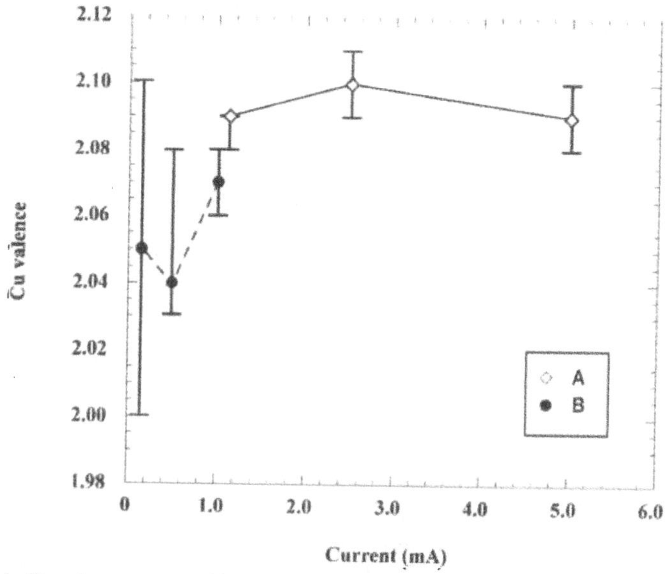

Fig. 5. Change in Cu valence measured by micro-coulometry using 5.0 mg (B) and 10 mg (A) of Y123 and varying the coulometric current from 0.15 to 5.0 mA

When the coulometric current is 2.5 and 5.0 mA, the average Cu valence of $YBa_2Cu_3O_{6.71}$ is calculated to be 2.09-2.10 ± 0.01 with good reproducibility. This Cu valence of 2.09-2.10 is ~0.04 smaller than that of 2.14, which was determined by conventional coulometry. To correct for the difference, an experiment for determining the amount of correction was carried out.

As described above, the average Cu valence measured by micro-coulometry was about 0.05 smaller than that determined by conventional coulometry. Since deposition of Cu on the cathode is observed during the measurement, the decrease in the Cu valence can possibly be explained by the side reaction on the cathode;

$$Cu^{2+} + 2e^{-} \rightarrow Cu, \tag{9}$$

which decreases the Cu^{2+} ion in the solution. From Nernst's equation, a decrease in the Cu^{2+} ion makes the redox potential more negative. Finally, the coulometric time for micro-coulometry becomes longer, which decreases the calculated Cu valence. To estimate the decrease of the Cu valence by the side reaction, the Cu valence of CuO was measured. If the side reaction (9) on the cathode occurs, the measured Cu valence of CuO will be smaller than 2.0. By correcting for the difference between 2.0 and the measured Cu valence, the correct Cu valence can be calculated.

The average Cu valence of CuO measured by micro-coulometry (i = 1.0 and 2.5mA) was 1.97, which was 0.03 smaller than the true valence, 2.0. Therefore, [measured Cu valence + 0.03] should be the correct Cu valence. The corrected Cu valence of $YBa_2Cu_3O_{6.71}$ is therefore (adding 0.03 to the measured value) 2.12-2.13 ± 0.01, which agrees very well with 2.14 ± 0.03.

Micro-coulometry was examined as a means to decrease the required $YBa_2Cu_3O_{6.71}$ mass from 10 mg to 5.0 mg. The results are listed in series B of Fig. 5. As the $YBa_2Cu_3O_{6.71}$ weight is decreased, the weight of CuCl is decreased from 2.0 mg to 1.0 mg while keeping the ratio of the CuCl vs. $YBa_2Cu_3O_{6.71}$ at 1:5. The current was set between 0.2 mA to 1.0 mA. As shown in B of Fig. 5, the average Cu valence measured by micro-coulometry was 2.05-2.07, which is about 0.08 smaller than the true Cu valence of 2.14. The average Cu valence of CuO measured by micro-coulometry with 5 mg of $YBa_2Cu_3O_{6.71}$ (i = 1.0 mA) was 1.95, which was 0.05 smaller than the true valence of 2.0, which can be explained by the side reaction (10). Therefore, [measured Cu valence + 0.05] would be the corrected Cu valence. The corrected Cu valence of $YBa_2Cu_3O_{6.71}$ is therefore 2.12 ± 0.01, which agrees well with 2.14.

Next, the reproducibility is discussed. As the current is reduced from 1.0 mA to 0.2 mA, the error range increases from ± 0.01 to ± 0.05. The expansion of the error range can be explained by the increase in the coulometric time. If a small amount of oxygen dissolves in the solution during coulometry, the coulometric time will increase because of parallel oxidation reaction on the anode.

CONCLUSION

The micro-coulometry system was successfully developed for measuring the Cu valence using only 5.0-10 mg of $YBa_2Cu_3O_{6.71}$.

Ar flow rate into the solution was optimized over the practical range from 0.02 to 0.1 l/min, with the result that an Ar flow of 0.1 l/min gave the smallest error range, ± 0.01. By measuring the dissolved oxygen content during coulometry, it was found that Ar flowing before and during coulometry is required to minimize the dissolved oxygen and keep the redox potential stable.

When using 10 mg of $YBa_2Cu_3O_{6.71}$, the Cu valence was determined to be 2.12-2.13 ± 0.01, which agrees well with the real valence of 2.14. Finally, micro-coulometry using only 5.0 mg of $YBa_2Cu_3O_{6.71}$ was tried. The Cu valence was successfully determined to be 2.12 ± 0.01 when the Ar flow rate, and the coulometric current were, 0.1 l/min, and 1.0 mA, respectively.

ACKNOWLEDGEMENT

The authors thank Jeffrey Willis of Los Alamos National Laboratory for his helpful discussions. This work was partly supported by the fund of Yamagata Prefecture.

REFERENCES

[1] S. Kambe, Y. Murakosi, R.Sekine, M.Kawai, K. Yamada, S.Ohsima and K.Okuyama, Physica C **178**, 71 (1991)

[2] M Karppinen, A Fukuoka, L Niinisto and H Yamauchi, Supercond. Sci. Technol. **9**, 121 (1996).

[3] Y. Yasukawa, H. Yamauchi and M. Karppinen, Applied Physics Letters, **81**, 502 (2002).

[4] R. J.Cava, A. W. Hewat, E. A. Hewat, B. Batlogg, M. Marezio, K. M. Rabe, J. J. Krajewski, W. F. Peck Jr and L. W. Rupp Jr, Physica C **165**, 419 (1990).

Electroceramic Materials and Applications
Edited by R. W. Schwartz

IODINE INTERCALATION AND DEINTERCALATION OF A Sr-FREE Bi-BASED CUPRATE: $Bi_2(La,Na)_2CuO_z$

Yoshiaki Tanaka, Makoto Nakanishi, Tatsuo Fujii and Jun Takada
Department of Material Chemistry, Graduate School of Natural Science and Technology,
Okayama University
Tsushima-naka 3-1-1,
Okayama 700-8530, Japan

Yoshihiro Kusano
Department of Applied Arts and Design, Kurashiki University of Science and the Arts
Nishinoura 2640, Tsurajima-cho, Kurashiki
Okayama 712-8505, Japan

Yasunori Ikeda
Institute for Chemical Research, Kyoto University
Gokasyo, Uji
Kyoto 611-0011, Japan

ABSTRACT

Iodine intercalation and deintercalation of a Sr-free Bi-based cuprate, $Bi_2(La,Na)_2CuO_z$, have been investigated. It has been confirmed by powder X-ray diffraction and thermogravimetric analysis that the stage-1 iodine-intercalated compound $Bi_2(La,Na)_2CuO_zI_x$ with x=1.0 was formed. As in the case of a $Bi_2(Sr,La)_2CuO_z$ (Bi-2201) specimen, expansion of the c-length of the stage-1 iodine-intercalated compound was 3.6 Å, compared with one half that of host compound. After deintercalation of the intercalated iodine, the stage-1 iodine-intercalated compound returned reversibly to the host compound. The powder X-ray diffraction pattern showed the formation of the intermediate phase containing a certain amount of iodine, not assigned to the stage-2 intercalated compound, during the deintercalation. Neither host compound nor iodine-intercalated compound exhibited superconductivity down to 2 K.

INTRODUCTION

Since the early 1990's, intercalation of halogens or interhalogen compound such as iodine, bromine and iodine monobromide into the Bi-based high temperature superconductors, $Bi_2Sr_2Ca_{n-1}Cu_nO_{2n+4}$ (n=1, 2, 3), which have a weakly coupled Bi-O bilayer in their crystal structures, has been investigated[1-6]. It is well-known that the intercalation of iodine into the Bi-based superconductors leads to the expansion of the c-length by about 3.6 Å and the superconductive properties for the host compounds are changed[1,2,7-9]. The latter could be understood in terms of the enhancement of the two-dimensionality due to the lattice expansion along c-axis for the iodine-intercalated compound and hole-doping into the CuO_2 plane deriving from the charge transfer from the iodine, which exists in the state of triiodide ion in the Bi-O bilayer[10], to the host compound[1-3,8,9]. This suggests that the iodine intercalation is a reliable hole-doping technique for hole-deficient Bi-based copper oxides, especially, in the Bi-2201 series possessing just one CuO_2 plane per unit cell.

Sr-free Bi-based cuprates, $Bi_2(La,A)_2CuO_z$ (A=Na, K), have been synthesized for the first time by Sasakura[11]. These compounds can be considered as Bi-based cuprates replaced all the Sr site of the Bi-2201 superconductor by lanthanum and alkaline metal (Na, K). But they exhibit no superconductivity probably because the hole concentration is not high enough to give rise to the superconducting transition. In order to induce the superconductivity in this system, the iodine intercalation could be the reliable hole-doping technique. Induced superconductivity in the Bi-based cuprates with no alkaline-earth elements will provide the important information on the superconductivity research.

In this study, we first tried the iodine intercalation into the Sr-free Bi-based cuprate, $Bi_2(La,A)_2CuO_z$ A=Na, and surveyed the crystal structure and magnetic properties. Here, we will not report about the case of A=K because no iodine-intercalated single-phase sample was obtained. We also elucidated the reversibility of the crystal structure upon iodine deintercalation, from the stage-1 iodine-intercalated compound.

EXPERIMENTAL PROCEDURE

Polycrystalline sample was prepared by conventional solid state reaction using powders of Bi_2O_3, La_2O_3, Na_2CO_3, and CuO. A mixture having a nominal composition of $Bi_2La_{1.5}Na_{0.65}CuO_z$ was pressed into pellet. The pellet lapped with an Ag-sheet was calcined at 780°C for 3 h in air and then quenched to room temperature. The calcined product was then ground, repressed into a pellet and sintered at 780°C for 15 h in air, then quenched to room temperature.

Iodine intercalation was carried out by a vapor-phase reaction of the host sample of $Bi_2La_{1.5}Na_{0.65}CuO_z$ with excess elemental iodine at 200°C for 50 h in a vacuum-sealed pyrex tube[1]. Iodine deintercalation of the stage-1 iodine-intercalated compound was done through the heat treatment ranging from 200°C to 400°C for 5 h with a heating rate of 1°C/min in flowing N_2. During the deintercalation the weight changes of the samples were measured.

Powder X-ray diffraction (XRD) measurements (Rigaku: RINT2000 with Cu $K\alpha$ radiation) were performed to identify the product phases and determine the lattice parameters for all the samples. Thermogravimetric analysis (TGA) was made using a Rigaku TG 8120 to determine the amount of iodine intercalated into the host compound. DC magnetic measurements were carried out for powdered samples down to 2 K under 100 Oe using a SQUID magnetometer (Quantum Design: MPMS-XL) after the zero-field cooled (ZFC) mode.

RESULTS AND DISCUSSION

Iodine intercalation

Fig. 1 (a) and (b) show powder X-ray diffraction patterns for $Bi_2La_{1.5}Na_{0.65}CuO_z$ and $IBi_2La_{1.5}Na_{0.65}CuO_z$, respectively. Obtained space group and lattice parameters are summarized in Table I. The single-phase $Bi_2La_{1.5}Na_{0.65}CuO_z$ was obtained without observable impurity phase as seen in Fig. 1 (a). On the other hand, after the iodine intercalation, the stage-1 iodine-intercalated $Bi_2La_{1.5}Na_{0.65}CuO_z$ was successfully prepared. All the (00l) diffraction peaks were systematically shifted to lower angle and the c parameter was increased from 23.77/2 Å of the host compound to 15.49 Å. These results were consistent with the cases of Bi-based high temperature superconductors containing Sr[1,7]. The space group of iodine-intercalated compound changed from Bbmb of the host compound to Pma2[7,12]. All the diffraction peaks were indexed on the basis of tetragonal lattice and no formation of impurity phases were revealed. The amount of iodine intercalated into $Bi_2La_{1.5}Na_{0.65}CuO_z$ was determined from the weight changes between before and after the intercalation. The estimated value was almost x=1.0 in $I_xBi_2La_{1.5}Na_{0.65}CuO_z$. This result also indicates that the intercalated compound has a stage-1 intercalation.

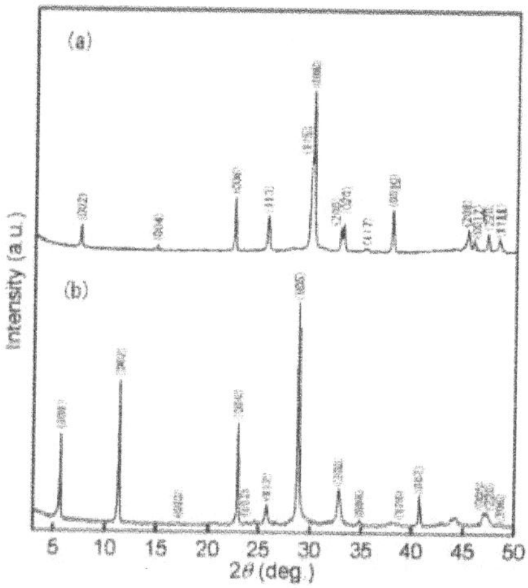

Fig. 1 Powder X-ray diffraction patterns for (a) $Bi_2La_{1.5}Na_{0.65}CuO_z$ and (b) $IBi_2La_{1.5}Na_{0.65}CuO_z$.

Table. 1 Space group and lattice parameters for $Bi_2La_{1.5}Na_{0.65}CuO_z$ and $IBi_2La_{1.5}Na_{0.65}CuO_z$.

Compound	Space group	Lattice parameters		
		a (Å)	b (Å)	c (Å)
$Bi_2La_{1.5}Na_{0.65}CuO_z$	*Bbmb*	5.472	5.431	23.77
$IBi_2La_{1.5}Na_{0.65}CuO_z$	*Pma*2	5.458	-*	15.49**

* The crystal system of the iodine-intercalated compound changes from ororthorhombic system of the host compound to tetragonal system.
** The unit cell of the iodine-intercalated compound changes to one half that of host compound.

Fig. 2 shows temperature dependence of DC magnetic susceptibility for (a) $Bi_2La_{1.5}Na_{0.65}CuO_z$ and (b) $IBi_2La_{1.5}Na_{0.65}CuO_z$. Both samples showed paramagnetism obeying the Curie-Weiss law in low temperature region. After iodine intercalation, the magnitude of susceptibility decreased, particularly, in low temperature region.

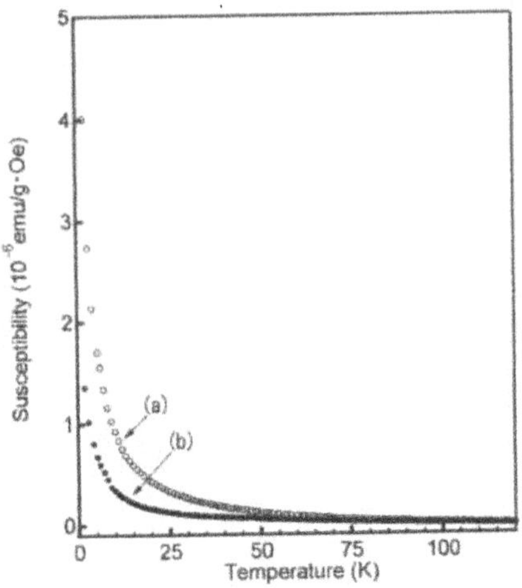

Fig. 2 Temperature dependence of DC magnetic susceptibility for (a) $Bi_2La_{1.5}Na_{0.65}CuO_z$ and (b) $IBi_2La_{1.5}Na_{0.65}CuO_z$ in zero-field cooled (ZFC) mode under 100 Oe.

The decrease in susceptibility could be understood as a reduction of the spin localized on Cu atoms of the CuO_2 planes resulting from hole-doping through the iodine intercalation. Neither sample exhibited superconducting behavior down to 2 K, but the result implies that the $Bi_2La_{1.5}Na_{0.65}CuO_z$ somewhat approached superconductivity through iodine intercalation.

Iodine deintercalation

Fig. 3 shows powder X-ray diffraction patterns for stage-1 iodine-intercalated compound before and after heat treatment at 200-400°C for 5 h in flowing N_2. As seen in the figure, each of the diffraction peaks of iodine-intercalated samples successively shifted to higher angle with increasing the heating temperatures up to 250°C. Then the peak shifting of the diffraction lines was saturated above 250°C and the secondary phase without iodine was observed. Finally each of the diffraction peaks returned reversibly to that of the host compound, $Bi_2La_{1.5}Na_{0.65}CuO_z$, by heating at 400°C. Unfortunately, Stage-2 iodine-intercalated compound was not obtained through the iodine deintercalation from the stage-1 compound, whereas formation of the intermediate phase containing a certain amount of iodine was observed at 250°C.

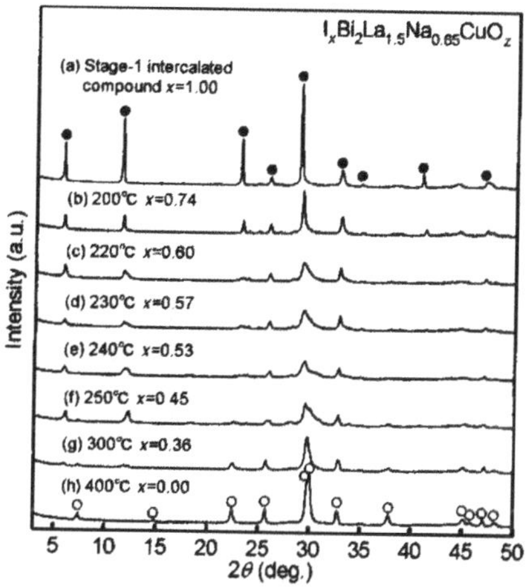

Fig. 3 Powder X-ray diffraction patterns for (a) $IBi_2La_{1.5}Na_{0.65}CuO_z$ and $I_xBi_2La_{1.5}Na_{0.65}CuO_z$ after heat treatment at (b) 200°C, (c) 220°C, (d) 230°C, (e) 240°C, (f) 250°C, (g) 300°C and (h) 400°C. ○: $Bi_2La_{1.5}Na_{0.65}CuO_z$, ●: $IBi_2La_{1.5}Na_{0.65}CuO_z$.

CONCLUSIONS

We have succeeded in the iodine interacalation into the Sr-free Bi-based cuprate, $Bi_2(La,Na)_2CuO_z$. Amount of the intercalated iodine and an expansion of c parameter were estimated as x=1.0 in $I_xBi_2La_{1.5}Na_{0.65}CuO_z$ and 3.6 Å, respectively, as well as the Bi-based superconductors containing Sr. The stage-1 iodine-intercalated compound exhibited no trace of superconductivity down to 2 K. Whereas, judging from a reduction in magnitude of the susceptibility, the iodine intercalation implied the charge transfer between the intercalated iodine and the CuO_2 planes. After the deintercalation of the intercalated iodine at the moderate heat treatment temperature, the stage-1 iodine-intercalated compound reversibly returned to the host compound, through the intermediate phase.

ACKNOWLEDGEMENTS

The authors are grateful to Prof. Masayuki Torii of the Okayama University of Science for SQUID measurements.

REFERENCES

[1]X. -D. Xiang, S. McKernan, W. Vareka, A. Zettl, J. Corkill, T. W. Barbee, and M. L. Cohen, "Iodine intercalation of a high-temperature superconducting oxide," *Nature*, **348**, 145 (1990).

[2]X. D. Xiang, A. Zettl, W. A. Vareka, J. L. Corkill, and M. L. Cohen, "Metallization of the resistivity tensor in $Bi_2Sr_2CaCu_2O_x$ through epitaxial intercalation," *Phys. Rev. B*, **43**, 11496 (1991).

[3]X. D. Xiang, W. A. Vareka, A. Zettl, J. L. Corkill, and M. L. Cohen, "Epitaxial intercalation of the Bi-Sr-Ca-Cu-O superconductor series," *Phys. Rev. Lett.*, **68**, 530 (1992).

[4]Y. Koike, T. Okubo, A. Fujiwara, T. Noji, and Y. Saito, "Trial of intercalation of Br and Li into $Bi_2Sr_2Ca_{n-1}Cu_nO_{2n+4}$ (n = 1, 2, 3)," *Solid State Commun.*, **79**, 501 (1991).

[5]T. Okubo, A. Fujiwara, Y. Koike, T. Noji, and Y. Saito, "Intercalation of Br and Li in $Bi_2Sr_2Ca_{n-1}Cu_nO_{2n+4}$," *Physica C*, **185-189**, 847 (1991).

[6]M. Mochida, Y. Koike, K. Sasaki, A. Fujiwara, T. Noji, and Y. Saito, "Synthesis and superconductivity of IBr-intercalated $Bi_2Sr_2CaCu_2O_8$," *Physica C*, **212**, 191 (1991).

[7]N. Kijima, R. Gronsky, X. D. Xiang, W. A. Vareka, A. Zettl, J. L. Corkill, and M. L. Cohen, "Crystal structure of stage-1 iodine-intercalated superconducting $IBi_2Sr_2CaCu_2O_x$," *Physica C*, **181**, 18 (1991).

[8]D. Pooke, K. Kishio, T. Koga, Y. Fukuda, N. Sanada, M. Nagoshi, K. Kitazawa, and K. Yamafuji, "Charge transfer associated with iodine intercalation in $Bi_2Sr_2Ca_{n-1}Cu_nO_y$ n=1−3," *Physica C*, **198**, 349 (1992).

[9]A. Fujiwara, Y. Koike, K. Sasaki, M. Mochida, T. Noji, and Y. Saito, "Two effects of iodine intercalation on T_c in $Bi_2Sr_2Ca_{1-x}Y_xCu_2O_8$: Two-dimensionality and charge transfer," *Physica C*, **208**, 29 (1993).

[10]E. Faulques, and R. E. Russo, "Raman spectra of iodide species in intercalated $IBi_2Sr_2CaCu_2O_{8+\delta}$," *Solid State Commun.*, **82**, 531 (1992).

[11]H. Sasakura, "New members of Sr-free Bi-2201 cuprates: $Bi_2(La,A)_2CuO_z$ (A = Na and K)," *Physica C*, **292**, 125 (1997).

[12]Y. Muraoka, M. Kikuchi, H. Nameki, S. Awaji, R. Suzuki, N. Kobayashi and Y. Syono, "Effects of iodine intercalation on the superconducting properties of the $Bi_2Sr_{2-x}La_xCuO_z$ compound," *Physica C*, **215**, 402 (1993).

Electroceramic Materials and Applications
Edited by R. W. Schwartz

Dielectric, Ferroelectric, Piezoelectric and Electrooptic Materials

Electroceramic Materials and Applications
Edited by R. W. Schwartz

PROGESS IN FERROELECTRIC DOMAIN ENGINEERING AT THE MICRO/NANOSCALE

Robert W. Eason, Sakellaris Mailis, Collin L. Sones, Alexander J. Boyland, Alistair C. Muir, Tleyane J. Sono, Jeffrey G. Scott, Christopher E. Valdivia, Iain T. Wellington.

Optoelectronics Research Centre
University of Southampton
Southampton, Hampshire, SO17 1BJ, UK.

ABSTRACT

Ferroelectric materials such as lithium niobate (LN) or lithium tantalate (LT) are examples of an extremely versatile class of optical crystals. In bulk single crystal, single domain format, these crystalline hosts find numerous applications in nonlinear optics, optical storage, photorefraction, surface acoustic wave devices, optical waveguides, piezoelectric and pyroelectric devices and electro-optic modulation. Single domain crystals can be subsequently engineered via spatially selective poling to yield domain structures whose size can lie in the region of a few tens of μm to sub-μm, for applications and device fabrication that are impossible to implement in single domain geometry.

This paper discusses our progress to date in micro- and nanostructuring of such materials, for applications in nonlinear optics, switching and deflection, and 3-dimensional sculpting for possible MEMS use. The techniques and benefits are discussed of using both light-assisted and direct optical poling for achieving controllable domains that can be irregular or periodic, bulk or surface, at sizes that approach the 100 nm scale. For surface inversion, domain features can be produced that lack the otherwise characteristic crystal symmetry imposed hexagonal shapes observed in conventional electric field poling.

INTRODUCTION

Ferroelectric materials play an extremely important role across many scientific and technological disciplines due to their wide range of intrinsic properties such as piezoelectricity, pyroelectic, electro-optic and photorefractive behaviour and are currently finding widespread use in periodically poled geometries for efficient quasi-phase-matched interactions[1]. Two widely used and commercially available materials are lithium niobate (LN) and lithium tantalate (LT) which are usually purchased in the form of single domain, pre-poled wafers of typically up to 3 inch diameter and thickness of order 1 mm or below.

For both LN and LT that have been congruently grown, the coercive, or poling, field at room temperature is ~21 kV mm^{-1} for forward poling[2], a value which is high, but manageable for producing periodically poled structures at periods of order 4-5 μm or larger in bulk crystals of thicknesses ~1 mm. Another variant of these crystals, termed stoichiometric, have poling fields of order one hundredth[3], of this value and there is currently great interest in the routine commercial availability of high quality stoichiometric crystal materials. As well as their much reduced coercive fields, stoichiometric crystals have far fewer intrinsic defects such as lithium vacancies, and nonlinear optical experiments have shown that they possess much greater resistance to optical damage by almost four orders of magnitude as compared to their congruent counterparts[4].

These crystals can be poled using electric fields to produce domain patterns that can be bulk (throughout the entire crystal thickness) or surface (some few μm depth) in nature, and such spatially selective domain patterning adds greatly to their functionality and areas of application.

While periodically poled lithium niobate (PPLN) in particular has found application in nonlinear optical harmonic and parametric generation, achieving high quality samples with periods of less than the ~4 μm value is experimentally challenging. It is therefore of considerable importance to investigate methods for realising μm scale periodicity, and below, for applications in tunable Bragg structures, filters and gratings, backward wave parametric generation, and other areas such as optical memory that involve domain patterning at the experimental limit which is of order ~10 nm.

This paper discusses recent progress in both physical structuring, using the technique of laser direct writing, and also domain engineering, to produce spatially selective domain patterns at feature sizes of ~100 nm and above, in both congruent and MgO doped stoichiometric materials. A generic end goal of our work to date may be summarised by the following schematic shown in figure 1 which concerns the idea of a functional lab-on-a-chip, using an active material such as LN rather than glass, polymers or silicon. Once the toolbox of techniques for structuring in 3D is complete, then arbitrary designs are possible for the manipulation of objects such as biological cells, micro-beads, or single molecules around the surface of such a chip. The added value of using a functional material such as LN compared to glass lies in the ability to integrate electrodes into the structure, thereby accessing the material's intrinsic piezoelectric, pyroelectric and electro-optic effects.

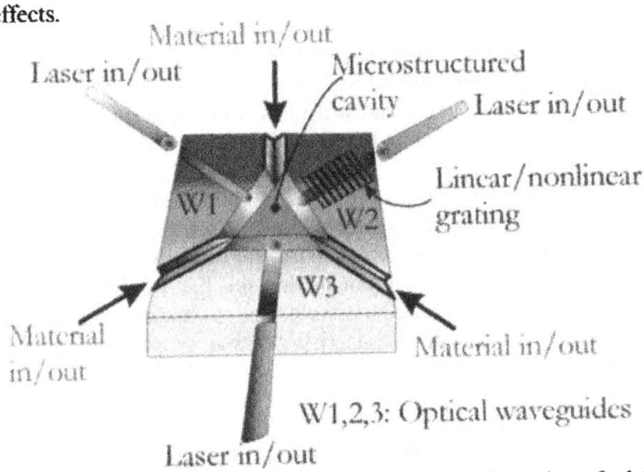

Figure 1. Schematic of physical structuring processes required for fabrication of a lab-on-a-chip.

PHYSICAL STRUCTURING PROCESSES

A range of physical structuring processes exists for generating surface profiles such as gratings, holes, vias and optical waveguides. Of prime importance however is that the processed crystal has a high quality finish, and that the number of processing steps is minimized. While laser ablation for rapid processing of LN may satisfy the second criterion, problems of direct laser machining such as debris, cracking and redeposition in single crystal materials are difficult to overcome. Secondly, conventional photolithographic processing which requires cleanroom access involves numerous sequential steps such as cleaning, photoresist spinning, patterning, baking, development, with subsequent diffusion, etching or other methods of spatially selective material removal or overgrowth. While these processes are to a large extent routine, and may be automated, there is clearly an opportunity here for faster single-step processing technology using for example laser direct writing and two such techniques are briefly discussed here.

Surface UV irradiation plus wet etching.

For writing of gratings, two-beam interferometry or the use of phase masks are the preferred routes. If UV light is used, and the energy density is kept below the ablation threshold, then although little physical damage may be evident, the exposed regions may have undergone latent damage and these areas can be subsequently removed in a smooth and precise manner with an etchant such as hydrofluoric acid (HF). This technique has been successfully used to form surface gratings with a period of 360 nm onto both the + z face of LN, and x-cut titanium indiffused LN channel waveguide structures[5]. The method has the advantage of rapid processing, as a single 20 ns duration pulse from an excimer laser is all that may be required, and subsequent etching reveals a faithful replica of the initial irradiation pattern. Precise depth control of the grating profile is somewhat harder to achieve however leading to a consequent reduction in the performance (linewidth and grating strength) of such devices in comparison to, for example, fiber Bragg gratings.

Direct writing of channel waveguides

A second technique that we have been exploring involves a process by which a scanned focused c.w. UV laser operating at a wavelength of 244 nm can be used to rapidly heat the surface of a LN crystal, inducing localized Li ion diffusion which can lead to the formation of optical waveguiding structures[6]. Since the first report of this technique, we have subsequently modeled the process using both analytical and finite difference techniques to determine the maximum temperature rise, spatial and temporal temperature gradients and consequent Li ion diffusion achieved in both lateral and depth directions[7]. The inset of figure 2 shows the results of an analytical model we have developed, which indicates that a transient temperature rise of almost 1000 °C is possible via scanned UV laser heating. Noteworthy is the fact that the extremely small absorption depth of ~30 nm at this wavelength produces a distinct difference in temperature profiles for surface (x or y) compared with depth (z) directions. The difference in temperature gradients greatly affects the Li ion diffusion process, and diffusion does not occur to any great extent in the depth direction, as shown in figure 2. Even though the index change associated with Li diffusion away from the heated regions is high enough to produce guided wave structures, as have been produced experimentally, the modeling performed so far indicates that the diffusion depths achievable are insufficient to support even single mode guidance. This discrepancy, which may have its origins in either the fundamental model assumptions or details associated with some unknown physical parameters, is currently under further investigation.

DOMAIN BASED STRUCTURING

An alternative technique that we have been pursuing for several years concerns the use of micro and nanostructuring based on domain engineering. For some applications, the domain patterning produced is subsequently converted into a corresponding topographical structure as in case for PPLN, the presence of the domain structure alone is all that is required. We show below a simple example of the application of a single domain boundary in deflection and switching, using the principle of voltage controllable total internal reflection (TIR) at an interface between opposite polarity domains.

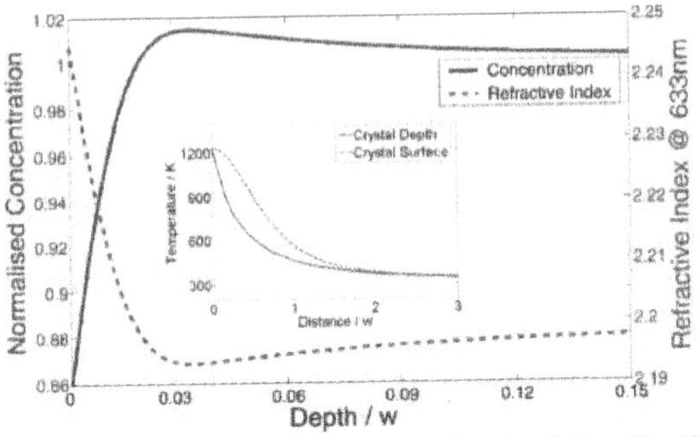

Figure 2. Modelled lithium concentration and associated refractive index distribution along the depth (z) direction for identical beam parameters as in the inset of this figure. Note that the induced Li concentration does not extend very far into the crystal depth, and peaks at a distance of order 0.03 of the incident laser spot size. The depth axis has been normalized relative to the beam width. Inset: Modelled temperature distributions on the surface (x or y directions) and in depth (z) for UV beam writing parameters of spot size w = 3.25 μm, incident laser power P = 30 mW and writing speed v = 0.83 mm s^{-1}.

TIR deflection and switching.

Shown in figure 3 is a schematic of a miniature TIR device that we have fabricated in congruent LN, with dimensions of 5 × 13 × 0.3mm in x, y, z directions[8]. One half of the wafer has been re-poled to produce a boundary between anti-parallel domains. On application of a voltage across this boundary region, the local refractive index is increased on one side of the boundary, and decreased on the other. A laser beam incident on the boundary from the elevated index side will experience deflection via Snell's law proportional to the applied voltage, and subsequently TIR when the induced index change exceeds the value required for the particular value of incident angle.

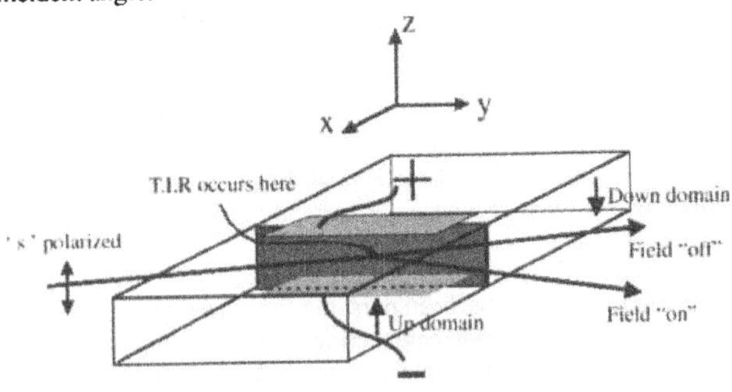

Figure 3. Schematic of TIR deflection and switching device with poled anti-parallel domains.

In practice, for *s* polarized light incident at an angle of ~88°, a voltage of ~1kV is required for TIR to occur, utilizing the r_{33} electro-optic coefficient. Deflection will also occur for *p* polarized light, which can access the r_{13} coefficient, but this requires an applied voltage roughly three times higher. If the exit facet of the device is cut at an appropriate angle, Snell's law can magnify the final deflection achievable to produce a ~10° swing of the exiting beam, and this can be further extended to 20° by reversing the polarity of the applied field across the interface. For an extended device length of 50 mm, modeling shows that the device is capable of achieving 300 resolvable spots. Further extensions of this technique allow both RGB colour separation and also 2-dimensional scanning operation.

Cantilevers

Micro-structuring in LN has been achieved via a process of spatially selective poling, crystal bonding and subsequent etching in HF. Such techniques permit full 3-dimensional structuring[9] in single crystal material which is a fundamental requirement if device implementation requires subsequent use of electro-optic or piezoelectric coefficients. Shown in figure 4 is an example of such sculpting used to produce a cantilever of transverse and height dimensions ~50 μm, and length of 5 mm. The gap between the cantilever and the substrate is afew μm, and during focused ion beam writing of electrodes on the side of the structure, without the use of a compensating electron flood beam to neutralize any surface charging, the cantilever experienced deflection to a height of ~15 μm, showing that such single crystal geometries are capable of flexure and bending, and hence suitable for MEMS based applications.

Figure 4. SEM image of a free-standing cantilever in LN (dimensions of ~50 μm×~50 μm ×5 mm).

Surface poling (overpoling).

As well as poling in bulk materials, where domain inversion is required throughout the full crystal thickness of ~500 μm to 1 mm, for waveguide geometries inversion is strictly only required throughout the waveguide depth of a few μm. As noted earlier, fabrication of small period gratings (Λ ~few μm) is far from routine, particularly when trying to achieve uniformity throughout a ~1 mm LN sample. Surface poling[10], alternatively termed overpoling, has been used recently to fabricate periodic structures with a domain size of ~0.5 μm, and period of 1 μm. Using a grating period of 2.5 μm, first order quasi-phase-matched harmonic generation from λ= 825 nm to λ=412.5 nm has been achieved in a Ti:indiffused LN channel waveguide[11], producing an overall output power of 6 mW, and an efficiency of ~10%.

Overpoling refers to a technique in which conventional photolithographic periodic patterning in photoresist is followed by the application of an electric field such that the charge delivered during poling exceeds the nominal charge Q calculated via:

$$Q = 2\,A\,P_s\,EF \qquad (1)$$

where A is the area to be poled, P_s is the spontaneous polarization for LN (0.72μC mm^{-2}) and EF is an empirical poling factor.

For an EF of ~1, normal bulk poling is achieved, while for an EF of >10 for example, the entire crystal poles, with domain wall spreading occurring throughout the crystal volume. For intermediate values (EF~4-6), while the bulk crystal experiences poling throughout its volume, regions directly below the photoresist do not, due most likely to trapped charge directly beneath the insulating photoresist layer. These regions of unpoled material can extend to depths of some few μm to perhaps 10 μm, dependent on both the periodicity chosen and the value of EF used. Figure 5 shows an example of such surface domains, taken via scanning force microscopy (SEM) on a polished and etched y face. For this example a periodicity of 6.5μm was used, and it can be seen that although some domains extend to beyond 10 μm depth, the degree of depth uniformity is poor. For waveguide applications however, as noted earlier, all that matters is that the minimum depth exceeds the waveguide dimension.

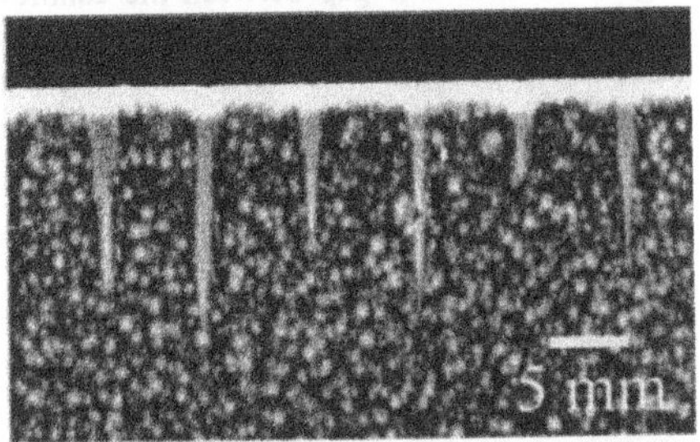

Figure 5. SEM image of an etched y-face showing typical surface domain profile (for Λ = 6.5 μm).

Even shorter period (Λ = 1.0 μm) surface domains have been realized, with a minimum domain feature size of ~0.5 μm, as shown in figure 6. A problem which is evident in this technique however, and can be clearly seen in figure 5, is that the mark/space ratio of domain inverted material differs from the ideal 50/50 value.

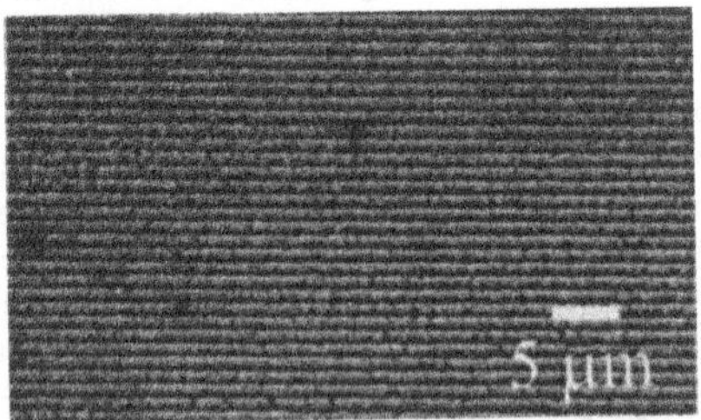

Figure 6. Surface poled periodic domain structure with domain widths of ~0.5 μm.

While we have tried to address this problem, success to date in achieving the 50/50 ratio is limited, and remains a problem for routine high quality surface domain structuring. For SHG experiments therefore, the trade-off that exists concerns the mark-space ratio efficiency reduction versus the efficiency increase achieved by operating in first order.

Nanotips.

The final example in this section on domain-based structuring relates to the fabrication of arrays of domain inverted tips on the surface of a LN wafer[12]. Applications for this technique lie in the area of SHG, atomic force microscopy and piezoelectric nano-devices. Figure 7 shows an SEM of a typical array, formed via photolithographic patterning, poling and etching in HF at a temperature of 60 °C. Tips with the characteristic 3-fold symmetry form due to the differentialnature of y-face etching. As etching proceeds in the $-z$ direction, the simultaneous y-face etching will result in the formation of pyramids, shown from a top view in figure 7. Also seen in the inset is a highly magnified view of a single tip, from which it is apparent that the tip apex is essentially of nm scale.

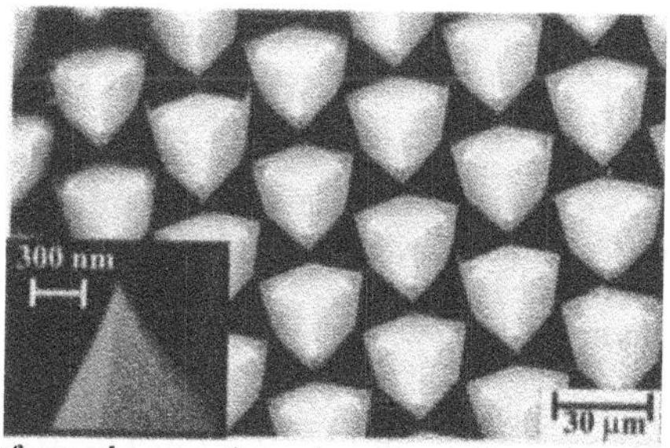

Figure 7. SEM picture of nanotip array formed on the $-z$ face of LN. Tip dimensions are ~30 μm at their base, and ~100 μm high. Inset shows that the tip apex is of ~nm dimensions.

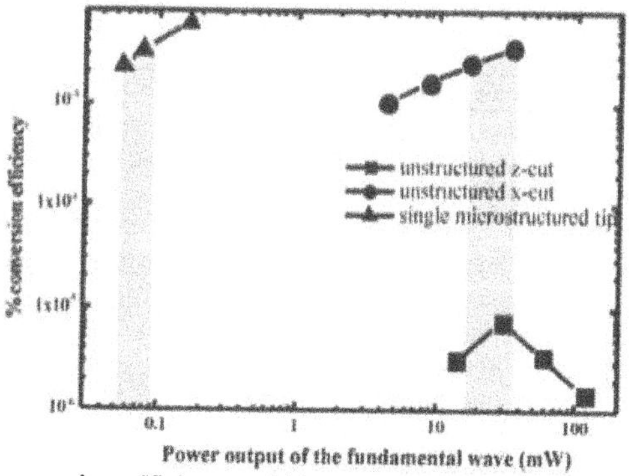

Figure 8. SHG conversion efficiency within nanostructured LN tips compared to unstructured x- and z-cut crystal samples.

Experiments have been performed using ~130 fs laser pulses at 800 nm to generate SHG at 400 nm from these tips, to investigate the efficiencies achievable from interactions within LN in a non-phase matched geometry under the extremely high power densities present in the vicinity of the tip. Figure 8 shows our first results, from which it can be seen that the harmonic conversion efficiency is dramatically enhanced, compared to that achievable with either unstructured x- or z - cut LN samples. This enhanced efficiency may be due to a combination of factors such as increase of the light intensity within the tip, access to nonlinear coefficients unavailable in the unstructured material, surface enhanced or surface phase-matched interaction, or even modification of the basic ferroelectric structure itself within these extreme crystal geometries. Whatever the predominant cause, highly efficient SHG has been achieved in an otherwise non-phase matched geometry. These investigations are ongoing, and material characterization is underway to determine whether the crystal structure within such tips has been modified with respect to that of bulk material.

LIGHT-INDUCED ETCH FRUSTRATION, LIGHT-ASSISTED POLING AND ALL-OPTICAL POLING.

We now address the area of laser light interaction with LN for investigation of the effects of light-induced frustrated etching (LIFE), light-assisted poling (LAP), and all-optical poling (AOP), using c.w. ion lasers operating in the UV and visible (244 nm, 488 nm, 514 nm, 532 nm), ns UV pulsed lasers (excimer, frequency quadrupled YAG and doubled dye) and finally 100 fs pulsed Ti:sapphire (from 305 nm to 800 nm). The work here spans several years, and we have investigated a range of ferroelectric hosts including congruent LN and LT (CLN, CLT), Fe:doped congruent LN (Fe:CLN), MgO:doped congruent LN (MgO:CLN) and MgO:doped stoichiometric (MgO:SLN). Table (I) summarises the experiments to date, and the abbreviations used.

Table I. Classification of light-assisted interactions in ferroelectric host materials.

Laser	Wavelengths	*CLT*	*CLN*	*Fe:CLN*	*MgO:CLN*	*MgO:SLN*
c.w.	UV	LAP	**LIFE**[b]	LIFE		
	visible	LAP	LAP	**LIFE**[a]	LAP	**LAP**[d]
pulsed	UV ns		**LIFE**[c] **AOP**[f]	LIFE	AOP	
	UV, visible fs		**LAP**[e]		**LAP**[e]	

(The abbreviations in **bold** and superscripts refer to subsequent sub-headed sections in the text)

Light induced frustrated etching (LIFE[a])

This has been investigated under two dissimilar experimental regimes. The first experiments were conducted with Fe:doped CLN only[13], with Fe dopant concentrations varying between 0.01 wt% and 0.2 wt%. The experimental arrangement involved the exposure of the –z face for periods ranging from 30 s up to one hour to an incident c.w. laser beam at wavelengths of 488 nm or 532 nm in the presence of an HF acid environment.

Initially, these experiments were performed in the expectation that this photoelectrochemical etching procedure would produce a dramatically enhanced etch rate, as had been seen previously for a range of semiconductor materials. In fact, the reverse effect was observed, namely the presence of the light totally inhibited any etching at the highest focused light intensity used (~1 kW cm^{-2}), and partially frustrated the etching at intensities between 1 kW cm^{-2} and ~1 W cm^{-2}. Within this partially frustrated regime[14], a spatially complex and visually dramatic patterning resulted, forming line features that followed the intrinsic crystallographic axes, with linewidths of order ~0.5 μm. An example of these surface features is shown in figure 9.

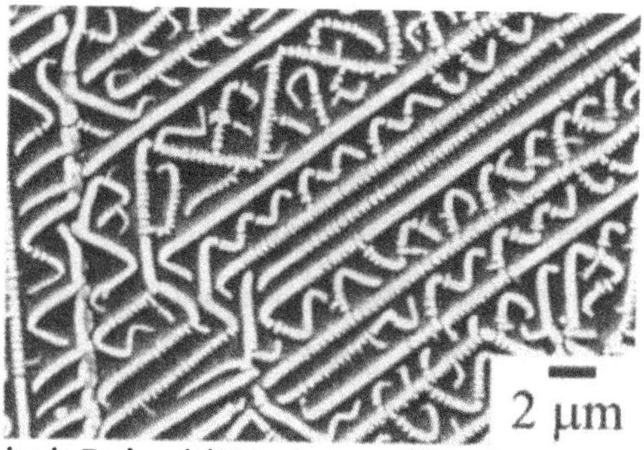

Figure 9. LIFE patterning in Fe:doped CLN using c.w. 532 nm laser light.

Although these results are artistically appealing, it has proved impossible to produce any useful degree of surface ordering. The intrinsic scale-lengths of the LIFE process appear to be almost ideal however, falling as they do in the sub-μm region. It has further proved difficult to uniquely classify the nature of these raised non-etched line features. While the temptation to refer to them as domains is considerable, there is as yet no clear evidence that this is the case, unlike our results for all-optically poled CLN which are discussed in a later section.

Light induced frustrated etching (LIFE^{b}).

The second set of experiments that fall within the LIFE category have been performed using c.w. focussed UV light at 244 nm from a frequency doubled Ar-ion (FRED)laser. The set-up is essentially identical to that used for direct writing of channel waveguides in the –z face of undoped LN as described earlier. Laser powers of between 20 mW and 50 mW at a focused spot size of 1.7 μm were used, at a reasonably fast scanning speed of 50 mm s^{-1} to ensure that no significant surface damage or melting occurred. The results shown in figure 10 reveal that the laser-exposed regions proved to be highly etch resistant[15]. One dimensional ridge features of height 1.7 μm as well as 2-dimensional square lattice patterns were written with good uniformity and reproducibility. Micro-Raman studies of the etch resistant structures revealed no departure from identical unexposed material, indicating that the UV laser writing process had neither damaged nor melted the crystal surface. As with the results for LIFE^{a} above, it is tempting to conclude that these features are laser-written inverted domains, but there is no direct evidence for this however. An equally plausible explanation might involve surface charge modification. Following laser writing, if the sample was annealed at a temperature of ~200 °C for one hour, then subsequent HF exposure did not reveal any etch-resistant features. Techniques such as scanning force microscopy (SFM), discussed in a subsequent section, cannot be easily used here once the sample has been etched. All such scanning probe techniques are also sensitive to sample topography, and both of these LIFE^{a,b} techniques result in highly textured topographical surfaces. The technique is simple, direct and fast however, and can result in feature sizes of a few μm.

Figure 11. SEM of –z face after scanned laser exposure and 1 hour of HF etching. The separation between each track was 9 μm.

Light induced frustrated etching (LIFE[c])

The final example of this surface structuring process can be seen in figure 11 below. A range of experiments was performed, using between one and ten laser pulses, followed by HF etching. A TEM grid that had multiple hexagonal openings of ~50 μm width, was loosely attached to the –z face of CLN samples[16]. Following exposure, the samples were etched in HF, and then examined by SEM. The level of detail and self-ordering in the surface topography is extremely interesting, suggesting that a similar process may be operating to that seen in LIFE[a], but in a time frame of ns rather than in minutes or hours. The surface features (as seen in the inset of figure 11b) are remarkably uniform, at a similar size of ~0.5 μm to previous LIFE[a] results. In the present case, it is also extraordinary that the etch frustration process appears to be inordinately sensitive to the local irradiance.

The etch pattern in figure 11b shows the result of exposure from a near-field (Fresnel) diffraction pattern produced by lack of perfect contact between the mask and the LN crystal surface. Although the variation of light intensity towards the edges of the irradiated area can only be slight (few % to perhaps few 10's of %), this variation is clearly sufficient to produce an equivalent modulation of the frustration effect. Indeed, towards the centre of the irradiated area, where the intensity modulation can only be minimal, it is still apparent that the frustration modulation and ordering persists. It would be very surprising if the frustration phenomena could be that sensitive to local intensity, or perhaps intensity gradient, but this is what appears to happen.

Even though the precise explanation is not yet apparent, it is clear that such a technique may be eminently suitable for precision surface topography manipulation.

Light-assisted poling (LAP[d])

There have been several reports to date of the role of incident light in modifying coercive fields for both forward and backward poling, and there appear to be at least two somewhat contradictory processes occurring. The first of these LAP processes concerns the modification of the recovery of the coercive field in CLT following a forward poling step[17]. Directly after a forward poling (some few seconds), the field required for reverse poling is considerably lower than the field needed to reverse pole perhaps one hour after the initial forward poling step. The internal field that plays an important role in such congruent crystals takes time to adjust itself to the new direction of spontaneous polarization. The recovery process is far from simple, and at least three distinct rate constants have been identified in the recovery mechanism, with time constants ranging from ~1s to more than one year. After one hour for example, the field required

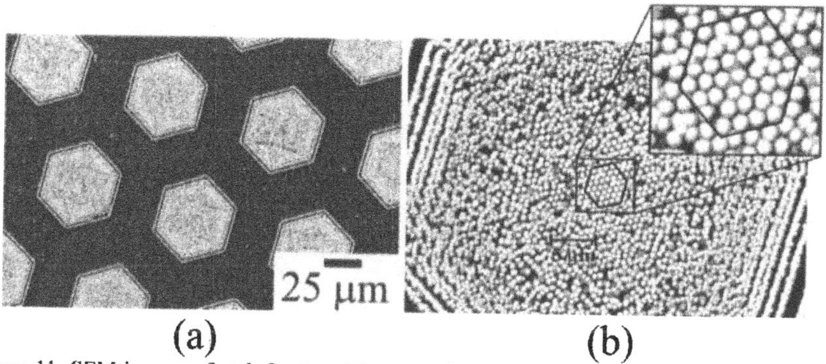

Figure 11. SEM images of etch frustrated hexagonal patterns induced on the −z face of LN via exposure through a TEM grid using a KrF excimer laser operating at 248 nm, 20 ns pulse duration.

for reverse poling in CLT remains some 35% lower than for the initial forward poling process.

During this coercive field recovery process, light in the visible spectral region, specifically at the Ar ion wavelengths of 488 nm and 514 nm, which is incident on the crystal, can enhance this rate of recovery. If the incident light is patterned, or structured, then by judiciously choosing the value of field applied during illumination, selected unilluminated areas can undergo reverse poling, whereas illuminated areas, where the coercive field has recovered faster, will not. Achieving the balance between poling and not poling in unilluminated and illuminated areas respectively is clearly difficult, particularly considering that all such local coercive fields are themselves time-dependent. Periodic patterning has been achieved however via this technique with a periodicity of ~6 μm, but over a very limited spatial extent, using 351nm UV illumination from an Ar ion laser.

The second of these processes appears to be the direct opposite of the first: during re-poling, incident light in the near-UV[18] or visible region[19] can dramatically reduce the local coercive field, by as much as 34% and 90% respectively in either MgO:CLN or MgO:SLN. Figure 12 shows a plot of the ratio of coercive field with and without illumination as a function of the irradiance for three crystal samples: CLN, MgO:CLN and MgO:SLN. For the latter two crystal types, the reduction is dramatic, and saturates at a value of only 10% of the field required for an unilluminated sample. One difference to note in these experiments however is that the field recorded here is not the conventionally defined coercive field, but the 'nucleation field', defined as the field at which the first evidence of poling occurs. The laser wavelength of 514 nm was chosen for these experiments as this line contained the highest power, but all other available Ar laser wavelengths in the visible of near UV produced substantially similar results. These experiments were greatly assisted through the use of a purposely designed transparent sample cell, shown in figure 13, which allowed the application of an electric field while simultaneously illuminating the LN sample with two different laser wavelengths: one for control of the coercive field and a second for viewing domain nucleation and growth.

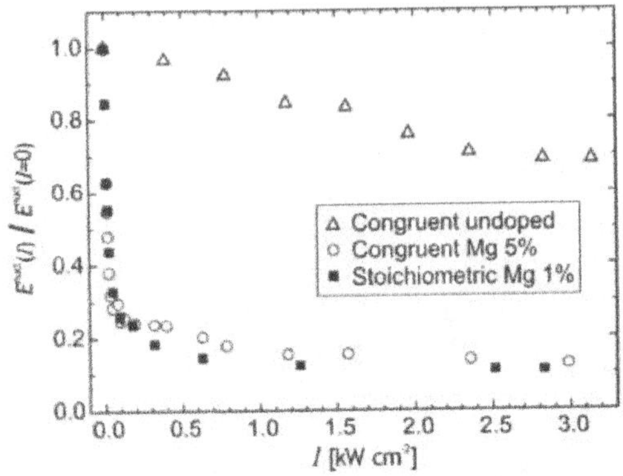

Figure 12. Nucleation field versus incident c.w. Ar ion laser irradiance at a wavelength of 514 nm.

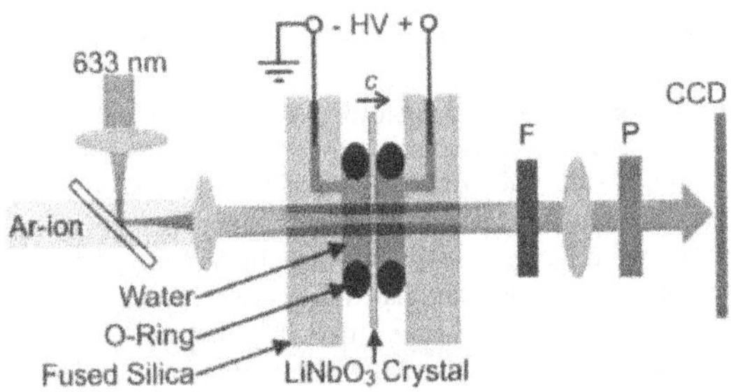

Figure 13. Transparent sample cell for simultaneous application of optical and electric fields.

Light-assisted poling (LAP[e]).

At the other extreme of laser exposure conditions, recent results have been obtained using a Ti:sapphire 100 fs laser source to investigate the process of LAP under very high peak power density. Figure 14 shows the result after etching in HF, in which a ~100 μm diameter domain has been formed in MgO:CLN at the remarkably low electric field of 100 V mm^{-1}, a value of less than 2% of the normal unilluminated nucleation field. These domains are not as easily visualized as those for the LAP[d] process, and are only revealed via HF etching after the LAP process. The laser average laser power was ~100 μW, for ~100 fs duration pulses at a repetition rate of 1 kHz. Light of wavelength 334 nm was used for figure 14, although a range of other wavelengths between 305 nm and 800 nm also process efficient LAP processing.

For the lowest wavelength of 305 nm, the incident power could be reduced to the extremely low value of 100 nW, a factor or 1000 less than used for figure 14. These experiments are ongoing, and hold considerable promise for further precise LAP experiments.

A final result in this fs LAP section is shown in figure 15, which was recorded in CLN. The letters were written by manually moving the position of the incident beam, hence the precision is less than can ultimately be achieved.

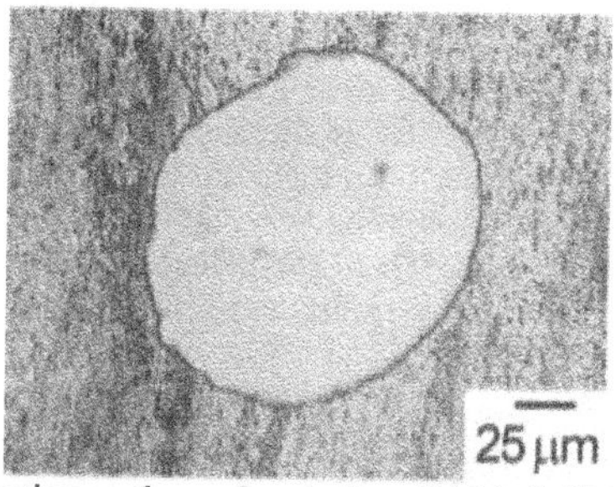

Figure 14. LAP domain formation on the –z face in 5-mol% MgO:CLN.

The domains are bulk, not surface as for other techniques, and an interesting and useful point concerns the lack of evidence of any crystal- imposed hexagonal symmetry. Deleterious post-formation domain spreading did not occur, indicating a very promising route for further study in formation and build-up of arbitrary 2-d domain patterns.

Figure 15. Domain patterning via fs LAP in CLN at a bias field of 16 kV mm^{-1}. Average power was 120 mW at a pulse repetition rate of 250 kHz.

All optical poling (AOP[f]).

The final technique in this section relates to domain formation using direct laser addressing only, without the requirement for any simultaneous externally applied electric field[20]. If this approach can be perfected, and controllable domain features produced, this would represent a significant step in precision domain engineering.

Three laser sources have been used for this technique. The first was a Q-switched 10ns duration frequency quadrupled diode-pumped Nd:YAG operating at $\lambda = 266$ nm, with good beam quality and up to 5 mJ output energy per pulse. The second was a KrF excimer laser, with poor beam quality, operating at $\lambda = 248$ nm and up to 300 mJ per pulse. Finally, a frequency-doubled Q-switched Nd:YAG laser pumping a dye laser, which was subsequently also frequency doubled, producing mJ level output at ~7 ns duration, but tuneable over the wavelength region of ~292 nm to 329 nm. This last source was valuable for its tunability, as the output wavelength straddles the absorption edge of undoped CLN which lies around >310 nm. In all cases, selected parts of the incident beams were chosen to maximise available uniformity.

Figure 16a shows a representative example of a laser-exposed sample subsequently HF etched, in which the typical three-fold symmetry is immediately apparent. Exposures were made in undoped and MgO-doped congruent lithium niobate using fluences ranging from 100 mJ/cm^2 up to 10 J/cm^2, applying 1 to 1000 pulses. In all instances where any ablation or laser-induced surface topography was observed, similar UV-induced surface domains were formed. Figure 16b shows a magnified view of the central area, from which it is apparent that this patterning technique produces features that are far from random. We have used SFM techniques to investigate these patterns, and conclude from studies of laser-exposed, but unetched samples, that they are surface domains. Using focussed ion beam milling to probe the sub-surface region, the domain depths extend to almost 2 μm. These UV-induced surface domains are uniform, regular, of linewidths as small as 100 nm, and curiously grow along the –y directions only, never in the +y direction.

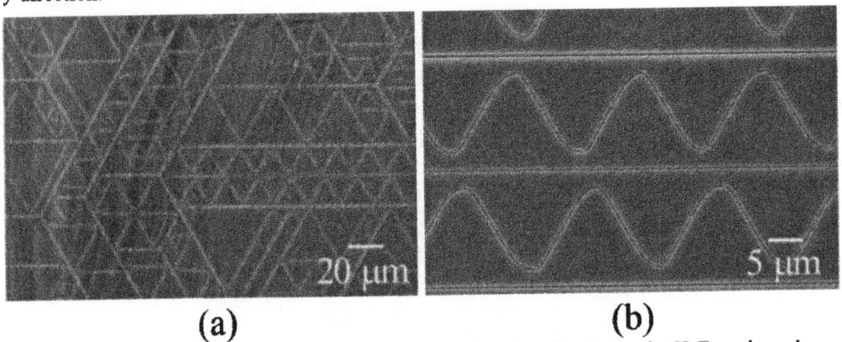

Figure 16. MgO-doped congruent lithium niobate exposed to 2 pulses of a KrF excimer laser at λ = 248 nm. (a) Large area self-organized UV-induced surface domains which extend across the entire laser-exposed region. (b) ~200 nm domain lines deflect away from other lines during growth, reliably reproducing similar features at each turn.

Detailed examination of figure 16b shows that the triangular domain regions situated between the three horizontal domain lines are incredibly self-similar. During growth, they have exactly the same closest-approach to the previously formed straight domains. Their shape on turning away from these lines is effectively identical. Their gradient and degree of curvature are identical. There is clearly ordered behaviour here, and from their growth habit, it is most likely that growth is electrostatically controlled. Fourier transformation shows some degree of preferred spatial feature size, and further work is in progress here on analysis of this behaviour.

The final part of this section concerns ordered periodic line features, and we have used phase masks to structure the incident laser light to try to achieve this. In figure 17 below, a phase mask with a period of ~750 nm was used to examine ordered domain growth along the y axis. It is clear that domain formation has only occurred at every fifth or sixth intensity maximum produced from the phase mask, another indication of perhaps an underlying electrostatic control mechanism. As earlier, these studies are being continued, to both elucidate the mechanism, and improve the fidelity, uniformity and spatial resolution.

CONCLUSION

Our progress to date in physical and domain engineering in doped lithium niobate has been presented. Applications where appropriate have been given and some recent experiments on laser-assisted and direct laser-written structures have been described. Domain features as small as

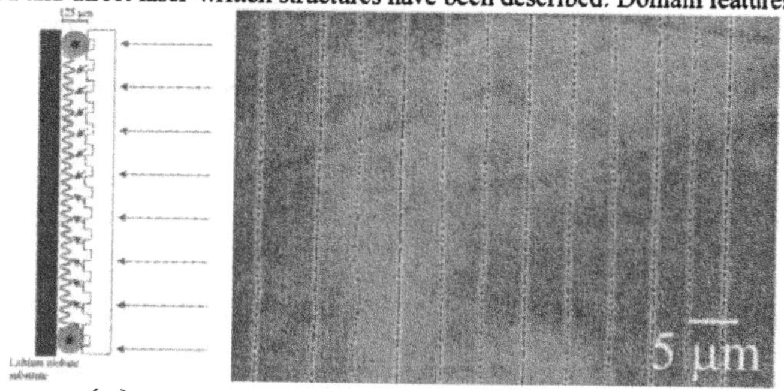

(a) (b)

Figure 17. +z face exposure from frequency doubled dye laser at a fluence of ~250 mJ cm^{-2} at λ= 297.5 nm. The contrast in figure (b) has been stretched to show underlying intensity maxima resulting from the phase mask producing a period Λ = 726 nm.

100nm in width have been achieved and progress made towards ordered, periodic arrays. While there is still more work required, laser-assisted techniques are seen to hold considerable promise for both domain patterning and also materials surface processing.

ACKNOWLEDGMENTS

The authors are grateful to the following for collaborative work within this programme: D. A. Scrymgeour and V. Gopalan, Penn State University, USA, for PFM work on surface domain analysis. T. Jungk, E. Soergel, M. C. Wengler, K. Buse, University of Bonn, Germany, for supply of the transparent poling cell, and DC-EFM domain analysis. I. Clark, CCLRC Rutherford Appleton Labs, UK, for laser access within the LSF laser programme. The Engineering and Physical Science Research Council, UK, (EPSRC) for research funding.

REFERENCES

[1] R. L. Byer, *J.Nonliner Opt. Phy. Mater.*, "Quasi-phasematched nonlinear interactions and devices," **6(4)**, 549-92 (1997)

[2] V. Gopalan, T. E. Mitchell, Y. Furukawa, and K. Kitamura, "The role of nonstoichiometry in 180° domain switching of $LiNbO_3$ crystals," *Appl. Phys. Lett.*, **72(16)**, 1981-3 (1998).

[3] L. Tian, V. Gopalan, and L. Galambos, "Domain reversal in stoichiometric $LiTaO_3$ prepared by vapour transport equilibration," *Appl. Phys. Lett.*. **85(19)**, 4445-47 (2004).

[4] Y. Furukawa, K. Kitamura, S. Takekawa, K. Niwa, and H. Hatano, "Stoichiometric Mg:$LiNbO_3$ as an effective material for nonlinear optics," *Opt. Lett.*, **23(24)**, 1892-94 (1998).

[5] S. Mailis, G. W. Ross, L. Reekie, J. A. Abernethy, and R. W. Eason, "Fabrication of surface relief gratings on lithium niobate by combined UV laser and wet etching," *Electron. Lett.*, **36(21)**, 1801-3 (2000).

[6] S. Mailis, C. Riziotis, I. T. Wellington, P. G. R. Smith, C. B. E. Gawith, and R. W. Eason, "Direct UV writing of channel waveguides in congruent lithium niobate single crystals," *Opt. Lett.*, **28(16)**, 1433-5 (2003).

[7] A. C. Muir, G. J. Daniell, C. P. Please, I. T. Wellington, S. Mailis, and R. W. Eason, "Modelling of UV laser-induced optical waveguide formation via lithium ion thermal-diffusion in $LiNbO_3$," to be submitted to Appl. Phys. A.

[8] A. J. Boyland, S. Mailis, J. M. Hendricks, P. G. R. Smith, and R. W. Eason, "Electro-optically controlled beam switching via total internal reflection at a domain-engineered interface in $LiNbO_3$," *Opt. Commun.*, **197(1-3)** 193-200 (2001).

[9] C. Sones, S. Mailis, V. Apostolopoulos, I. E. Barry, C. B. E. Gawith, P. G. R. Smith, and R. W. Eason, "Fabrication of piezoelectric micro-cantilivers in domain engineered $LiNbO_3$ single crystals," *J. Micromech. Microeng.*, **12(1)**, 53-7 (2002).

[10] A. C. Busacca, C. L. Sones, V. Apostolopoulos, R. W. Eason, and S. Mailis "Surface domain engineering in congruent lithium niobate single crystals: a route to sub-micron periodic poling," *Appl. Phys. Lett.*, **81(26)**, 4946-8 (2002).

[11] A. C. Busacca, C. L. Sones, R. W. Eason, and S. Mailis, "First-order quasi-phase-matched blue light generation in surface-poled Ti-indiffused lithium niobate waveguides," *Appl. Phys. Lett.*, **84(22)**, 4430-2 (2004).

[12] S. Mailis, P. G. Lagoudakis, M. Paturzo, J. D. Mills, J. Feldmann, and R. W. Eason, "Second harmonic generation enhancement in lithium niobate micro-tips," *CLEO/QELS 2005*, CMN4, (2005).

[13] A. J. Boyland, S. Mailis, I. E. Barry, R. W. Eason, and M. Kaczmarek, "Latency effects and periodic structures in light induced frustrated etching of Fe:doped $LiNbO_3$," *Appl. Phys. Lett.*, **77(18)**, 2792-4 (2000).

[14] J. G. Scott, A. J. Boyland, S. Mailis, C. Grivas, O. Wagner, S. Lagoutte, and R. W. Eason, "Self ordered sub-micron structures in Fe-doped $LiNbO_3$ formed by light-induced frustration of etching (LIFE)," *Appl. Surf. Sci.*, **230(1-4)**,138-50 (2004).

[15] S.Mailis, C.Riziotis, P.G.R.Smith, J.G.Scott, R.W.Eason, "Continuous wave ultra violet radiation induced frustration of etching in lithium niobate single crystals," *Appl. Surf. Sci.*, **206(1-4)**, 46-52 (2003).

[16] S. Mailis, C. L. Sones, J. G. Scott, and R. W. Eason, "UV laser-induced ordered surface nanostructures in congruent lithium niobate single crystals," *Appl. Surf. Sci.*, **247(1-4)**, 497-503 (2005).

[17] P. T. Brown, G. W. Ross, R. W. Eason, and A. R. Pogosyan, "Control of domain structures in lithium tantalate using interferometric optical patterning," *Opt. Commun.*, **163**, 310-6 (1999).

[18] M. C. Wengler, B. Fassbender, E. Soergel, and K. Buse, "Impact of ultraviolet light on coercive field, poling dynamics and poling quality of various lithium niobate crystals from different sources," *J. Appl. Phys*, **96 (5)**, 2816-20 (2004).

[19] C. L. Sones, M. C. Wengler, C. E. Valdivia, S. Mailis, R. W. Eason, and K. Buse, "Light-induced order-of-magnitude decrease in the electric field for domain nucleation in MgO-doped lithium niobate crystals," *Appl. Phys. Lett.*, **86(21)**, 212901 (2005).

[20] C. E. Valdivia, C. L. Sones, J. G. Scott, S. Mailis, R. W. Eason, D. A. Scrymgeour, V.Gopalan, T. Jungk, E. Soergel, and I. Clark, "Nanoscale surface domain formation on the +z face of lithium niobate by pulsed UV laser illumination," *Appl. Phys. Lett.*, **86(2)**, 022906 (2005).

Electroceramic Materials and Applications
Edited by R. W. Schwartz

DOMAIN WALL ENGINEERING IN LEAD-FREE PIEZOELECTRIC MATERIALS FOR ENHANCED PIEZOELECTRIC PROPERTIES

Satoshi Wada, Koichi Yako, Keisuke Yokoo, Hirofumi Kakemoto and Takaaki Tsurumi
Department of Metallurgy & Ceramics Science, Tokyo Institute of Technology,
2-12-1 Ookayama, Meguro, Tokyo 152-8552, Japan

ABSTRACT

For the [111] poled barium titanate ($BaTiO_3$) single crystals with the engineered domain configuration, it was clearly observed that the piezoelectric properties increased with decreasing domain sizes. To explain the phenomenon, the multidomain single crystals were regarded as the composite of (a) a distorted domain wall region and (b) a normal tetragonal domain region. Using a 2-phase model, the piezoelectric properties from the domain wall were estimated. As a result, ultrahigh piezoelectric constants over 8,000 pC/N were expected from the domain wall region. Moreover, this study suggested that it is possible to obtain the lead-free piezoelectric materials with the d_{31} and d_{33} over 1,000 pC/N, when the domain sizes can decrease below 1 μm.

INTRODUCTION

Recently, lead-free ferroelectrics have become highly attractive materials from the viewpoint of a solution to environmental problems.[1] However, as compared with $Pb(Zr,Ti)O_3$ (PZT) ceramics,[2] their ferroelectric related properties were very poor, and therefore it is difficult to replace the PZT ceramics. Many researchers have attempted to improve the piezoelectric properties of lead-free ferroelectrics such as bismuth layer-structure ferroelectrics, $BaTiO_3$ and potassium niobate ($KNbO_3$) by chemical modification, but no significant improvement has been achieved.[1-4] Recently, however, chemically modified $KNbO_3$ and sodium niobate ($NaNbO_3$) solid solution ceramics were reported as new lead-free piezoelectrics with the similar piezoelectric properties to PZT ceramics.[5,6] Thus, there are still some possibilities to obtain high performance lead-free piezoelectrics by the chemical modification. On the other hand, there is also a possibility that a new ferroelectrics with a new chemical formula can show a high piezoelectric performance, but although it took over 50 years after discovery of $BaTiO_3$, there are no new ferroelectrics with high piezoelectric properties.

To achieve much higher piezoelectric properties than those of PZT ceramics, the domain engineering technique should be applied to lead-free ferroelectric materials. Domain engineering is an important technique for obtaining enhanced piezoelectric properties in ferroelectric single crystals. In [001] oriented rhombohedral $Pb(Zn_{1/3}Nb_{2/3})O_3$ (PZN) – lead titanate ($PbTiO_3$ = PT) single crystals, ultrahigh piezoelectric activities were found by Park and Shrout[7,8] and Kuwata *et al.*[9,10] and these ultrahigh piezoelectric properties were originated from the application of the domain engineering techniques, *i.e.*, engineered domain configuration.[11-12]

$BaTiO_3$ crystals are the typical lead-free ferroelectrics and there have been a number of reports on its domain engineering.[13-16] Park *et al.*[14] reported that the [001] poled orthorhombic $BaTiO_3$ crystals with the engineered domain configurations exhibited d_{33} of over 500 pC/N and k_{33} of over 85%. These piezoelectric properties are much higher than those of "soft" PZT ceramics.[2] Moreover, for the tetragonal $BaTiO_3$ crystals with the engineered domain configurations, it was recently found that the piezoelectric properties significantly improved with decreasing domain sizes.[15-16] These results suggested that the domain walls in the engineered domain configuration could contribute significantly to the piezoelectric properties.

The objective in this study is to explain the domain size dependence of the piezoelectric properties for the $BaTiO_3$ crystals with engineered domain configurations. For this objective, the piezoelectric properties of the $BaTiO_3$ crystals were investigated as a function of domain size. Finally, the effect of 90° domain walls on the piezoelectric properties is discussed.

EXPERIMENTAL DETAILS

$BaTiO_3$ single crystals were grown by a TSSG method at Fujikura, Ltd.[17] These crystals were oriented along [111] direction using a back-reflection Laue method. For the piezoelectric measurement using the 31 resonators, $BaTiO_3$ single crystals were sized into 4.0x1.2x0.4 mm^3 (4.0mm // [1-10], 1.2mm // [11-2], 0.4mm // [111]) by cutting and polishing using fine Al_2O_3 powders. Gold electrodes were sputtered on both sides (4.0x1.2 mm^2). Poling treatment for the preparation of $BaTiO_3$ crystals with different domain sizes was performed on the basis of the previous reports of domain size dependence on the electric-fields and the temperatures.[15-16]

On the other hand, for the piezoelectric measurement using the 33 resonators, the $BaTiO_3$ single crystals were sized into 0.8x0.8x2.0 mm^3 (0.8mm // [1-10], 0.8mm // [11-2], 2.0mm // [111]) by cutting and polishing. Gold electrodes were sputtered on both sides (0.8x0.8 mm^2). The $BaTiO_3$ crystals were set in the micro silicone-oil bath on the heater of cryostat, and the various electric-fields were applied just above Curie temperature.

After the poling treatment, their domain configurations of the both resonators were investigated using the polarizing microscope (Nikon, Labophoto2-POL). Finally, their piezoelectric properties were measured using an impedance analyzer (Agilent, HP-4294A) by a conventional resonance-antiresonance method[18].

RESULTS AND DISCUSSION

Domain size dependence of piezoelectric properties using the 31 resonators

By the poling treatment at the similar electric-fields and temperatures reported for $BaTiO_3$,[15-16] the 31 resonators of $BaTiO_3$ crystals with different domain sizes were prepared. The average domain sizes in the engineered domain configuration (Fig. 1) were changed from 40

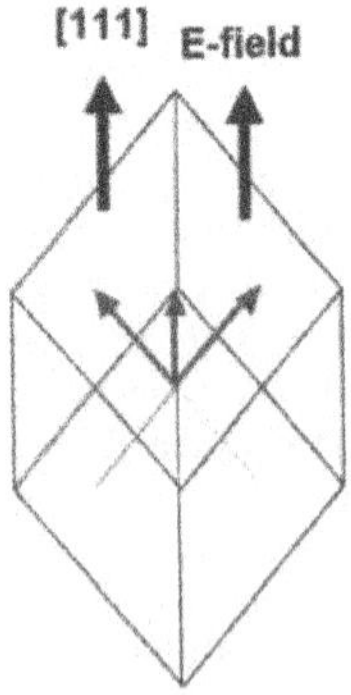

Fig. 1 Schematic engineered domain configuration in [111] poled tetragonal $BaTiO_3$ crystals.

μm to 5.5 μm. The domain configurations for all resonators prepared in this study were composed of the same 90° domain walls,[19] and the difference between these domain configurations was just domain size, i.e., domain wall density as shown in Fig. 2.

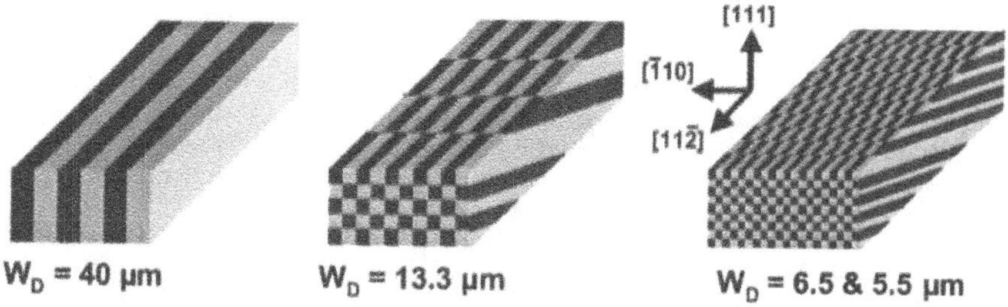

Fig. 2 Schematic 31 resonators composed of the engineered domain configuration with different domain sizes.

Table I. Domain size dependence of piezoelectric properties for the 31 $BaTiO_3$ resonators.

$BaTiO_3$ single crystals	ε_{33}^T	s_{11}^E (pm^2/N)	d_{31} (pC/N)	k_{31} (%)
[001] a) (single-domain)	129	7.4	-33.4	---
[111] b) (single-domain)	---	---	-62.0	---
[111], charged (domain size of 80 μm)	1,299	10.9	-85.3	24.1
[111], charged (domain size of 50 μm)	2,117	7.80	-98.2	25.7
[111], charged (domain size of 40 μm)	2,185	7.37	-97.8	25.9
[111], charged + neutral (domain size of 20.0 μm)	2,117	8.30	-102.7	26.0
[111], charged + neutral (domain size of 15.0 μm)	2,186	8.20	-112.5	28.2
[111], charged + neutral (domain size of 13.3 μm)	2,087	7.68	-134.7	35.7
[111], charged + neutral (domain size of 12.0 μm)	1,921	8.20	-137.6	36.8
[111], charged + neutral (domain size of 10.0 μm)	2,239	9.30	-140.5	32.8
[111], charged + neutral (domain size of 8.0μm)	2,238	9.10	-159.2	37.5
[111], charged + neutral (domain size of 7.0 μm)	2,762	9.30	-176.2	36.9
[111], charged + neutral (domain size of 6.5 μm)	2,441	8.80	-180.1	41.4
[111], charged + neutral (domain size of 5.5 μm)	2,762	9.58	-230.0	47.5
"soft" PZT ceramics c) $Pb_{0.988}(Ti_{0.48}Zr_{0.52})_{0.976}Nb_{0.024}O_3$	1,700	16.4	-171.0	34.4

These piezoelectric properties were measured at 25 °C using a weak AC electric-field of 125 mV/mm. Table I shows the domain size dependence of piezoelectric-related properties using these 31 resonators. As a reference, a calculated d_{31} piezoelectric constant for [111] oriented $BaTiO_3$ single-domain crystal using the material constants reported by Zgonik *et al.*[20] was also listed. With decreasing domain size, all piezoelectric-related properties increased significantly. Especially, the 31 resonator with a domain size of 5.5 μm showed much higher d_{31} of −230 pC/N and k_{31} of 47.5 % than those (d_{31} of −171 pC/N and k_{31} of 34.4 %) reported for soft PZT ceramics.[2]

To date, it was considered that the highest piezoelectric property must be obtained for the single-domain crystals, and it is impossible for the material constants to be beyond the single-domain crystals' values. However, this study revealed that the 90° domain walls in the engineered domain configuration significantly contributed to piezoelectric properties, and caused much higher values than those from single-domain crystals.

In general, under the high electric-field exposure, the domain walls can move very easily, and this domain wall motion made an intrinsic contribution of the domain walls themselves to the piezoelectric properties very unclear. However, in the engineered domain configuration as shown in Fig. 1, it should be noted that the 90° domain walls cannot move with or without unipolar DC electric-field exposure.[13-14] This means that in the engineered domain configuration, the 90° domain walls can exist very stably with and without unipolar electric-field exposure. Therefore, the intrinsic contribution of the domain walls to the piezoelectric properties was clarified for the first time using the engineered domain configuration. In other words, the engineered domain configuration is considered as domain-wall engineering among the domain engineering techniques.[21]

Domain size dependence of piezoelectric properties using the 33 resonators

By the poling treatment at various electric-fields and temperatures,[15-16] the 33 resonators of $BaTiO_3$ crystals with different domain sizes were prepared. The average domain sizes in the engineered domain configuration (Fig. 1) were changed from 100 μm to 14 μm. For the 31 resonator, minimum domain size was 5.5 μm while for the 33 resonator, minimum domain size was 14 μm. This difference caused from the difference of applied electric-field. This is because that in this study, domain sizes decreased with increasing applied electric-fields. The domain configurations for all resonators prepared in this study were composed of the same 90° domain walls,[19] and the difference between these domain configurations was just domain size as shown in Fig. 3.

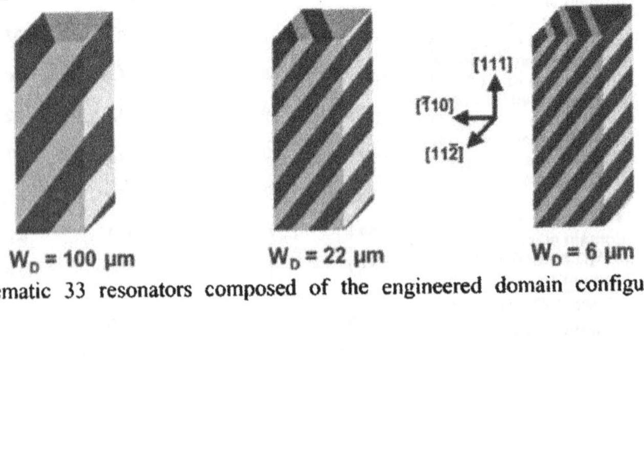

Fig. 3. Schematic 33 resonators composed of the engineered domain configuration with

different domain sizes.

These piezoelectric properties were measured at 25 °C using a weak AC electric-field of 50 mV/mm. Table II shows the domain size dependence of piezoelectric-related properties using these 33 resonators. As a reference, a calculated d_{33} piezoelectric constant for [111] oriented $BaTiO_3$ single-domain crystal using the material constants reported by Zgonik *et al.*[20] was also listed. With decreasing domain size, all piezoelectric-related properties increased. However, for the 33 resonators, with decreasing domain sizes, the enhancement of piezoelectric properties was not so large. Now, we cannot explain this difference between the 31 and 33 resonators, but it may be originated from the difference of these domain configurations (Figs. 2 and 3).

Table II. Domain size dependence of piezoelectric properties for the 33 $BaTiO_3$ resonators.

$BaTiO_3$ single crystals	ε_{33}^T	s_{33}^E (pm²/N)	d_{33} (pC/N)	k_{33} (%)
[001] a) (single-domain)	---	---	90	---
[111] a) (single-domain)	---	---	224	---
[111], neutral (domain size of 100 μm)	1,984	10.6	235	54.4
[111], neutral (domain size of 60 μm)	1,959	10.7	241	55.9
[111], neutral (domain size of 22 μm)	2,008	8.8	256	64.7
[111], neutral (domain size of 15 μm)	2,853	6.8	274	66.1
[111], neutral (domain size of 14 μm)	1,962	10.8	289	66.7
[111], neutral (domain size of 6 μm)	2,679	10.9	331	65.2

Calculation of piezoelectric properties contributed from the 90° domain walls

It is well known that a region near the 90° domain walls is gradually distorted to relax the strain between domains with different polar directions as shown in Fig. 4.[22-26] Moreover, for $BaTiO_3$ crystals, the crystal structure near the 90° domain walls gradually changed from normal tetragonal with *c/a* ratio of 1.011 to tetragonal with *c/a* ratios closed to 1.0. Thus, it can be expected that the region near the 90° domain walls with pseudo-cubic structure exhibits the material constants of $BaTiO_3$ single crystals near the Curie temperature as reported by Budimir *et al.*[27] In this study, the $BaTiO_3$ crystals with the engineered domain configuration were regarded as a composite of (a) normal tetragonal region and (b) distorted domain wall region. On the basis of this 2-phases model, a volume fraction of the distorted domain wall region on the normal tetragonal region was estimated as follows. For a simplification of calculation, one dimensional model was applied.[28] First, a thickness of the distorted 90° domain wall region was assumed using various sizes from 1 to 100 nm.[22-26] Next, using this thickness of the distorted 90° domain wall region (W_{DW}) and the domain size (W_D), the volume fraction of the distorted 90° domain wall region (F) was estimated using the following equation,

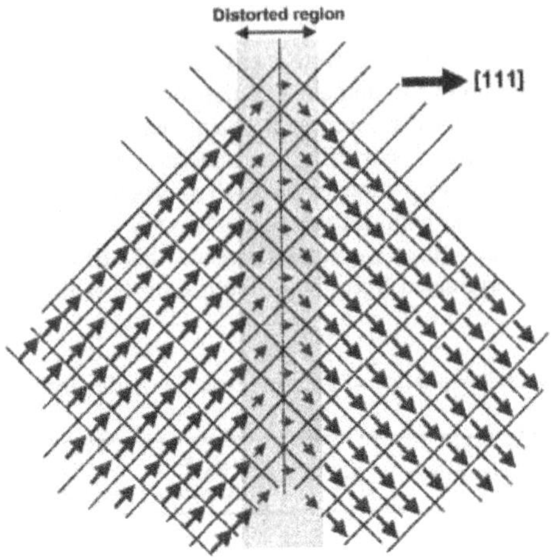

Fig. 4 Schematic structure near the 90° domain walls for $BaTiO_3$ crystals.

$$F = \frac{W_{DW}}{W_D} \qquad (1).$$

W_D can be measured from the experiment in this study while W_{DW} is unknown values. Thus, we must use valid W_{DW} values for the above calculation. To date, many researchers tried to

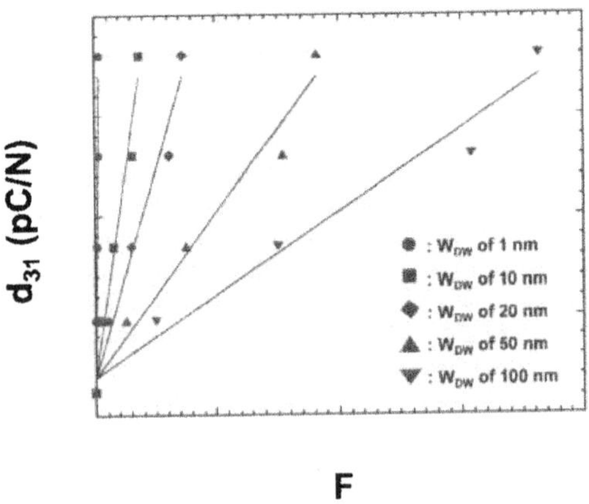

Fig. 5 Relationship between d_{31} and F calculated using various W_{DW} from 1 to 100 nm.

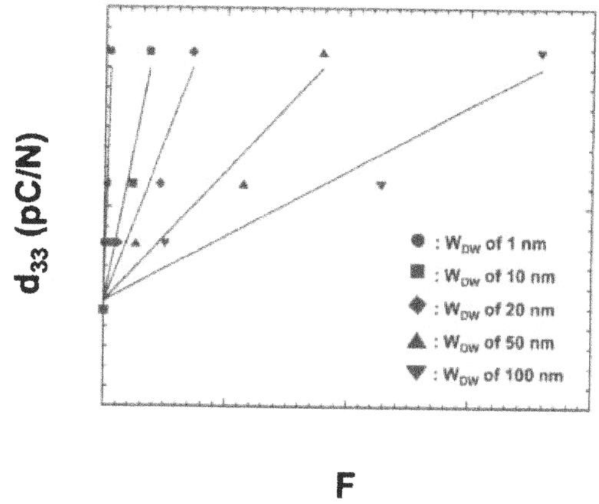

Fig. 6 Relationship between d_{33} and F calculated using various W_{DW} from 1 to 100 nm.

clarify the domain wall thickness using LGD theories and TEM observation, and their estimated values are distributed from 1 to 10 nm.[22-26] Recently, with developing measurement equipments and theories, some new methods were proposed to estimate the domain wall thickness.[28-30] As a result, the domain wall thickness was related to point defect, and the defect was responsible for the broadening of the domain wall thickness.[29] This means that it is very difficult to determine the 90° domain wall thickness for $BaTiO_3$. Thus, in this study, F values were calculated using the various domain wall thickness from 1 to 10 nm. Using the F values, the relationship between d_{31} and F was plotted in Fig. 5 while that between d_{33} and F was also plotted in Fig. 6. In Figs. 5 and 6, a slope of the line means the piezoelectric constants expected from the distorted 90° domain wall region. If the 90° domain wall thickness can be estimated as 10 nm, d_{31} and d_{33} was expressed using the following equation, respectively,

$$d_{31} = 82676 * F + 69 \quad (2),$$

$$d_{33} = 81744 * F + 227 \quad (3).$$

This suggested that if the 90° domain wall thickness is 10 nm, d_{31} and d_{33} from the distorted 90° domain wall region can be estimated as 82,676 and 81,744 pC/N, respectively. On the other hand, if the 90° domain wall thickness is 20 nm, d_{31} and d_{33} from the distorted 90° domain wall region can be also estimated as 41,338 and 40,872 pC/N, respectively. Moreover, if the 90° domain wall thickness is 100 nm, d_{31} of 8,268 and d_{33} of 8,174 can be expected from the distorted 90° domain wall region. The above discussion revealed that the piezoelectric constants expected from the distorted 90° domain wall region are significantly high (over 8,000 pC/N as a minimum value).

This study showed that the distorted 90° domain walls can contribute significantly to the piezoelectric properties. When the domain wall density increases more and more, what piezoelectric constants we can expect? The domain size dependence of d_{31} and d_{33} can be also expressed as follows,

$$d_{31} = \frac{826760}{W_D} + 69 \quad (4),$$

$$d_{33} = \frac{817440}{W_D} + 227 \quad (5).$$

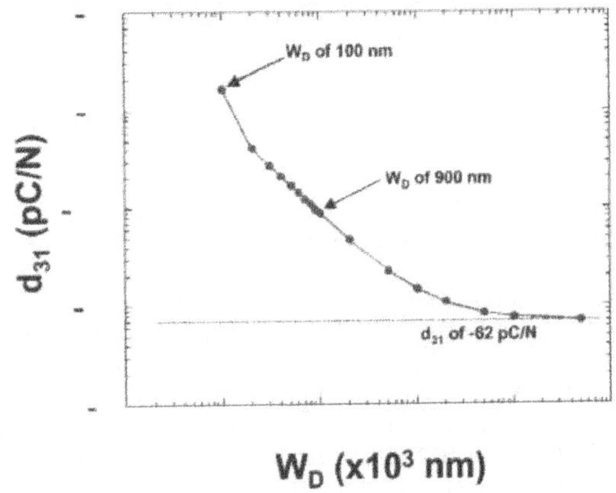

Fig. 7 Relationship between d_{31} and W_D calculated using the equation (4).

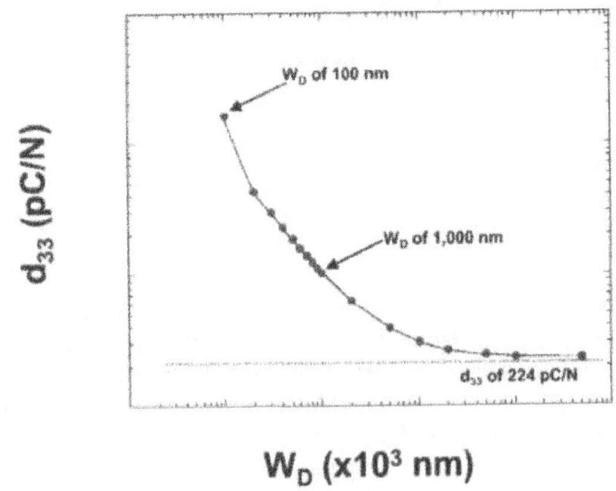

Fig. 8 Relationship between d_{33} and W_D calculated using the equation (5).

It should be noted that these equations (4) and (5) are independent of W_{DW}. Thus, using the various W_D values, the relationship between d_{31} and W_D was plotted in Fig. 7 while that between d_{33} and W_D was also plotted in Fig. 8. In Figs. 7 and 8, above the W_D of 20 μm, piezoelectric constants were almost constant at the calculated single-domain's values while below the W_D of 10 μm, piezoelectric constants drastically increased with decreasing domain sizes. Moreover, when the domain size decreased down to 1 μm, both d_{31} and d_{33} became to around 1,000 pC/N.

Park and Shrout reported that [001] poled PZN single crystal exhibited the ultrahigh d_{33} of 1,100 pC/N,[31] and the domain size in this PZN crystal was observed at around 1 μm.[11] They also reported that for [111] poled PZN single-domain crystal had the d_{33} of just 83 pC/N[31] similar to that of $BaTiO_3$. Therefore, it can be expected that when the domain size of around 1 μm is induced into the [111] poled $BaTiO_3$ crystals, high performance lead-free piezoelectrics with ultrahigh piezoelectric constants over 1,000 pC/N can be created.

CONCLUSION

For the $BaTiO_3$ single crystals with the engineered domain configuration, it was clearly observed that the piezoelectric properties increased with decreasing domain sizes. To explain the phenomenon, these multidomain single crystals were regarded as the composite of the distorted domain wall region and normal tetragonal domain region. Using the 2-phases model, the piezoelectric properties from the domain wall were estimated simply. As a result, ultrahigh piezoelectric constants over 8,000 pC/N as a minimum value were expected from the domain wall region. Moreover, to obtain the $BaTiO_3$ single crystals with the d_{31} and d_{33} over 1,000 pC/N, the domain sizes below 1 μm must be required. We believe that the significant contribution of the distorted domain wall region to piezoelectric properties is a universal concept for all of ferroelectric crystals. According to this concept, we can create new lead-free piezoelectric materials with ultrahigh piezoelectric performance.

ACKNOWLEDGEMENT

We would like to thank Mr. O. Nakao of Fujikura Ltd. for preparing the TSSG-grown $BaTiO_3$ single crystals with excellent chemical quality. We would like to thank Dr. Y. Ishibashi of Aichi-shukutoku University, Dr. D. Damjanovic of EPFL, Dr. A. J. Bell of University of Leeds and Dr. L. E. Cross of Pennsylvania State University for their helpful discussions about the domain wall contribution to the piezoelectric properties. We would like to thank Dr. J. Erhart and Dr. J. Fousek of Technical University of Liberec for their helpful discussions about the analysis of the domain configuration. We also would like to special thank Dr. T.R. Shrout and Dr. S.-E. "Eagle" Park of Pennsylvania State University for giving us a great opportunity to study the engineered domain configurations. This study was partially supported by (1) a Grant-in-Aid for Scientific Research (**16656201**) from the Ministry of Education, Culture, Sports, Science, and Technology, Japan, (2) the Japan Securities Scholarship Foundation, (3) the Toray Science Foundation, (4) the Kurata Memorial Hitachi Science and Technology Foundation, (5) the Electro-Mechanic Technology Advanced Foundation, (6) the Tokuyama Science Foundation and (7) the Yazaki Memorial Foundation for Science and Technology.

REFERENCES

[1]M. Demartin Maeder and D. Damjanovic: *Piezoelectric Materials in Devices*, ed. N. Setter (N. Setter, Lausanne, 2002) p.389.

[2]B. Jaffe, W. R. Cook, Jr. and H. Jaffe: *Piezoelectric Ceramics*, (Academic Press, New York, 1971) p.135.

[3]F. Jona and G. Shirane: *Ferroelectric Crystals*, (Dover Pub., New York, 1993) p.108.

[4]Y. Zu: *Ferroelectric Materials and Their Applications*, (North-Holland, New York, 1991) p. 101.

[5]Y. Saito, H. Takao, T. Tani, T. Nonoyama, K. Takatori, T. Homma, T. Nagaya and M. Nakamura: Nature (2004) in press.

[6]Y. Guo, K. Kakimoto and H. Ohsato: Appl. Phys. Lett. **85** 4121 (2004).

[7]S .-E. Park and T. R. Shrout: Mater. Res. Innov. **1** 20 (1997).

[8]S.-E. Park and T. R. Shrout: J. Appl. Phys. **82** 1804 (1997).

[9]J. Kuwata, K. Uchino and S. Nomura: Ferroelectrics **37** 579 (1981).

[10]J. Kuwata, K. Uchino and S. Nomura: Jpn. J. Appl. Phys. **21** 1298 (1982).

[11]S. Wada, S.-E. Park, L. E. Cross and T. R. Shrout: J. Korean Phys. Soc. **32** S1290 (1998).

[12]S. Wada, S.-E. Park, L. E. Cross and T. R. Shrout: Ferroelectrics **221** 147 (1999).

[13]S. Wada, S. Suzuki, T. Noma, T. Suzuki, M. Osada, M. Kakihana, S.-E. Park, L. E. Cross and T. R. Shrout: Jpn. J. Appl. Phys. **38** 5505 (1999).

[14]S.-E. Park, S. Wada, L. E. Cross and T. R. Shrout: J. Appl. Phys. **86** 2746 (1999).

[15]S. Wada and T. Tsurumi: Br. Ceram. Trans. **103** 93 (2004).

[16]S. Wada, K. Yako, H. Kakemoto, J. Erhart and T. Tsurumi: Key Eng. Mater. **269** 19 (2004).

[17]S. Ajimura, K. Tomomatsu, O. Nakao, A. Kurosaka, H. Tominaga and O. Fukuda: J. Opt. Soc. Am. B **9** 1609 (1992).

[18]EMAS-6100: Standard of Electronic Materials Manufacturers Association of Japan (1993).

[19]J. Fousek: Czech. J. Phys. B **21** 955 (1971).

[20]M. Zgonik, P. Bernasconi, M. Duelli, R. Schlesser, P. Gunter, M. H. Garrett, D. Rytz, Y. Zhu and X. Wu: Phys. Rev. B **50** 5941 (1994).

[21]J. Fousek, D. B. Litvin and L. E. Cross: J. Phys.: Condens. Matter. **13** L33 (2001); J. Fousek and L. E. Cross: Ferroelectrics **293** 43 (2003).

[22]D. Damjanovic: Rep. Prog. Phys. **61** 1267 (1998).

[23]Y. Ishibashi and E. Salje: J. Phys. Soc. Jpn. **71** 2800 (2002).

[24]N. Setter: *Piezoelectric Materials in Devices*, ed. N. Setter (N. Setter, Lausanne, 2002) p.1.

[25]B. Meyer and D. Vanderbilt: Phys. Rev. B **65** 104111 (2002).

[26]M. Budimir, D. Damjanovic and N. Setter: J. Appl. Phys. **94** 6753 (2003).

[27]Y. Ishibashi: Private Communication (2004).

[28]H. Chaib, F. Schlaphof, T. Otto and L. M. Eng: J. Phys.: Condens. Matter **15** 1 (2003).

[29]D. Shilo, G. Ravichandran and K. Bhattacharya: Nature Mater. **3** 453 (2004).

[30]T. Tsuji, H. Ogiso, J. Akedo, S. Saito, K. Fukuda and K. Yamanaka, Jpn. J. Appl. Phys. **43** 2907 (2004).

[31]S.-E. Park and T. R. Shrout: IEEE Trans. Ultrason., Ferroelectr. & Freq. Control. **44** 1140 (1997).

Electroceramic Materials and Applications
Edited by R. W. Schwartz

OBSERVATIONS OF 180° FERROELECTRIC DOMAINS THROUGH LITHIUM NIOBATE USING BRAGG AND LAUE X-RAY TOPOGRAPHY WITH THE APPLICATION OF ELECTRIC FIELDS

Terrence Jach
Chemical Science and Technology Laboratory
National Institute of Standards and Technology
Gaithersburg, MD 20899-8371

Stephen Durbin
Department of Physics
Purdue University
West Lafayette, IN 47907

Sungwon Kim, Venkatraman Gopalan
Materials Research Institute and Dept. of Materials Science and Engineering,
Pennsylvania State University
University Park, PA 16802

ABSTRACT

We report on the visualization of domain wall strains in congruent $LiNbO_3$ using Bragg and Laue x-ray topography. The Laue geometry allows imaging through the entire crystal. Using monochromatic synchrotron undulator radiation and a magnifying x-ray camera, we are able to image domains created by electric field poling, as well as the strain associated with the subsequent application of an electric field to the domains in real time. Electric fields below the coercive field applied along the c-axis produce unexpected and complex strain patterns which are precursors to 180° domain wall switching. The strains observed with field off and on in the Laue topographs differ significantly from the surface strains observed previously with Bragg topography.

INTRODUCTION

The creation of antiparallel ferroelectric domains of various shapes and sizes in $LiNbO_3$ is a key technology in nonlinear optics[1] and electrooptics[2]. These applications are sensitive to the underlying physics of ferroelectric domain walls, both their local structure and dynamics. Strain distributions will ultimately determine the minimum size of usable domains in these devices. While numerous techniques have been in use for domain wall observation[3-5], much remains to be determined on a microscopic scale. In previous work, we reported observing the changes in domain behavior under the application of an external electric field by means of x-ray Bragg topography[6]. Using intense synchrotron radiation, we were able to obtain diffraction images from crystals in real time. Strains were determined in samples with and without applied fields[7]. The Bragg topograph is highly sensitive to slight displacements both parallel and normal to the surface as a result of stress associated with the presence of domains (Fig. 1). By means of ray tracing, we obtained quantitative information about displacements near the surface with and without an applied field. These observations led to the conclusion that strains are maximized in the vicinity of 180° domain walls but extend out for distances of the order of 100 μm.

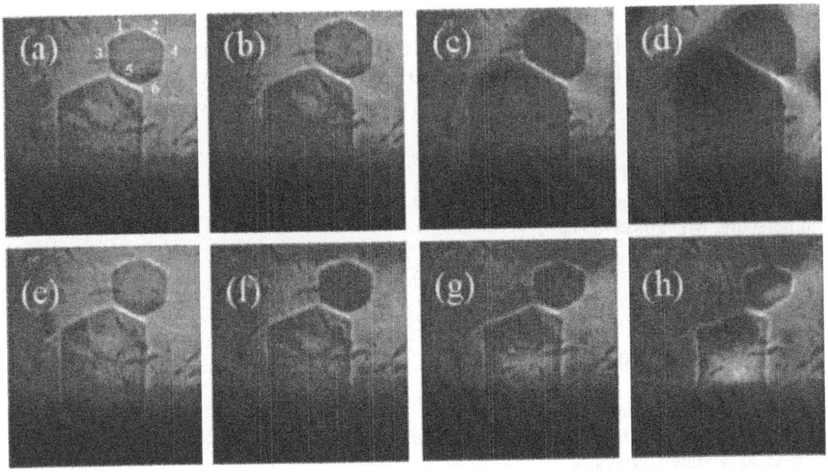

Fig. 1. Domains observed in a (0 0 0 12) Bragg topograph in $LiNbO_3$ for positive applied voltages (forward bias) of (a) 0 V, (b) +1500 V, (c) +3000 V, and (d) +4500 V; for negative applied voltages (reverse bias) of (e) 0 V, (f) –1500 V, (g) -3000 V, and (h) –4400 V. (see ref. 7)

It was equally desirable to study internal strains deep within the crystal. We report carrying out an investigation of the strains normal to the c-axis of congruent $LiNbO_3$ using Laue topography. In this case, we used x-ray reflections from crystal planes normal to the surface, and imaged x-rays which were transmitted through the crystal by dynamical diffraction at small Bragg angles. This also permitted us to observe beams in the undiffracted and diffracted directions. The observations allow us to see anisotropic strains associated with antiparallel domains that are present in the lithium niobate crystals, but not visible in Bragg topographs.

EXPERIMENTAL METHOD

The experiments were carried out at the 4-ID Beamline at the Advanced Photon Source at Argonne National Laboratory. The general experimental layout is depicted in Fig. 2. The beamline makes use of a Type A undulator followed by a symmetric Si (111) monochromator. The undulator and monochromator were tuned to a photon energy of 8.53 keV. The monochromatic flux was of the order of 10^{13} photons/s after passing through slits which limited the beam size to approximately 3.0 mm horizontally × 0.6 mm vertically.

Because of the small undulator source size and the high energy of operation of the Advanced Photon Source, the angular spread of radiation incident on any element of a diffracting crystal is very small, 1.4 μrad. It is because of the small source size and great distance of the undulator that further collimation of the incident beam was not necessary. This means that an asymmetric crystal monochromator is not required for x-ray topography—a distinct advantage of third generation light sources.

The ferroelectric specimens were typically 3 cm × 3 cm sections of congruent $LiNbO_3$ wafers that were Z-cut and polished to a thickness of 300 μm. The samples were further prepared by deposition of a 100 nm thick amorphous carbon film on both surfaces, to which electrodes were attached. The conductive films were deposited over 1 cm^2 in the center of the crystal, to leave an adequate nonconducting path length on either side. The crystals were completely poled as obtained, then they were partially reverse-poled above the coercive field at room temperature prior to the experiment. This created domains visible in phase contrast optical microscopy that extended through the thickness of the wafer. The samples were mounted in the configuration shown in Fig. 2. A voltage source was attached, and it was possible to monitor the current through the crystals.

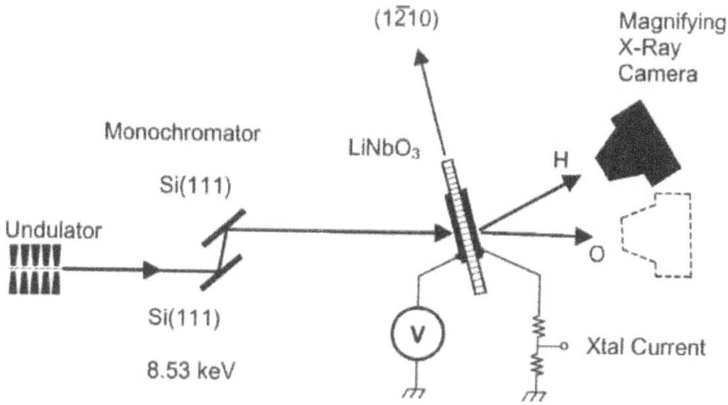

Figure 2. Experimental configuration employed on the 4-ID beamline for Laue x-ray topography. H is the diffracted beam direction, and O is the undiffracted beam direction.

$LiNbO_3$ is a trigonal crystal with *3m* symmetry. We examined Laue topographs made from the $(1\bar{2}10)$ Bragg reflection at an angle θ_B=16.396°. The natural width of the Bragg reflection is 33.56 μrad. Strains that change the lattice angle locally by a large fraction of that amount or that change the lattice constant locally by 2×10^{-5} will be readily visible. The presence of two separate wave solutions in Laue topographs leads to an interference phenomenon known as "Pendellösung" that indicates strain in the bulk of the crystal as well[8].

The Laue topographic images were recorded with a magnifying x-ray camera. The camera consists of a Gd oxysulfide sputtered fluorescent film deposited on a magnifying optical taper that is coupled to a cooled CCD detector with 12-bit readout accuracy. The fluorescent film is relatively insensitive to the third harmonic radiation from the monochromator. The combination was capable of producing images with 6 μm resolution over a field of 3 mm × 3 mm.

With the undulator flux and minimum filtering at the camera, we produced images with exposures of 70 ms-800 ms.

The crystal specimens were mounted on an insulated stage in a 6-circle goniometer. Each run began with a rocking curve over the selected portions of the crystal. Images were then taken as the electric field was either raised monotonically or cycled manually to greater and greater fields of alternating polarity. This was done because the breakdown limit for a specimen with a given set of electrodes was unknown ahead of time. The maximum field that was applied before breakdown was 22 kV/mm, which is close to the coercive field in congruent $LiNbO_3$. The maximum applied field was limited by carriers generated by the intense x-ray beam in the crystal, and breakdown occurred in the area illuminated by the beam when the electric field was sufficient to provoke an avalanche in the carriers.

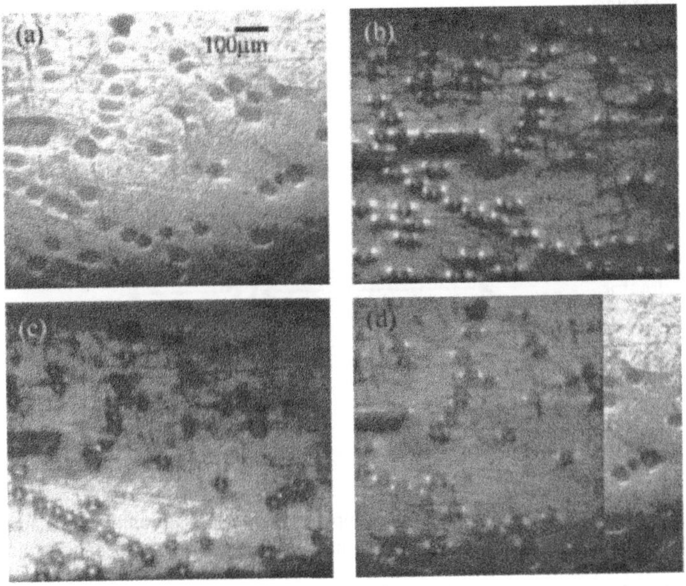

Fig. 3 (a) Bragg topograph from the (0 0 0 12) reflection showing hexagonal domains; (b) $(1\bar{2}10)$ Laue topograph through the same region in the crystal in the H direction showing areas of high strain; (c) Laue topograph showing the diffracted beam in the O direction; (d) superposition of (a) and (b).

RESULTS

We started by examining Bragg and Laue topographs obtained from the same crystal sample in zero field. Fig. 3 (a) shows a topograph of the Bragg diffraction along the (0 0 0 12) direction at a Bragg angle θ_B= 38.97°. Previous reverse poling has created a large number of small, hexagonal domains. Fig. 3(b) shows the H (diffracted direction) $(1\bar{2}10)$ Laue topograph of the same region. The image shows a large number of dislocations, some apparent domains, and a large number of features that show high contrast. Each feature consists of a small dark region along whose outline are a few bright spots corresponding to very high transmission through the crystal. Two of the spots on each site are very intense, and a third spot is visible which is less intense. Fig. 3(c) shows the Laue topograph as seen by the beam diffracted in the O (incident) direction. This topograph also shows the features but with regions of high intensity undiffracted into the H beam.

Fig. 3(d) shows a superposition of Figs. 3(a) and 3(b), with changes in scale on Fig. 3(b) to account for the differences in projection between the two Bragg angles. The overlapping regions between the two images show a complete registration between the hexagonal domains visible in the Bragg topograph and the strong features in the Laue topograph. The regions of high transmission in the Laue topograph are aligned along the $[10\bar{1}0]$ axis of the crystal. They are evidently due to a large gradient in strain that occurs preferentially along this axis across the domains. This strain is visible through the bulk but not on the surface of the domains. It is also highly localized at the domain walls. The presence of a single region of high transmission in the direction of the O beam as seen in Fig. 3(c) in the center of each domain indicates a symmetric distortion occurring across each domain. A finite element analysis has shown that strain buildup at the domain walls can distort the domain symmetrically so that three spots of high diffracted intensity would develop in the H beam of the Laue topograph at vertices of the domain[9].

Fig. 4 shows the effect of an electric field applied to the crystal. The region viewed in Fig. 4(a) at V=0 shows both large and small 180° domains. The reverse-bias potential across the crystal was varied from 0 to -2000 V (0 to -6667 V/mm). As the potential was increased [Fig. 4 (b) and (c)], the topograph developed light and dark bands which can be attributed to the fringes associated with Pendellösung. These are interference fringes that develop between the absorptive and nonabsorptive solutions of waves propagating through the crystal at the diffraction condition[8]. The significance of the fringes is that they indicate where distortions in the lattice will cause intensity to be displaced from the H direction into the O direction and vice-versa. The fringes are sensitive to displacements of the lattice by less than the size of a unit cell. The fringes disappeared when the field was removed, although at some of the higher field values, time-delayed effects were observed. These corresponded to a change in the fringe patterns which exhibited hysteresis when the field was increased and also decreased. We attribute these effects to charge trapping in crystal defects that accumulated and leaked off with time. As long as the field did not exceed values which resulted in permanent damage to the crystal, the Laue topograph was ultimately observed to return to the original appearance.

At high fields [Fig. 4(d)], collective strains emerged along strings of domains which indicated coupling of the strains from individual domains. These resulted in a wholesale distortion of domain images that was reversible as well.

CONCLUSIONS

High-resolution Laue x-ray topography from 180° domains in congruent lithium niobate shows internal strains in zero applied field at the domain wall which were not visible in previous Bragg topography. These strains appear to be strongest in a uniaxial direction which, in this case,

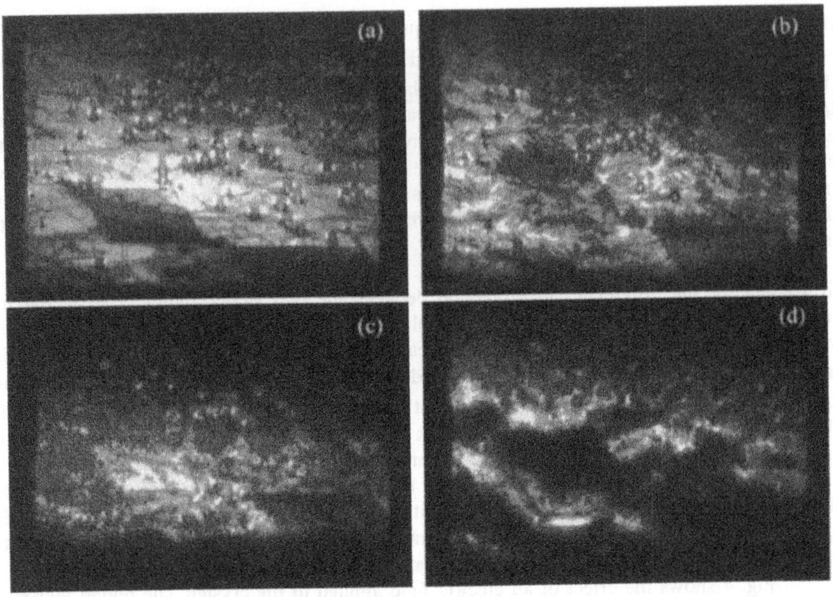

Fig. 4 Laue topographs of the $(1\bar{2}10)$ diffracted beam (H) for negative applied voltages (reverse bias) of (a) -0 V, (b) -300 V, (c) -900 V, (d) -2000 V.

was aligned along the $[10\bar{1}0]$ crystal axis. The strains are symmetric with respect to the domain boundaries indicating a high degree of curvature at the domain walls associated with the presence of the domains.

When an electric field was applied along the c-axis, the Laue topographs developed a pattern of fringes due to Pendellösung that are indicative of the distribution of strains in the crystal. We expect that a detailed analysis of the Pendellösung fringe patterns will lead to additional information on the distribution of strains in the ferroelectric crystals under the applied fields. Collective strains which run along patterns of domains at large applied fields were also observed. These strains are precursors of the domain growth that occurs at the coercive field. Some time delays were observed in the relaxation of the strains, indicating that charging effects at trapping sites are present.

ACKNOWLEDGMENTS

We thank Dr. G. Srajer and the SRI-CAT for assistance on the 4-ID Beamline at the Advanced Photon Source. Use of the Advanced Photon Source was supported by the U.S. Department of Energy, Office of Science, Office of Basic Energy Sciences, under Contract No. W-31-109-Eng-38. The Penn State authors would like to acknowledge support from the National Science Foundation Grant Nos. DMR-0103354,-0507146, and -9984691

REFERENCES

[1]R. G. Batchko, V. Y. Shur, M. M. Fejer, and R. L. Byer, "Backswitch poling in lithium niobate for high-fidelity domain patterning and efficient blue light generation," *Appl. Phys. Lett.* **75**, 1673-1675 (1999).

[2]K. T. Gahagan, V. Gopalan, J. M. Robinson, Q. X. Jia, T. E. Mitchell, M. J. Kawas, T. E. Schlesinger, and D. D. Stancil, "Integrated electro-optic lens/scanner in a $LiTaO_3$ single crystal," *Appl. Opt.* **38**, 1186-1190 (1999).

[3]E. Willibald, H. Oppolzer, and E. Born, "X-Ray Diffraction Topographic Characterization of Crystal Defects in Lithium Niobate Single Crystals," *Siemens Forsch.- u. Enwickl.-Ber.* **11**, 271-274 (1982).

[4]H. Cerva, P. Pongratz, and P. Skalicky, "Lattice defects in single-crystal lithium niobate, I. Transmission electron microscopy,"*Phil. Mag. A* **54**, 185-197 (1986).

[5]T. J. Yang and U. Mohideen, "Nanoscale measurement of ferroelectric domain wall strain and energy by near-field scanning optical microscopy," *Phys. Lett. A* **250**, 205-210 (1998).

[6]T. Jach, S. Kim, S. Durbin, V. Gopalan, and D. Bright , "Real-Time Studies of Strains at Ferroelectric Domain Walls under an Applied Field," *Fundamental Physics of Ferroelectrics* (edited by R. E. Cohen), **CP262,** American Institute of Physics (2002).

[7]T. Jach, S. Kim, V. Gopalan, S. Durbin, and D. Bright,"Long-range Strains and the Effects of Applied Field at 180° Ferroelectric Domain Walls in Lithium Niobate," *Phys. Rev. B*, **69**, 064113-1—064113-9 (2004).

[8]B. W. Batterman and H. Cole, "Dynamical Diffraction of X-rays by Perfect Crystals," *Rev. Mod. Phys.* **36**, 681-717 (1964).

[9]S. Kim, "Optical, Electrical and Elastic Properties of Ferroelectric Domain Walls in $LiNbO_3$ and $LiTaO_3$," *Ph. D. Thesis,* Dept. of Intercollegiate Graduate Program in Materials, Pennsylvania State University (2003).

References

Electroceramic Materials and Applications
Edited by R. W. Schwartz

DOMAIN ENGINEERING FERROELECTRIC CRYSTALS FOR NONLINEAR OPTICS

Peter Powers
University of Dayton
300 College Park
Dayton, OH 46469-2314

ABSTRACT

The ability to domain engineer bulk ferroelectric crystals has translated to engineering nonlinear optical interactions. By periodically inverting the domain structure of a ferroelectric crystal one can optimize for nonlinearity, interaction length, and design wavelength. This capability has led to a wide range of new applications from low-threshold devices to THz generation. The impact of these crystals on the field of nonlinear optics is in its infancy and will continue to have strong impact in the coming years. This talk will cover the underlying nonlinear optics of domain-engineered ferroelectrics, crystal design considerations, comparisons between ferroelectrics, and applications of these crystals to devices.

INTRODUCTION

Ferroelectric crystals are widely used for nonlinear optical applications. All of these crystals have intrinsic properties, such as a high nonlinearity, that make them attractive. However, perhaps the most attractive feature is that the crystal structure can be altered by the application of an electric field. Such alteration gives the possibility of engineering the crystal for nonlinear optical applications. In the case of crystals discussed in this talk, if a field is applied above the coercive field then the crystal domain can be flipped by 180°. This flip of the domain does not change the linear optical properties, such as the index of refraction, but it does change the nonlinear optical properties. In the case of the commonly used crystal, lithium niobate, the sign of some of the nonlinear susceptibilities (such as χ_{33}) changes. This modification of the nonlinear properties while keeping the linear properties unchanged is the attribute that allows for engineered nonlinear optical crystals. This paper gives a brief overview of how to exploit this domain engineering for second order nonlinear interactions followed by a discussion of practical design considerations and nonlinear optical device applications.

Nonlinear optical engineering: quasi-phase matching

The ability to flip a crystal domain with an applied field can be used to achieve so-called quasi-phase matching for a given NLO interaction. To visualize this effect consider a difference frequency mixing process with the two incident beams polarized along the z-axis of a crystal. If we consider the output at the difference frequency that is also aligned along the z-axis, then the amplitude of the nonlinear polarization at the difference frequency is given by

$$P_z^{(2)} = 2\varepsilon_0 \chi_{zzz}^{(2)} E_z(\omega_1) E_z(\omega_2) \tag{1}$$

where ε_o is the permittivity of free space, $\chi_{zzz}^{(2)}$ is the second order nonlinear susceptibility, $E_z(\omega_1)$ and $E_z(\omega_1)$ are the complex field amplitudes for the two inputs. This nonlinear polarization in turn generates the freely propagating difference frequency. In general, the freely propagating field is not in phase with the nonlinear polarization that is driving it. This phase mismatch inhibits efficient generation of the desired mixed signal. The distance over which the generated field builds up is referred to as the "coherence length." The ability to modify the crystal structure of a ferroelectric crystal can be used to keep the nonlinear polarization and the freely propagating field in phase even beyond one coherence length. To understand this, we look at what happens if we flip the crystal in this example about the x-axis. The nonlinear polarization then becomes

$$P_z^{(2)} = -2\varepsilon_0 \chi_{zzz}^{(2)} E_z(\omega_1) E_z(\omega_2) \qquad (2)$$

where the only change from equation (1) is to the sign of the nonlinearity. The significance of this statement is that we now have some control over the nonlinear polarization's phase and we can use this control to keep the freely propagating field and the nonlinear polarization quasi-phase matched. The basic idea is that we can let the freely propagating field build up for one coherence length at which point energy would then start to flow from the generated field back into the input fields. However, if we invert the crystal at this point, thus changing the sign of the nonlinearity, then we also flip the phase of the nonlinear polarization such that energy continues to flow from the pump fields into the generated field. This re-phasing then allows the interaction to build up for another coherence length after which the crystal would need to be flipped to attain further build up. Such a scheme is called quasi-phase matching.

Figure 1 shows the idea in terms of interacting fields' k-vectors. The phase mismatch of a process in a bulk medium is given by Δk. The efficiency of the process is, in the small signal regime, proportional to

$$\eta = \mathrm{sinc}^2\left(\frac{\Delta k L}{2}\right) \qquad (3)$$

where L is the crystal length. A large mismatch results in low conversion efficiency. This phase mismatch can be compensated for by the addition of a "grating vector," k_g that comes about from the addition of a periodic inversion of the sign of the nonlinearity. The addition of the grating vector to the process has the effect of making the effective $\Delta k=0$.

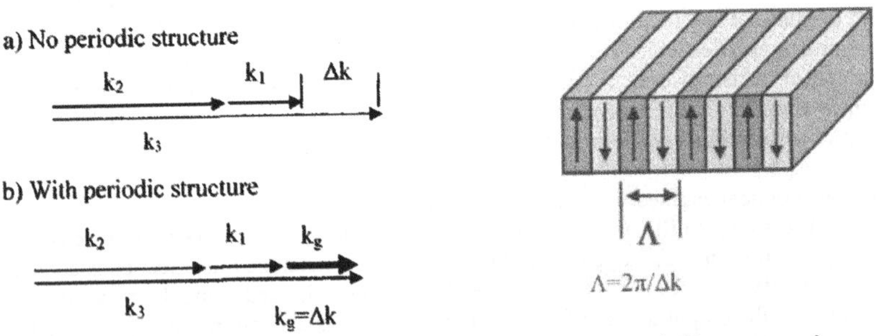

Figure 1. Phase matching diagram a) for the case where no periodic structure is present and b) including the effect of a periodically poled structure. When $k_g=\Delta k$ then the process is quasi-phase matched. Also shown is a representation of a periodically poled crystal with arrows indicating the direction of the spontaneous polarization.

The grating vector is a simple function of the periodicity of the structure and so in principle any second order nonlinear interaction can be accommodated by simply fabricating a periodic structure that matches the phase mismatch of the interaction. An example of a novel structure is shown in Figure 2 where the pattern is one with a continuous change in periodicity from one side of the crystal to the other.[1] The reason for fabricating such a structure is to attain broad tunability. By simply translating the crystal with respect to the interacting beams, different periodicities are sampled and hence one can compensate for the phase mismatches for different interactions.

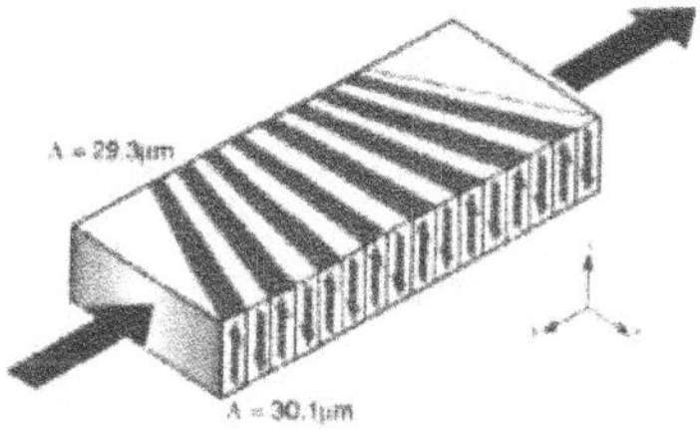

Figure 2. An exaggerated view of a fan-out poled crystal. By translating the pump beam to access different regions of the crystal, different periodicities are sampled making the device highly tunable. The arrows indicate the direction of the spontaneous polarization. The QPM interaction occurs when all the interacting beams propagate in the x-direction and are polarized along the z-axis.

ELECTRIC FIELD POLING

The ability to electric field pole ferroelectric crystals now comes into play for fabricating such a structure as shown in Figure 2. The technique of making periodically poled ferroelectric crystals has been developed to the point where commercially poled crystals are now off-the-shelf items. Periodic poling is achieved using standard photolithographic techniques to place a periodic contact on the + surface of the ferroelectric crystal. A periodic pattern is achieved by, Schmidt applying a positive field that is above the coercive field of the crystal.

Once a lithographic pattern is placed on the crystal, the application of a poling field is accomplished with several techniques. The most common is to pole using a direct connection to the patterned contact using an electrolytic solution or a metal contact,[2] but corona discharge,[3] direct electron beam domain reversal,[4] and AFM poling[5] are other techniques that have also been employed. Many studies have been conducted to optimize the poling process. Depending on the crystal, techniques such as UV assisted poling,[6,7] elevated temperature poling,[8] and programming the poling voltage waveform[9] have all been used to improve the poled structures.

Poled ferroelectrics for nonlinear optics have been demonstrated in a large number of crystals. Lithium niobate is the most widely used crystal. This is due to its relative inexpensive cost and the maturity of the crystal growth process for lithium niobate. However other important considerations come into play where other ferroelectrics may be more favorable. Table I gives a list of several ferroelectrics used for nonlinear optical applications along with information about the poling properties of the crystals. The choice of which crystal to use is dictated by several factors in addition to the poling properties. Simple considerations such as the damage threshold and transparency range of the crystal are common reasons to consider different crystals. However, more subtle considerations such as poling quality and ability to pole a large thickness must also be considered. With congruent lithium niobate (CLN), the poling quality near the +z surface (where the poling originates) is usually better than the –z surface. This is important when considering large aperture crystals since this effect can lead to a change in duty cycle from one side of the crystal to the other. Such a duty cycle variation then leads to a gradation in the effective nonlinearity as shown in Figure 3.[10] To test the poling quality across a 2.0 mm periodically poled stoichiometric lithium tantalate (PPSLT) crystal we focused a beam to a diameter of ~300 μm over the length of the 1cm long crystal. Since the beam diameter was much smaller than the aperture we were able to sample the poling quality as a function of distance away from the original +z surface. The transmitted pulse energy was monitored to ensure that the beam did not clip on the top or bottom surface. The poling quality was estimated by measuring the output energy of an optical parametric generator, and from the figure we see a marked drop-off in efficiency away from the +z surface and hence a drop in poling quality.

Table I.

Crystal	Coercive field* (kV/mm)	Comments
Congruent $LiNbO_3$ (CLN)	21 [11]	
Stoichiometric $LiNbO_3$ (SLN)	0.2 [12] 6.6 [13] 4-5 [14]	
Congruent $LiTaO_3$ (CLT)	20 [15]	
Stoichiometric$LiTaO_3$ (SLT)	1.7 [15,16]	
Mg:SLN	3.5 [13]	
Mg:CLN	6 [3,17]	
$KTiOPO_4$ (KTP)	~2 Flux [18] Hydothermal [19]	Low temperature poling, or Rb doped to reduce ionic conduction
$RbTiOAsO_4$ (RTA)	2.4 [19]	Room temperature poling
$KTiOAsO_4$ (KTA)	5 [20]	Low-temperature poled
$RbTiOPO_4$ (RTP)	3.5 [18]	

* In some cases the value given is the poling field used to pole the crystal

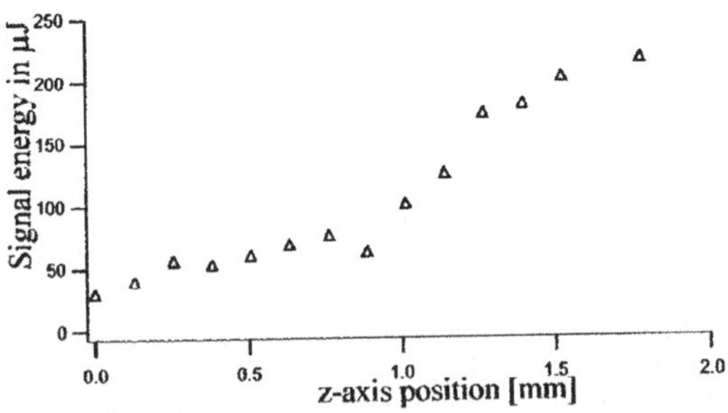

Figure 3. Output energy of an optical parametric generator from one surface to the other. The zero position is the position of the beam under the original –z surface where clipping does not occur.

The coercive field of a crystal limits the poling thickness, for example in the case of CLN, the high coercive field limits the poling thickness to roughly 1mm or less. Hence for thicker samples, a crystal with a lower coercive field must be used. Indeed, SLT crystals as well as KTP crystals have been poled with thicknesses up to 3 mm.[21]

Once the crystal is poled, its intrinsic properties must be considered. For example in PPLN one must consider photorefraction (PR), green-induced IR absorption (GRIIRA), and optical damage. Simple fixes such as heating to reduce PR are possible, but more involved work to improve these properties has centered on doping the crystal with MgO which reduces PR and GRIIRA while increasing the optical damage threshold. In the KTP family of crystals grey-tracking due to the formation of color centers is an issue.[22]

DEVICES

Applications of ferroelectric crystals to nonlinear optical devices have been common even before the use of QPM structures. Birefringently phasematched (BPM) interactions with lithium niobate, KTP, and many of the other crystals listed in Table I have been in use well before these crystals were periodically poled. Some features of BPM are worth noting in that they make a contrast to QPM devices. The natural birefringence of these crystals in BPM phasematching leads, in general, to walkoff. The walkoff effect is such that interacting beams in the crystal will physically separate given enough crystal length which limits the conversion efficiency. In addition, the BPM process dictates the crystal orientation and beam polarizations to accommodate phase matching. Hence in order to phasematch for a given interaction, the process may couple to a less than optimal nonlinearity and the polarization combination may be such that walkoff is present. With QPM we can choose a crystal orientation with no walk-off and a polarization combination that couples to the maximum nonlinearity. This paper will focus on devices based on QPM interactions which take advantage of ferroelectric crystal's ability to be poled.

Most devices used for frequency conversion applications are based on second order, $\chi^{(2)}$ coupling. These processes can be summarized using the energy conservation relation for a three-wave process:

$$\omega_1 + \omega_2 = \omega_3 \tag{4}$$

where by convention $\omega_3 > \omega_2 \geq \omega_1$ and the subscript refers to one of the interacting waves. A host of second order nonlinear devices, such as second harmonic generators, are reported in the literature. In this paper we will focus on a subset of these devices and in particular parametric interactions. These consist of optical parametric oscillators (OPO), optical parametric generators (OPG), and optical parametric amplifiers (equivalent to difference frequency generators). The impact of poled ferroelectrics on this subset of possible interactions carries over to other types of nonlinear devices as well.

Because of the high nonlinearity accessible to QPM interactions and because of the long interaction lengths (without walkoff), devices have much lower thresholds than their BPM counterparts. For example continuous wave OPO's, before the advent of QPM crystals such as PPLN, could not operate except by resonating two of the three fields (called a doubly resonant oscillator). With PPLN the threshold of the much more stable singly resonant OPO is accessible to moderate powered pump lasers on the order of 5W. Because of the ability to engineer ferroelectric crystals with different periodicities on a single crystal, we are able to make the OPO device wavelength-versatile. Figure 4 shows the setup for a PPLN OPO using a fan-out PPLN crystal.[1] The OPO is tuned by translating the crystal with respect to the pump beam and the resonator beam. This translation allows us to quasi-phase match over the range dictated by the range of periodicities on the crystal.

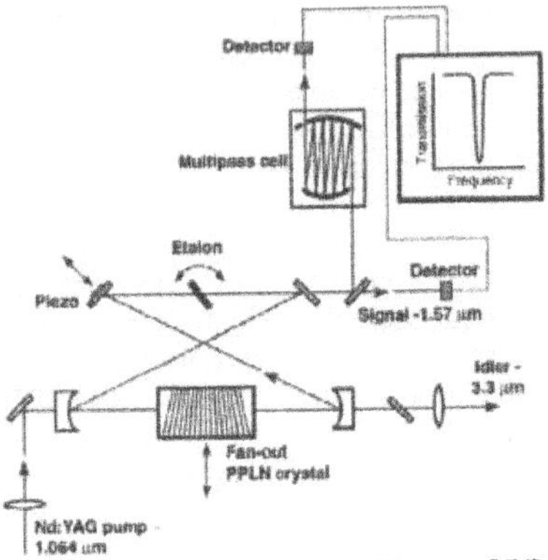

Figure 4. Experimental setup for a broadly tunable cw OPO based on a fan-out PPLN crystal.

The setup in Figure 4 also shows diagnostics for a simple transmission measurement. A measurement of the absorption spectrum of CO_2 was made to show the ability to use the fan-out PPLN crystal to access any wavelength within its theoretical tuning range. Figure 5 shows the CO_2 transmission measurement.[1] Beyond demonstrating random-access tunability, the measurement also shows the utility of using this source for spectroscopy. The measurements shown here were made using one of the outputs of the OPO which is near 1.5 μm, however the other OPO output, as dictated by equation (4) is near 3μm. Typically these outputs are called the "signal" and the "idler." This second output, or the idler, is also useful for spectroscopic applications occurring in the 3 μm range.

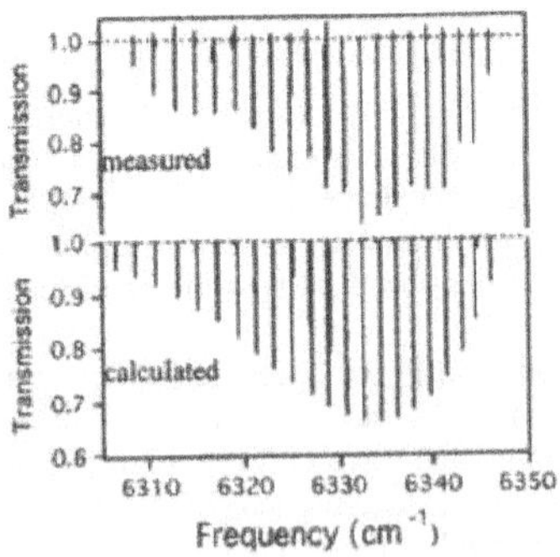

Figure 5. Measured spectrum of the CO_2 combination-band transition, showing the measured transmission and the calculated line positions.

Pulsed OPO devices based on QPM crystals have also shown improved performance in terms of much lower thresholds. Figure 6 shows the output signal and idler energy as a function of pump energy for a monolithic OPO.[23] A monolithic OPO is one where the crystal end-faces are plane-parallel polished and are sometimes coated making for a very simple device. In Figure 6 the OPO had no reflective coating so that the reflectivity was simply the Fresnel reflection from the crystal itself (on the order of 14%). This figure shows that even for a very low finesse cavity that the OPO can reach threshold. Other experiments using smaller pump beam diameters than used here (1 mm) demonstrate thresholds on the order of 50 μJ.

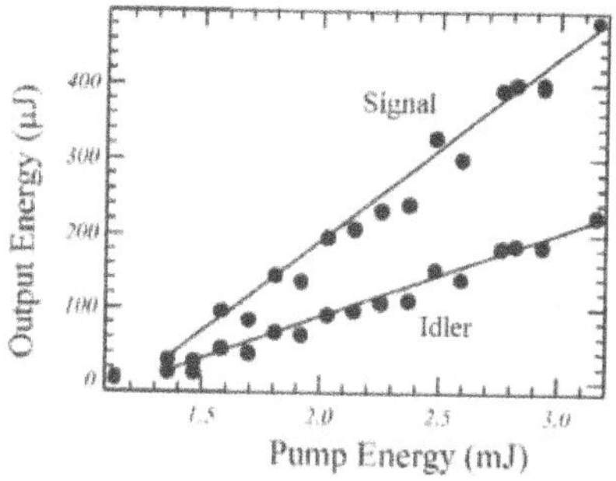

Figure 6. Output signal energy versus input pump energy for a 12 mm uncoated monolithic PPLN OPO. The crystal facets were plane-parallel polished.

Another parametric device that has benefited from the high accessible nonlinearity and long interaction length is the optical parametric generator (OPG). Previous to crystals such as PPLN, OPG devices were efficient only for high peak-power sources and required psec and fsec pulse durations to get the required peak power. With PPLN crystals with lengths on the order of 2 cm, the OPG process is efficient with nsec pulses. The same is true with other periodically poled crystals such as PPRTA. Figure 7 shows simple setup for an OPG, indicating that a pump laser is simply focused through the device. Unlike the monolithic OPO mentioned above, the end-faces of the OPG are anti-reflection coated to prevent resonator modes. The conversion efficiency for such an OPG device is typically on the order of 15% to the signal and idler. However, the free-running OPG typically has a bandwidth that is too large for many spectroscopic measurements, but the bandwidth can be controlled using a seeding scheme. A seeding technique is one where a second narrow bandwidth beam is co-aligned with the pump. If this narrow bandwidth beam is tuned to be within the signal bandwidth, it can serve to control the signal bandwidth. We have used two techniques to seed. The first is shown in Figure 7 where the output of a first stage OPG is filtered with an etalon whose output then seeds a second stage OPG.[24]

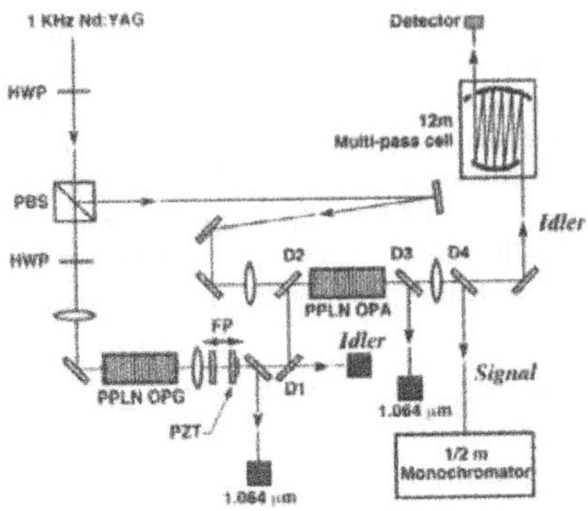

Figure 7. Experimental schematic. HWP, half wave plate; PBS, polarizing beam splitter; FP, air-spaced plane-parallel Fabry-Perot etalon; D1 - D4, dichroic mirrors to separate the signal, idler, and pump; other abbreviations defined in the text.

A second technique is to replace the filtered OPG with another laser source, in our case we have used a tunable diode laser. The diode laser seeding has shown that powers on the order of 1mW are sufficient to control the OPG bandwidth for a 1 mJ, 3 nsec pump pulse. Figure 8 shows the unseeded and seeded bandwidths from the OPG setup.[24] The advantage of this technique is that since the OPG bandwidth is relatively large, we can seed over that same large bandwidth and obtain a tunable narrow bandwidth seeded output.

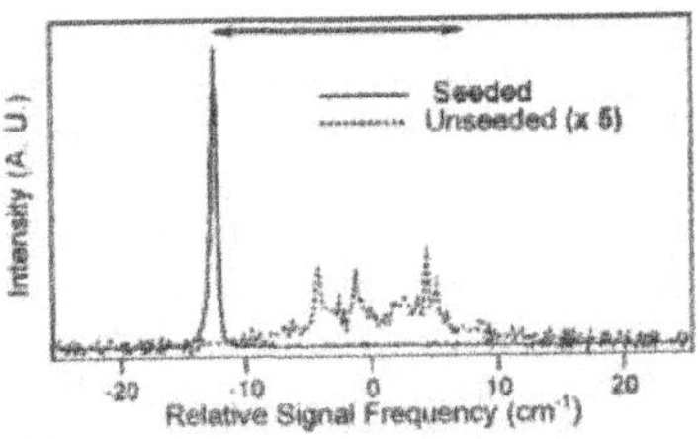

Figure 8. Seeded and unseeded OPG outputs. The unseeded spectrum is multiplied by 5 for visibility. The arrow indicates the tuning range that is possible without changing the crystal temperature or QPM periodicity. The resolution of the monochromator is less than the linewidth of the seeded OPG output.

This tunability is an important feature when considering spectroscopic applications. The ability to tune over a relatively large frequency range without changing the periodicity or crystal temperature allows one to rapidly scan over several spectral features. This is a great advantage when trying to identify a given species from the measured spectrum. Such a scan is shown as the inset to Figure 9.[25] This measurement was a cavity ring-down measurement of a variety of species present in trace amounts. The inset is an expanded view of two scans showing that each scan can cover many spectral features.

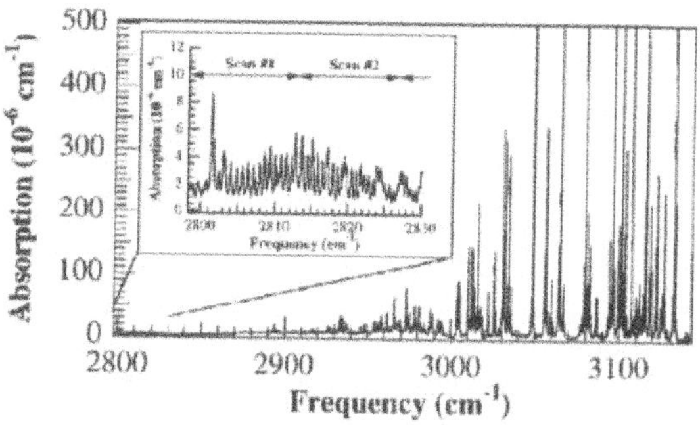

Figure 9. Demonstration of frequency tunability: the 350 cm^{-1} cavity ringdown spectrum was obtained by combination of 27 smaller continuous scans, each with a unique PPLN poling period and temperature. Inset: an expanded view of two scans.

By fabricating a crystal with several QPM periodicities or by using a continuously varying periodicity as with the fan crystal, an even broader tuning range is achieved by piecing scans together. Figure 9 shows a ring-down scan over 350 cm^{-1} which is the result of 27 scans pieced together.

We have used this same technique to attain seamless and broad tunability in the THz region. For this setup we used two seeded OPG's operating at wavelengths separated by a frequency in the THz region. By DFG mixing the outputs in a DAST crystal we then obtained the THz difference frequency. Other researchers have used periodically poled crystals directly to generate THz frequencies both in the ultrafast regime and in the quasi-cw regime.

The broad tunability of these OPO's and OPG's are attractive for other applications such as remote sensing and laser radar. To accommodate such applications, the output pulse energies need to scale up significantly to reach reasonable ranges. We have carried out a series of experiments on scaling up the pulse energy in both OPO and seeded OPG devices using single-periodicity crystals. In particular we have focused on monolithic devices with the goal of attaining a rugged system. In both cases of a monolithic OPO and an OPG, scaling up in energy introduces new behavior based on the high nonlinearity of the QPM crystals. Scaling up energy requires an increase in the beam size to avoid damaging the crystal, however most QPM crystals have thickness limited to under 3 mm, with most crystals available at ~1mm. Hence these crystals require pumping with an elliptical beam to increase the energy throughput. The output energy of OPO's did increase with elliptical pumping, however, as shown in Figure 10 there is an accompanying rapid increase of the beam divergence as the beam diameter increases.[23,26,27] Other measurements of the monolithic OPO showed a similar increase in spectral bandwidth with aperture scaling.

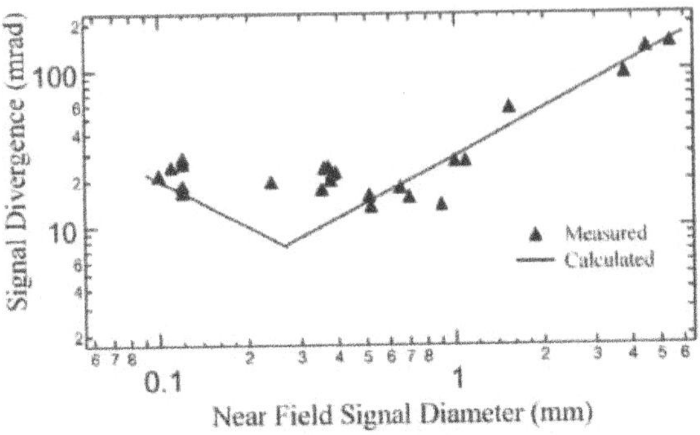

Figure 10. Output divergence increases rapidly as the pump beam size is increased.

The reason for the increased beam divergence and bandwidth is a consequence of non-collinear phasematching processes. As the transverse pump size increases, the overlap of noncollinear processes with the pump is increased. The good overlap is coupled with a higher parametric gain since the noncollinear processes tune towards degeneracy where the parametric gain peaks.

To control the divergence and spectral properties, we went to a seeding scheme. To avoid a competition between resonator modes and the seeding, we also switched to an OPG setup where the end-faces of the crystal were anti-reflection coated, and for one crystal the output face was Brewster polished to eliminate any spurious cavity modes. Figure 11 shows the output energy of a PPLN and a PPRTA unseeded OPG when pumped with an elliptical beam.[28] The pump beam was measured to be 4.4 mm by 180 μm (FWHM). Similar to the monolithic OPO's the conversion efficiency is high. but when unseeded the beam divergence and the bandwidth is high.

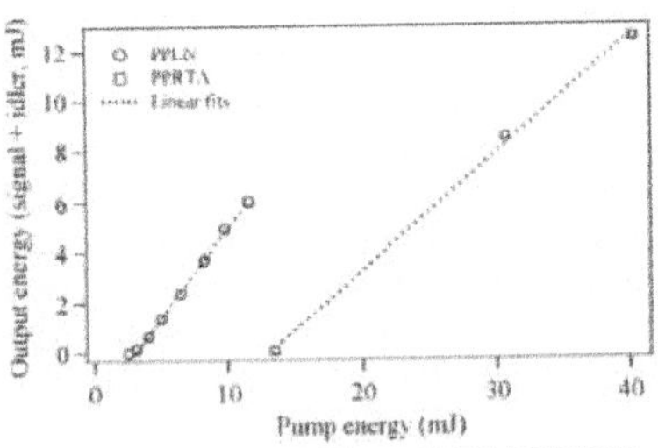

Figure 11. OPG energy output for PPRTA and PPLN OPG's.

By introducing a seed beam into the experiment we investigated the ability of the seed beam to control the divergence and bandwidth. In our attempts with PPLN, we saw that the high gain of the PPLN was such that noncollinear processes could compete with the seeded mode.

Figure 12 shows that seeding is effective for low pump energies, but as the pump is increased, the unseeded background begins to dominate.[28]

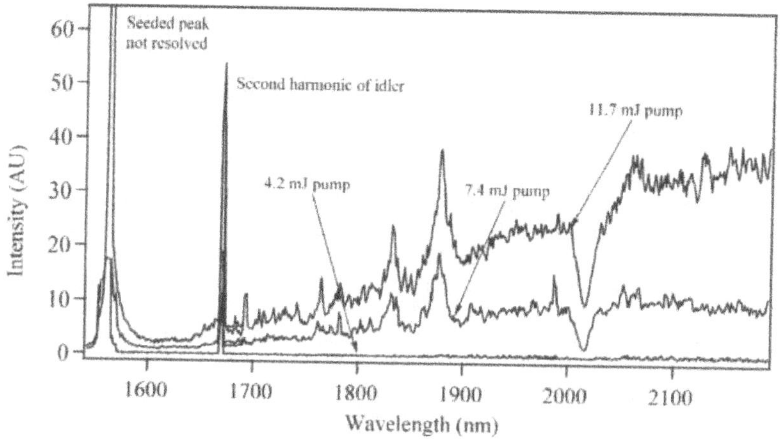

Figure 12. High energy seeded OPG for a range of pump energies. A broadband unseeded background is present when pumping at high energies.

The inability to seed PPLN in the high energy regime results from the high nonlinearity of the PPLN and the overlap of the noncollinear beams with the pump. We tested this hypothesis by going to PPRTA which has a lower nonlinearity than PPLN. With this crystal we were able to seed effectively with large elliptical beams up to the damage threshold of the crystal. The result of effective seeding is to control the bandwidth, and as Figure 13 shows, it also controls the beam divergence.[28] Note that since the OPG is pumped with an elliptical beam in the near field, that the far field pattern should be a vertical stripe (as it is for the seeded case).

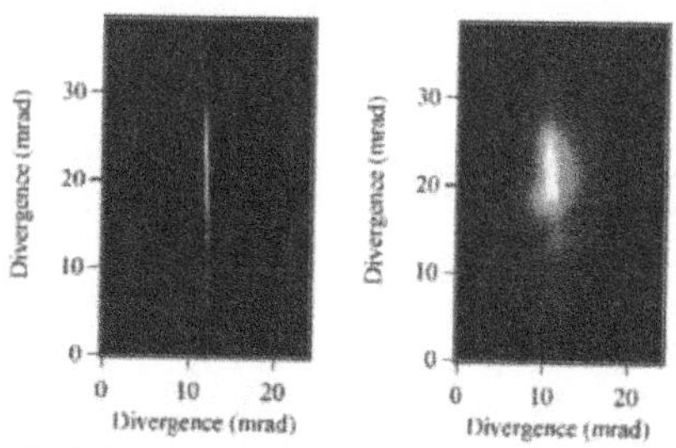

Figure 13. Seeded (left) and unseeded (right) far-field beam profiles of an elliptically-pumped PPRTA OPG.

The best solution for high energy seeded devices is not to switch to a different crystal with a lower nonlinearity, rather it is possible to use the domain engineering capability of ferroelectric crystals to use a QPM pattern which has a lower nonlinearity. We are currently pursuing work that involves using long PPLN crystals with a low nonlinearity by using 3rd order QPM or by reducing the duty cycle of the poled structure. The result of the lower nonlinearity is that noncollinear processes will not have enough gain to build up before they steer out of the pump beam. Only processes that are nearly collinear will see sufficient gain to be efficient. In this regime we expect the seeding to be effective, making possible diffraction limited and transform limited outputs. Such a system would then have excellent qualities for remote sensing applications in terms of beam quality, but also in terms of system simplicity.

RECEIVER DEVICES

Most of the above work is focused on source development. The engineering capabilities of ferroelectric crystals can be extended to other applications as well. We have recently begun an effort to improve the sensitivity of mid-IR receiver systems by frequency converting the mid-IR to near-IR telecommunications wavelengths which can then take advantage of mature detection technology.

Detectors in the mid-IR (3 to 5 μm region) make use of liquid nitrogen cooled detectors and newer thermo-electrically (TE) cooled detectors. However, cryogenic detectors are not always practical in a setting outside of the laboratory. The TE-cooled detectors are more convenient to work with and can have high sensitivities, but there is a trade-off between sensitivity and speed.[29] Faster detectors trade off noise for speed. For applications such as range-resolved LIDAR, a fast detector is needed, and ideally we would like to have noise properties similar to near-IR or visible detectors. Our approach is shown in Figure 14. In this scheme a mid-IR beam is frequency converted to the 1.55 μm range where telecom detectors and infrastructure are well developed. In our experiment, the mid-IR signal was a 2nsec pulse at 3 μm and was frequency converted to 1.55 μm. The 1.55 μm signal was then coupled into a fiber and detected with an avalanche photodiode (APD). Since the frequency conversion process is nearly instantaneous, this approach preserves the timing information on a LIDAR return while being able to use high speed near-IR detectors with much more favorable noise properties when compared to direct detection in the mid-IR. Synchronization between the return LIDAR signal and the pump laser are ensured by using a cw pump so that whenever a return signal is present we are guaranteed to have a pump.

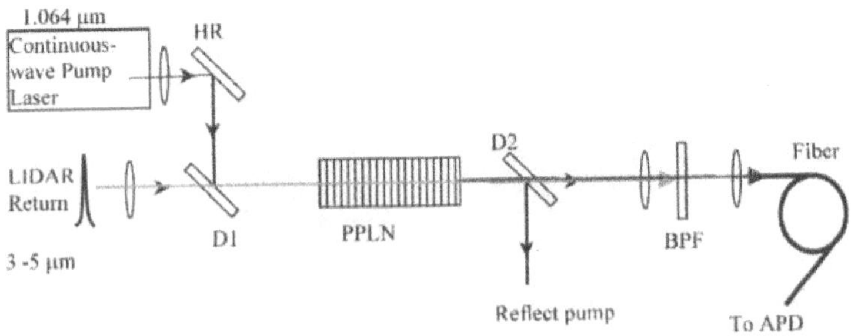

Figure 14. Receiver for LIDAR system using a frequency conversion scheme based on PPLN.

Our initial measurements show that we have attained a noise-equivalent power (NEP) equivalent to the fast direct detectors in the mid-IR while being able to operate at much higher speeds. The NEP of our measurement is 4 nW/√ (Hz) with a detection bandwidth of 250 MHz (limited by the oscilloscope used). This compares well to state-of-the-art direct detectors the fastest of which have an NEP of 4 nW/√(Hz) but have a detection bandwidth of only 8 MHz. We are working to improve this scheme by increasing the power of the pump laser and by improving the coupling of the return signal with the pump. Further improvement could be made by working on the timing of the pump and the LIDAR return so that a high-peak power pump laser could be timed to overlap with the return. Such a scheme could even result in amplification of the frequency converted beam.

CONCLUSION

Domain engineered ferroelectric crystals are beginning to have a significant impact on the field of nonlinear optics. Periodically poled crystals in particular have made possible high efficiency nonlinear frequency conversion and low threshold nonlinear optical devices. Broad tunability covering the entire IR transparency range of the crystal is possible with OPO and OPG devices. The results presented here show how this tunability can be coupled with a narrow bandwidth suitable for many spectroscopic applications. Scaling up to higher energies is also possible with these crystals by using an elliptical pumping scheme, and provided that the domain engineering is done right, the output at high energy has a narrow bandwidth and a low divergence. Such a high energy source is attractive for remote sensing applications. The same properties that make domain engineered crystals attractive for tunable sources are being exploited to work on detector systems. The high nonlinearity of periodically poled crystals makes frequency converting a signal from the mid-IR to the near-IR or visible a practical possibility. This paper has gone over just a few of the devices and applications enabled by domain engineered crystals. The ability to engineer the crystals is a significant advantage which is also being exploited in numerous other applications.

REFERENCES

1 P. E. Powers, T. J. Kulp, and S. E. Bisson, "Continuous tuning of a cw PPLN OPO using a fan-out grating design," Optics Letters **23**, 159-161 (1998).

2 L. E. Myers, R. C. Eckardt, M. M. Fejer, R. L. Beyer, and W. R. Bosenburg, "Quasi-phase-matched 1.064-µm-pumped optical parametric oscillator in bulk periodically poled $LiNbO_3$," *Opt. Lett.* **21**, 591 (1996).

3 A. Harada and Y. Nihei, "Periodically poled MgO-LiNbO3 by corona discharge method," *Appl. Phys. Lett.* **69**, 2629 (1996).

4 C. Restoin, V. Couderc, C. Darraud-Taupiaca, J.L. Decossasa, J.C. Vareillea, A. Barthélémy, and J. Hauden, "Periodically Poled Lithium Niobate by Electron Beam: Irradiation Conditions and Second Harmonic Generation," *Applications of Photonic Technology 4*, Roger A. Lessard, George A. Lampropoulos, Editors, Proceedings of SPIE Vol. 4087, 571 (2000).

5 A. Agronin, Y. Rosenwaks, and G. Rosenman, "Ferroelectric domain reversal in LiNbO3 crystals using high-voltage atomic force microscopy," *Appl. Phys. Lett.* **85**, 452 (2004).

6 For a comparison of CLN and Mg:CLN see, M. C. Wengler, B. Fassbender, E. Soergel, and K. Buse, "Impact of ultraviolet light on coercive field, poling dynamics and poling quality of various lithium niobate crystals from different sources," J. Appl. Phys. **96**, 2816 (2004).

7 M. Fujimura, T. Sohmura and T. Suhara, "Fabrication of domain-inverted gratings in MgO:LiNb03 by applying voltage under ultraviolet irradiation through photomask at room temperature," *Elect. Lett.* **39**, 719 (2003).

8 H. Ishizuki, I. Shoji and T. Taira, "Periodical poling characteristics of congruent MgO:LiNbO3 crystals at elevated temperature," *Appl. Phys. Lett.* **82**, 4062-4 (2003).

9 R.G. Batchko, G.D. Miller, V. Ya. Shur, E. L. Rumyantsev, M.M. Fejer, and R.L. Byer, "Domain patterning in lithium niobate using spontaneous backswitching,"Part of the SPIE Conference on Laser Material Crystal Growth and Nonlinear Materials and Devices, San Jose, California **SPIE 3610**, 36 (January 1999).

10 E. M. Vershure, "Fabrication and direct comparison of periodically poled materials for mid-IR generation," M. S. Thesis, University of Dayton, May 2005.

11 M. Yamada, N. Nada, M. Saitoh, and K. Watanabe, "First-order quasi-phase matched $LiNbO_3$ waveguide periodically poled by applying an external field for efficient blue second harmonic generation," *Appl. Phs. Lett.* **62**, 435-436 (1993).

12 A. Grisard, E. Lallicr, K. Polgar and A. Peter, "Low electric field periodic poling of stoichiometric lithium niobate," *Elect. Lett.* **25**, 651-653 (2000).

13 L. Huang, D. Hui, D. J. Bamford, S. J. Field, I. Mnushkina, L. E. Myers, J. V. Kayser, "Periodic poling of magnesium-oxide-doped stoichiometric lithium niobate grown by the top-seeded solution method," *Appl. Phys. B* **B72**, 301-6 (2001).

14 V. Gopalan, T. E. Mitchell, Y. Furukawa, and K. Kitamura, "The role of nonstoichiometry in 180° domain switching of $LiNbO_3$ crystals," *Appl. Phys. Lett.* **72**, 1981-1983 (1998).

15 Y. Furukawa, K. Kitamura, V. Gopalan, T.E. Mitchell, G. Foulon, A. Alexandrovski, R.K. Route, and M.M. Fejer, "Stoichiometric $LiTaO_3$ as new nonlinear material for bulk quasi-phasematching," *Conference on Lasers and Electro-Optics*, Pp. 167–168 (1999).

16 T. Hatanaka, K. Nakamura, T. Taniuchi, H. Ito, Y. Furukawa, and K. Kitamura "Quasi-phase-matched optical parametric oscillation with periodically poled stoichiometric LiTaO3." Opt. Lett. **25**, 651-653 (2000).

17 For a comparison of CLN and Mg:CLN see, M. C. Wengler, B. Fassbender, E. Soergel, and K. Buse, "Impact of ultraviolet light on coercive field, poling dynamics and poling quality of various lithium niobate crystals from different sources," J. Appl. Phys. **96**, 2816 (2004).

18 H. Karlsson and F. Laurell, "Electric field poling of flux grown KTiOPO4," Appl. Phys. Lett. **71**, 3474 (1997).

19 Z. W. Hu, P. A. Thomas, J. Webjorn, and G. M. Loiacono, "Domain inversion in RbTiOAsO4 using electric field poling," *J. Phys. D: Appl. Phys.* **29** 1681–1684 (1996).

20 G. Rosenman, A. Skliar, Y. Findling, P. Urenski, A. Englander, P. A. Thomas, and Z. W. Hu, "Periodically poled KTiOAsO4 crystals for optical parametric oscillation" J. Phys. D: Appl. Phys. **32,** L49–L52 (1999).

21 J. Hellstrom, V. pasiskevicius, F. Laurrell, and H. Karlsson, "High-power optical parametric oscillation in large-aperture periodically poled $KTiOPO_4$," *Opt. Lett.* **24**, 1233 (1999).

22 M. P. Scripsick, D. N. LoIacono, J. Rottenberg, S. H. Goellner, L. E. Halliburton, and F. K. Hopkins, "Defects responsible for gray tracks in flux-grown KTiOPO4," *Appl. Phys. Lett.* **66**, 3428 (1995).

23 M. J. Missey, "Characterization of periodically poled lithium niobate monolithic optical parametric oscillators and generators," Ph. D. Dissertation, University of Dayton (1999).

24 P. E. Powers, K. W. Aniolek, T. J. Kulp, B. R. Richman, and S. E. Bisson, "Periodically poled lithium niobate optical parametric amplifier seeded with the narrow-band filtered output of an optical parametric generator", *Opt. Lett.* **23**, 1186-1889 (1998).

25 K. W. Aniolek, R. L. Schmitt, T. J. Kulp, B. A. Richman, S. E. Bisson, P. E. Powers, "Microlaser-pumped periodically poled lithium niobate optical parametric generator–optical parametric amplifier," *Opt. Lett.* 25, 557 (2000).

26 M. J. Missey, V. Dominic, P. E. Powers, and K. L. Schepler, "Aperture scaling effects with monolithic periodically poled lithium niobate optical parametric oscillators and generators," *Opt. Lett.* **25**, 248-250 (2000).

27 M. J. Missey, V. Dominic, P. E. Powers, and K. L. Schepler, "Periodically poled lithium niobate monolithic nanosecond optical parametric oscillators and generators," *Opt. Lett.* **24**, 1227-1229 (1999).

28 P. E. Powers, P. K. Bojja, R. Alkuwari, E. M. Vershure, K. L. Schepler, "High Energy Seeded Optical Parametric Generation with Elliptical Beams in Quasi-Phase-Matched Crystals," accepted for publication in *J. Opt. Soc. Am. B*, publication date, October (2005).

29 J. Piotrowski, "Uncooled operation of IR photodetectors," *Opto-Electronics Review* **12**, 111 (2004).

Electroceramic Materials and Applications
Edited by R. W. Schwartz

ROLE OF EXTRINSIC DEFECTS IN FERROELECTRIC DOMAIN INVERSION OF LITHIUM NIOBATE

Volkmar Dierolf and Christian Sandmann
Lehigh University, Physics Department
16 Memorial Drive East
Bethlehen, PA 18015

ABSTRACT

We apply confocal luminescence microscopy to investigate the interaction between ferroelectric domain walls both statically and dynamically. In our technique, we use rare earth ions such as Er^{3+} and probe the changes in the emission spectra of these probe-defects. The changes that occur can be interpreted in terms of changes in internal electric fields and the arrangement of charge compensating defects. With this technique we are able to precisely image the domain wall regions and to measure changes that occur within a moving domain wall.

INTRODUCTION

The ferroelectric, lithium niobate ($LiNbO_3$), has found wide application due to its favorable acousto-optical, electro-optical, and nonlinear optical properties. For many applications such as periodically poled pattern for nonlinear frequency conversion and electro-optical steering elements it is crucial to create precise ferroelectric domain patterns. With the most common technique, using electrodes patterned by optical lithography, periodically poled components with period lengths down to 3μm could be fabricated [1]. Smaller structures, however, are difficult to achieve with present techniques. For the investigation of novel domain inversion techniques, which potentially offer the feasibility to create smaller structures, it is vital to understand the interaction between domain walls and intrinsic defects present in all compositions of the materials except the perfectly stoichiometric, which has not been realized in $LiNbO_3$. It has been shown that even a small number of defects drastically influence the shape of spontaneously formed domains and the coercive fields required [2]. Unfortunately, it is virtually impossible to detect these intrinsic defects (e.g.: Li-vacancies, Nb-antisites) on a fast time scale and with high spatial resolution. In order to circumvent this problem, we developed an approach that uses intentionally introduced rare earth defects, e.g. Er^{3+} ions, as local probes for the domain inversion process. In earlier studies [3] we have shown that the emission of the Er^{3+} ion is different for the as grown and domain inverted part of the sample. Scanning over the sample in a raster mode, while continuously measuring and evaluating the emission spectra, allows 3D imaging of domain structures with high spatial resolution and free of topographic artifacts. We used this imaging technique to visualize domain structures produced with our light-induced domain writing technique [4]. In this paper we will review details of our detection technique and show that it is possible to probe the Er^{3+} defect and its interaction with a domain wall in real time during a domain inversion event. This offers the potential to study the defect/domain wall interaction associated with a moving domain wall and allows determining the quantities such as the dimension of the domain wall regions during domain growth. This real-time diagnostic technique can be used in an active feedback system for light-induced domain inversion. First results in this direction will be presented in this paper.

EXPERIMENTAL SET-UP

For our experiments, we developed a sample holder, which allows the application of high electric fields and at the same time close access for a home-built confocal luminescence microscope (CLM) that allows optical spectroscopy with high spatial resolution (500nm). For the optical excitation of the Er^{3+} ions, we use the 488nm emission line of an Ar^{+}-laser or a tunable dye laser (440nm-460nm). In either case the laser is coupled into a single-mode optical fiber. For detection of the emission at a wavelength of about 550nm we use a nitrogen cooled CCD detector attached to a grating spectrometer. The use of a CCD detector allows the detection of the emission spectrum and the transmitted red light from the sample illumination at once with a repetition rate of up to 140Hz. The red illumination allows the observation of the sample with a CCD camera attached to the scanning head of the CLM and at the same time does not overlap with the Er^{3+} emission. A schematic and a picture of the experimental setup are shown in Figure 1. As sample material we used congruent $LiNbO_3$ z-cut crystals of 0.5mm thickness doped with Er^{3+} ions with a concentration of $1 \cdot 10^{-4}$mol in the melt (Deltronix) and also nearly stoichiometric $LiNbO_3$ z-cut crystals with the same Er concentration obtained from Dr. K. Polgar.

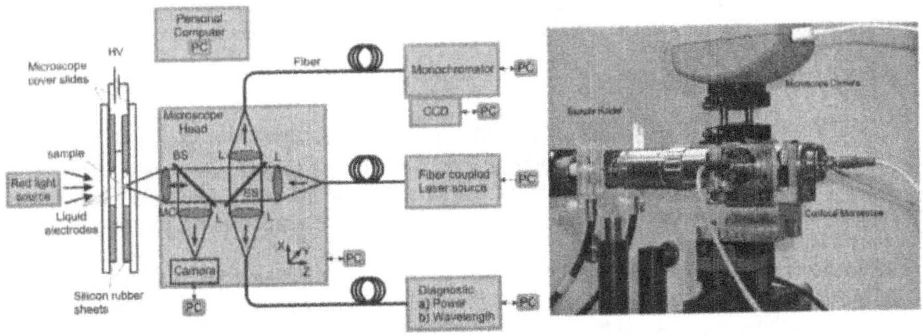

Figure 1: (left) Schematic of the experimental setup; BS: Beamsplitter, L: Lens, MO: Microscope objective. (right) Picture of the microscope

EXPERIMENTAL RESULTS AND THEIR DISCUSSION

Properties of the Er-ion probe ion

Our imaging and real time diagnostic schemes rely on a good understanding of rare earth ion defects in the ferroelectric material and the changes that can be observed in their emission during a domain inversion process. For the Er^{3+} ion this knowledge has been accumulated over several years.

Rare earth ions are incorporated into $LiNbO_3$ mainly onto a Li-site [5] but in several different configurations that are different due to perturbations from defects that are required for local charge compensation. These "sites" can easily be distinguished using our combined excitation-emission spectroscopy technique, as shown in Fig. 2 [6]. Changes in local environment can be simulated by application of external perturbations such as electric fields and hydrostatic pressure. They lead to site-specific shifts that are indicated in Fig. 2 by arrows. *This sensitivity is the basis for using the rare earth ions as probes.* We come back to this capability in a more quantitative way in the next section.

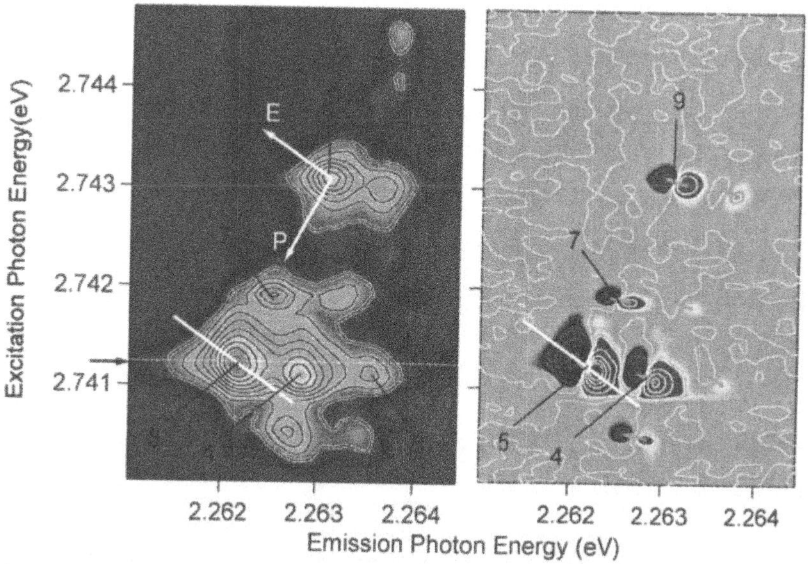

Fig. 2: *Contour/Image plot of combined excitation emission data for Er^{3+} in nearly stoichiometric $LiNbO_3$ at 4K.*
(Left) Emission intensities. In the chosen range of the spectrum only one excitation and emission transition is expected. The number of peaks indicates the number of different Er environments. Several sites are labeled in accordance with Ref. [6]. The excitation energy used for Fig. 3 is indicated by a dotted line. The white arrows indicate the shifts that occur under application of an electric field (E) and hydrostatic pressure (P). (Right) Differences between emission intensities with or without application of a field along the z-axis of the sample (10kV/mm).

In terms of ferroelectric domain inversion, we make the following observations in the spectra for stoichiometric and congruent samples (see Fig. 3):

1. The relative peak height for emission from different sites is changed. More precisely, one of the sites (labeled 5 in Fig. 2) is unchanged while all the others are reduced. The unchanged defect site is most prominent in stoichiometric samples with low Er^{3+} concentration and is attributed to an Er^{3+} ion that has no local charge compensation.

2. For the emission from a particular site, the emission is shifted. This is most apparent for site 5 since its peak is at the low energy end of the spectrum.
3. The changes that are induced by domain inversion are reversed when the domain is reverted to its original orientation.
4. Annealing of the sample at 200°C leads to almost identical spectra for both before and after domain inversion. These spectra, however, are different from the ones before annealing.
5. Under application of an electric field the emission intensity is changed.

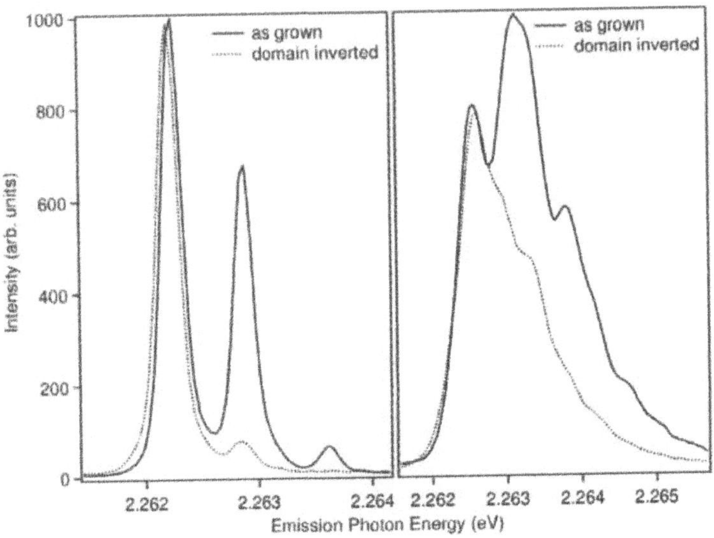

Fig. 3: Left: Emission spectra of Er^{3+} doped stoichiometric $LiNbO_3$ for an excitation photon energy of E_{exc}=2.74225eV for as-grown and domain inverted samples at low temperature T=4K. Right: Single emission spectra of Er^{3+} doped congruent $LiNbO_3$ for an excitation photon energy of E_{exc}=2.7425eV for as-grown and domain inverted samples at low temperature T=4K.

We base the following discussion of the Er-defect structure on a model for the Nb antisite, $(Nb_{Li})^{4+}$, defect that has been proposed earlier to describe the influence of stoichiometry on the coercive fields and the asymmetry in the hysteresis loops [7,8].

The niobium antisite $(Nb_{Li})^{4+}$, is surrounded by three lithium vacancies $(V_{Li})^-$ in the nearest lateral neighborhood and one more independent (V_{Li})– vacancy along the c-axis axial to the $(Nb_{Li})^{4+}$ antisite. During domain inversion the $(Nb_{Li})^{4+}$ antisite defect moves through the oxygen plane to the neighboring oxygen octahedron and also the four Li-vacancies move through their respective oxygen plane. However, the fourth Li vacancy determines the orientation of the defect dipole moment, which will be now anti-parallel to the dipole moment of the crystal. Due to the low mobility of lithium ions at room temperature the axial Li – vacancy is not able to move to the other side of the $(Nb_{Li})^{4+}$ antisite. The result is a frustrated defect state (Fig. 4, center) Annealing above 200° C increases the lithium mobility [7] and the axial Li-vacancy can arrange accordingly to the $(Nb_{Li})^{4+}$ antisite and the dipole moment of the crystal (Fig 4 left). This model is able to quantitatively predict many observed features during domain inversion as discussed in Ref. [8] .

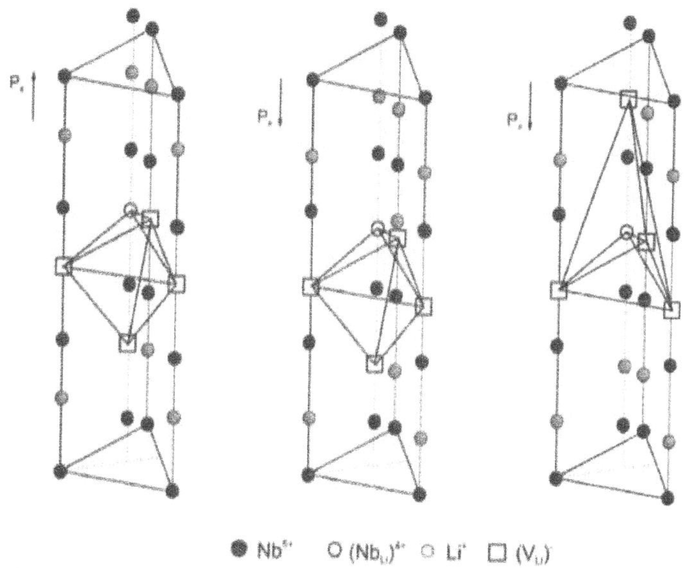

Fig. 4: *right: Schematic of a niobium antisite (Nb_{Li}) defect with lithium vacancies (V_{Li})• for charge compensation in $LiNbO_3$. b) Frustrated defect after domain inversion. c) Relaxed defect after annealing above 200 °C for a couple of minutes.*

The Er^{3+} defect in congruent and stoichiometric $LiNbO_3$ is somewhat similar to the niobium antisite, because the Er^{3+} defect is also positively charged and needs a negative charge compensation. Additionally, just as for the niobium antisite, the majority of Er^{3+} ions are located on a lithium site and a charge compensation by Li– vacancies is likely. Observations (3) and (4) above resemble the observation that lead to the $(Nb_{Li})^{4+}$, antisite model and suggest frustrated defect configurations might be created. In terms of the creation of a frustrated defect state it is not imperative that the charge compensation is generated by Li– vacancies. Any other charged defect, not mobile at room temperature, and creating together with the Er^{3+} ion a local defect dipole moment with a component parallel to the c-axis, will convert to a frustrated defect state during domain inversion. Therefore, based on the experimental evidence, a conversion of the Er^{3+} defect and its local charge compensation to a frustrated defect state seems likely for all sites which reduce in intensity during domain inversion and reestablish their initial intensity during domain inversion back to the as-grown state. The defect site that experiences only a change in spectral position is likely to be connected, as stated above, to an Er ion without nearby charge compensation. Such a defect structure has been found for Cr^{3+} already by detailed ENDOR measurements [9]. Its small shift is due to more distant frustrated defects. Making use of site-selective spectroscopy, our spectra will hence give us information about defects that are close and somewhat distant to the Er ion, simultaneously.

Calibration of Er^{3+}-ion probe ion

In order to use spectra of the Er ion as a quantitative probe for changes in the local intrinsic field, we need to calibrate their sensitivity. To this end, we applied several approaches. Most accurately, we determined the spectral shifts site selectively taking advantage of the ability of our combined excitation emission technique to follow individual sites under application of perturbations. The changes that can be produced experimentally (≈10kV/mm) are quite small and the shifts are best seen by taking the differences between spectra taken with and without field or, even better, opposite field direction. These differences are shown in Fig. 2 on the right. Since changes occur both in excitation and emission energies, the maxima and minima are not horizontal or vertical in these plots. In order to determine the shifts in both energies, we extended the well-known method of moments to this two dimensional case. The zeroth (A) corresponding to the total intensity, the first moment (E), corresponding to the emission peak position and the change in the latter (ΔE) are defined as

$$A = \int_{-\infty}^{\infty} I(E)dE$$

$$\overline{E} = \int_{-\infty}^{\infty} I(E)EdE \qquad (1)$$

$$\Delta E = \int_{-\infty}^{\infty} (E-\overline{E})\Delta I(E)dE$$

and can be calculated from the intensity, I(E), and intensity difference data ΔI(E) data. In our case, $\vec{E}$ and $\Delta\vec{E}$ are two-dimensional vectors with components for the excitation and emission energy. The direction of $\Delta\vec{E}$ is given in the difference data image by the connection between the maximum and minimum. The absolute value is obtained when the data are evaluated along a path in the direction of $\Delta\vec{E}$. An example of such a path is shown in Fig. 2. From the absolute shift value and the direction the components for the shift in emission and excitation energies can be calculated. We list in Table 1, the spectral shift factors, $V_{em} = \frac{\Delta E_{em}}{applied\ Voltage}$, in emission energy for two different sites and five transitions in the green spectral region.

Table:1 Spectral shift factors V_{em} (in $eV\frac{cm}{V}\cdot 10^{-10}$) expressing shifts in emission energy per voltage for several sites and emission transitions

Emission Energy	V_E for Site 5	V_E for Site 9
2.241 eV	0	0
2.249 eV	-2.6	-4.5
2.248 eV	-2.1	-3.4
2.256 eV	-3.1	-3.5
2.263 eV	-3.7	-3.8

For using these values in cases in which the excitation and emission is not selective, we note that the site-selectivity of the shift for some transitions is rather small and good accuracy can be achieved if the values for the most abundant site are used. However, this statement is no longer true, when the relative strength of the emission from different sites is changed as seen for instance under ferroelectric domain inversion. In such cases, the shifts have to be evaluated site-selectively. In our real time diagnostic experiments of the domain inversion (see below), we generally use 488nm excitation and work at room temperature. For these reasons, we are not site-selective (which is not an option for time resolved measurements anyway). In order to improve our accuracy for this case, we perform an in-situ calibration and measure spectra and evaluate the quantities of equation (1) while we are increasing the electric field.

Imaging of Static Domains and Domain Wall Regions

The changes in rare earth emission that we observe (and show in Fig. 3) can be used to look more closely at changes in local electric fields and changes in the defect arrangement across a domain wall. Due to the more narrow lines and the dominance of the main defect site, the results are most conclusive for the near stoichiometric samples with rather low Er concentration. They will be presented here (see Fig. 5). For these experiments we freshly prepared inverted domain regions and measured line scans across the domain wall. The recorded spectra are evaluated in terms of the peak position of the main site (Site 5) and the relative strength of the main site (Site 5) and its satellite (Site 4). The same experiment is repeated after the sample has been annealed for 5 hours at T=250 °C.

For the annealed case, the ratio of the two peaks is identical across the domain wall excluding defect rearrangement and hence the observed shifts can directly be related to changes in the intrinsic local fields. For the data taken before annealing, peak fitting is required to determine the shifts of the main peak. In both cases, the shifts can be translated into field changes using the calibration from Table 1. Compared to the as grown region, we can "measure" changes in the local intrinsic field of 60kV/cm within the wall and of 100kV/cm across the wall in a freshly domain inverted sample. The corresponding value for the annealed cases are 60kV/cm and 40kV/cm. The presence of differences between the areas in the latter case indicates that despite of the anneal procedure some intrinsic defects did not realign completely. We should also note, that the intrinsic fields are different before and after annealing, even in the as grown areas (ΔE= 80kV/cm) suggesting that different thermal equilibria of the defect configuration exist for different temperatures. This fact is also reflected in the observation that annealing leads to changes in the Er emission spectra for the as grown region. This may account for the observed dependence of coercive fields on the thermal history of the sample.

The results are not exclusive to stoichiometric materials and can be observed in congruent material as well. Figure 6 shows a line scan for the changes that occur across a domain pattern within a periodically poled waveguide device. It has been annealed to relief stress. Again, we see in the position of Site 5 sharp peaks at the domain walls and a difference for the as grown and the domain inverted region. The observed shifts are somewhat higher but the relative size of the effects is similar to the stoichiometric case.

We are turning now our attention to the dimensions over which the changes are occurring. The domain wall of $LiNbO_3$ is the cause of strong discrepancies between theoretical calculations and experimental results. Theoretical first principle calculations on the important class of oxygen octahedron ferroelectrics show that the ferroelectric wall between two anti-parallel domains should be atomically sharp of the order of 0.5nm [10]. In contrast, intrinsic wall widths in $LiNbO_3$ measured with atomic force microscopy revealed a width of 150nm [11] and the determined width with X-ray topography revealed ≤300nm [12]. Using a near-field scanning optical microscope [13], birefringence in the vicinity of the domain wall in a distance of up to 1μm was detected in $LiTaO_3$ (isostructural to $LiNbO_3$). These length scales were supported by recorded strain images with synchrotron radiation [14]. Additional measurements using synchrotron radiation and detecting strain in the vicinity of the domain wall under simultaneous application of external electric fields revealed an extension of the domain wall region to up to 100μm [15].

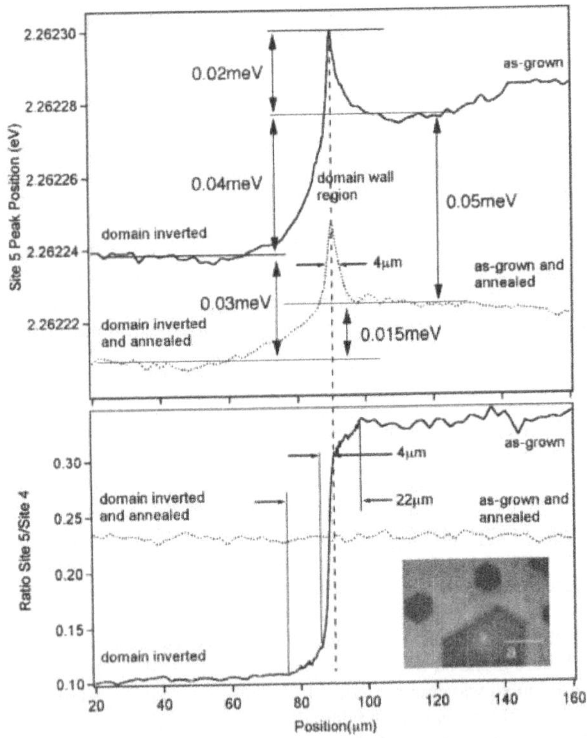

Fig. 5: *Evaluated emission data for a scan over a domain wall regions (see inset) for a nearly stoichiometric sample before and after annealing (5h, 250°C). Top: Spectral position of the emission from site 5 (same spectral range as in Fig. 2 and 3). Bottom: Intensity ratio of main and satellite peak. The data have been collected at 4K.*

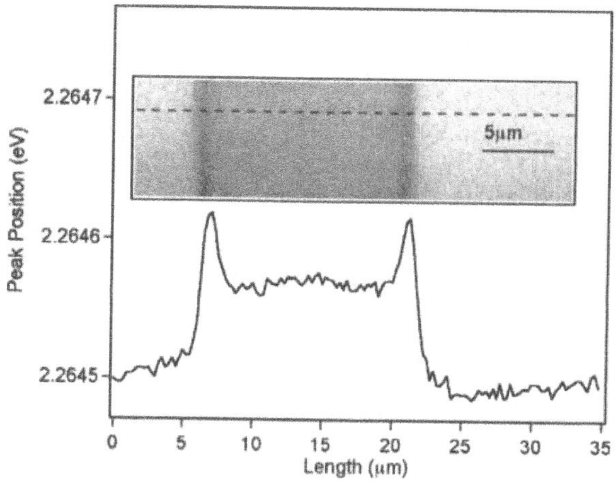

Fig. 6: *Evaluated emission data for a scan over the region of a periodically poled congruent $LiNbO_3$ device. The image in the inset has been obtained using the emission peak positions of area scan. The data have been collected at 4K.*

In our experiments, we find several length scales (all of them are well within our resolution limit):

(1) The sharp peak in the center of the domain wall region stretches over about 4μm.
(2) The transition of the peak positions across the domain wall occurs within about 20μm.
(3) The changes in relative peak height have a short-range component of about 4μm and another one with an extent of about 20μm.

In our interpretation, we assign (1) and (2) to modification of intrinsic defects that are not closely associated with the Er-probe ion and hence just modify the local field the probe ion experiences. The changes in (3), on the other hand, are due to rearrangements of defects close to the probe ions. The similarity of length scales suggests that in both cases the domain wall has a similar effect in the arrangements of the defect dipoles.

Real-time Observation of Moving Domain Wall Regions

The ability to measure spectra rapidly cannot only be exploited for scanning across an area but also to measure dynamic changes at one spot while a voltage above the coercive field is applied to the samples and a ferroelectric domain is growing through that spot. The growth of the domain is monitored simultaneously with a CCD camera such that the velocity of the domain can be determined.

Such real time measurements are performed at room temperature and hence do not allow the same spectral selectivity as the static data presented above. We will use the position of the whole band as measure for the changes that are occurring. In this way, it is not possible to distinguish between changes that are due to changes in nearby or distance intrinsic defects. As we have seen, above, the length scales are similar such that still valuable information can be obtained. First results applying this approach are shown in Fig. 7 for congruent $LiNbO_3$.

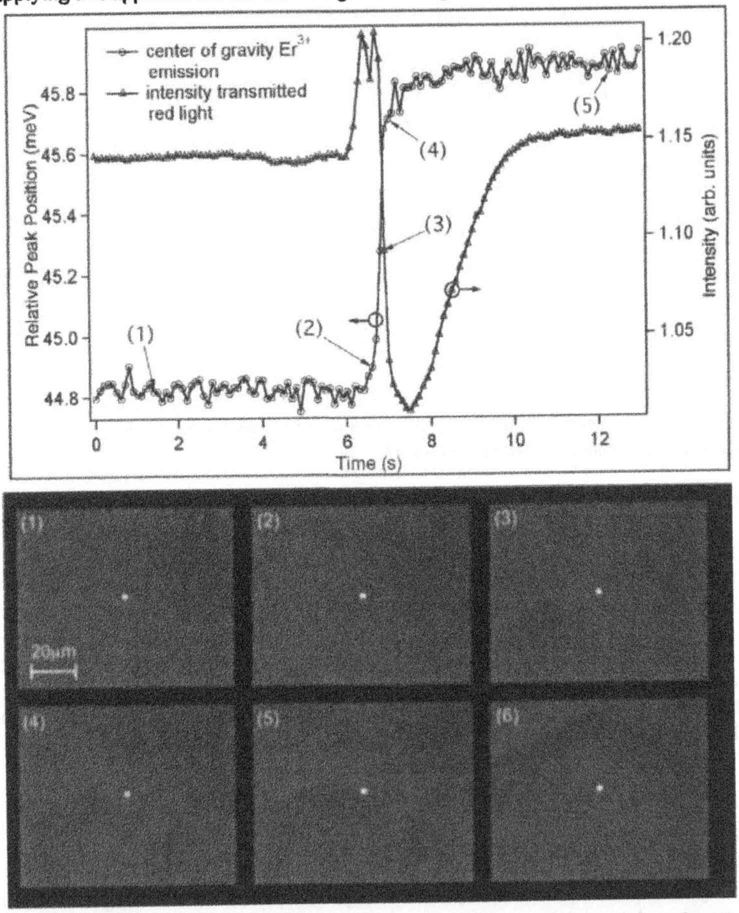

Fig. 7: *Evaluated emission data recorded at the spot as a function of time while a domain wall is moving though it. Top: Relative emission peak position and intensity of transmitted light. Bottom: Time sequence from camera observing the moving domain wall. The white dots indicate the point of observation.*

Along with the emission spectra, we recorded also the changes in the transmitted light giving us indication of the extent of the strain fields. As we can see, the changes in the peak positions and the transmitted light occur on quite different time scales. These first results will need further refinement that will enable us to determine domain wall width dynamically.

SUMMARY and OUTLOOK

We have reviewed our work on static imaging of ferroelectric domain walls and presented first results on real time observation of ferroelectric domain wall motion showing the power of using rare earth ions as the probe for intrinsic electric fields. The technique relies on fast measurement of whole spectra and their evaluation using moments. It can readily be applied to other intrinsic emission processes and Raman scattering. Experiments in that directions have been performed in our lab and show promising results. The method that we introduced is not limited to the spatial (500nm) and temporal resolution (4ms) that we achieved in our experiments. As long as emission intensity is sufficiently high, improvements in both can be achieved. In first experiments, we have already successfully adopted our probing approach to near-field optical microscopy that offers spatial resolution of 100nm. In order to further exploit the atomic scale information contained in our data, we are planning ligand-sensitive measurements such as EPR and ENDOR to further clarify the defect environment around the Er ion.

This work is supported by the Lehigh Center for Optical Technology, NSF (DMR- 0349632) and the ARO (W911NF-04-1-0323).

REFERENCES

[1] M. Yamada, N. Nada, M. Saitoh, and Watanabe, "First-order quasi-phase matched $LiNbO_3$ waveguide periodically poled by applying an external field for efficient blue second-harmonic generation", Appl. Phys. Lett. **62** (5), 435 (1993)

[2] V. Gopalan, T. E. Mitchell, Y. Furukawa, and K. Kitamura, The role of nonstoichiometry in 180 degrees domain switching of LiNbO3 crystals, Appl. Phys. Lett. 72, 1981 (1998).

[3] V. Dierolf, C. Sandmann, S. Kim, V. Gopalan, and K. Polgar, "Ferroelectric domain imaging by defect-luminescence microscopy", J. Appl. Phys. **93**, 2295 (2003)

[4] V. Dierolf and C. Sandmann, "Direct-write method for domain inversion patterns in $LiNbO_3$", Appl. Phys. Lett. **84**, 3987 (2004)

[5] L. Rebouta, M. F. da Silva, J. C. Soares, D. Serrano, E. Dieguez, F. Agullo-Lopez, and J. Tornero, "Nonaxial sites for Er in $LiNbO_3$," Appl. Phys. Lett., vol. 70, pp. 1070–1072, 1997.

[6] V. Dierolf and M. Koerdt, "Combined excitation-emission spectroscopy of Er ions in stoichiometric $LiNbO_3$: The site selectivity of direct and up conversion excitation processes," Phys. Rev. B,, vol. 61, pp. 8043–8052, 2000.

[7] C. C. Battle, S. Kim, V. Gopalan, K. Barkocy, M. C. Gupta, Q. X. Jia, and T. E. Mitchell, "Ferroelectric domain reversal in congruent $LiTaO_3$ crystals at elevated temperatures," Appl. Phys. Lett., vol. 76, pp. 2436–2438, 2000.

[8] S. Kim, V. Gopalan, K. Kitamura, and Y. Furukawa, "Domain reversal and nonstoi-

chiometry in lithium tantalate," J. Appl. Phys., vol. 90, pp. 2949–2963, 2001.

[9]G.Malovichko, V.Grachev, A.Hofstaetter, E.Kokanyan, A.Scharmann, O.Schirmer. "ENDOR study of Cr3+ centers substituting for lithium in lithium niobate". Physical Review, B65, 224116 (2002).

[10]J. Padilla, W. Zhong, and D. Vanderbilt, "First-principles investigation of 180° domain walls in $BaTiO_3$," Phys. Rev. B, vol. 53, pp. 5969–5973, 1996.

[11]J. Wittborn, C. Canalias, K. V. Rao, R. Clemens, H. Karlsson, and F. Laurell, "Nanoscale imaging of domains and domain walls in periodically poled ferroelectrics using atomic force microscopy," Appl. Phys. Lett., vol. 80, pp. 1622–1624, 2002.

[12] M. Drakopoulos, Z. W. Hu, S. Kuznetsov, A. Snigirev, I. Snigireva, and P. A. Thomas, "Quantitative X-ray Bragg diffraction topography of periodically domain-inverted $LiNbO_3$," J. Phys. D, vol. 32, pp. 160–165, 1999.

[13] T. J. Yang and U. Mohideen, "Nanoscale measurement of ferroelectric domain wall strain and energy by near-field scanning optical microscopy," Phys. Lett. A, pp. 205–210, 1998.

[14] S. Kim, V. Gopalan, and B. Steiner, "Direct X-ray synchrotron imaging of strains at 180° degree domain walls in congruent $LiNbO_3$ and $LiTaO_3$ crystals," Appl. Phys. Lett., vol. 77, pp. 2051–2053, 2000.

[15] T. Jach, S. Kim, V. Gopalan, S. Durbin, and D. Bright, "Long-range strains and the effects of applied field at 180 degree ferroelectric domain walls in lithium niobate," Phys. Rev. B, vol. 69, pp. 64113–64121, 2004.

Electroceramic Materials and Applications
Edited by R. W. Schwartz

EFFECTS OF EXCESS BISMUTH ADDITION ON FERROELECTRIC PROPERTIES AND MEMORY CHARACTERISTICS OF $Bi_{3.25}La_{0.75}Ti_3O_{12}$ THIN FILMS FABRICATED BY SPUTTERING

Koji Fukushima, Masafumi Kobune, Toru Yamaji, Jun Yoshizumi, and Tetsuo Yazawa
Graduate School of Engineering, University of Hyogo, Hyogo 671-2201, Japan

ABSTRACT

Bismuth lanthanum titanate (BLT) films with compositions of $\{(1-x)Bi_{3.25}La_{0.75}Ti_3O_{12} + xBi_2O_3\}$, where x=0.28-0.36, were successfully deposited on $Pt(111)/SiO_2/Si(100)$ substrates by rf-magnetron sputtering using a powder target. All the BLT films were confirmed to have a single-phase bismuth-layer perovskite structure without the presence of a secondary phase. The x=0.28 and 0.30 BLT films exhibited rounded grain morphologies with an average grain size of approximately 320 nm. On the other hand, BLT films with excess Bi_2O_3 of $x \geq 0.32$ significantly enhanced grain growth, exhibiting oval grain morphologies with an average grain size of approximately 500 nm × 200 nm. From the results of J-E characteristics, the x=0.30 BLT films showed the highest electrical insulation. The x=0.30 BLT film also showed the best hysteresis loop shape with a remanent polarization of $2P_r=31$ $\mu C/cm^2$ and a coercive field of $2E_c=170$ kV/cm. Measurements of fatigue characteristics showed that all the samples exhibited nearly fatigue-free behavior that resisted degradation even after 1×10^{10} cycles.

INTRODUCTION

Ferroelectric thin films have attracted significant recent attention due to practical applications in nonvolatile ferroelectric random access memories (FeRAMs).[1, 2] A number of materials have been investigated as potential ferroelectric materials for FeRAM device applications. In particular, $Pb(Zr_{1-x}Ti_x)O_3$ (PZT) thin films with various compositions incorporating different Zr/Ti ratios appear to the most promising materials for practical applications, and intensive research has been carried out on these materials. Although PZT has a large remanent polarization (P_r) of 20-90 $\mu C/cm^2$ [3,4], the simplified PZT capacitor structure, which is shown by the Pt/PZT/Pt structure, frequently results in a polarization fatigue phenomenon, wherein a significant decrease in pulse-derived switchable polarization (Q_{sw}) occurs after repetitive read/write cycles.[2,5] To solve the above issue, bismuth layer-structured ferroelectric (BLSF) materials such as $SrBi_2Ta_2O_9$ (SBT) and $Bi_4Ti_3O_{12}$ (BIT) have attracted much attention recently for FeRAM applications. It has been reported that these materials show fatigue-free characteristics up to 1.0×10^{10} cycles[6-8] in the simplified capacitor structure with top and bottom Pt electrodes described above. The crystal structures of BLSF materials are generally described as alternately stacked multiple pseudoperovskite units, which are responsible for the generation of electric dipoles by ionic displacements, and $(Bi_2O_2)^{2+}$ layer units.[3] The $Bi_4Ti_3O_{12}$ (BIT) base material is composed of a triple pseudoperovskite unit sandwiched between $(Bi_2O_2)^{2+}$ layers. The monoclinic BIT unit possesses lattice parameters of a=0.5450, b=0.54059, and c=3.2832 nm,[9] and exhibits spontaneous polarizations (P_s) of 50 and 4 $\mu C/cm^2$, along the a-and c-axes, respectively.[10,11] In films with the BIT-type structure, therefore, a large ferroelectric polarization can be expected by the appearance of the orientation which tilts even a little faces to an *a*-axis direction. Thus far, although the fabrication of the BLSF films has been demonstrated using sol-gel[12-14], pulsed laser deposition (PLD)[6,15] and metalorganic chemical vapor deposition (MOCVD) [16,17] processes, there have been hardly any reports on the fabrication as well as ferroelectric and memory characteristics of $Bi_{3.75}La_{0.75}Ti_3O_{12}$ (BLT) films formed by sputtering. Moreover, with respect to other BLSF materials such as $Sr_{0.8}Bi_xTa_{1.2}Nb_{0.9}O_{9+y}$ and Nd-doped $Bi_4Ti_3O_{12}$, Tsai et al.[18] and Gao et al.[19] have previously reported that excess Bi_2O_3 addition strongly affects the structure and physical properties of thin films derived from sputtering.

In the present study, BLT films with excess Bi_2O_3 were fabricated on Pt(111)/SiO_2/Si(100) substrates by rf-magnetron sputtering. The ferroelectric properties (polarization-electric field measurements) and memory characteristics (J-E characteristics and fatigue tests) of the fabricated Pt/BLT/Pt capacitors were investigated, and the optimal excess Bi_2O_3 content is discussed based on the previous physical properties.

EXPERIMENTAL PROCEDURE

Bismuth lanthanum titanate (BLT) films with compositions of $\{(1\text{-}x)Bi_{3.25}La_{0.75}Ti_3O_{12} + xBi_2O_3\}$ were deposited on Pt(111)/SiO_2/Si(100) substrates by high-temperature rf-magnetron sputtering using a powder target. The target powder was calcined in air at 750°C for 2 h. The excess Bi_2O_3 content in this target was varied in the range of 28-36 mol%. The thickness of the BLT film was approximately 300 nm. Top and bottom Pt electrodes with respective thicknesses of 100 and 200 nm were similarly fabricated by rf-magnetron sputtering. Top Pt electrodes were patterned by the photolithographic technique. The sputtering conditions used for fabrication of the BLT films, and the top and bottom Pt electrodes are summarized in Table I. The crystalline phases, the orientation and the crystallinity of the deposited BLT films were investigated by X-ray diffractometry (XRD; RINT2200, Rigaku). Energy-dispersive X-ray fluorescence (XRF) analysis was employed for chemical analysis of the deposited BLT films. The polarization-electric field (P-E) hysteresis loops and polarization fatigue characteristics were measured using a ferroelectric film test system (RT-66A, Radiant Tech.) and a function/arbitrary waveform generator (33120A, Hewlett-Packard). Polarization fatigue tests were performed using bipolar voltage pulses with amplitudes of 4.6-6.4 V at 500 kHz. The dc-voltage offset calculated from the measurement of the P-E hysteresis loop was set so as to equalize the positive and negative fields applied to the sample. The leakage current density-applied voltage (J-E) characteristics were measured in the applied field range of ±50 to ±800 kV/cm at 50 kV/cm steps and 5 s intervals using a programmable electrometer (617, Keithley). The surface morphology of the BLT films was characterized by atomic force microscopy (AFM).

Table I. Sputtering conditions for fabricating BLT films, and top and bottom Pt electrodes.

Item	BLT	Bottom Pt electrode	Top Pt electrode
Target	*BLT powder	Pt	Pt
Substrate temp. (°C)	600	214	R.T.
Input power density (W/cm^2)	1.2	0.3	1.2
Sputtering gas (Ar/O_2)	90/10	100/0	50/50
Gas pressure (Pa)	0.4	0.4	0.5
Deposition rate (nm/min)	6.3	7.2	100

*BLT powder target; $\{(1\text{-}x)Bi_{3.25}La_{0.75}Ti_3O_{12} + xBi_2O_3\}$
x=0.28, 0.30, 0.32, 0.34, 0.36

RESULTS AND DISCUSSION

Figure 1 shows the XRD patterns of BLT films with excess Bi_2O_3 deposited on Pt(111)/SiO_2/Si(100) substrates by rf-magnetron sputtering. All the BLT films were confirmed to have a single-phase bismuth-layer perovskite structure without an apparent secondary phase being detected. Moreover, the analyzed values of BLT films by the XRF measurements coincide well with the design values within the accuracy of measurement, indicating that any compositional fluctuation of the BLT films fabricated in the present study is negligible. A change in crystallinity and orientation of the BLT films with increasing excess Bi_2O_3 content was also ascertained. The x=0.30 sample had almost a preferential (117) orientation. However, samples with x≥0.32 exhibited a tendency for the peak intensity of (00l), (200) and (220) reflections to increase. The excess Bi_2O_3 content dependence of the full-width at half maximum (FWHM) of a rocking curve of the BLT(117) reflection is shown in Fig. 2. The value of FWHM, $\Delta\theta$, at x=0.32 was 3.7°, exhibiting excellent crystallinity, whereas those with excess Bi_2O_3 content of x≤0.30 and x≥0.34 increased up to 4.1-9.0°. In particular, the FWHM at x=0.36 exhibited a maximum ($\Delta\theta$=9.0°). These results suggest that the crystallinity and the orientation of the fabricated films are strongly affected by the excess Bi_2O_3 addition. It is considered that the excess Bi_2O_3 of 0.30≤x≤0.32 successfully compensates for the Bi constituent which is volatilized from the film surface during the fabrication of the films, yielding an improvement in the crystallinity of the films. In contrast, the crystallinity of the films with excess Bi_2O_3 of x≥0.34 declined significantly. This could be attributed to the existence of excess Bi_2O_3 taken in the BLT crystal lattice

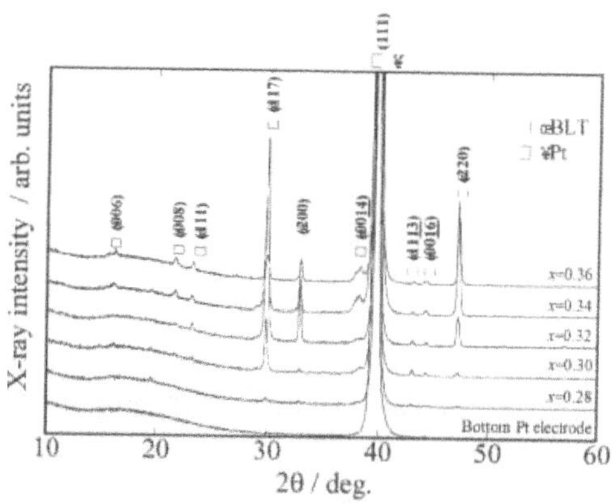

Fig. 1. XRD patterns of BLT films with compositions of {(1-x)$Bi_{3.25}La_{0.75}Ti_3O_{12}$ + xBi_2O_3}, where x=0.28-0.36 deposited on Pt bottom electrode with $\Delta\theta$=3.0°.

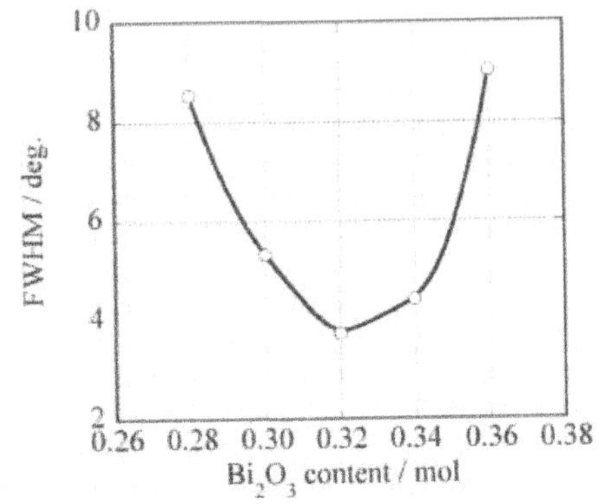

Fig. 2. Excess Bi_2O_3 content dependence of the full-width at half maximum of a rocking curve of the BLT(117) reflection.

Morphological changes of the BLT films were analyzed using AFM over an area of 2 μm × 2 μm. AFM images of BLT films fabricated by varying the excess Bi_2O_3 content (x=0.28-0.36) are shown in Fig. 3. The x=0.28 and 0.30 BLT films exhibited rounded grain morphologies with an average grain size of approximately 320 nm. Furthermore, BLT films with excess Bi_2O_3 of x≥0.32 showed significant enhancement in grain growth, exhibiting oval grain morphologies with an average grain size of approximately 500 nm × 200 nm. These results demonstrate that the surface morphology of BLT films is significantly affected by the excess Bi_2O_3 content.

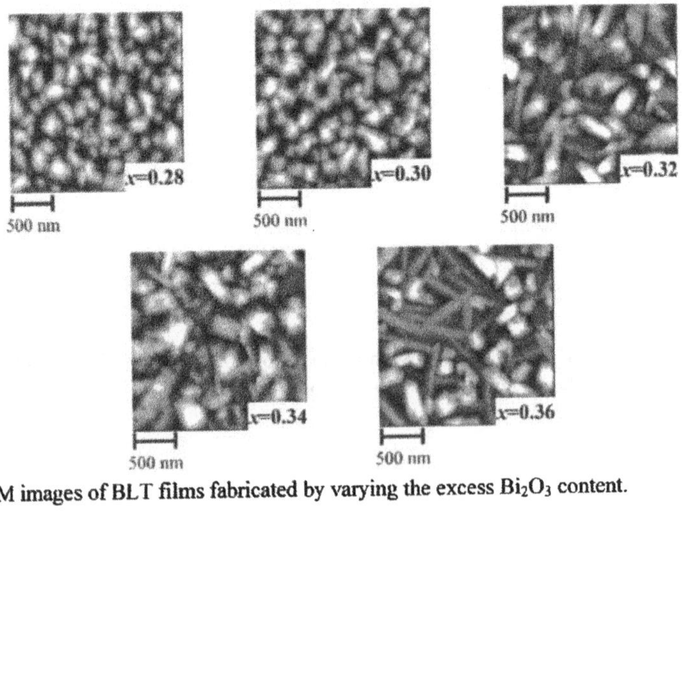

Fig. 3. AFM images of BLT films fabricated by varying the excess Bi_2O_3 content.

Leakage current density-electric field (J-E) characteristics of BLT samples with different excess Bi_2O_3 content are shown in Fig.4. The J-E characteristics of the x=0.30-0.36 films indicated good symmetrical behavior. Since the BLT films with various excess Bi_2O_3 content, except for x=0.30, show almost similar J-E characteristics, these samples have the level electrical insulation. On the other hand, the x=0.30 BLT films apparently show the best electrical insulation among the four x values. This result suggests that the excess Bi_2O_3 compensates for A-site vacancies that form due to the volatilization of Bi which results in leakage current. As another reason for its high electric insulation, the x=0.30 film is considered to have a preferential (117) orientation showing relatively high electrical insulation in the orientations that bismuth-layer-structured ferroelectrics (BLSF) can be expected to take.[20] In addition, Bi_2O_3 is well known as a good conductor of electricity, yielding high electrical conductivity with increasing temperature.[21] Therefore, the excess Bi_2O_3 becomes a leak path as the excess Bi_2O_3 content increases. The appearance of (200) and (220) orientations showing low electrical insulation results in the deterioration of the electrical insulation.

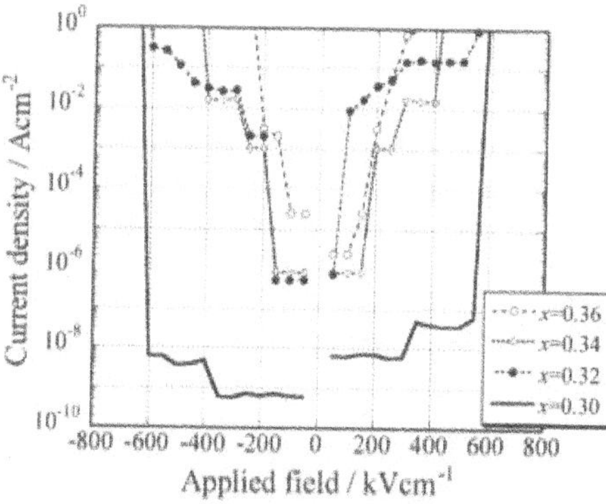

Fig.4. Leakage current density-electric field (J-E) characteristics of BLT samples with different excess Bi_2O_3 contents.

Figure 5 shows the P-E hysteresis loops for the x=0.30-0.36 BLT samples. All the BLT samples exhibited comparatively good symmetric hysteresis loop shapes with a remanent polarization of $2P_r$=14-31 μC/cm^2 and a coercive field of $2E_c$=123-170 kV/cm. In particular, the x=0.30 BLT film yielded the best hysteresis loop shape with a remanent polarization of $2P_r$=31 μC/cm^2 and a coercive field of $2E_c$=170 kV/cm. On the contrary, the three BLT films with x=0.32, 0.34 and 0.36 exhibited unfavorable characteristics with a relatively small $2P_r$=14-16 μC/cm^2 and a relatively large $2E_c$=123-170 kV/cm for ferroelectric applications. Moreover, as can be easily seen from the XRD results, the P-E hysteresis loops of the x=0.28 BLT film almost exhibited paraelectric behavior because of its amorphous nature. These results may be explained by presuming that Bi deficiency results in the amorphous structure, whereas sufficient Bi_2O_3 addition produces a deterioration in the crystallization temperature, as has been reported in detail elsewhere.[17, 22] Based on these results, the optimal excess Bi_2O_3 content is determined to be x=0.30.

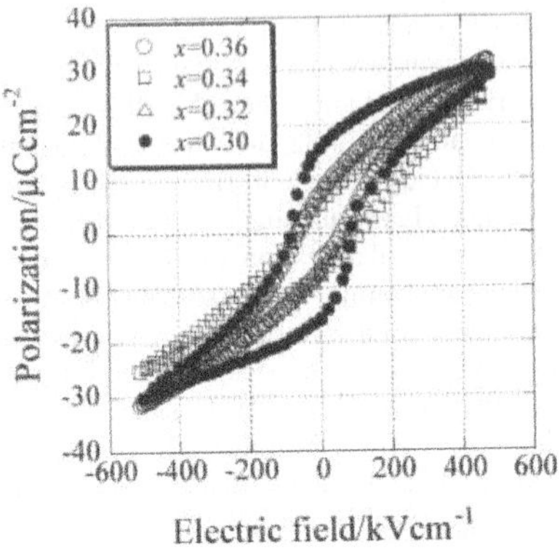

Fig. 5. P-E hysteresis loops of BLT thin films with compositions of {(1-x)$Bi_{3.25}La_{0.75}Ti_3O_{12}$ + xBi_2O_3, where x=0.30-0.36.

Figure 6 shows the polarization fatigue characteristics of the BLT films with different values of excess Bi_2O_3 content as a function of the switching cycles. The experimental data in the present study were measured using an applied electric field of E=2.5E_c. All the samples exhibited almost fatigue-free behavior that resisted degradation even after 1×10^{10} cycles. These results can be explained by the fact that the added excess Bi_2O_3 compensates sufficiently for Bi vacancies which is the main cause of fatigue.

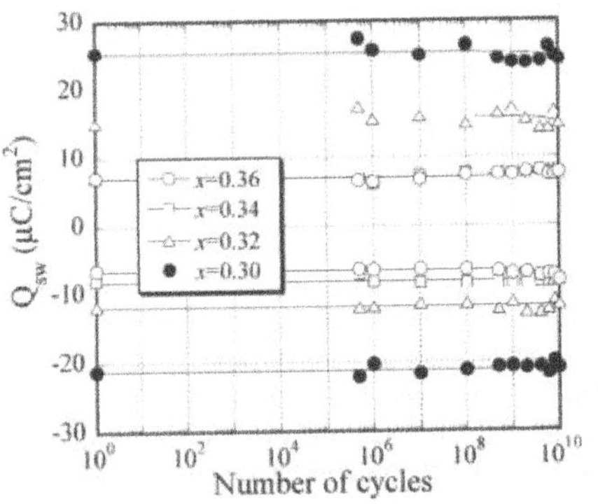

Fig. 6. Polarization fatigue characteristics of the BLT films with different excess Bi_2O_3 contents.

CONCLUSIONS

Bismuth lanthanum titanate films with compositions of $\{(1-x)Bi_{3.25}La_{0.75}Ti_3O_{12} + xBi_2O_3\}$, where x=0.28-0.36, were successfully deposited on $Pt(111)/SiO_2/Si(100)$ substrates by rf magnetron sputtering using a powder target. The results of this study can be summarized as follows.

(1) All the BLT films were confirmed to have a single-phase bismuth-layer perovskite structure without the presence of a secondary phase.

(2) The x=0.28 and 0.30 BLT films exhibited rounded grain morphologies with an average grain size of approximately 320 nm. The BLT films with excess Bi_2O_3 of $x \geq 0.32$ yielded significantly enhanced grain growth, exhibiting oval grain morphologies with an average grain size of approximately 500 nm × 200 nm.

(3) From the results of J-E characteristics, the x=0.30 BLT films yielded the best electrical insulation properties.

(4) The x=0.30 BLT film showed the best hysteresis loop shape with a remanent polarization of $2P_r=31\ \mu C/cm^2$ and a coercive field of $2E_c=170$ kV/cm.

(5) Measurements of fatigue characteristics indicated that all the samples exhibited nearly fatigue-free behavior that resisted degradation even after 1×10^{10} cycles.

REFERENCES

[1] J. F. Scott and C. A. Paz de Araujo, "Ferroelectric Memorys," *Science*, **246**, 1400-1405 (1989).

[2] C. A. Paz de Araujo, J. D. Cuchiaro, L. D. McMillan, M. C. Scott and J. F. Scott, "Fatigue-free ferroelectric capacitors with platinum electrodes," *Nature*, **374**, 627-629 (1995).

[3] B. H. Park, B. S. Kang, S. D. Bu, T. W. Norh, J. Lee and W. Jo, "Lanthanum-Substituted for use in non-volatile memories," *Nature*, **401**, 682-684 (1999).

[4] H. Morioka, G. Asano, T. Oikawa, and H. Funakubo, "Large remanent polarization of 100% polar-axis-oriented epitaxial tetragonal $Pb(Zr_{0.35}Ti_{0.65})O_3$ thin films" *Appl. Phys. Lett.*, **82**, 4761-4763 (2003).

[5] K. Fukushima, M. Kobune, Y. Nishioka, T. Yamaji, T. Yazawa, H. Fujisawa and, M.Shimizu, "Invesitigation of Annealling Programs by Hot Isostatic Pressing for Crystallization of Amorphous $Pb(Zr_{0.3}Ti_{0.7})O_3$ Thin Films," *Trans. Mater. Res. Soc. Jpn.*, **30**, 97-100 (2005).

[6] J.-K. Lee, C.-H. Kim, H.-S. Suh, and K.-S. Hong, "Correlation between internal stress and ferroelectric fatigue in $Bi_{4-x}La_xTi_3O_{12}$ thin films," *Appl. Phys. Lett.*, **80**, 3593-3595 (2002).

[7] U. Chon, G.-C. Yi, and H. M. Jang, "Fatigue-free behavior of highly oriented $Bi_{0.35}La_{0.75}Ti_3O_{12}$ thin films grown on $Pt/Ti/SiO_2/Si(100)$ by metalorganic solution decomposition," *Appl. Phys. Lett.*, **78**, 658-660 (2001).

[8] D. Bao, T.-W. Chiu, N. Wakiya, K. Shinozaki, and N. Mizutani, "Structural and electrical characteristics of chemical-solution-derived $(Bi,La)_4Ti_3O_{12}$ thin films with various Bi_2O_3 template layers" *J. Appl. Phys.*, **93**, 497-503 (2003).

[9] A. D. Rae, J. G. Thompson, R. L. Whithers, and A. C. Willis, "Structure refinement of commensurately modulated bismuth titanate, $Bi_4Ti_3O_{12}$," *Acta Crystallogr. Sect. B*, **46**, 474-487 (1990).

[10] H. Irie, M. Miyayama, and T. Kudo, "Structure dependence of ferroelectric properties of bismuth layer-structured ferroelectric single crystals," *J. Appl. Phys.* **39**, 4089-4094 (2001).

[11]M. Soga, Y. Noguchi, M. Miyayama, H. Okino, and T. Yamamoto, "Domain structure and polarization properties of lanthanum-substituted bismuth titanate single crystals," *Appl. Phys. Lett.*, **84**, 100-102 (2004).

[12]N. Sugita, M. Osada, and E. Tokumitsu, "Characterization of Sol-gel Derived $Bi_{4-x}La_xTi_3O_{12}$ Films," *Jpn. J. Appl. Phys.*, **41**, 6810-6813 (2002).

[13]R. E. Melgarejo, and M.S. Tomar, "Large ferroelectric response in Bi4-xNdxTi3O12 films prepared by sol-gel process," *Appl. Phys. Lett.*, **81**, 2611-2613 (2002).

[14]N. Sugita, and E. Tokumitsu, "Crystallization of Bi4-xLaxTi3O12 films prepared by the sol-gel technique on IrO2/Ir multi-layered electrode," *Trans. Mater. Res. Soc. Jpn.*, **28**, 153-156 (2003).

[15]J. S. Zhu, D. Su, X. M. Lu, H. X. Qin, Y. N. Wang, D.Y. Wang, H. L. W. Chan, K. H. Wong, and C. L. Choy, "La-doped effect on the ferroelectric properties of $Bi_4Ti_3O_{12}$-$SrBi_4Ti_4O_{15}$ thin film fabricated by pulsed laser deposition," *J. Appl. Phys.*, **92**, 5420-5424 (2002).

[16]T. Kojima, T. Watanabe, and Hiroshi Funakubo, "Ferroelectric properties of lanthanide-substituted $Bi_4Ti_3O_{12}$ epitaxial thin films grown by metalorganic chemical vapor deposition," *J. Appl. Phys.*, **93**, 1707-1712 (2003).

[17]T. Higuchi, M. Nakamura, Y. Hachisu, M. Saitoh, T. Hattori, and T. Tsukamoto, "Ferroelectric Properties of $Bi_4Ti_3O_{12}$ Thin Films Prepared on TiO_2 Anatase Layer," *Jpn. J. Appl. Phys.*, **43**, 6585-6589 (2004).

[18]H.-M. Tsai, P. Lin, and T.-Y. Tseng, "Effect of bismuth content on the properties of $Sr_{0.8}Bi_xTa_{1.2}Nb_{0.9}O_{9+y}$ ferroelectric thin films," *J. Appl. Phys.*, **85**, 1095-1099 (1999).

[19]X. Gao, Z. Zhou, J. Xue, and J. Wang, "Effects of Excess Bi_2O_3 on the Ferroelectric Behavior of Nd-Doped $Bi_4Ti_3O_{12}$ Thin Films," *J. Am. Ceram. Soc.*, **88**, 1037-1040 (2005).

[20]A. Fouskova, and L. E. Cross, "Dielectric Properties of Bismuth Titanate," *J. Appl. Phys.*, **41**, 2834-2838 (1970).

[21]"bismuth oxide," *Rikagakujiten* (Dictionary of physics and chemistry), 515 (1978) [in Japanese].

[22] T. Hayashi, D. Togawa, W. Sakamoto, and S. Hirano, "Effects of Bi_2O_3 seeding layer on crystallinity and electrical properties of CSD-derived $Bi_{4-x}La_xTi_3O_{12}$ ferroelectric thin films," *J. Eur. Ceram. Soc.*, **24**, 1621-1624 (2004).

Electroceramic Materials and Applications
Edited by R. W. Schwartz

SCANNING PROBE METHODS FOR CHARACTERIZATION OF ELECTRICAL PROPETIES IN NANO-MATERIALS

Takuji Takahashi
Institute of Industrial Science, University of Tokyo
4-6-1 Komaba, Meguro-ku,
Tokyo 153-8505
Japan

ABSTRACT

We present Kelvin probe force microscopy (KFM) for accurate measurements of surface potential as well as magnetic force microscopy (MFM) for non-contact and non-destructive evaluation of current in fine structures.

As a new approach in KFM, we have proposed and demonstrated an intermittent bias application method with a sampling extraction of the electrostatic force, in which electrostatic force variation due to an oscillation of KFM cantilever are avoidable and very sensitive extraction of the electrostatic force are realized. Owing to them, circular potential distribution around self-assembled InAs quantum dots was experimentally observed, and the results suggested that the achieved spatial resolution was better than 10 nm.

To realize, on the other hand, the non-contact and non-destructive measurement of current, we have investigated MFM detection of current-induced magnetic fields. The magnetic field around a current path of a few hundred nm in width was correctly observed when we appropriately eliminated an influence of the electrostatic force by adjusting the tip potential. We confirmed that the spatial resolution in this method was better than 200 nm.

INTRODUCTION

Nano-materials are very promising for future device application. In order to characterize their electrical properties, we need special equipments with high spatial resolution at nano-meter scale. Since scanning probe methods are suited for such a purpose, we have examined several scanning probe methods as the characterizing tools of the electrical properties.

Among them, surface potential determination by Kelvin probe force microscopy (KFM)[1] and current evaluation by magnetic force microscopy (MFM) will be presented in this paper. In the former, we are aiming at accurate potential measurements by introducing both intermittent bias application method[2] to suppress variation of the electrostatic force and sampling extraction method[3] to achieve sensitive detection of the electrostatic force because the electrostatic force plays a very important role for determining the surface potential. Actually the surface potential measured around self-assembled InAs quantum dots indicates that the stable potential measurements of good quality were realized and that the achieved spatial resolution was better than 10 nm. In the latter one, on the other hand, current-induced magnetic fields around fine current paths have been investigated by MFM. When unnecessary electrostatic force acting between the MFM tip and the current path was adequately suppressed[4-7], the current-induced magnetic fields were successfully detected. The magnetic fields observed around T-junction of current paths or around one pair of parallel paths showed the validity of this method, and they also indicated that the estimated spatial resolution was better than 200 nm.

SURFACE POTENTIAL DETERMINATION BY KELVIN PROBE FORCE MICROSCOPY

By means of KFM, which is based on AFM, surface potential of the nano-materials can be studied in high spatial resolution. In KFM, electrostatic force acting between the KFM tip and the sample is used to determine the surface potential, but this electrostatic force has a nature as a long-range force. Since the actual KFM system is normally operated in an intermittent contact mode of AFM, in which the cantilever is mechanically oscillated for controlling the tip height, we should pay attention to the variation of the electrostatic force due to this cantilever oscillation to realize the accurate potential determination. We previously investigated the influence of the variation of electrostatic force through both experiments[8,9] and numerical simulations[9,10].

To realize more accurate KFM measurements, we propose an intermittent bias application method to suppress the electrostatic force variation due to the cantilever oscillation as well as a sampling extraction method to achieve very sensitive detection of the electrostatic force[2,3].

Principle of Intermittent Bias Application Method

The principle of the intermittent bias application method we propose is illustrated in Fig. 1. The cantilever in KFM is mechanically driven at a frequency of f_1 for observing topography on the sample surface as shown in Fig. 1(a). In the conventional KFM, a sinusoidal ac bias, which has a different frequency f_2 from f_1, is applied between the sample and the tip to generate the electrostatic force and the f_2 component in the electrostatic force is nullified by adjusting dc offset bias additionally applied to determine the contact potential difference[1]. In the intermittent bias application method, on the other hand, we use spiky biases as illustrated in Fig. 1(b) instead of the sinusoidal waveform. Since these spiky biases have a sinusoidal envelope at a frequency of f_2, it is possible to extract the f_2 component in the electrostatic force, and from this component we can adjust the dc offset bias in the same way as in the conventional KFM.

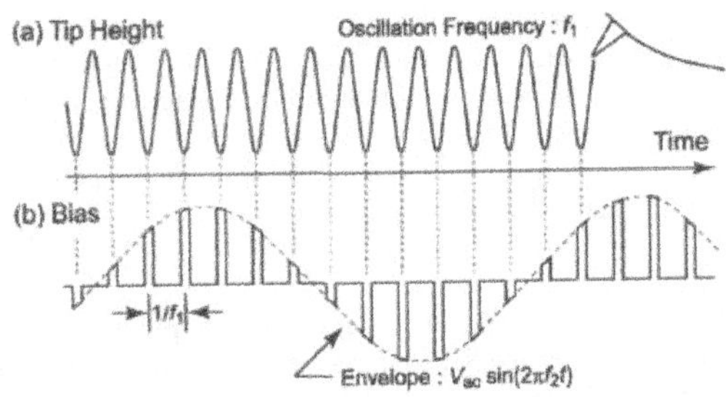

Fig. 1 Schematic illustration of (a) tip movement during cantilever oscillation in KFM and (b) spiky biases intermittently applied synchronized with the cantilever oscillation. An envelope of the spiky biases is coincident with a sinusoidal waveform to be used for extracting the cantilever bending due to electrostatic force.

When a phase, that is, a timing of bias application, is properly tuned, we can realize the situation that the spiky biases are applied at exact moments when the tip reaches the closest position to the sample surface during the oscillation. At those moments, the electric field between the tip and the sample is mostly concentrated beneath the tip apex[8], and therefore the improvement in the spatial resolution is expectable. Moreover, the electrostatic force used for potential measurements acts at a equivalently constant tip height, and therefore the variation in the electrostatic force due to the cantilever oscillation can be suppressed.

Principle of Sampling Extraction of Electrostatic Force

Figure 2 shows a model of a sample-and-hold (S/H) operation[11,12]. The cantilever deflection signal, which includes both a mechanical vibration component (for topography) and an electrostatic force component (for potential), is transmitted into the signal input of the S/H circuit. At the same time, the trigger signal synchronized with the tapping frequency f_1 is sent to the control input of the S/H circuit. If f_2 is appropriately chosen to be much lower than f_1, a trace of the clipped signals includes an f_2 component due to the electrostatic force, as shown in Fig. 2(a). Then, this f_2 component is extracted by a lock-in amplifier and used for potential determination. Note here that we can clip the deflection signal at any desired moment in the oscillation period by tuning a time delay for sampling, that is, sampling phase θ, as shown in Fig. 2(b).

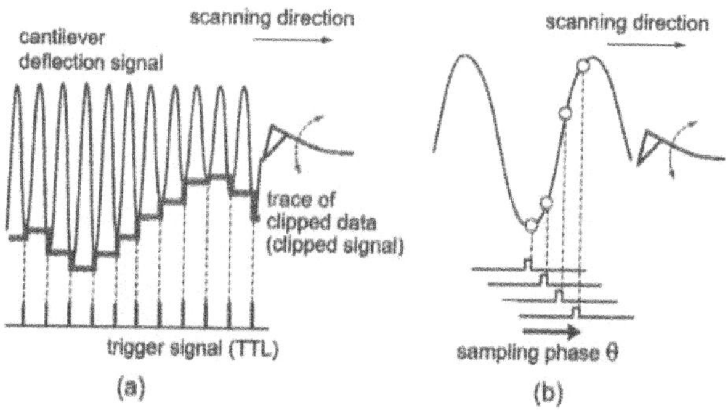

Fig. 2 (a) Model of a sample-and-hold (S/H) operation and (b) function of sampling phase θ. The cantilever deflection signal is clipped by the trigger signal synchronized with the tapping frequency f_1. A trace of the clipped signals includes an f_2 component due to the electrostatic force. By tuning θ, the deflection signal can be clipped at any desired moment in the oscillation period.

Prior to the potential measurements, we experimentally searched the optimum sampling phase suited for the sensitive detection of the electrostatic force. Figures 3(a) and 3(b) show the f_2 components of cantilever bending, which represent the electrostatic forces acting on the cantilevers with the sharp tip and the dull tip, respectively, as a function of the sampling phase. Here, the sampling phase θ corresponds to the tip position where the deflection signal was clipped. For example, θ = 0 and 180° represent the bottom and top positions in the cantilever oscillation, respectively. Although the details of the features appearing in this figure are discussed elsewhere[2], the most important feature is that very sensitive extraction of the electrostatic force is realized when we detect the cantilever bending at θ = 90° regardless of the tip curvature. Therefore the sampling phase is fixed at 90° in the following experiments presented in this section.

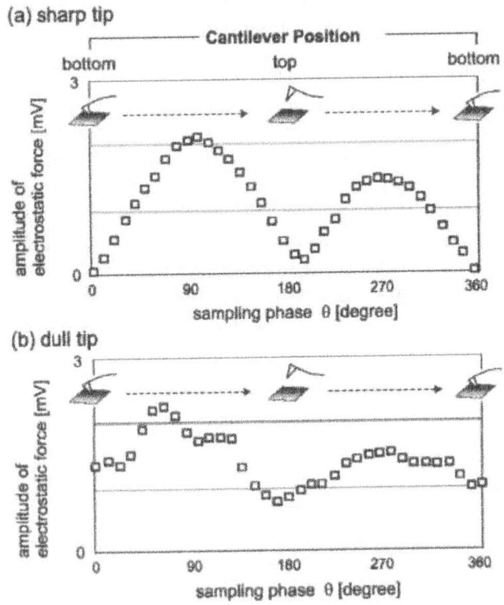

Fig. 3 Extracted electrostatic force at a frequency of f_2 at various sampling phases θ, obtained with (a) sharp tip and (b) dull tip. The sampling phase corresponds to the tip position where the deflection signal was clipped; θ = 0, 90, 180 and 270° represent the bottom, intermediate, top and next intermediate positions in the cantilever oscillation, respectively. The obtained value of the f_2 component became the largest at the intermediate position after tapping (θ = 90°).

Potential Measurements around InAs Quantum Dots

Our KFM system to realize the intermittent application of bias and the sampling extraction of electrostatic force is schematically illustrated in Fig. 4. A function generator which drives the oscillatory motion of cantilever also gives two trigger signals for the intermittent bias application and the electrostatic force sampling as well as the reference signal to detect the cantilever oscillation amplitude for controlling the tip height in the same way as the conventional AFM. We used a homemade controller for the potential feedback[13].

We performed surface potential measurements around self-assembled InAs quantum dots grown on an n-type (001) GaAs substrate, where three-dimensional dot structures were formed due to lattice mismatch between InAs and GaAs[14-16]. Here, the InAs dots themselves and an underlying GaAs buffer layer of 270 nm in thickness were undoped, while carrier density in the GaAs substrate was 2×10^{18} cm^{-3}. Owing to both the thick undoped buffer layer and the high carrier density in the substrate, we consider that the lateral undulation of the depletion zone in the n-type GaAs substrate due to the existence of the InAs dots on the top surface is negligibly small.

In our KFM, we used the Si cantilever coated with Pt whose spring constant was about 2 N/m, and its tip curvature was around 20 nm. Typical parameters for f_1, f_2, and V_{ac} were 75 kHz, 1.5 kHz, and 0.5 $V_{p\text{-}p}$, respectively. The whole measurements were performed in dry N_2 gas at 1 atm and at room temperature. Figure 5(a) shows a surface topography of the sample, and Figs. 5(b) and 5(c) are potential images sequentially obtained by the intermittent bias method and the normal sinusoidal bias method, respectively. As shown in this figure, quality of the potential image taken with the intermittent bias is improved very well, and circular distribution of potential around the dots are clearly recognized. This improvement is attributable to both suppression of the electrostatic force variation and concentration of the electric field just beneath the tip end owing to the intermittent application of bias at the bottom position in the cantilever oscillation.

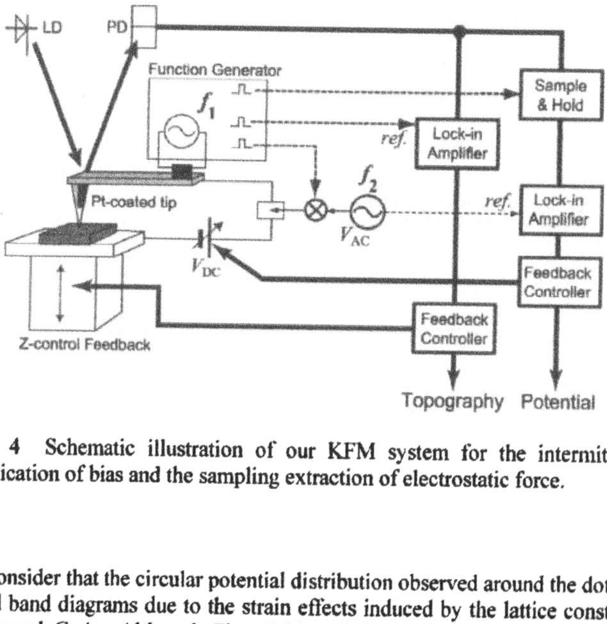

Fig. 4 Schematic illustration of our KFM system for the intermittent application of bias and the sampling extraction of electrostatic force.

We consider that the circular potential distribution observed around the dot resulted from the modulated band diagrams due to the strain effects induced by the lattice constant difference between InAs and GaAs. Although Figs. 5(b) and 5(c) show slight asymmetry in potential distribution around individual dots, we attribute it to the asymmetry of metal coating on the KFM tip and/or the asymmetry of the tip shape itself. Moreover, we confirmed that the potential images became almost featureless when we reduced the modulation bias amplitude V_{ac} down to several mV, and therefore it is considered that the residual effect on the potential measurements due to topography is negligible.

In order to discuss the spatial resolution, on the other hand, we took the line profile of the potential image observed by the intermittent bias method as shown in Fig. 6 along a solid line indicated in the inset. From this figure, very abrupt change within 10 nm was recognized, and therefore the achieved spatial resolution by this method is estimated to be better than 10 nm. This value is much smaller than the oscillation amplitude of the cantilever (typically, 200 nm in peak-to-peak), and is very close to the value of tip curvature (typically, 20 nm). We believe this is owing to the sufficient suppression of the electrostatic force variation, and then the tip curvature determines the spatial resolution.

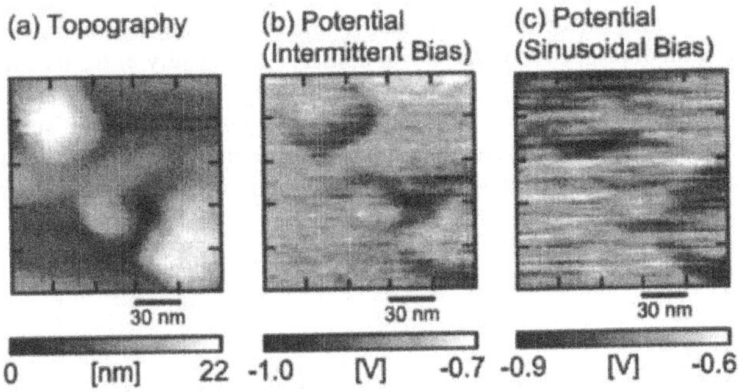

Fig. 5 (a) Topographic image around self-assembled InAs quantum dots, and potential images sequentially observed (b) by the intermittent bias method and (c) by the conventional sinusoidal bias method. Clear potential distribution was obtained in (b) owing to the suppression of the electrostatic force variation compared with (c).

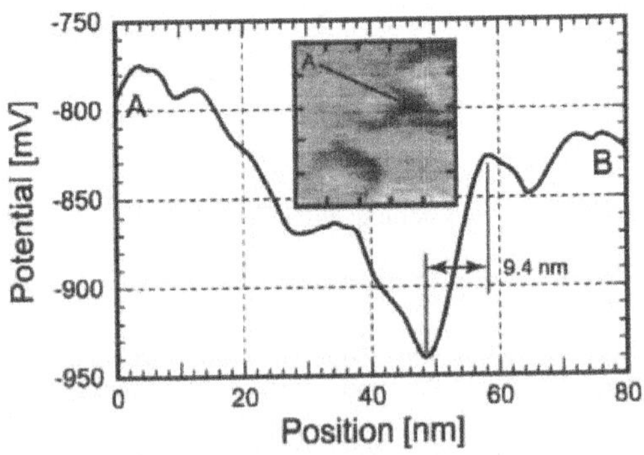

Fig. 6 Line profile of potential distribution along a solid line A-B in the inset obtained by the intermittent bias method. Very abrupt change within 10 nm was recognized, suggesting that the spatial resolution in this method is better than 10 nm.

Summary

We have proposed the intermittent bias application method to suppress the electrostatic force variation due to the oscillatory motion of the tip as well as the sampling extraction method for very sensitive detection of electrostatic force in KFM. By those methods. we observed the potential distribution around the InAs quantum dots, and the result indicates that the quality of the potential images were improved very well compared with the conventional KFM and that the estimated spatial resolution in potential measurements was better than 10 nm.

CURRENT EVALUATION THROUGH MAGNETIC FIELD DETECTED BY MAGNETIC FORCE MICROSCOPY

Current evaluation in individual paths of current networks consisting of, for instance, molecular wires, quantum wires, and carbon nanotubes will become important as decreasing device sizes. For such a purpose, nondestructive ammeter with high spatial resolution as well as high sensitivity is strongly required. As one candidate for such an ammeter, we have proposed and demonstrated a novel operation method of MFM, by which current-induced magnetic fields around fine current paths were detected simultaneously with topography in very high spatial resolution[4-7]. We have examined to detect the magnetic field gradients around the fine current paths by MFM, and have investigated the validity of this method as well as its spatial resolution.

Methods for Current-induced Magnetic Field Detection

Our MFM system is based on the intermittent contact mode of AFM as illustrated in Fig. 7. The cantilever displacement is monitored by an optical-beam-deflection sensor, and a vertical displacement signal due to a mechanical vibration is used for the height control of the tip. On the other hand, the MFM cantilever is deformed due to the magnetic field induced by the ac current flowing through the current path. From a torsional displacement of the MFM cantilever synchronized with the ac current, magnetic force signals are extracted as their amplitudes (R) and phase differences (θ) by a lock-in technique. Owing to this setup, we can obtain magnetic force images simultaneously with a topography. We strongly emphasize here that our MFM is sensitive to the gradients of the magnetic field in a perpendicular direction to a cantilever beam in principle. It is also important for the correct detection of the magnetic force that unnecessary effects from the electrostatic force acting between the MFM tip and the sample should be eliminated by adjusting the offset bias by the potential feedback which is used in Kelvin probe force microscopy[1,8].

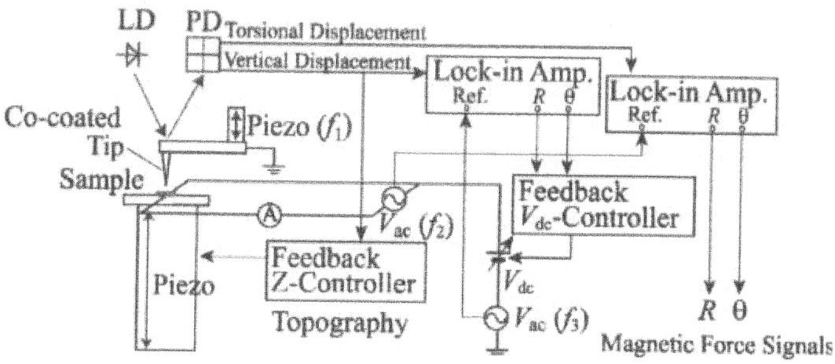

Fig. 7 Experimental setup for current-induced magnetic force detection method by magnetic force microscopy.

We used the Si cantilever whose spring constant was about 3 N/m, and the tip was coated with Co. To control the magnetic interaction between the tip and the magnetic field intentionally, the tip was magnetized by a permanent magnet prior to each measurement. The MFM measurements were performed in dry N_2 gas at 1 atm and at room temperature.

We fabricated two samples of current paths as illustrated in Fig. 8, which consist of double metal layers (Au on Ti) fabricated by electron beam lithography and lift-off processes. The path widths in the sample in a T-junction configuration shown in Fig. 8(a) are 200, 300 and 400 nm for upper, lower and left-hand paths, respectively, while two parallel paths of 500 nm in width have a 200 nm gap in between as shown in Fig. 8(b).

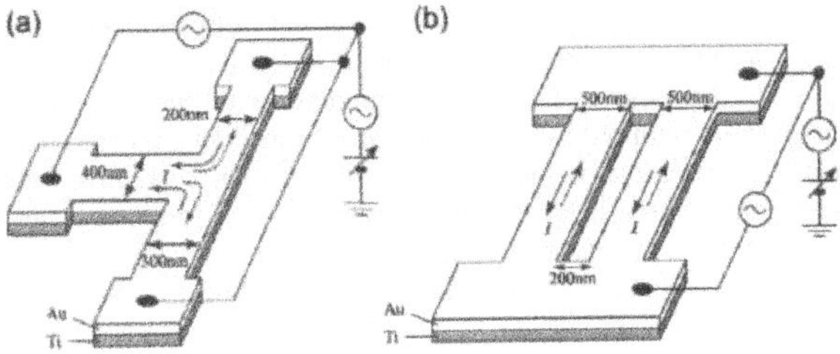

Fig. 8 Schematic illustrations of (a) T-junction and (b) parallel configurations of current paths fabricated by the electron beam lithography and the lift-off processes.

Validity of Current-induced Magnetic Field Detection Method

To verify the validity of the method we propose, the current-induced magnetic fields around the T-junction of current paths were observed. The obtained magnetic force images as well as the topographic image are shown in Fig. 9, where the cantilever alignment against the current paths is illustrated in the inset. Here, the ac bias frequency was about 350 kHz which was tuned to be coincident with the torsional resonant frequency of the cantilever in order to enhance the sensitivity in the magnetic force detection.

As shown in Fig. 9(b), the magnetic force signals were clearly recognized at both edges of the upper and lower paths, while almost no signal was detected around the left-hand path. This result is attributable to the fact that our method is sensitive to the gradients of the magnetic field in a perpendicular direction to a cantilever beam as mentioned before. We also confirmed that the peak value in amplitude is nearly proportional to the current value, which indicates the good quantitativeness of this method.

In addition, Fig. 9(c) shows that the phases at left and right sides of the upper or lower paths are inverted and that their relationship is also inverted between the upper and lower paths. We can attribute these results to the opposite directions of the magnetic field gradient for the left side and the right side of each path and to the opposite directions of the current flow between the upper and lower paths, respectively.

From these results, we have confirmed that our method successfully detected the current-induced magnetic field.

At the right sides of the upper and lower paths, both the amplitude and phase signals seem to be less distinct than those at the left sides, and we attribute it to the non-uniformity in the tip magnetization probably due to the non-uniform coating of the magnetic material near the tip apex.

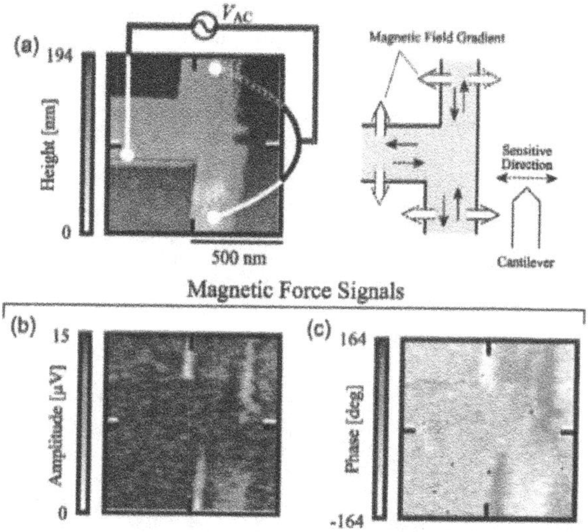

Fig. 9 (a) A topographic image and (b) amplitude and (c) phase images of magnetic force signals observed around the T-junction of current paths. The inset indicates the configuration of cantilever against the current paths during measurements. Total current flowing in the left-hand path was 200 μA_{rms} and nearly half of it flowed in each of the upper and lower paths.

Spatial Resolution

As shown in Fig. 9(b), the magnetic force signal was clearly identified around the current path of 200 nm in width, which indicates that the spatial resolution of our method is better than 200 nm. In order to confirm the spatial resolution from another point of view, on the other hand, we observed the magnetic fields around one pair of the parallel current paths with 200 nm separation illustrated in Fig. 8(b). As shown in Fig. 10, the specific features such as amplitude enhancement and phase inversion were clearly recognized around the both sides of each current path, and almost no interference was observed. This result also support our estimation that the spatial resolution is better than 200 nm.

Summary

We have proposed and demonstrated a current evaluation method through the detection of the current-induced magnetic field by MFM. As a result, we have confirmed the validity of our method including the correct detection of the interaction between the magnetic field and the MFM tip as well as the linear response to the current values. The spatial resolution is estimated to be better than 200 nm. These results open the possibility of spatially resolved and quantitative current mapping through the magnetic field detection by MFM.

CONCLUSIONS

We have presented on nano-probing techniques to characterize electrical properties of nano-materials with high spatial resolution, including Kelvin probe force microscopy for potential determination and magnetic force microscopy for current evaluation. They revealed the wide ability of nano-probes for electrical characterizations at nanometer scale.

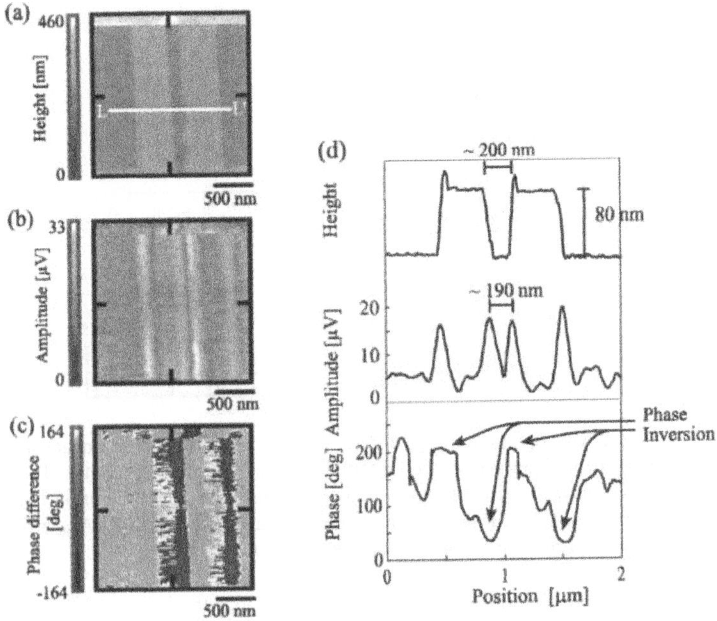

Fig. 10 (a) A topographic image, (b) amplitude and (c) phase images of magnetic force signals observed around the parallel current paths, and (d) their line profile along Line L-L' indicated in (a). Current value flowing in each path was about 80 μA_{rms}.

ACKNOWLEDGEMENTS

The author would like to express sincere thanks to Prof. H. Sakaki of University of Tokyo for fruitful discussions and to Dr. S. Ono and Mr. D. Saida of University of Tokyo for performing the KFM and MFM experiments presented in this paper.

The author also would like to appreciate Prof. Y. Wada, Dr. K. Tsutsui and Dr. T. Edura of Waseda University and Dr. M. Takeuchi of Institute of Physical and Chemical Research (RIKEN) for providing samples used in this study.

This work was partly supported by a Grant-in-Aid from Ministry of Education, Culture,

Sports, Science and Technology of Japan (MEXT), Japan, and by Focused Research and Development Project for the Realization of the World's Most Advanced IT Nation, IT Program, MEXT, Japan.

REFERENCES

[1]M. Nonnenmacher, M. P. O'Boyle and H. K. Wickramasinghe, "Kelvin Probe Force Microscopy," *Appl. Phys. Lett.*, **58**, 2921-2923 (1991).

[2]S. Ono and T. Takahashi, "Sample-and-hold Operation in Kelvin Probe Force Microscopy," *Jpn. J. Appl. Phys.*, **44**, 6213-6217 (2005).

[3]S. Ono and T. Takahashi, "Intermittent Bias Application in Kelvin Probe Force Microscopy for Accurate Determination of Surface Potential," to be published in *Jpn. J. Appl. Phys.*

[4]D. Saida and T. Takahashi, "Current-induced Magnetic Field Detection by Magnetic Force Microscopy around a GaAs/AlGaAs Mesa Stripe," *Jpn. J. Appl. Phys.*, **42**, 4874-4877 (2003).

[5]T. Takahashi and D. Saida, "Magnetic Field Detection for Current Evaluation by Magnetic Force Microscopy," *Ultramicroscopy*, **100**, 293-299 (2004).

[6]D. Saida and T. Takahashi, "Magnetic Field Observation around a Current Flow by Magnetic Force Microscopy," *Jpn. J. Appl. Phys.*, **43**, 4643-4646 (2004).

[7]D. Saida, T. Edura, K. Tsutsui, Y. Wada and T. Takahashi, "Investigation of the Spatial Resolution in the Current-induced Magnetic Field Detection by Magnetic Force Microscopy," to be published in *Jpn. J. Appl. Phys.*

[8]S. Ono, M. Takeuchi, T. Noda and T. Takahashi, "Surface Potential Imaging on InAs Quantum Dots and InAs Thin Films by Kelvin Probe Force Microscopy Operated in High Vacuum," *Jpn. J. Appl. Phys.*, **42**, 4869-4873 (2003).

[9]T. Takahashi and S. Ono, "Tip-to-sample Distance Dependence of an Electrostatic Force in KFM Measurements," *Ultramicroscopy*, **100**, 287-292 (2004).

[10]S. Ono and T. Takahashi, "Lateral Averaging Effect on Surface Potential Measurements on InAs Dots Studied by Kelvin Probe Force Microscopy," *Jpn. J. Appl. Phys.*, **43**, 4612-4614 (2004).

[11]T. Takahashi and S. Ono, "Sample-and-hold Imaging for Fast Scanning in Atomic Force Microscopy," *Jpn. J. Appl. Phys.*, **43**, L582-L584 (2004).

[12]T. Takahashi and S. Ono, "Sample-and-hold Atomic Force Microscopy for Fast Operation," *Ultramicroscopy*, **105**, 42-50 (2005).

[13]T. Takahashi and T. Kawamukai, "Phase Detection of Electrostatic Force by AFM with a Conductive Tip," *Ultramicroscopy*, **82**, 63-68 (2000).

[14]J. M. Moison, F. Houzay, F. Barthe, L. Leprince, E. André and O. Vatel, "Self-organized Growth of Regular Nanometer-scale InAs Dots on GaAs," *Appl. Phys. Lett.*, **64**, 196-198 (1994).

[15]D. Leonard, M. Krishnamurthy, S. Fafard, J. L. Merz, and P. M. Petroff, "Molecular-beam Epitaxy Growth of Quantum Dots from Strained Coherent Uniform Islands of InGaAs on GaAs," *J. Vac. Sci. Technol.*, **B 12**, 1063-1066 (1994).

[16]A. Madhukar, Q. Xie, P. Chen and A. Konkar, "Nature of Strained InAs Three-dimensional Island Formation and Distribution on GaAs(100)" *Appl. Phys. Lett.*, **64**, 2727-2729 (1994).

Electroceramic Materials and Applications
Edited by R. W. Schwartz

DIELECTRIC PERMITTIVITY MAPPING WITH NON-CONTACT MICROWAVE PROBE FOR DIELECTRIC MATERIALS

Hirofumi Kakemoto*, Song-Min Nam, Satoshi Wada and Takaaki Tsurumi
Graduate School of Engineering, Tokyo Institute of Technology
2-12-1 Ookayama, Meguro, Tokyo 152-8552, Japan

ABSTRACT

The high frequency reflection intensity was measured for Al_2O_3 and $SrTiO_3$ substrates at room temperature using a non-contact microwave probe as a function of distance between sample and probe. The difference of reflection intensity for Al_2O_3 and $SrTiO_3$ substrates was observed in the region where the distance of 0.2mm between sample and probe, and it was caused from dielectric permittivities of samples. The reflection coefficient of sample was estimated in comparison with results of electromagnetic simulation using finite differential time domain method. The reflection intensity for Al_2O_3 and $SrTiO_3$ substrates was transformed to dielectric permittivity at reflection intensity minimum point. The *in-plane* reflection intensity mapping for Al_2O_3 substrate/Cu-plate was carried out at 8.5 and 9.4GHz using reflection intensity minimum point, and the difference of reflection intensity between Al_2O_3 and Cu-plate was observed reflecting their dielectric permittivity.

INTRODUCTION

Nowadays, dielectric devices, such as multi-layered ceramics capacitors (MLCCs), have been miniaturized and used at high frequency region. MLCCs are characterized from their total capacitance, and also should be measured in view of local dielectric properties due to realizing more miniaturization. Recently, the local dipole moment behavior in miniaturized MLCCs has been reported about their core-shell structure.[1] However, at high frequency region, general dielectric measurement techniques are difficult for investigating the local dielectric properties. The uneven dipole moment behavior in dielectric materials is well known, [2,3] and local dipole moment behavior should be also investigated in high frequency region.

Therefore a high frequency measurement method for the local dielectric properties should be developed. Recently, near field scanning microwave microscopy was developed by Gao et al. and Steinhauer et al. using a resonator and contact probe, and local dielectric properties are investigated at high frequency region.

Also, non-contact probe preventing resonance effect and obtaining wide frequency range is useful for the high frequency (f) measurement method, however the impedance of measurement system is changed as the functions of f in input electric (E) field and electrical length (l) between probe and sample (i.e. changed as a function of phase: $\theta=\beta l=2\pi fl/c$).[4-9]

In this study, the microwave reflection (r) intensity using non-contact probe was measured in high f with deciding the accurate θ for impedance of measurement system and sample (Z_L). The r intensity was successfully transformed to dielectric permittivity (ε_r), and *in-plane* r intensity mapping was carried out using r intensity minimum point (non-contact state).

EXPERIMENTAL

Reflection intensity measurement

Figure 1 shows the measurement system and the simulation model for electromagnetic (EM) field analysis. The measurement system is shown in **Fig.1(a)**. The measurement system was constructed from an oscillator generated by a Gunn diode (8.5-11GHz), directional coupler, detector (diode), test-probe (Cu, 0.99mmϕ, Micro Denshi Co. Ltd.) and micro-probe (Pt, 6μmϕ, KEYCOM Corp.). The phase shifter (0-360°) was inserted between oscillator and directional coupler. The frequency (f) was changed from 8.5GHz to 11GHz. The test-probe was formed from semi-rigid coaxial cable, and sized 8mm for length (l_o). The micro-probe was sized also l_o=8mm, and it was attached to coaxial cable.

The *XYZ*-stage was moved by d.c. servomotors (Chuo precision industrial Co.Ltd.) for setting to each axes, and they are controlled by a personal computer using GP-IB interface. The sample was selected for Cu-plate (reference, size: 40x40x2.0mm^3), Al_2O_3(1000) (ALO) and $SrTiO_3$(100) (STO) substrates (Shinko-sha Co.Inc., size: 10x10x0.5mm^3). The reverse surface of ALO and STO was deposited an Au film as the ground, and they were put on Cu-plate. The r was measured at room temperature as a function of distance (d) between probe and sample. In addition, the scanning for *XY*-plane was carried out for samples.

Analysis

The EM field analysis was carried out using finite differential time domain (FDTD) method using MAGNA/TDM software (CRC solutions Corp.). **Figure 1(b)** shows the simulation model. The size and physical quantities were input to the simulation model, and then E field was loaded to simulation model. The reflection coefficient (Γ) and Z were calculated using results of EM field distribution.

a) b)

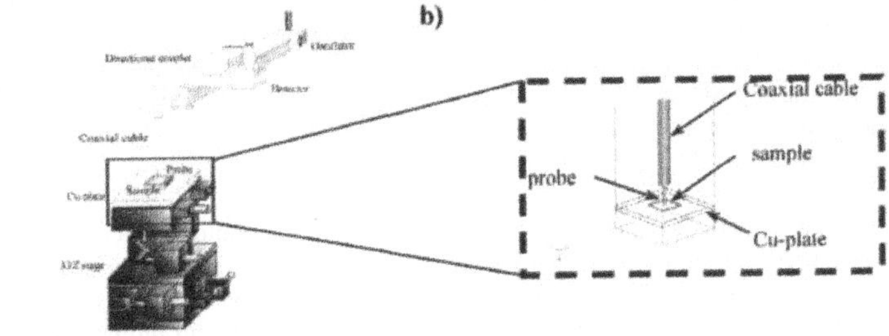

Fig.1 The measurement system (a) and simulation model for electromagnetic field analysis (b).

PRINCIPLE

The probe acts as a quarter wavelength (λ/4) dipole antenna with the non-contact state for the surface of sample. The r intensity properties of dipole antenna were investigated in 1941, and r intensity minimum was observed with the configuration toward to the ground. The r intensity and impedance of λ/4 dipole antenna were calculated.[10] The schematic illustration of λ/4 dipole antenna is shown in **Fig.2(a)**. The impedance of λ/4 dipole antenna (probe) (Z_{probe}) is expressed as

$$Z_{probe} = -j(\beta\eta/2\pi d_0)[1/j\beta d_0 - 1/\beta^2 d_0^2]\exp(-j\beta d_0) l_0^2, \quad (1)$$

where η is $(\mu/\varepsilon)^{0.5}$; μ: permeability, d_0 is distance between probe (dipole antenna) and its image in the sample, and l_0 is the length of λ/4 dipole antenna.

The equivalent circuit in this measurement system is shown in **Fig.2(b)**. The impedance of air-gap ($Z_{air\text{-}gap}$) as a function of electrical length (l) follows transmission theory, and is denoted as

$$Z_{air\text{-}gap}(l) = Z_0[Z_L - jZ_0\tan\beta l]/[Z_0 - jZ_L\tan\beta l], \quad (2)$$

where Z_0 is characteristic impedance (50Ω). The $Z_{air\text{-}gap}$ is also changed as a function of θ. Therefore, the total impedance (Z_{total}) is $Z_{total} = Z_{probe} + Z_{air\text{-}gap}$.

The reflection coefficient (Γ) at l=0 is defined as

$$\Gamma = r/i = [Z_L - Z_0]/[Z_L + Z_0]. \quad (3)$$

[9].

The electromagnetic (EM) field distribution irradiated from the probe was also analyzed by computer simulation software using finite differential time domain (FDTD) method. The simulation model, as shown in **Fig.1(b)**, using FDTD method is followed by Yee algorism.[11] In the simulation model, skin depth of sample's surface was considered, and cell size for sample in simulation model was created as 10 times smaller than other parts. The EM field distribution was obtained with input Gaussian pulse as the E field to FDTD simulation model in **Fig.1(b)**, the plane wave response in r EM field was observed in simulation model at r minimum point. In experimental, the r intensity minimum point is also ascribed to be formed the plane wave. The irradiation area of EM field to the sample (S) was observed from above simulation.

The adjustment of θ is very important for obtaining accurate Z_L at l=0, because Γ is transformed to the dielectric permittivity (ε_r) with inputting θ, as follows,

$$\varepsilon_r = 2|\Gamma|\sin\theta/[\omega C_0 Z_0(|\Gamma|^2 + 2|\Gamma|\cos\theta + 1)], \quad (4)$$

where C_0 is $\varepsilon_0 S/d$.[12]

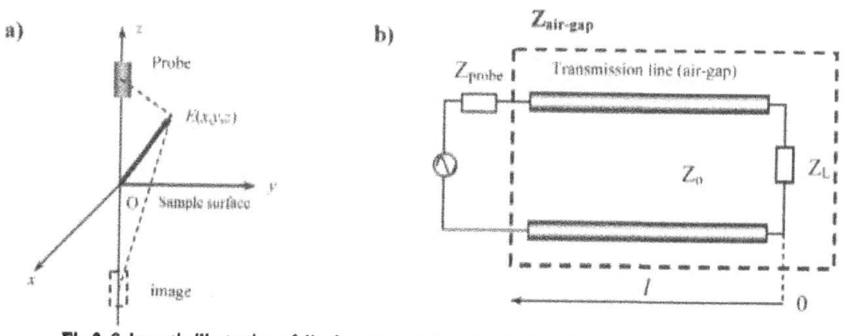

Fig.2 Schematic illustration of dipole antenna (a), and equivalent circuit in this measurement system (b).

RESULTS AND DISCUSSION

Reflection intensity for Cu-plate

Figure 3 shows the r intensity for Cu-plate using test probe as a function of distance (d) between probe and Cu-plate, and θ at measurement f region. At first, r intensity was measured as a function of d, as shown in **Fig.3(a)**. The measurement was done without θ shifter. The probe fully acts as a dipole antenna. The r intensity showed the minimum values at d=0mm for 8.6GHz and d=0.2mm for 9.4GHz, and r intensity at other f was increased with decreasing d in **Fig.3(a)**. From obtained r intensities at 8.6GHz and 9.4GHz in **Fig.3(a)**, we consider the factor about appearing r intensity minimum, i.e., the probe length, quarter wavelength ($\lambda/4$) of EM wave and θ at measurement f. Firstly, probe length and $\lambda/4$ of EM wave at measurement f were considered. Although the probe length is as same as $\lambda/4$ of EM wave at 9.4GHz (8mm), $\lambda/4$ of EM wave at 8.6GHz is, to the contrary, 8.72mm. Hence, the probe length is not related to the appearing of r intensity minimum. Secondary, in order to measure the θ at d=0.2mm at measurement f, θ shifter was equipped for measurement system. **Figure 3(b)** shows the r intensity versus θ at measurement f range. Although r intensity showed the different values because of oscillator's f dependence, r intensity was periodically changed with increasing θ. Therefore, the resonance phenomena in this measurement do not occur. The peaks of r intensity at 8.6GHz and 9.4GHz were found about 150° and 330° (#1, #2), and the curve shapes of them are accordance with periodically in **Fig.3(b)**. However, r intensity at 8.6GHz is slightly shifted to lower θ. The minimum r intensity at 9.4GHz in **Fig.3(a)** was measured at θ=300° (#3). If peak #1 in **Fig.3(b)** is consider as θ=0°, peak #3 is $3\pi/2$.

Therefore the θ is the main factor of appearing r intensity minimum. The θ of EM wave incidence to sample at 8.6GHz and 9.4GHz is assumed to be $n\pi/2$ from **Fig.3(b)**. The r intensity at 9.4GHz shows the minimum value at d=0.2mm, however r intensity at 8.6GHz appears the minimum value at d=0mm because of slight shifting of θ. Therefore, it is possible that r intensity minimum is appeared at other f with adjusting θ.

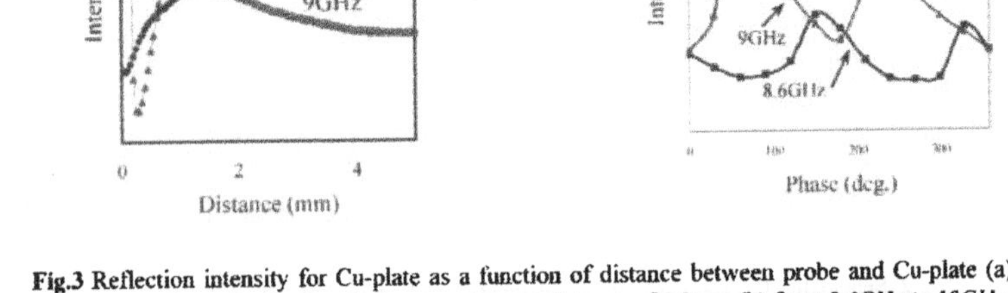

Fig.3 Reflection intensity for Cu-plate as a function of distance between probe and Cu-plate (a), and reflection intensity for Cu-plate versus phase at distance of 0.2mm (b) from 8.6GHz to 10GHz.

Figure 4 shows the *r* intensity for ALO and STO at 9.4GHz as a function of *d*. The *r* intensity was increased with increasing ε_r of sample (ALO:ε_r=10, STO:ε_r =310). The EM wave was attained and reflected at reverse sides of sample, and reflected EM wave was then detected by the non-contact probe. In the **Fig.4**, the positions of showing *r* intensity minimum values for ALO and STO did not shift, because optical lengths for ALO and STO, denoted as $(\varepsilon_r)^{0.5}$x(0.5mm), were smaller than $\lambda/4$. Hence θ at *d*=0.2mm in **Fig.4** is kept to $n\pi/2$. Therefore it is important for keeping the θ as $n\pi/2$ that EM wave is conducted to electrode (Cu-plate) or transmitted to dielectric materials (ALO, STO).

Various *f* dependence of dielectric material can be measured with non-contact probe. In order to investigate *r* intensity property, the EM analysis was carried out using simulation model in **Fig.1(b)**.

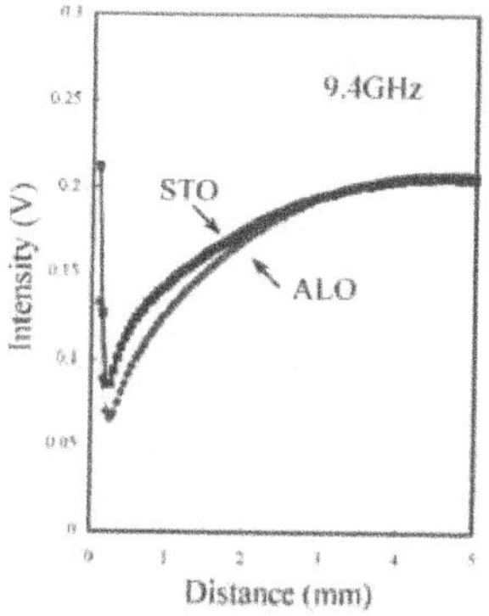

Fig.4 Reflection intensity for Al_2O_3 and $SrTiO_3$. The minimum point of reflection intensity at 9.4GHz was appeared at distance of 0.2mm between probe and sample, and difference of reflection intensity of samples was clearly observed.

Results of electromagnetic analysis

The time domain *E* field distribution of ALO (ε_r=10) was calculated by EM analysis. **Figure 5** shows the *E* field distribution of test-probe and ALO (a), and result of calculation for Γ as a function of *d* (b). The *E* field distributes to space and into sample predicted in **Fig.2** and eq.(2), as shown in **Fig.5(a)**. The *E* field distribution to space behaves as a dipole antenna, and other part of *E*-filed distribution was irradiated to sample. The area (*S*) of *E* field irradiated to sample was difficult to be calculated from **Fig.5(a)**. However, from *E*-field distribution, Γ and Z were calculated by EM analysis software.

Figure 5(b) shows Γ as a function of *d*. The value of Γ at *d*=0mm is much large, and curve shape between 0.2mm and 5mm is slightly smooth. As the results, it is slightly difficult for analyzing. However Γ is also presents minimum value at *d*=0.2mm. The Γ and Z_{total} were calculated to be 0.34 and 98.7Ω at *d*=0.2mm, respectively. The relationship between Γ and Z_{total} =Z_L as shown in **Fig.2(b)** is expressed in eq.(3) at *l*=0. The Γ and Z_L at *d*=0.2mm in **Fig.5** are well accordance with eq.(3).[9]

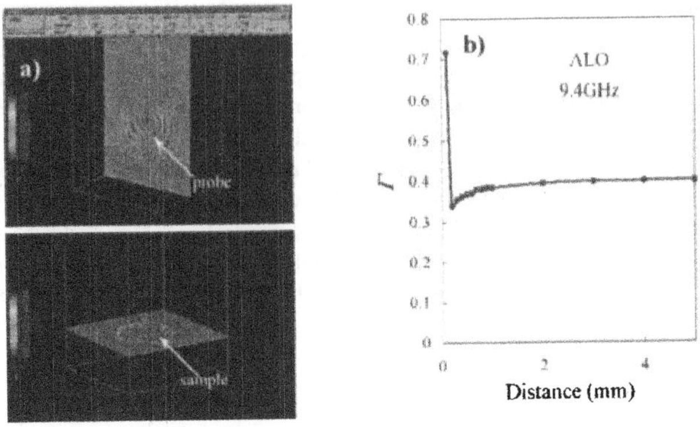

Fig.5 Time domain electric field distribution around probe and sample (a), and reflection intensity of Al_2O_3 versus distance calculated by electromagnetic analysis (b).

Reflection intensity and reflection coefficient of ALO and STO

Figure 6 shows the *r* intensity (**Fig.3**) and Γ(**Fig.4**) properties of ALO. The solid line is experimental *r* intensity, and open circle is calculated value of EM analysis. At d=0.2mm, Γ of ALO is estimated to be 0.34. From results in section 4.2, Z matching was found at d=0.2mm. In addition, Γ of STO is also estimated to be 0.36. The Γ of sample was successfully estimated in comparison between experimental and analysis data. The transformation from *r* intensity to Γ involves background-calibration, and it is solved in comparison with experimental value (*r* intensity) and EM analysis (Γ). The obtained results indicate that d=0mm is not a corrective standard point for measure the *r* intensity but d=0.2mm is a corrective one. In addition, the *d* at minimum *r* intensity can be considered as the *l* (=0mm) between probe and sample.

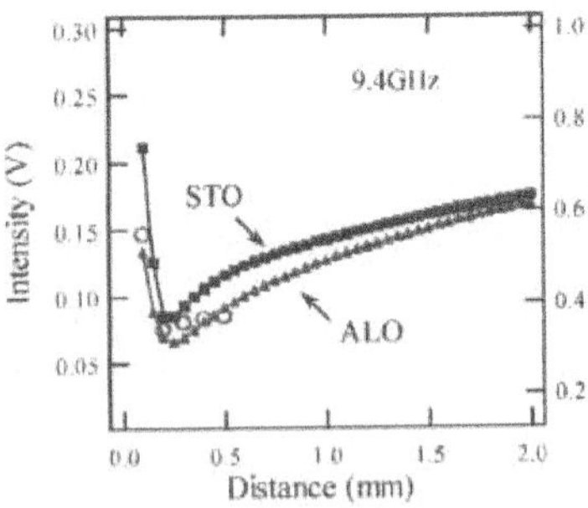

Fig.6 Plots of reflection intensities (experimental) and coefficient (analysis) of Al_2O_3.

Transform to dielectric permittivity

From the result in section "*Reflection intensity and reflection coefficient of ALO and STO*", Z in eq.(3) indicates impedance of sample ($Z_L=1/j\omega C$, where C is capacitance of sample). The C of sample is $C=\varepsilon_r\varepsilon_o S/(0.5\text{mm})$, where S is effective are of transmitted E-field in the sample.

Figure 7 shows the Γ as a function of S (a), and S versus ε_r (b). The S was estimated to be 1.33mm^2 (ALO) and 0.06mm^2 (STO) in **Fig.6(a)**. As shown in **Fig.6(b)**, S was decreased with increasing ε_r. The transmitted E-field in sample is decreased with increasing ε_r, and r intensity became higher value, as shown in **Fig.3**. From above all discussions, the ε_r at d=0.2mm was estimated using eq.(4), $\varepsilon_r = 2|\Gamma|\sin\theta/[\omega C_o Z_o(|\Gamma|^2+2|\Gamma|\cos\theta+1)]$, where $\theta=90^o$, $\omega=2\pi f$, $C_o=\varepsilon_o S/(0.5\text{mm})$ and $Z_o=50\Omega$.[12]

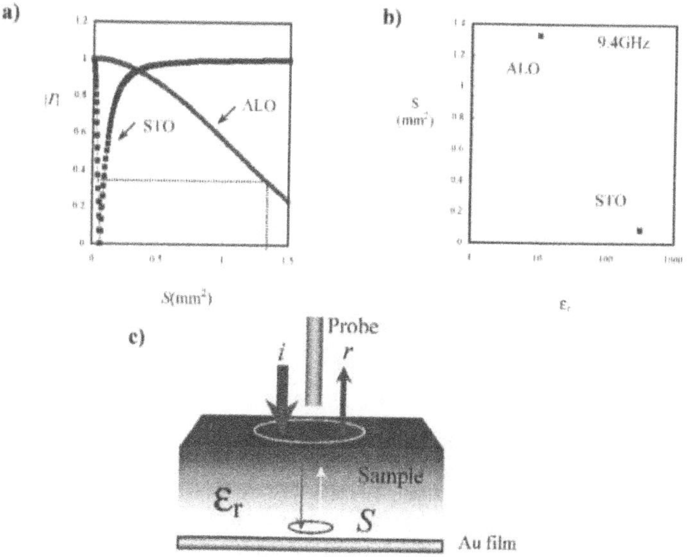

Fig.7 Reflection coefficient versus effective area of transmitted electric field in samples (a), and dielectric permittivity versus effective area (b), and their images (c).

Figure 8 shows the log plot of ε_r versus l at 9.4GHz. The ε_r of ALO and STO is estimated to be 8.9 and 290. The ε_r in **Fig.8** is accordance with the ε_r of samples ($\delta\varepsilon_r^{ALO}$=10-8.9, $\delta\varepsilon_r^{STO}$=310-290).

We developed a new dielectric measurement method for high frequency using a non-contact probe. The measurement method prevents resonance phenomena, and measures accurate ε_r of dielectric materials. In measurement data, ε_r was estimated to slightly lower value. It was caused from irradiation loss as the result of using the non-contact probe. In addition, the measurement of dielectric loss (tanδ) will be done in future study.

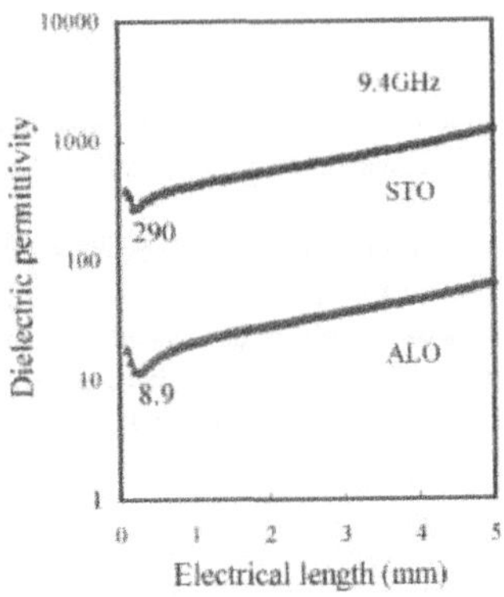

Fig. 8 Dielectric permittivity of samples versus electrical length.

x-y plane mapping using un-contact probe

Figure 9 shows the *r* intensity mapping of ALO at 9.4GHz using **test-probe**. The obtained *r* intensity mapping shows the homogeneous distribution of *r* reflecting ε_r of ALO.

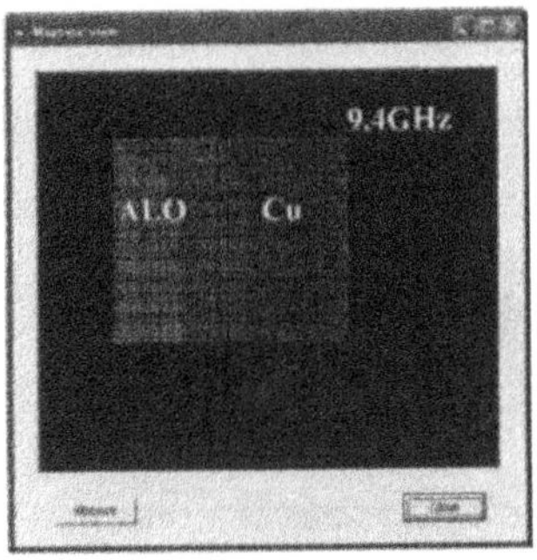

Fig. 9 Reflection intensity mapping of Al_2O_3/Cu-plate using test-probe operated at 9.4GHz.

Figure 10 shows *r* intensity versus *d* for ALO and Cu-plate (a) and also the *r* intensity mapping for ALO/Cu-plate (b) operated at 8.5GHz using **micro-probe**. Although initial adjustment is difficult for micro-probe more than test-probe, *r* intensity minimum point for Cu-plate and ALO were observed in **Fig.10(a)**. However the position of *r* intensity minimum point of ALO was shifted to larger *d* value (the adjustment is needed). In-plane mapping for ALO/Cu-plate, as shown in **Fig.10(b)**,the diameter of probe was decreased, and scanning range became 5 times for axis smaller than test-probe.

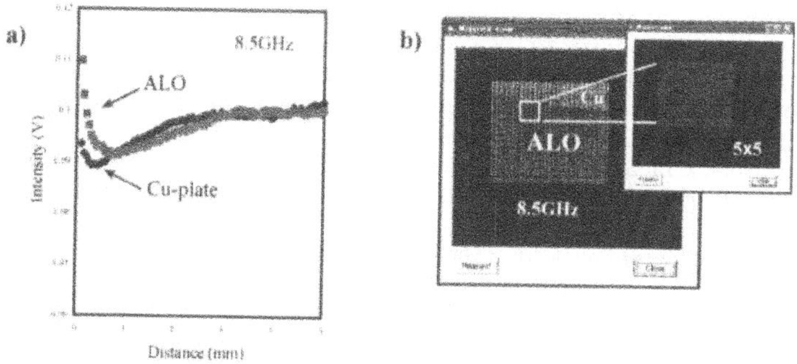

Fig. 10 Reflection intensity versus distance (a) and mapping (b) for ALO/Cu-plate using micro-probe operated at 8.5GHz.

CONCLUSION

A high frequency dielectric measurement method is proposed using a non-contact probe. The reflection intensity measurement at high frequency was carried out for dielectric materials. The reflection intensity showed the minimum value in the region where the distance of 0.2mm between sample and probe. The reflection intensity minimum point presented an impedance matching one. The dielectric permittivity of samples was calculated by using reflection intensity minimum point value and result of electromagnetic analysis. The *in-plane* reflection intensity mapping for Al_2O_3 substrate/Cu-plate was carried out using reflection intensity minimum point, and the difference of reflection intensity between Al_2O_3 and Cu-plate was observed reflecting their dielectric permittivity.

ACKNOWLEDGMENT

The author would like to thank Y.Suzuki (KEYCOM Corp.) for designing the probe, and M.Saruhasi (CRC solution Corp.) for useful discussions of electromagnetic analysis. This work was supported by Japan Society Promotion of Science (Project No.16760542).

* Corresponding author

REFERENCES

[1] H. Chazono and H. Kishi, "dc-Electrical Degradation of the BT-Based Material for Multilayer Ceramic Capacitor with Ni internal Electrode: Impedance Analysis and Microstructure," *Jpn. J. Appl. Phys.* **40**, 5624-5629 (2001).
[2] A. J. Moulson and J. M. Herbert: "Electroceramics, Properties, Applications. 2nd edition" Jhon Wiley & Sons (2003).
[3] Y. Cho, S. Kazuta and K. Matsuura, "Scanning Nonlinear Dielectric Microscopy with Nanometer Resolution," *Appl. Phys. Lett.* **75**, 2833-2835 (1999).
[4] B. W. Hakki and P. D. Coleman, *IEEE Trans. Microwave Theory Tech.* **8**, 381 (1960).
[5] Y. Kobayashi and M. Katoh, *IEEE Trans. Microwave Theory Tech.* **33**, 586 (1985).

[6] C. Gao and X.-D. Xiang, "Quantitative Microwave Near-field Microscopy of Dielectric properties," *Rev. Sci. Instrum.* **69**. 3846-3851 (1998).
[7] D. E. Steinhauer, C. P. Vlahacos, C. Canedy, A. Stanishevsky, J. Melngailis, R. Ramesh, F. C. Wellstood, and Steven M. Anlage, "Imaging of Microwave Permittivity, Tenability, and Damage Recovery in (Ba,Sr)TiO_3 thin films," *Appl. Phys. Lett.* **75**, 3180-3182 (1999).
[8] H. Kakemoto, J. Li, S-M. Num, S. Wada and T. Tsurumi, "Dielectric Spectra of $BaTiO_3$-Based Materials Measured by Impedance Analyzer up to 1GHz," *Jpn. J. Appl. Phys.* **42**, 6143-6148 (2001).
[9] F. Okada:"Microwave Engineering"Gakken-sha, Tokyo (1993) [in Japanese].
[10] Y. Kato:"Denpa Ku-chu Senro no Kenkyu"Kyoritsu-shuppan, Tokyo (1941) [in Japanese].
[11] T. Uno:"Electromagnetic field and antenna analysis using FDTD method" Corona-sha, Tokyo (1998) [in Japanese].
[12] S. S. Stuchly, M. A. Rzepecka, and M. F. Iskander, "Permittivity Measurements at Microwave Using Lumped Elements," *IEEE Trans. Instrum. Meas.* **23**, 56-62 (1974).

Electroceramic Materials and Applications
Edited by R. W. Schwartz

DIELECTRIC ANOMALY IN PEROVSKITE ARTIFICIAL SUPERLATTICES

Takaaki Tsurumi*, Takakiyo Harigai, Hirohumi Kakemoto and Satoshi Wada
Graduate School of Science and Engineering, Tokyo Institute of Technology
2-12-1 Ookayama, Meguro-ku, Tokyo 152-8552, Japan

ABSTRACT

$BaTiO_3/SrTiO_3$, $SrZrO_3/SrTiO_3$ and $BaTiO_3/BaZrO_3$ artificial superlattices were fabricated by the molecular beam epitaxy process. X-ray diffraction (XRD) reciprocal space mapping measurement was performed using high-resolution XRD. In all artificial superlattices, the superlattices with the 10-periodic structure showed the clear satellite peaks in the XRD pattern and were mostly distorted in the direction of film thickness. Superlattices with the 10-periodic structure showed a high dielectric permittivity of ε_r=33,000 or artificially induced ferroelectricity. It was clarified that the anisotropic lattice distortion introduced by the strains due to the lattice mismatch was the origin of the unique dielectric characteristics of artificial superlattices.

INTRODUCTION

Perovskite-type oxide thin films show various properties, such as ferroelectricity, superconductivity and so on, and are considered to be important for new electric devices. In recent years, oxide artificial superlattices have been studied extensively. Artificial superlattices, which multi-layered structures consist of two different lattice constants, have the potential to drastically improve material properties or to show unknown functionalities. In hetero-structured artificial superlattices, large lattice distortion could be introduced in the in-plane direction by the stacking layers. So, material properties of artificial superlattice might be controlled artificially.

$BaTiO_3$(BTO)/$SrTiO_3$(STO) system have been intensively investigated in terms of calculations and experimental results. The lattice constant of BTO bulk is larger about 3% than that of STO bulk. Tabata *et al.* [1] has reported that the pressure induced by the STO layers is estimated to be approximately 400-500 MPa. A large number of researches concerning BTO/STO system [1-4] have succeeded in fabricating artificial superlattices. Their superlattices show different behaviors from solid solution $Ba_xSr_{1-x}TiO_3$(BST). In our previous studies [5,6], dielectric properties of BTO/STO superlattices with several kinds were measured using the planar type electrode structure. It was clarified that the dielectric permittivity in which the in-plane field direction depended on the stacking period of superlattices and was maintained up to high frequency region. Even though it is considered that the stress introduced due to lattice mismatches at the interface might play an important role to material properties, the origin of these material properties has not been clearly understood at present. So, it is necessary to elucidate the mechanism of the dielectric properties of oxide artificial superlattices.

Therefore, this research aimed at investigating the influence of the superlattice structure in the perovskite-type artificial superlattices on their dielectric properties. The artificial superlattices with the ferroelectric/paraelectric (A-site modulated, BTO/STO) system as well as those with the paraelectric/paraelectric systems (B-site modulated $SrZrO_3$(SZO)/STO) and the ferroelectric/paraelectric (B-site modulated, BTO/$BaZrO_3$(BZO)) were prepared in order to investigate the relationship between their structures and dielectric properties.

EXPERIMENTAL PROCEDURES

Artificial superlattices were fabricated on STO (001) single crystal substrates by molecular beam epitaxy (MBE) method at the substrate temperature of 600°C. Strontium and barium metals were evaporated from the Knudsen cells, and titanium and zirconium metals were evaporated from electron-beam guns. Electron cyclotron resonance plasma was irradiated on the surfaces of the growing films for generating the oxygen plasma, and the working pressure in the vacuum chamber was 10^{-6} Torr. The structure of superlattices is represented by a formula $[(BTO)_m/(STO)_m]_n$, where the subscript *m* indicates the period of layers (m = 1, 10, 20, and 40) and *n* is the number of repeating periods. The total film thickness was fixed at 80 unit cells of the primitive perovskite lattice. XRD reciprocal space mappings around the STO substrate (103) were performed using a high-resolution x-ray diffractometer (BRUKER AXS, Discover8). The capacitance and the complex admittance of the superlattices with interdigital electrodes were measured at room temperature by an impedance analyzer (Agilent Tech., 4294A) at frequencies up to 110 MHz. The dielectric permittivity was determined from the complex admittance using electromagnetic field analysis. In addition, charge *vs.* voltage (Q-V) hysteresis curves were measured using a standardized ferroelectric test system (Radiant Tech., RT-6000).

RESULTS AND DISCUSSION

A. Lattice distortion of artificial superlattice

The XRD reciprocal space mappings measured with the high-resolution x-ray diffractometer were shown in Fig. 1. As for the BTO/STO artificial superlattices, the satellite peaks were clearly observed in the sample with the periodicity of 10 unit cells (call 10-periodic structure hereafter). The satellite peaks observed in the 20-periodic structure were not clear in comparison with those of the 10-periodic structure. In the 40-periodic structure, the position of broad peaks observed on the reciprocal space map indicated that the a-axis lattice constant was expanded from that of the STO substrate. This means that the crystal lattice in the film was relaxed from the restriction of the substrate. The a-axis lattice constant of the 40-periodic structure was larger than that of the STO substrate, while the a-axis lattice constant of the 1-periodic structure almost agreed with the STO substrate. As for the SZO/STO artificial superlattices, the results showed the similar behavior to the case of the BTO/STO artificial superlattices. On the other hand, the BTO/BZO artificial superlattice with 10-periodic structure showed different tendency compared with the other artificial superlattices. Clear satellite peaks of this 10-periodic structure were observed, but the satellite peak positions in the reciprocal map were deviated from the a-axis length of the STO

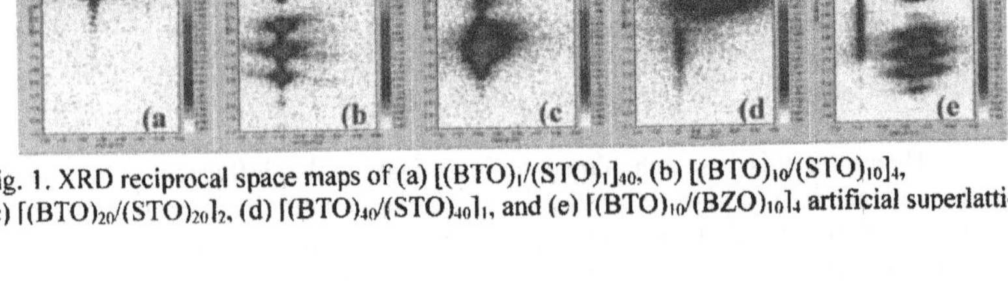

Fig. 1. XRD reciprocal space maps of (a) $[(BTO)_1/(STO)_1]_{40}$, (b) $[(BTO)_{10}/(STO)_{10}]_4$, (c) $[(BTO)_{20}/(STO)_{20}]_2$, (d) $[(BTO)_{40}/(STO)_{40}]_1$, and (e) $[(BTO)_{10}/(BZO)_{10}]_4$ artificial superlattice.

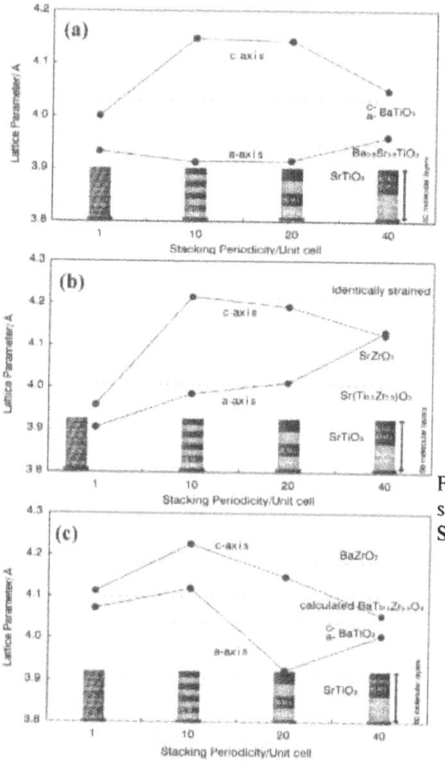

Fig. 2. Lattice constants of (a) BTO/STO, (b) SZO/STO, and (c) BTO/BZO artificial superlattices as a function of the periodicity.

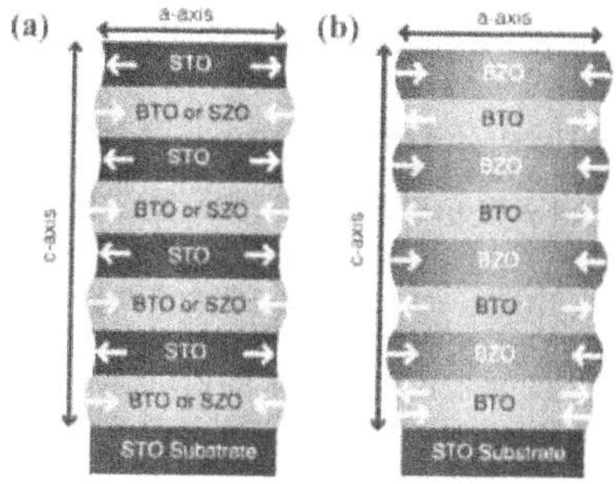

Fig. 3. Model of strains induced in each layer of the superlattices. (a) Schematic diagram of BTO/STO or SZO/STO, and (b) BTO/BZO artificial superlattice.

substrate. This indicates that the film growth is coherent in spite of less restriction from the substrate. It is also noticeable in the 20-periodic structure that these peak positions are consistent with the a-axis length of the STO substrate.

Figure 2 shows the lattice constants of the BTO/STO, SZO/STO, and BTO/BZO artificial superlattices calculated from the XRD reciprocal space maps by assuming that the crystal system of the superlattices was tetragonal. As reference values, the lattice constants of STO (cubic; 0.3905 nm[7]), BTO (tetragonal; 0.3994, 0.3994, 0.4033 nm[8]), BST (x=0.5, cubic; 0.3947 nm[9]), and BZO (cubic; 0.4182 nm[10]) of bulk state are plotted in the same figure. The a-axis lattice constant of the BTO/STO artificial superlattice with the 10-periodic structure was consistent with that of the substrate. The a-axis length increased with the periodicity from 20 to 40, but the c-axis length (lattice parameter along the film thickness) decreased with increasing periodicity above the 10-periodic structure. It is considered that the origin of this phenomenon stems from the strains in the superlattices caused by their different lattice constants. The lattice constant of BTO bulk is larger about 3% than that of STO bulk. Therefore, in the hetero-structured BTO/STO artificial superlattice, the material with a larger lattice constant (BTO) is under a compression stress in the in-plane direction, while the material with a smaller lattice constant (STO) is under a tensile stress. Since STO was used as the substrate, the BTO layer deposited on the substrate should be compressed in the in-plane direction and should be elongated in the direction of c-axis. After the deposition of the first

BTO layer, the STO layers deposited on it also give the compressive stress to the BTO layer as shown in Fig. 3(a), which gives rise to the reduction of a-axis length and the increase of c-axis length of the BTO lattice. However, the STO layers in the superlattice receive the tensile stress from the BTO layer in plane direction and deform its crystal lattice. In the artificial superlattice with the 10-periodic structure, the compressive stresses from the STO substrate and the STO layers are effectively applied to the BTO layers before the lattice relaxation of the BTO layers, therefore, the a-axis length in the whole film is almost consistent with that of the STO substrate. When the periodicity of the superlattice increased to 20 and 40, the crystal lattice of the BTO layers is relaxed before the deposition of the STO layers. This is the reason of the increase in the a-axis length in these superlattices. In the case of the 1-periodic structure, the effect of lattice distortion was expected to be maximum since it had the shortest stacking periodicity. The compressive stress from the STO layers seemed to be induced into the BTO layers most effectively and it was seemed that the induced lattice distortion was larger in all periodic structures. However, it was found that the c-axis length was markedly reduced. This phenomenon could not be explained by the internal stress in the superlattice. In this film, the superlattice was hardly formed because of the thermal diffusion between the BTO and the STO layers during the deposition at 600°C in the MBE process. Actually, the satellite peaks were not observed for this film. The lattice constant of the 1-periodic structure consequently had the similar value to that of BST(x=0.5) solid solution.

As for the SZO/STO artificial superlattices, the change in the lattice constants with the periodicity (Fig. 2(b)) showed the similar behavior to the case of the BTO/STO superlattice. However, the a-axis length increased more steeply with the periodicity than that of the BTO/STO superlattice. This means that the lattice relaxation was enhanced in this superlattice. The a-axis lattice constant of SZO is larger than that of BTO. This may be the reason of the difference between the two superlattices.

The c-axis length of BTO/BZO superlattice showed a maximum at the 10-periodic structure and it reduced with increasing periodicity, which was a similar behavior to other superlattices. However, the change in the a-axis length was totally different from those of other superlattices: the a-axis length of the BTO/BZO superlattice showed a maximum at the 10-periodic structure. This discrepancy is attributable to the difference in the structure of the BTO/BZO superlattice from those of others (Fig. 3(b)). The BTO/STO and SZO/STO superlattices involved the STO layer with the same lattice constant as the substrate, therefore, a compressive stress was induced in the BTO or SZO layers from the STO layers as mentioned above. On the contrary, the lattice constants of both BTO and BZO in the BTO/BZO superlattice are larger than the STO substrate. Since the lattice constant of BZO is larger than the BTO, in the BTO layers in this superlattice, a compressive stress was induced from the substrate and a tensile stress was induced from the BZO layers. The lattice distortion of BTO layer by the tensile stress from the BZO layer increased with decreasing thickness of the BTO layer. Consequently, the distortion was maximized in the 10-periodic structure to break the lattice matching between the substrate and superlattice. The increase in the a-axis length in the 10-periodic structure could be explained as mentioned above. In the case of the 20-periodic structure, the tensile stress induced to the BTO layer from the BZO layer decreased to keep the lattice matching between the substrate and superlattices, giving rise to the reduction of the a-axis length to that of the STO substrate. The a-axis length of the BTO/BZO superlattice was determined from the balance of the compressive stress from the substrate and the tensile/compressive stress induced into the BTO/BZO layers due to the lattice mismatch between the two materials as well as the lattice relaxation with increasing thickness of each layer. The lattice constants of the 1-periodic structure could not be explained by the above speculation. Because the lattice constant of the BZT(x=0.5) has not been reported so far, they were estimated by extrapolating observed lattice constants with different chemical composition [11]. The lattice constant of the 1-periodic structure was consistent with those estimated, indicating that the solid solution was formed by the thermal inter-diffusion between the BTO and BZO layers.

B. Lattice distortion and dielectric properties

Figure 4 shows the dielectric permittivity of the superlattices at 110 MHz as a function of stacking periodicity [6, 12]. It was found that the permittivity decreased with increasing periodicity over the 10-periodic structure in all superlattices. In particular, the 10-periodic BTO/STO artificial superlattice showed the extraordinarily high relative dielectric permittivity of 33,000 up to 110 MHz.

Figure 5 shows the Q-V hysteresis curves of the superlattices measured at 132.5 Hz using interdigital electrodes. In the BTO/STO superlattice, hysteresis loops were not observed for all periodic structures. The SZO/STO superlattice system shows ferroelectricity, nevertheless both of SZO and STO are paraelectric materials [12-15]. This ferroelectricity depended on the periodicity of the superlattice. As for the BTO/BZO superlattice, hysteresis loops were observed only for the 10-periodic structures.

In the SZO/STO superlattice with the 10-periodic structure, a clear hysteresis was observed because the anisotropic lattice distortion was largest in this structure as shown in Fig. 2(b). The ferroelectricity reduced with reducing anisotropic lattice distortion. In the BTO/STO superlattice, ferroelectricity was not observed in the 10-periodic structure in spite of the large anisotropic lattice distortion. The c/a ratios obtained from the XRD reciprocal space maps were almost same values, 1.058 and 1.060, for the SZO/STO and BTO/STO superlattices with the 10-periodic structure. The reasons of the difference in two superlattices are not clear at present. However, it must be pointed out that the electric field was mainly applied in the film plane in the measurement using planar electrodes [6], so that the spontaneous polarization should rotate 90 degrees to give a hysteresis loop in the Q-V curve. The a-axis of the BTO/STO superlattice was almost consistent with that of STO substrate (Fig. 2(a)), which means that the crystal lattice was not relaxed. We are thinking that this may restrict the rotation of the spontaneous polarization. It is also interesting that the superlattice consisting of paraelectric/paraelectric materials showed ferroelectric nature but that of ferroelectric/paraelectric materials showed the paraelectric nature. This indicates that the anisotropic lattice distortion is more effective to induce the ferroelectricity than the ferroelectric nature of BTO.

In this study, we found that the superlattices with the 10-periodic structure showed anomalous behaviors in dielectric properties: i.e. the BTO/STO superlattice showed an extremely high dielectric permittivity, while the BZO/BTO and SZO/STO superlattices showed ferroelectricity. In the 10-periodic structures, the strains due to the lattice mismatch between consisting layers were most effectively induced into the superlattices before the lattice relaxation occurring with the increase of layer thickness, which may be the reason of these peculiar properties.

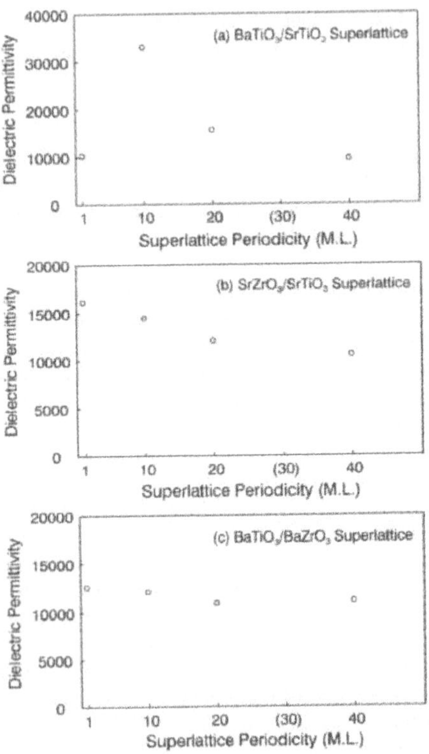

Fig. 4. Dielectric permittivity of (a) BTO/STO, (b) SZO/STO, and (c) BTO/BZO artificial superlattice at 110MHz as a function of periodicity

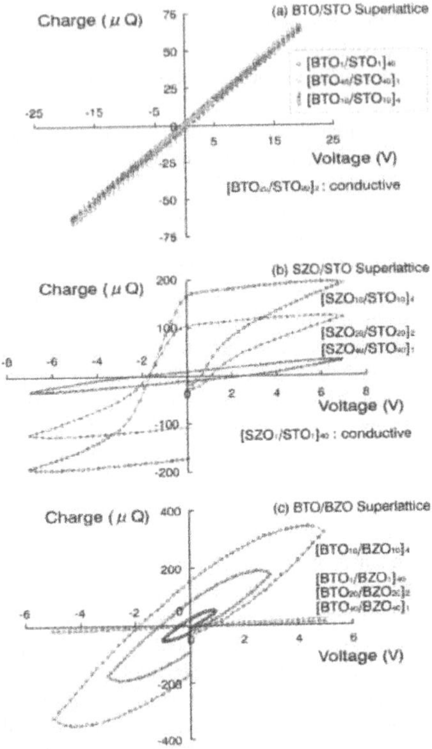

Fig. 5. Charge (Q) *vs.* voltage (V) curves of (a) BTO/STO, (b) SZO/STO, and (c) BTO/BZO artificial superlattices.

CONCLUSIONS

In this study, to investigate the origin of the unique dielectric properties of the perovskite-type artificial superlattice, the BTO/STO, SZO/STO, and BTO/BZO superlattices were fabricated by MBE method and XRD reciprocal space mapping measurement around the STO substrate (103) was performed using high-resolution XRD. The superlattices with 10-periodic structure were most distorted in the direction of c-axis, and the lattice distortion decreased with increasing superlattice periodicity above the 10-periodic structure. The lattice constant of the 1-periodic superlattice was close to the value of bulk state of the solid solution. The superlattices with the 10-periodic structure showed anomalous behaviors in dielectric properties. The BTO/STO artificial superlattice with the 10-periodic structure showed an extraordinarily high relative dielectric permittivity of 33,000. The SZO/STO superlattice with the 10-periodic structure showed a clear hysteresis loop, although the superlattice consisted of the paraelectric/paraelectric combination. These results could be understood by a lattice strain effect induced between the hetero-layers in the superlattices because the strains due to the lattice mismatch between the consisting layers were most effectively induced into the superlattices with the 10-periodic structure.

REFERENCES

[1] H. Tabata, H. Tanaka, and T. Kawai, "Formation of artificial $BaTiO_3/SrTiO_3$ superlattices using pulsed laser deposition and their dielectric properties", Appl. Phys. Lett. 65, 1970 (1994).
[2] K. Iijima, T. Terashima, Y. Bando, K. Kamigai, and H. Terauchi, "Atomic layer growth of oxide thin films with perovskite-type structure by reactive evaporation", J. Appl. Phys. 72, 2840 (1992).
[3] T. Tsurumi, T. Suzuki, M. Yamane, and M. Daimon, "Fabrication of Barium Titanate/Strontium Titanate Artificial Superlattice by Atomic Layer Epitaxy", Jpn. J. Appl. Phys., Part 1 33, 5192 (1994).
[4] T. Tsurumi, T. Miyasou, Y. Ishibashi, and N. Ohashi, "Preparation and Dielectric Property of $BaTiO_3$-$SrTiO_3$ Artificially Modulated Structures", Jpn. J. Appl. Phys., Part 1 37, 5104 (1998).
[5] T. Tsurumi, T. Ichikawa, T. Harigai, H. Kakemoto, and S. Wada, "Dielectric and Optical Properties of $BaTiO_3/SrTiO_3$ and $BaTiO_3/BaZrO_3$ Superlattices", J. Appl. Phys. 91, 2284 (2002).
[6] T. Harigai, D. Tanaka, H. Kakemoto, S. Wada, and T. Tsurumi, "Dielectric Properties of $BaTiO_3/SrTiO_3$ Superlattices measured with Interdigital Electrodes and Electromagnetic Field Analysis", J. Appl. Phys. 94, 7923 (2003).
[7] R. J. Nelmes, G. M. Meyer and J. Hutton, Ferroelectrics 21, 461 (1978).
[8] H. T. Evans Jr, Acta Crystallographica 14, 1019 (1961).
[9] Wong-Ng, W., McMurdie, H., Paretzkin, B., Hubbard, C., Dragoo, A., NBS (USA). ICDD Grant-in-Aid (1988).
[10] H. D. Megaw, Proc. Phys. Soc. 58, 133 (1946).
[11] T. N. Verbitskaia, G. S. Zhdanov, Iu. N. Venevtsev and S. P. Soloviev, Sov. Phys.-Cryst. 3, 182 (1958).
[12] T. Harigai, D. Tanaka, S. M. Nam, H. Kakemoto, S. Wada, K. Saito and T. Tsurumi, "Preparation and Dielectric Properties of $SrZrO_3/SrTiO_3$ Superlattices", Jpn. J. Appl. Phys. 43, 6530 (2004).
[13] T. Harigai, D. Tanaka, S. M. Nam, H. Kakemoto, S. Wada and T. Tsurumi, "Ferroelectric behavior in $SrZrO_3/SrTiO_3$ Arificial Superlattices", Trans. Mater. Res. Soc. Jpn. 29, 1211 (2004).
[14] T. Tsurumi, T. Harigai, D. Tanaka, H. Kakemoto and S. Wada, "Anomalous dielectric and optical properties in perovskite-type artificial superlattices", Science and Technology of Advanced Materials 5, 425 (2004).
[15] T. Tsurumi, T. Harigai, D. Tanaka, S. M. Nam, H. Kakemoto, S. Wada, and K. Saito, "Artificial Ferroelectricity in Perovskite Superlattices", Appl. Phys. Lett. 85, 5016 (2004).

Electroceramic Materials and Applications
Edited by R. W. Schwartz

EFFECTS OF BISMUTH DEFICIENCY ON PIEZOELECTRIC PROPERTIES OF $(Bi_{0.5-x}Na_{0.5}TiO_3)_{0.94}(BaTiO_3)_{0.06}$ CERAMICS

Jiro Abe, Masafumi Kobune, Tetsuo Yazawa, and Hiroshi Nishioka
Department of Material Science Engineering, Graduate School of Engineering, University of Hyogo, 2167, Shosha, Himeji, 671-2201, Japan

Yasuhiro Nakai, and Shigeru Osaka
Kyocera corporation, Kokubu Plant 1-1 Yamashita-cho, Kokubu, Kagoshima, 899-4396, Japan

ABSTRACT

High-density ceramics with compositions of $(Bi_{0.5-x}Na_{0.5}TiO_3)_{0.94}(BaTiO_3)_{0.06}$ (BNBT-6), where x=0-0.03, were fabricated to develop useful lead-free piezoelectric materials with high mechanical quality factors (Q_m). Effects of Bi deficiency on physical properties of the fabricated ceramics were investigated. In the microstructures of the samples with Bi-deficiency concentration of $x \geq 0.015$, no appreciable grain boundaries were observed. This is due to the liquid-phase sintering occurred by the Bi-deficiency concentration of $x \geq 0.015$. The mechanical quality factors for radial vibration mode and for thickness vibration mode exhibited the maxima at x=0.022 and 0.024 that were 674 and 128, respectively. The temperature coefficients of resonant and antiresonant frequencies (TCF) of the BNBT-6 samples with x=0.019-0.030 indicated almost constant, being TCF_{fr1}=-400 ppm/°C for the fundamental resonant frequency and TCF_{fa1}=-320 ppm/°C for the fundamental antiresonant frequency, respectively. This technology that makes deficiency of a very small quantity of Bi to the BNBT-6 composition produces a substantial improvement in Q_m characteristics without lowering other piezoelectric parameters.

INTRODUCTION

In recent years, activities for reducing an environmental load on a global scale have been materialized. Particularly, the appearance of high-performance electronic parts having no harmful substances has been greatly expected. $Pb(Zr, Ti)O_3$ (PZT) that is a binary system solid solution between lead titanate ($PbTiO_3$) and lead zirconate ($PbZrO_3$) and the ternary system ceramics which can be synthesized by incorporating ferroelectric $Pb(M_1, M_2)O_3$ relaxors with PZT described above have been widely used in practical piezoelectric applications. However, as PZT generally contains harmful lead of approximately 60 mass% in PZT, the diffusion of lead to a natural environment is feared when scrapping them. To solve this problem, the development of high-performance and lead-free piezoelectric ceramics is strongly required. Bismuth sodium titanate ($Bi_{0.5}Na_{0.5}TiO_3$; abbreviated as "BNT") as one of lead-free piezoelectric ceramics has recently attracted much attention.[1] The composition that shows morphotropic phase boundary (MPB) in the same material system was found out, exhibiting a high piezoelectricity. Especially, it is well known that BNT-based solid solution ($(Bi_{0.5-x}Na_{0.5}TiO_3)_{1-a}(BaTiO_3)_a$; abbreviated as "BNBT-100a") ceramics which indicate MPB at a=0.06 show an excellent piezoelectricity.[2] Takenaka et al.[3,4] reported that the mechanical quality factor, Q_m of BNBT-6 ceramics with 0.5 mass% MnO_2 becomes approximately three times higher than that of BNBT-6 ceramics without additives. According to their report, the MnO_2 addition was effective in improving Q_m characteristics without deteriorating other piezoelectric parameters.

In the present study, we attempt to improve the piezoelectric properties by making a deficiency in a part of compositional constituents. There have as yet been no approaches from such an angle on the improvement in the piezoelectric properties. The piezoelectric ceramics with compositions of $(Bi_{0.5-x}Na_{0.5}TiO_3)_{0.94}(BaTiO_3)_{0.06}$, where x=0-0.03, are fabricated by making deficiency of a very small quantity of Bi in the compositional formula previously described. The chemical constituents, microstructure, piezoelectric properties and the temperature dependence of resonant and antiresonant frequencies of the fabricated BNBT-6 ceramics are investigated in detail. The influence of Bi-deficiency concentration on the physical properties of the BNBT-6 ceramics is discussed based on the compositional analysis.

EXPRIMENTAL PROCEDURE

Reagent-grade, Bi_2O_3, Na_2CO_3, TiO_2 and $BaCO_3$ were used as starting materials. All compositions studied here were prepared by conventional mixed oxide method. The calcination temperatures of various BNT powders and $BaTiO_3$ powder synthesized were 800°C and 900°C, respectively. After the obtained various BNT powders and BT powder were weighted according to the compositional formula, $BNT_{0.94}BT_{0.06}$, the mixtures with 2 mass% polyvinyl alcohol (PVA) solution were ground for 12 h in ethanol using a planetary zirconia ball mill. After drying the powders at 105°C for 1.5 h, they were sieved using a 120 mesh (opening; 125 μm) screen. The obtained granules were pressed at approximately 175 MPa into disks of 12.0 mm in diameter and then the disks were sintered at 1200°C for 2 h in air. The crystalline phase and the species of crystal defects were identified using an X-ray diffractometer (XRD; RINT2200, Rigaku) and a Raman spectrometer (T64000, Jobin-Yvon). Real chemical constituents were analyzed by fundamental parameter method using a fluorescent X-ray analyzer (RIX2000, Rigaku)[5]. The microstructure of the samples fractured along the direction perpendicular to the surface of the sample was observed using a scanning electron microscope (SEM, JSM-5310; Jeol). The samples with silver top and bottom electrodes were poled at room temperature under a DC field of 60 kV/cm in a silicone oil. Piezoelectric properties of the poled samples were measured by resonant-antiresonant frequency method based on EMAS-6100[6] using an impedance analyzer (HP4194A, Hewlett-Packard). The temperature dependence of resonant and antiresonant frequencies was investigated at temperature range of -20-80°C.

RESULT AND DISCUSSION

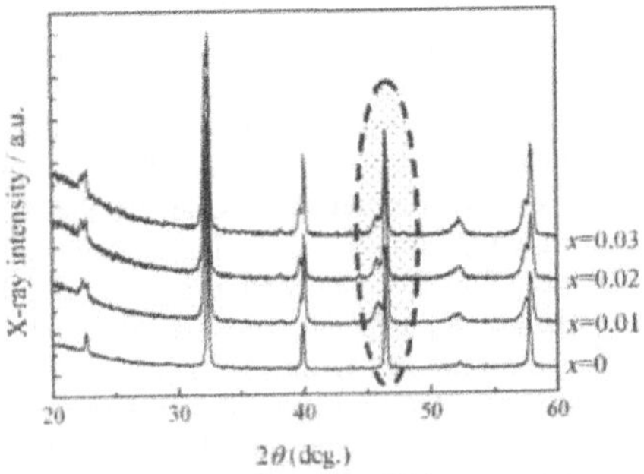

Fig. 1. XRD patterns of $(Bi_{0.5-x}Na_{0.5}TiO_3)_{0.94}(BaTiO_3)_{0.06}$ ceramics as a function of Bi-deficiency concentration, x.

Figure 1 shows the XRD patterns of the polished $(Bi_{0.5-x}Na_{0.5}TiO_3)_{0.94}(BaTiO_3)_{0.06}$ ceramics as a function of Bi-deficiency concentration, x. It has been previously reported that the stable structure of a pure BNBT-6 ceramic at room temperature is a rhombohedral phase.[2] The BNBT-6 samples with Bi deficiency studied here are considered to have the tetragonal structure since the peaks observed at 40, 47 and 58°split into two, suggesting particularly clear split peaks which correspond well to (002) and (200) at around 47°.[7]

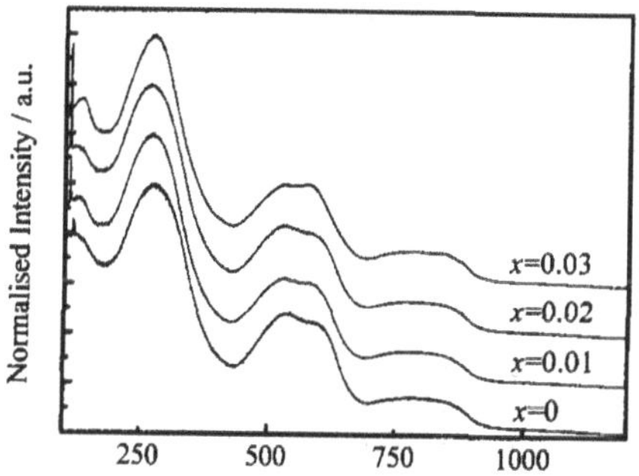

Fig. 2. Raman spectra of $(Bi_{0.5-x}Na_{0.5}TiO_3)_{0.94}(BaTiO_3)_{0.06}$ ceramics as a function of Bi-deficiency concentration, x.

To investigate the species of crystal defects caused by the Bi-deficiency concentration, Raman spectroscopy was employed by using a 514 nm line (50 mW) of an argon ion laser. Raman spectra[8] of the polished $(Bi_{0.5-x}Na_{0.5}TiO_3)_{0.94}(BaTiO_3)_{0.06}$ ceramics as a function of Bi-deficiency concentration, x are shown in Fig. 2. As shown, no appreciable change in crystal structure was ascertained as well as the XRD results. Moreover, all peaks of the observed bands in Raman spectra were accurately separated. Here, the relative peak area, α of the band observed at approximately 600 cm^{-1} can be defined as follows:

$$\alpha = S_{600cm-1} / S_{all} \qquad (1)$$

where $S_{600cm-1}$ and S_{all} denote the peak area of the band observed at approximately 600 cm^{-1} and all the peaks of the bands measured in this study, respectively.[9] The relative peak area, α of the band observed at approximately 600 cm^{-1} decreased slightly from 0.019 to 0.016 as the deficiency concentration increased. The sintering of BNBT-6 ceramics with the same composition was also examined as the preliminary experiment in an oxidation atmosphere and a non-oxidation atmosphere. This peak is consequently considered to be closely connected with the bond between oxygen and metal, based on the fact that α of samples sintered in a non-oxidation atmosphere becomes lower than that of samples sintered in an oxidation atmosphere. Because the decrease in α with increasing Bi-deficiency concentration can be explained by the fact that the Bi deficiency produces the formation of additional oxygen vacancies.

Figure 3 shows a linear relationship between the designed values and the analyzed values. In the range of $0 \leq x \leq 0.025$, the relationship between the designed values and the analyzed values exhibit a straight line with correlation coefficient, r=99.5. After this, piezoelectric properties of BNBT-6 ceramics are discussed based on the Bi content due to the compostional analysis.

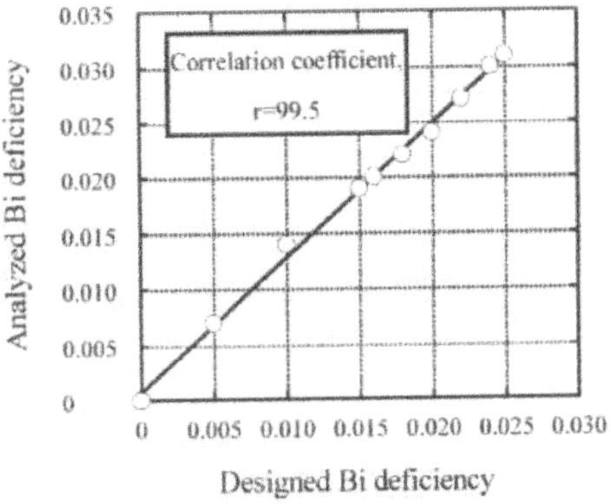

Fig. 3 Relationship between analyzed values and designed values on Bi-deficiency concentration, x, in $(Bi_{0.5-x}Na_{0.5}TiO_3)_{0.94}(BaTiO_3)_{0.06}$ ceramics.

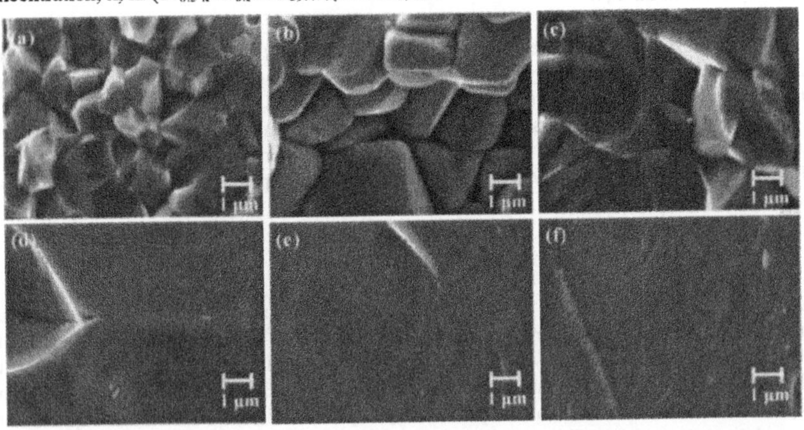

Fig. 4. Cross-sectional SEM photographs of $(Bi_{0.5-x}Na_{0.5}TiO_3)_{0.94}(BaTiO_3)_{0.06}$ ceramics with and without Bi deficiency: (a) x=0, (b) x=0.007, (c) x=0.014, (d) x=0.019, (e) x=0.024 and (f) x=0.031.

Figure 4 shows the SEM photographs of the fractured surface for $(Bi_{0.5-x}Na_{0.5}TiO_3)_{0.94}(BaTiO_3)_{0.06}$ ceramics with and without Bi deficiency. In x=0 sample, several grains which exhibit a rhombohedral structure[2] with an average grain size of approximately 2.3 μm can be observed appreciably, whereas in the case of $0.007 \leq x \leq 0.014$ the shape of the grains changed from rhombohedral to tetragonal and the average grain size became around 1.8 times larger than that of x=0 sample. This result suggests that the phase transition point between the rhombohedral phase and the tetragonal phase exists in the vicinity of x=0.007. Furthermore, for the samples with $x \geq 0.019$, no appreciable grain boundaries were observed as a result of increase in Bi- deficiency concentration. This is attributable to the fact that the liquid phase occurred by the Bi-deficiency concentration of $x \geq 0.019$ played an important role in promoting the densification of the microstructures.

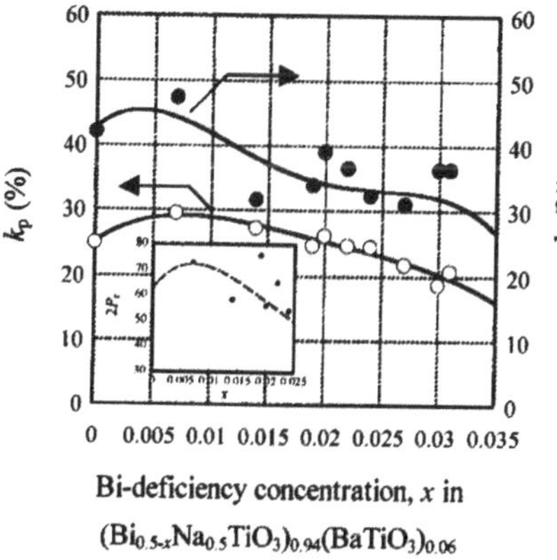

Fig. 5. Electromechanical coupling factors, k_p and k_t of $(Bi_{0.5-x}Na_{0.5}TiO_3)_{0.94}(BaTiO_3)_{0.06}$ samples as a function of Bi-deficiency concentration, x.

The Bi-deficiency concentration dependence of electromechanical coupling factors, k_p and k_t for $(Bi_{0.5-x}Na_{0.5}TiO_3)_{0.94}(BaTiO_3)_{0.06}$ ceramics are shown in Fig. 5. Here, k_p and k_t indicate the efficiency to transform electric energy into mechanical energy in the radial vibration mode and in the thickness vibration mode for thin circular plates, respectively. As shown, x=0.007 samples showed the maximum k_p and k_t values of the ten samples studied here that were 29.4% and 47.3%, respectively. This result corresponds well to the decision of the phase transition point by the previously reported SEM observation. In addition, in the case of $x \geq 0.010$, the behaviors of k_p and k_t show the tendencies to decrease gradually with increasing Bi-deficiency concentration. Since k_p has almost proportional relation with a remanent polarization, P_r, as previously reported by Sakata et al.[10], it can also be seen from the inset of Fig. 5 that oxygen vacancies additionally formed by Bi deficiency produce the restraint in the movement of ferroelectric domain walls as observed in the previous Raman spectra, resulting in a lowering of k_p and k_t.[11, 12]

Mechanical quality factor, Q_m for each vibration mode indicates the sharpness in the vicinity of resonant and antiresonant frequencies. Figure 6 shows the mechanical quality factors of $(Bi_{0.5-x}Na_{0.5}TiO_3)_{0.94}(BaTiO_3)_{0.06}$ ceramics as a function of Bi-deficiency concentration, x. Both Q_m values significantly increased from x=0.015, exhibiting the maxima at x=0.022 and 0.024 that were 674 for radial vibration mode and 128 for thickness vibration mode, respectively.It was found from these results that the respective Q_m values for radial mode and thickness mode correspond to around 3.5 times and 3.3 times as large as those of x=0 BNBT-6 samples. As predicted from the previous SEM observations, the samples with $x \geq 0.015$ indicated a distinct change in microstructures as the volume of grain boundary decreased, showing a significant improvement in Q_m.

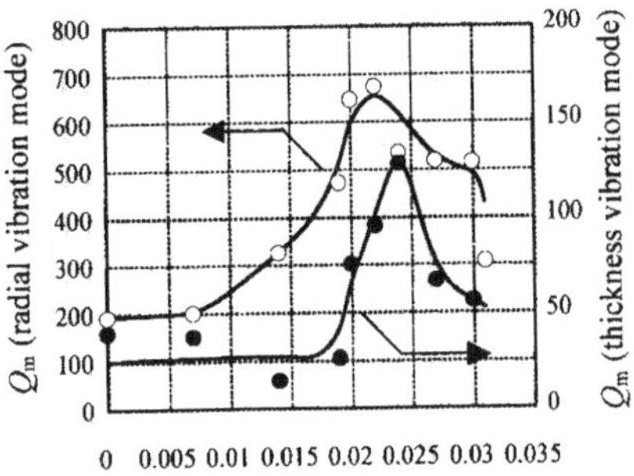

Fig. 6. Mechanical quality factors of $(Bi_{0.5-x}Na_{0.5}TiO_3)_{0.94}(BaTiO_3)_{0.06}$ samples as a function of Bi-deficiency concentration, x.

Figure 7 shows the temperature dependence of resonant and antiresonant frequencies for radial vibration mode of the $(Bi_{0.5}Na_{0.5}TiO_3)_{0.94}(BaTiO_3)_{0.06}$ ceramics without Bi deficiency. As shown, both resonant and antiresonant frequencies tend to decrease almost linearly with increasing temperature in the range -20-80°C. From the slope of this straight line obtained by the method of least squares, the temperature coefficients of resonant and antiresonant frequencies can be estimated. The temperature coefficients of resonant and antiresonant frequencies (TCF)[13, 14] of $(Bi_{0.5-x}Na_{0.5}TiO_3)_{0.94}(BaTiO_3)_{0.06}$ ceramics with and without Bi deficiency are shown in Fig. 8. The TCF values of the BNBT-6 samples with x=0.019-0.030 indicated almost constant, being TCF_{fr1}=-400 ppm/°C for the fundamental resonant frequency and TCF_{fa1}=-320 ppm/°C for the fundamental antiresonant frequency, respectively. This improvement in temperature coefficient is thus considered to be due to the densification of microstructures. In conclusion, the present BNBT-6 ceramics with x=0.022-0.024 exhibited the highest Q_m and excellent temperature characteristics of resonant and antiresonant frequenies.

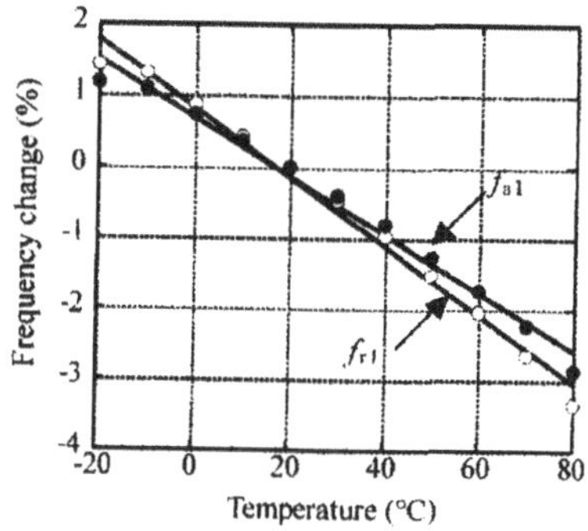

Fig. 7. Temperature dependence of resonant and antiresonant frequencies for radial vibration mode of $(Bi_{0.5}Na_{0.5}TiO_3)_{0.94}(BaTiO_3)_{0.06}$ samples without Bi deficiency. Here, f_{r1} and f_{a1} denote the resonant frequency and the antiresonant frequency, respectively.

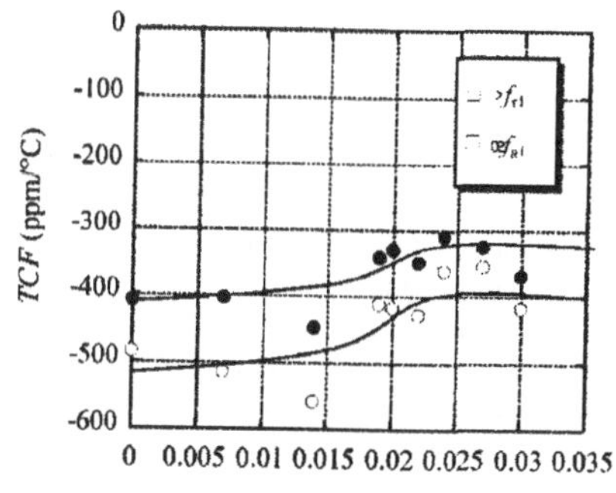

Fig. 8. Temperature coefficients of resonant and antiresonant frequencies of $(Bi_{0.5-x}Na_{0.5}TiO_3)_{0.94}(BaTiO_3)_{0.06}$ samples with and without Bi deficiency. Here, f_{r1} and f_{a1} denote the resonant frequency and the antiresonant frequency, respectively.

This technology that makes deficiency of a very small quantity of Bi to the BNBT-6 composition produced a substantial improvement in Q_m characteristics a without deteriorating other piezoelectric parameters. The proposed technology can hence be expected to be highly useful in promising lead-free piezoelectric materials such as $SrBi_2Ta_2O_9$,[15] $NaNbO_3$[16] and $(Sr_{0.6}Ba_{0.4})NaNb_5O_{15}$.[17]

CONCLUSION

The results of this study can be summarized as follows.

(1) In the microstructures of the samples with Bi-deficiency concentration of $x \geq 0.015$, no appreciable grain boundaries were observed. This is due to the liquid-phase sintering occurred by the Bi-deficiency concentration of $x \geq 0.015$.

(2) The mechanical quality factors, Q_m for radial vibration mode and for thickness vibration mode exhibited the maxima at x=0.022 and 0.024 that were 674 and 128, respectively.

(3) The TCF values of the BNBT-6 samples with x=0.019-0.030 indicated almost constant, being TCF_{fr1}=-400 ppm/°C for the fundamental resonant frequency and TCF_{fa1}=-320 ppm/°C for the fundamental antiresonant frequency, respectively.

(4) This technology that makes deficiency of a very small quantity of Bi to the BNBT-6 composition produced a substantial improvement in Q_m characteristics without deteriorating other piezoelectric parameters.

REFERENCES

[1]T. Takenaka, and K. Sakata, "New Piezo- and Pyroelectric Sensor Materials of $(BiNa)_{1/2}TiO_3$-based Ceramics." *Sensors and Materials Letter*, **3** 123-131 (1997).

[2]T. Takenaka, K. Maruyama, and K. Sakata, "$(Bi_{1/2}Na_{1/2})TiO_3$-$BaTiO_3$ System for Lead-Free Piezoelectric Ceramics." *Jpn. J. Appl. Phys.*, **30** 2236-2239 (1991).

[3]N. Kaewkamnerd, T. Takenaka, K. Sakata, and K. Toda, "The effect of Mn on Piezoelectric Characteristics of $(Bi_{1/2}Na_{1/2})_{0.94}Ba_{0.06}TiO_3$ Ceramics." Abstracts of The Autumn Meeting The Inst. Elec. Infor. Commu. Engi. 5-9 (1992).

[4]H. Nagata, and T. Takenaka, "Additive Effects on Electrical Properties of $(Bi_{1/2}Na_{1/2})TiO_3$ Ferroelectric Ceramics." *J. Eur.Ceram. Soc.*, **21** 1299-1302 (2001).

[5]Y. Kataoka, "Standardless X-ray Fluorescence Spectrometry." *The Rigaku Journal*, **6** 33-40 (1989).

[6]*Electrical Testing Method of Piezoelectric Ceramic Vibrators*, pp. 21-95 in EMAS-6100. Edited by Electronic Materials Manufacturers Association of Japan, (1993).

[7]A. Sasaki, T. Chiba, Y. Mamiya, and E. Otsuki,"Dielectric and Piezoelectric Properties of $(Bi_{0.5}Na_{0.5})TiO_3$-$(Bi_{0.5}K_{0.5})TiO_3$ Systems." *Jpn. J. Appl. Phys.*, **38** 5564-5567 (1999).

[8]C. Y. Kim, T. Sekino, and K. Niihara, "Synthesis of Bismuth Sodium Titanate Nanosized Powders by Solution/Sol-Gel Process." *J. Am. Ceram. Soc.*, **89** 1464-1467 (2003).

[9]A. Mineshige, T. Yasui, N. Ohmura, M. Kobune, S. Fujii, M. Inaba, and Z. Ogumi, "Oxygen Chemical Potential and Mixed Conduction in Doped Ceria under Influence of Oxygen Partial Pressure Gradient." *Solid State Ionics* 493-498. (2002)

[10]K. Sakata, K. Okazaki, "Space Charges Poling and Aging of $BaTiO_3$ Ceramics." *The Inst. Elec. Engi. Jpn.*, **80** 1460-1466 (1960).

[11]T. Izaki, H. Haneda, A. Watanabe, Y. Uchida, J. Tanaka, and S. Shirasaki, "Effects of Mn Ions on the Piezoelectric Property of $(Pb, La)(Zr, Ti)O_3$." *Jpn. J. Appl. Phys.*, **31** 3045-3047 (1992).

[12]T. Kamiya, T. Suzuki, T. Tsurumi and M. Daimon, "Effects of Manganese Addition on Piezoelectric Properties of $Pb(Zr_{0.5}Ti_{0.5})O_3$."; *Jpn. J. Appl. Phys.*, **31** 3058-3060 (1992).

[13]M. Kimura, T. Sawada, A. Ando, and Y. Sakabe, "Energy Trapping Characteristics of Bismuth Layer Structured Compound $CaBi_4Ti_4O_{15}$." *Jpn. J. Appl. Phys.*, **38** 5557-5560 (1999).

[14]M. Suzuki, H. Nagata, J. Ohara, H. Funakubo, and T. Takenaka, "$Bi_{3-x}M_xTiTaO_9$ (M = La or Nd) Ceramics with High Mechanical Quality Factor Q_m." *Jpn. J. Appl. Phys.*, **42** 6090-6093 (2003).
[15]K. Shoji, K. Shibata, T.Nakayama, and K. Sakata, "Preparation and Properties of $SrBi_2Ta_2O_9$ Ceramics II." *Jpn. J. Appl. Phys.*, **39** 5569-5572 (2000).
[16]L. A. Reznitchenko, A. V. Turik, E. M. Kuznetsova, and V. P. Sakhnenko, "Piezoelectricity in $NaNbO_3$ ceramics." *J. Phys. Condens. Matter.*, **13** 3875-3881 (2001).
[17]M. Kimura, T. Minamikawa, A. Ando, and Y. Sakabe, "Temperature Characteristics of $(Ba_{1-x}Sr_x)_2NaNb_5O_{15}$ Ceramics." *Jpn. J. Appl. Phys.*, **36** 6051-6054 (1997).

Electroceramic Materials and Applications
Edited by R. W. Schwartz

ELECTRICAL PROPERTY IMPROVEMENT OF $CaBi_4Ti_4O_{15}$ CERAMICS BY DOPING AND GRAIN ORIENTATION

Yongxiang Li, Jiangtao Zeng, Qingrui Yin
The State Key Lab of High Performance Ceramics and Superfine Microstructure, Shanghai Institute of Ceramics, Chinese Academy of Sciences
1295 Dingxi Road
Shanghai, 200050, P. R. China

ABSTRACT

$CaBi_4Ti_4O_{15}$ (CBT) is a very promising candidate for high temperature piezoelectric applications due to its high Curie temperature and high resistivity. Two approaches were carried out to improve the electrical properties, one is to modify their compounds incorporating other dopants into the A-site or B-site of the crystal lattice; the other is to make grain orientation (textured ceramic) by the multilayer grain growth (MLGG) processing. Nd, V and W doped CBT piezoelectric ceramics were prepared by conventional solid-state reaction process. The influences of these dopants on the sintering temperature, ceramic density and crystal structure were investigated. The improvement on the dielectric, ferroelectric and piezoelectric properties of the doped CBT ceramics were studied. The mechanism of decrease oxygen vacancies and domain structure change by the dopants were discussed. Screen-printing process and nanosized starting materials were used in our MLGG process to get a highly textured $CaBi_4Ti_4O_{15}$ piezoceramics without the need of templates. The degree of grain orientation and the grain morphologies were examined using XRD and SEM techniques. It was found that the particle size of starting materials and the multilayer process are the key factors for the grain growth with preferential orientation. A model to explain the mechanism of MLGG process was proposed.

INTRODUCTION

Bismuth layer-structured ferroelectrics (BLSFs) were first synthesized by Aurivillus in 1949 and they showed a very peculiar crystal structure. BLSFs compounds can be described by the formula $(Bi_2O_2)^{2+}(A_{m-1}B_mO_{3m+1})^{2-}$, where A denotes mono-, di- or trivalent ions with a large ionic radius, and B denotes tetra- or pentavalent ions with a small ionic radius [1-2]. The integer m ranges from 2 to 5. The BLSFs consist of $(Bi_2O_2)^{2+}$ layers interleaved with perovskite-like layer $(A_{m-1}B_mO_{3m+1})^{2-}$.

There are a large number of compounds belonging to BLSFs family which include $SrBi_2(TaNb)_2O_9$ (m=2), $Bi_4Ti_3O_{12}$ (m=3), $(Ca,Sr,Ba)Bi_4Ti_4O_{15}$ (m=4) and $(Ca,Sr,Ba)Bi_4Ti_5O_{18}$ (m=5).[3] This kind of materials have recently attracted a great deal of attention because of their low dielectric constant, high Curie temperature and high resistance to polarization fatigue, so they are considered to be a good candidate for high temperature piezoelectric and nonvolatile memory applications. $SrBi_2Ta_2O_9$(SBT) and lanthanide doped $Bi_4Ti_3O_{12}$ (BIT) films can be used for non-volatile FeRAM[4-5] and donor doped BIT and $CaBi_4Ti_4O_{15}$ (CBT) ceramics can be used in high temperature applications[6-8]. Undoped BLSFs couldn't be practically used because of low ferroelectric and piezoelectric properties and high coercive field.

Many efforts have been paid to improve the ferroelectric and piezoelectric properties of BLSFs, such as A-site doping, B-site doping and grain orientation. Noguchi *et al*[9] found that the remanent polarization of SBT could be markedly improved by Bi^{3+} doping which is accompanied by cation vacancies. Lanthanide doping is an effective way to improve the ferroelectric and fatigue properties of BIT [10-11]. B-site doping is also important for BLSFs especially in BIT, donor doping such as Nb^{5+} and W^{6+} can greatly reduce the current leakage and enhance piezoelectric properties.

The BLSFs are highly anisotropic and their spontaneous polarization mainly lie on *a(b)* direction [12-13]. Randomly oriented grains in ceramics can be aligned along one direction by using grain orientation techniques, and their ferroelectric properties can be enhanced markedly without changing the Curie temperature. Conventionally used grain orientation methods include hot forge (HF) [14-15], templated grain growth (TGG) [16-17] and reactive templated grain growth (RTGG) [18-19]. Pressure should be used in HF process and anisotropic templates are needed for TGG and RTGG processes which make the process quite complicated.

$CaBi_4Ti_4O_{15}$ is a typical BLSF with m=4 and it has very high Curie temperature (Tc=790°C) because of the small radius of Ca^{2+} compared with those of Sr^{2+} and Ba^{2+}. However, it is suffered from low piezoelectricity. In recent years, several studies have been performed to improve the electrical properties of $CaBi_4Ti_4O_{15}$ ceramics by our group [20-22]. The effects of different dopants and the preparation of grain orientated CBT ceramics without using template are summarized in this paper.

EXPERIMENTAL

Nd^{3+} was chosen for A-site doping. There are two ways for A-site doping: Nd^{3+} substituting Bi^{3+} or Ca^{2+}. Therefore, there are two formulae $CaBi_{4-x}Nd_xTi_4O_{15}$ (x=0, 0.05, 0.10, 0.20) (CBNT-x) and $Ca_{1-3/2x}Nd_x\ _{1/2}Bi_4Ti_4O_{15}$ (x=0, 0.25, 0.50, 0.75) (CN BT-x), respectively. Here denotes the cation vacancy. Since Nd^{3+} has higher valence than Ca^{2+}, Nd^{3+} substitute will induce cation vacancies in A-site. V^{5+} and W^{6+} were chosen for B-site doping, there are two formulae $CaBi_4Ti_{4-x}V_xO_{15}$ (x=0, 0.01, 0.03, 0.05, 0.10, 0.15, 0.20) ($CBTV_x$) and $CaBi_4Ti_{4-x}W_xO_{15}$ (x=0, 0.025, 0.050 0.075) ($CBTW_x$), respectively.

Doped ceramics were prepared by the solid-state reaction method. The starting raw materials were analytical pure grade oxide and carbonate powders, Bi_2O_3, TiO_2, $CaCO_3$, Nd_2O_3, V_2O_5 and WO_3. The chemical stoichiometric amounts of the starting powders were thoroughly mixed with ethanol in a ball mill for 4 h, dried and calcined at 850-900°C for 3 h in an alumina crucible. After calcination, the ground and ball-milled powders were pressed into disks of 15 mm in diameter and about 2 mm in thickness. The pressed tablets were finally sintered at temperatures of 1100-1180°C for 4 h in a sealed alumina crucible. Pt electrodes were sputtered on both sides of the ceramic specimens.

$CaBi_4Ti_{3.95}V_{0.05}O_{15}$ ($CBTV_{0.05}$) ceramics were prepared by the screen-printing method. The starting materials were analytical grade oxide and carbonate powders: Bi_2O_3, $CaCO_3$ and TiO_2. Two kinds of TiO_2 powders were used in the experiment: nanosized TiO_2 (P25) and micrometer sized TiO_2. Stoichiometric amounts of the starting powders were thoroughly mixed with ethanol in a ball mill for 4 h, then dried and calcined at 850°C for 3 h in an alumina crucible. After calcination, the ground and ball-milled powders mixed with ethyl-cellulose and -terpineol organic vehicle for 3 h to obtain the screen-printing paste. The paste was composed with 38 wt.% of the calcined powders and 62 wt.% of the organic vehicle. The paste was screen-printed onto a glass substrate and then dried at 90°C, and repeated the process 20 times until the multi-layered thick film reaches about 100 μm. The film was removed carefully from glass substrate and cut into 12×12 mm^2, then, stacked 20 layers high, and pressed uniaxially at 100 MPa. The binder was burned out by heating the green-body in air at 1°C/min to 600°C. The laminates were cold isostatic pressed at 400 MPa and then sintered at temperatures of 1000°C to 1180°C in air.

Bulk densities of the sintered ceramics were measured by the Archimedes method. The crystal phase was determined by X-ray diffraction (XRD) analysis (D/max 2550V) using Cu K_{α_1}

radiation with a scan speed of 4°/min and a step width of 0.02°. The ferroelectric hysteresis loops were measured by a TF Analyzer 2000 FE-Module ferroelectric tester. Dielectric constant as a function of temperature was measured with an HP4284A LCR meter at 100 kHz. Piezoelectric constant d_{33} was measured by a quasi-static d_{33} meter (Model ZJ-3A, Institute of Acoustics, CAS, Beijing).

RESULTS AND DISCUSSION

(1) A-site doping

The XRD patterns of CBNT-x and CN BT-x ceramics confirmed that for CBNT-x ceramics, single phase BLSFs with m=4 ceramics were obtained when Nd^{3+} doping from 0.25 to 0.75, but for CN BT-x ceramics, single phase BLSFs with m=4 can only be obtained when $x \leq 0.1$. It is implied that Nd^{3+} could substitute more Bi^{3+} than Ca^{2+}. The ionic radius of $Nd^{3+}<Ca^{2+}<Bi^{3+}$, Nd^{3+} substituting Bi^{3+} will cause much structure distortion, but the solid solubility limit shows opposite tendency. The reason is that Nd^{3+} can also substitute the Bi^{3+} ions in $(Bi_2O_2)^{2+}$ layer. Osada [23] studied La^{3+} doped BIT ceramics by Raman spectra and found that La^{3+} could enter the $(Bi_2O_2)^{2+}$ layer when La^{3+} content is more than 1.0 mole. The distortion of crystal structure can be decreased by Nd^{3+} entering the $(Bi_2O_2)^{2+}$ layer, so Nd^{3+} have much higher solid solution capacity when substituting Bi^{3+}.

Figure 1 shows the dielectric constant and loss of CBNT-x and CN BT-x ceramics at room temperature. The dielectric constants of CBNT-x ceramics only increased a little than that of CBT ceramics, but the dielectric constants of CN BT-x ceramics are much higher than that of CBT ceramics at room temperature. The ionic polarizability of Nd^{3+} is much higher than that of Ca^{2+}, which led to an increased dielectric constant. The dielectric loss of samples increased for CN BT-x, but decreased for CBNT-x.

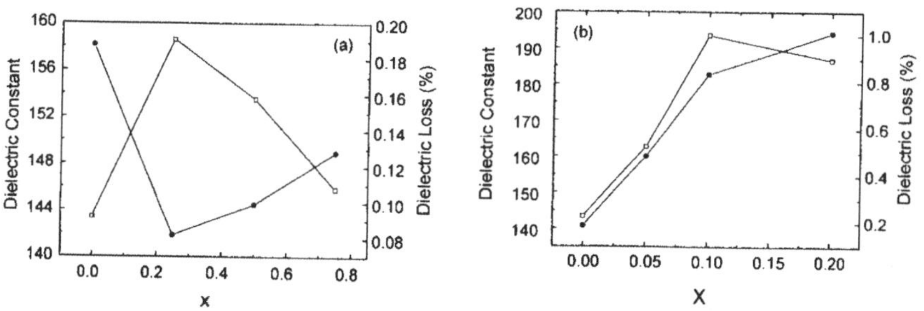

Fig. 1. Dielectric constant and loss of CBNT-x (a) and CN BT-x (b) at room temperature as a function of Nd^{3+} dopant.

The temperature dependence of dielectric constant and loss of CBNT-x and CN BT-x ceramics are shown in Fig.2. It can be seen that there are different dielectric behaviors for two types doped ceramics. For CBNT-x ceramics, the Curie temperature and the dielectric constant at Tc decreased obviously with the increase of Nd^{3+} content, the Tc decreases linearly with Nd^{3+} contents which follows the formula Tc=791-150x. For CN BT-x ceramics, the Curie temperature was slightly changed by Nd^{3+} doping and the dielectric constants at *Tc* for different Nd^{3+} contents are nearly the same. There is another dielectric peak at 120°C below the Curie when Nd^{3+} substituting Ca^{2+} which can't be detected in CBNT-*x* ceramics. It can also be noticed that the dielectric peaks become more distinct with the increase of Nd^{3+} content. It is well known that there is another phase transition in BLSFs with even number (m=2, 4) and the transition belongs to first-order, though the phase transition is difficult to be observed in dielectric constant curve in most cases. It can be noticed also in Fig. 2 that the ceramics with Nd^{3+} substituting Bi^{3+} have much lower dielectric loss than that of ceramics with Nd^{3+} substituting Ca^{2+}. We know that Bi_2O_3 is volatile that will cause oxygen vacancies in the crystal lattice and increase the dielectric loss. Nd^{3+} substituting Bi^{3+} will decrease the content of Bi^{3+} and hence decrease the dielectric loss.

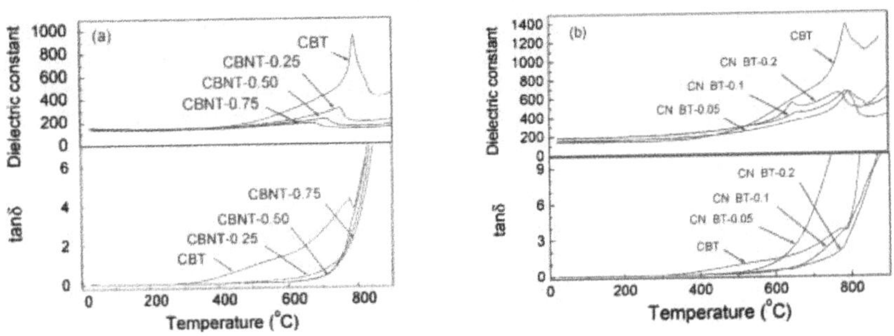

Fig. 2. Temperature dependence of dielectric constant and loss of CBNT-*x* (a) and CN BT-*x* ceramics (b).

Figure 3 shows the *P-E* hysteresis loops of CBNT-*x* and CN BT-*x* ceramics. It can be seen that Nd^{3+} substituting Bi^{3+} will increase the remanent polarization but the coercive field increase at the same time. Nd^{3+} substituting Ca^{2+} will not only increased polarization but also decreased the coercive field. Nd^{3+} substituting Bi^{3+} decreases the leakage current of the CBNT-x ceramics and increases the polarization. The cation vacancy caused by Nd^{3+} substituting Ca^{2+} increases the mobility of ferroelectric domains in CN BT-*x* ceramics and greatly enhance the *Pr* and decrease the *Ec*.

Figure 4 shows the piezoelectric constant of CBNT-*x* and CN BT-*x*. It can be seen that Nd^{3+} substituting Bi^{3+} only slightly increase the piezoelectric constant d_{33} but the piezoelectric constant d_{33} was greatly increased by Nd^{3+} substituting Ca^{2+}. The large remanent polarization and low coercive field are the reasons for the enhancement of piezoelectricity.

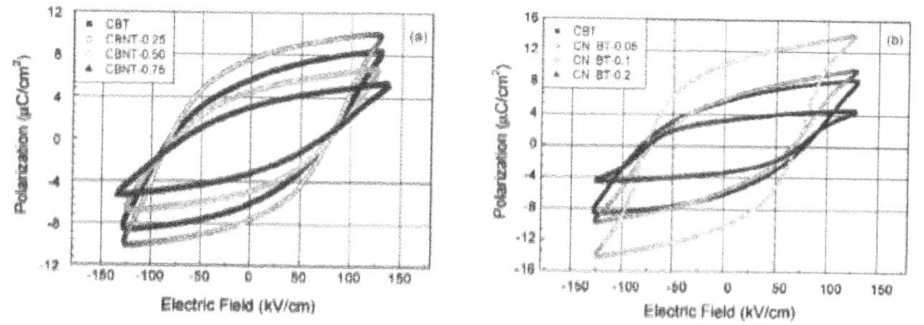

Fig. 3. P-E hysteresis loops of CBNT-x (a) and CN BT-x (b) ceramics.

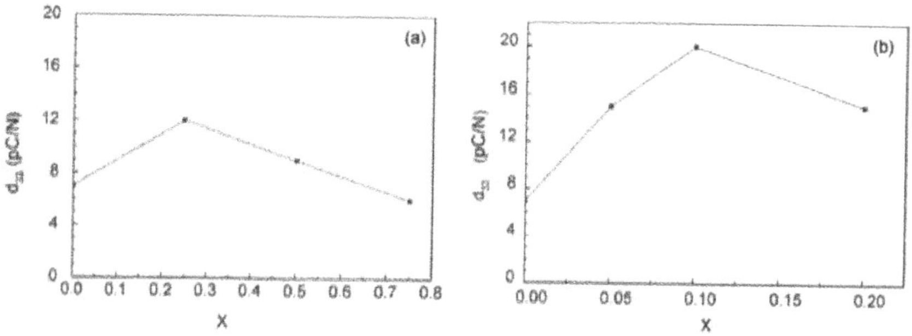

Fig. 4. Piezoelectric constant d_{33} of CBNT-x (a) and CN BT-x (b) as a function of Nd^{3+} dopant.

(2) B-site doping

The XRD patterns of $CBTV_x$ and $CBTW_x$ ceramics indicated that the solid solubilities for V^{5+} and W^{6+} in CBT are about 0.1 and 0.05, respectively. It is well known that V^{5+} has much smaller radius than W^{6+}, but the solubility of V^{5+} is larger than of W^{6+}, so the ionic valence in this case is more important than ionic radius.

Figures 5 shows the temperature dependence of dielectric constant and loss of $CBTV_x$ and $CBTW_x$ ceramics. It can be seen that B-site doping increases the temperature stability of dielectric constant and decreases the dielectric loss at high temperature. Regarding to the optimal doping content of the ceramics with the lowest dielectric loss, for $CBTV_x$ ceramics, the optimal doping content is about 0.05, but for $CBTW_x$, it is about 0.025. Oxygen vacancy is the main source for dielectric loss. With the high valence dopants such as V^{5+} and W^{6+}, the oxygen vacancy concentration of the ceramics is reduced, thus decreased the dielectric loss.

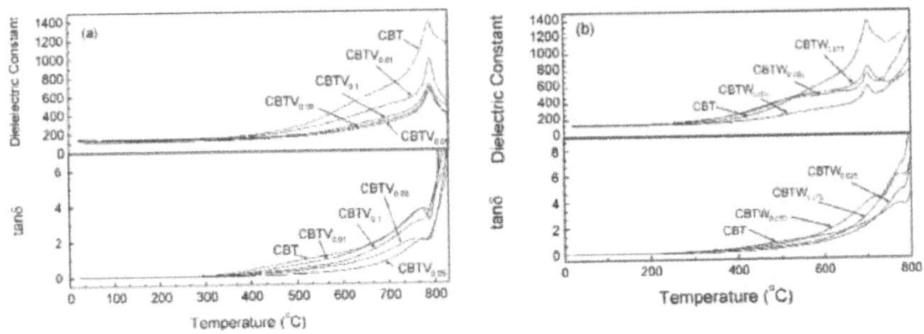

Fig. 5. Temperature dependence of dielectric constant and loss of $CBTV_x$ (a) and $CBTW_x$ (b) ceramics.

Figure 6 shows the *P-E* hysteresis loops of CBT, $CBTV_{0.05}$ and $CBTW_{0.025}$ ceramics. It can be seen that V^{5+} doping only slightly increase the remanent polarization and decrease the coercive field, but W^{6+} doping greatly decrease the remanent polarization and coercive field as well. Though CBT and BIT have similar crystal structure, it is quite different from V^{5+} and W^{6+} doped BIT in which the *Pr* was greatly enhanced due to the different defects. Normally, there are high concentration oxygen vacancies in BIT and make it a *p*-type conductor, while CBT, $CBTV_{0.05}$ and $CBTW_{0.025}$ ceramics showed an *n*-type conductor. In BLSFs, the Bi^{3+} in A-site is very important to its conductivity. The BLSFs systems with all the A-site occupied by Bi^{3+} such as $Bi_4Ti_3O_{12}$, Bi_3NbTiO_9 and $Bi_5Ti_3FeO_{15}$ are highly conductive and high valency doping in B-site have significant influence on the high temperature conductivity. When the A-site is only partly occupied by Bi^{3+}, the ceramics will have rather low conductivity. The concentration of oxygen vacancies in CBT is less than that in BIT, so high valency ion doping have little influence on dielectric and ferroelectric properties.

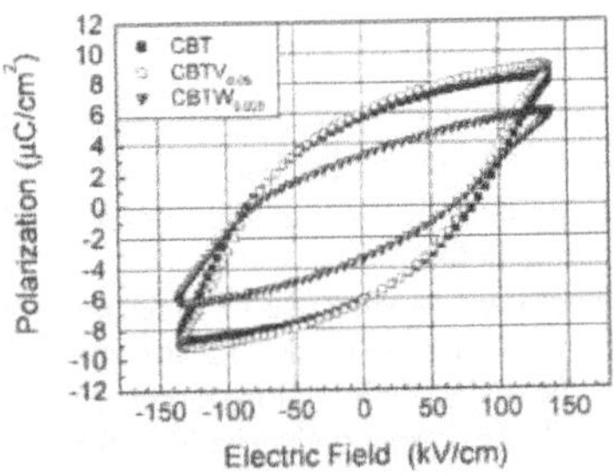

Fig. 6. *P-E* hysteresis loops of CBT, $CBTV_{0.05}$, and $CBTW_{0.025}$ ceramics.

Figure 7 shows the piezoelectric constant of $CBTV_x$ and $CBTW_x$ ceramics. V^{5+} and W^{6+} dopants could increase the piezoelectricity, but the piezoelectric constant is much lower than of

A-site doped CBT ceramics. The improvement of piezoelectricity is due to the decrease of coercive field since B-site doping couldn't increase the remanent polarization effectively.

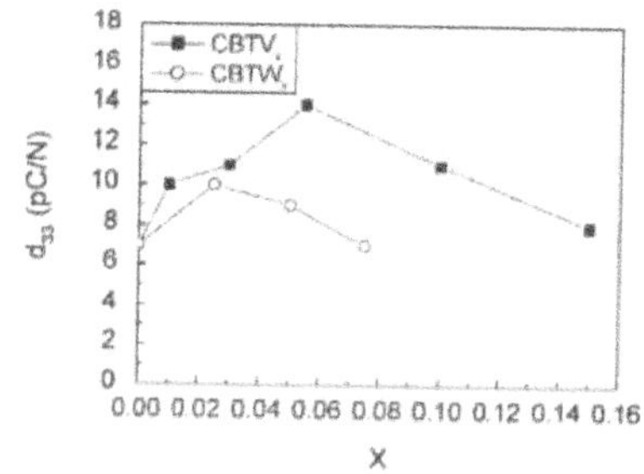

Fig. 7. Piezoelectric constant d_{33} of $CBTV_x$ and $CBTW_x$ ceramics.

(3) Grain orientation

Figure 8 shows the XRD patterns for screen-printed ceramics, the plane perpendicular and parallel to screen-printing plane and the comparison with the XRD pattern of $CBTV_{0.05}$ powders by solid-state reaction. The differences of three patterns can be clearly seen. The *(119)* peak is the strongest peak for $CBTV_{0.05}$ powder, but it decreased markedly and all *(00l)* reflections increased in the parallel plane. The *(00l)* reflections can hardly be observed in the perpendicular plane and *(020)* reflection became the strongest peak instead. This indicates that the ceramic grains were oriented parallel with the *c*-axis, which is perpendicular to the screen-printing plane.

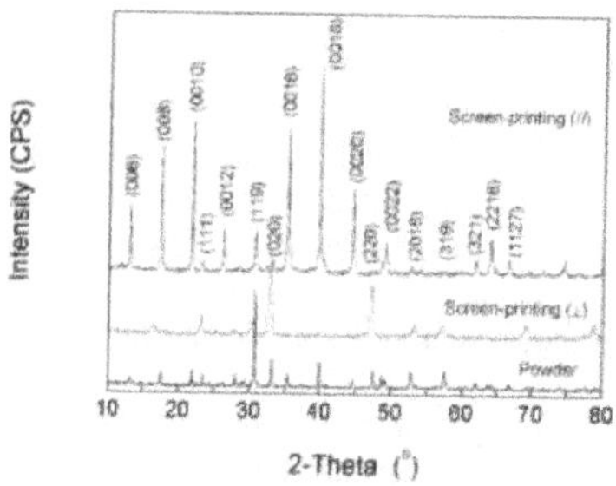

Fig. 8. The XRD pattern of the $CBTV_{0.05}$ ceramics prepared by screen-printing method on the surfaces parallel (//) and perpendicular (□) to the original sheet plane, compared to the XRD pattern of the $CBTV_{0.05}$ powder by conventional solid-state processing.

The degree of grain orientation can be calculated by Lotgering factor *f*, which is defined as: $f=(p-p_0)/(1-p_0)$, where $p=\Sigma I(00l)/\Sigma I(hkl)$, $p_0=p$ for randomly oriented sample. *f* varies from 0 to 1. The reflection peaks between 2θ=10° and 50° were used to calculate p and p_0. In this way, f is as high as 96% for $CBTV_{0.05}$ ceramics prepared by the screen-printing method, which is comparable to the samples prepared by hot forge and templated grain growth methods.

Our study showed that the size of starting material has great influence on grain orientation. Fig. 9 shows the XRD patterns of $CBTV_{0.05}$ ceramics prepared by different sizes of TiO_2 powders. Using nanosized raw materials, highly oriented CBT ceramics were obtained, but when using micrometer sized TiO_2 only randomly oriented ceramics were obtained by the same process. The result indicated that the grain oriented ceramics can only be obtained by using nanosized TiO_2 powders.

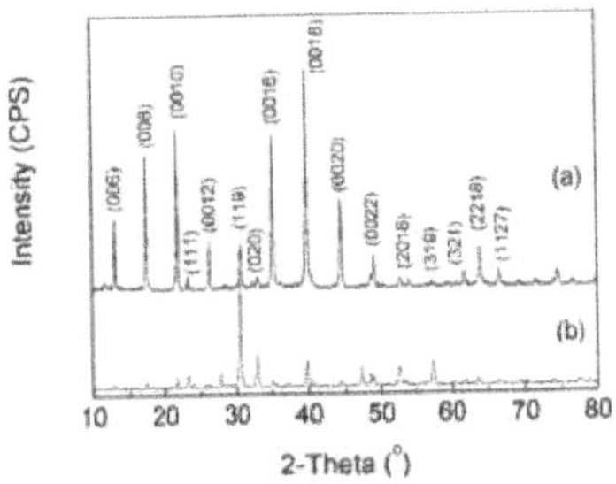

Fig. 9. XRD patterns of screen-printed ceramics with different TiO_2: (a) nanosized TiO_2, (b) micrometer sized TiO_2 powders.

In order to illustrate the size effect more clearly, thick films prepared by nanosized TiO_2 and micrometer sized TiO_2 were laminated together and sintering at various temperatures. Figs.10(a)-(c) show the SEM images of ceramics prepared by MLGG process. The microstructure showed large differences between two adjoining layers. The thickness of nanosized layer is about 15 μm and the thickness of micrometer sized layer is larger than 26 μm. Both layers are randomly oriented at 850°C, but the fracture surface of nanosized layer is quite rough compared to micrometer sized layer. Increasing the sintering temperature up to 900°C, a textured microstructure developed in nanosized layer, and there are small fraction of grains randomly oriented since *a(b)* plane can be seen in the layer. In micrometer sized layer only randomly oriented grains can be seen. When increase the temperature to 1050°C, the grain coarsened in both layers. The grain orientation degree in nanosized layer was higher than that sintered at 900°C.

Figure 11 shows the SEM images of dewaxed green pellets prepared by screen-printing using different TiO_2 powders. It can be seen that the shape of grains is plate-like that indicate layer-structured grains was synthesized after calcinating. Most of the grains are randomly oriented in Fig.11(a), but there are still some large particles which align along the screen-printing direction. Because the thickness of screen-printed film is very small, the large plate-like particles can be aligned along the screen-printing direction by shear stress in the process. These particles act as templates during sintering process. This process was similar to templated grain growth, but the template was synthesized during calcinating process, no need to prepare the templates by molten-salt or hydrothermal processes in advance. In the sample prepared by micrometer sized

TiO_2, because the screen-printed film is much thicker than that prepared by nanosized TiO_2, plate-like particles couldn't be aligned along the screen-printing direction by shear stress, hence the grain orientation could not be obtained.

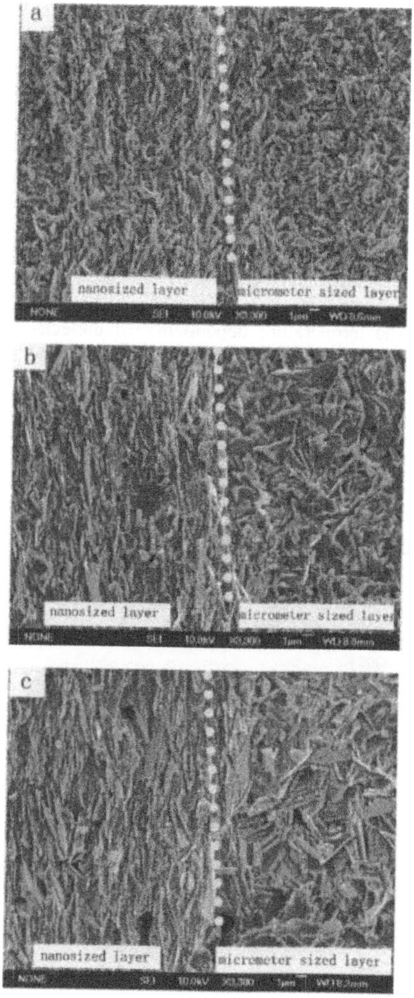

Fig. 10. SEM images of the laminates with different TiO_2 powders sintered at: (a) 850°C, (b) 900°C and, (c) 1050°C.

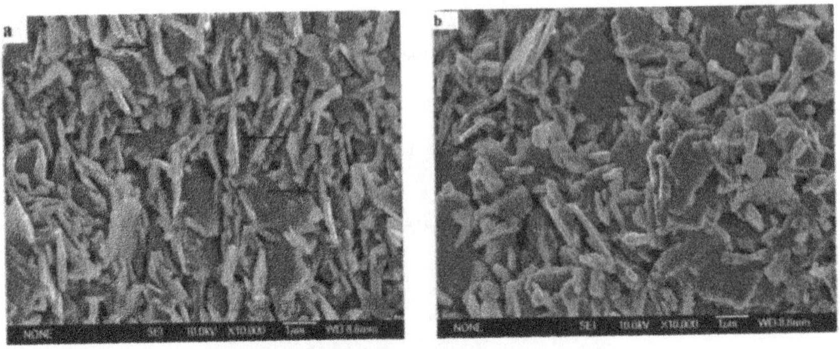

Fig. 11. SEM images of dewaxed green pellets prepared by screen-printing using different TiO_2 powders: (a) nanosized TiO_2 and, (b) micrometer sized TiO_2.

Figure 12 shows the temperature dependence of dielectric constant of screen-printed ceramics at two directions and randomly oriented ceramics. The inset is an enlargement of dielectric constant between R.T. to 350°C. It can be seen that the Curie temperature wasn't changed by grain orientation. The dielectric constant in perpendicular plane is higher than that in parallel plane, and the dielectric constant of randomly oriented ceramics lies between these two grain orientated planes. The difference of dielectric constant between two planes increases with the increase of temperature and reaches the maximum at Curie temperature. The dielectric constant in perpendicular plane is about 12 times larger than that in parallel plane. The value is very close to some single crystal of BLSFs. Irie *et al* studied the dielectric properties of $SrBi_4Ti_4O_{15}$ and found that the dielectric constant along *a(b)* direction is about ten times higher than that along the c axis at *Tc* [24].

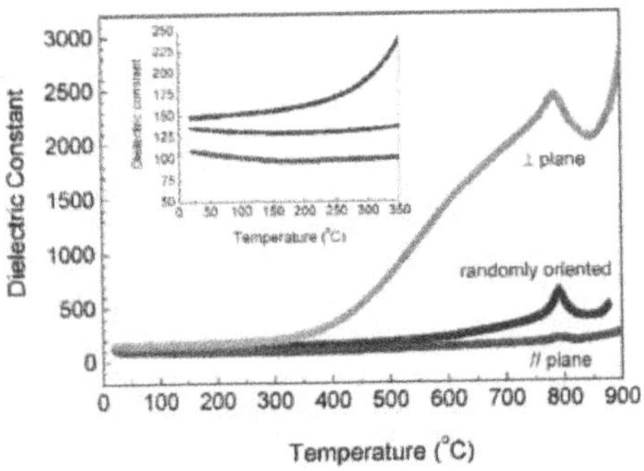

Fig. 12. Temperature dependence of dielectric constant of screen-printed $CBTV_{0.05}$ ceramics in two planes and the comparison with random oriented ceramics.

Figure 13 shows the temperature dependence of dielectric loss of screen-printed ceramics in two planes and randomly oriented ceramics. The dielectric loss in perpendicular plane is higher than that in parallel plane. The higher dielectric loss in perpendicular plane is caused by the high conductivity along *a(b)* direction. The dielectric loss of randomly oriented ceramics lies between these two planes at low temperature, but it is lower than those prepared by screen-printing process at high temperature. The reason for the low dielectric loss of the ceramics prepared by conventional method is that they usually have higher density than that prepared by screen-printing method.

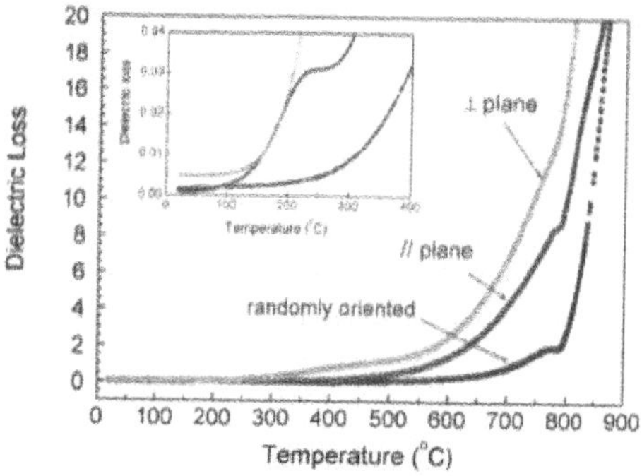

Fig.13. Temperature dependence of dielectric loss of screen-printed $CBTV_{0.05}$ ceramics in two planes and the comparison with random oriented ceramics.

Figure 14 shows a typical *P-E* hysteresis curve of $CBTV_{0.05}$ ceramics at a temperature of 150°C. It is very difficult to obtain saturated hysteresis loop because the leakage current at perpendicular direction is quite high, but the difference of the hysteresis in two planes are obvious. The remanent polarization in parallel plane is as low as 0.31 $\mu C/cm^2$, but it is much

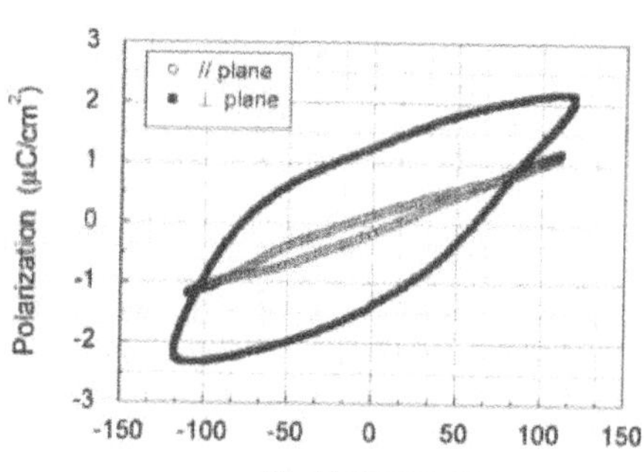

Fig. 14. *P-E* hysteresis loops of screen-printed $CBTV_{0.05}$ ceramics in two perpendicular planes.

higher in perpendicular plane (2.64 $\mu C/cm^2$). For even layered BLSFs, the spontaneous polarization lies along *a(b)* direction, and no polarization along the *c*-axis due to a mirror symmetry. $CBTV_{0.05}$ ceramics prepared by screen-printing have the *c*-axis perpendicular to the screen-printing plane, so the perpendicular plane was expected to have better properties.

CONCLUSIONS

A-site and B-site doping and grain orientation technique were used to improve the electrical properties of CBT ceramics. Nd^{3+} doped CBT ceramics were prepared by solid state-reaction. Nd^{3+} substituting Ca^{2+} or Bi^{3+} have different effects on the electrical properties. CBNT-x ceramics have lower dielectric loss than CNBT-x ceramics. CNBT-x ceramics have better ferroelectric and piezoelectric properties than CBNT-x ceramics. When Nd^{3+} substitutes Bi^{3+}, Nd^{3+} can enter the $(Bi_2O_2)^{2+}$ layer.

V^{5+} and W^{6+} doped CBT ceramics were prepared by solid state-reaction. High valence B-site dopants have very low solid solubility. Small amount of V^{5+} and W^{6+} doping can decrease the dielectric loss. V^{5+} doping slightly increased the remanent polarization, but W^{6+} doping markedly decreased the remanent polarization and coercive field. CBT showed an n-type conducting mechanism and V^{5+}, W^{6+} doping increased the conductivity at high temperature.

Grain oriented $CBTV_{0.05}$ ceramics were prepared by screen-printing method. The particle size of starting materials plays a crucial role in grain growth, highly oriented ceramics could only be obtained by using nanosized TiO_2 powders. Screen-printed $CBTV_{0.05}$ ceramics showed highly anisotropic dielectric and ferroelectric properties. Without the need of templates and pressure densification process, the screen-printing based MLGG is a new approach to fabricate grain-orientated piezoceramics.

ACKNOWLEDGEMENTS

This work was supported by the Ministry of Sciences and Technology of China through 973-project (2002CB613307), National Advanced Materials Committee of China (863-project no. 2001AA325070) and the Innovation project of Shanghai Institute of Ceramics (no. SCX200409).

REFERENCES

[1]R. E. Newnham, R. W. Wolfe, and J. F. Dorrian, "Structural Basis of Ferroelectricity in the Bismuth Titanate Family", *Mat. Res. Bull.*, **6**, 1029-40 (1971).

[2]R. E. Newnham, R. W. Wolfe, R. S. Horsey, and F. A. Diaz-colon, "Crystal Structure of (Sr, Ba)$Bi_2Ta_2O_9$", *Mat. Res. Bull.*, **8**, 1183-96 (1973).

[3]E. C. Subbarao, "A Family of Ferroelectric Bismuth Compound", *J. Phys. Chem. Solids*, **23**, 665-676 (1962).

[4]J. F. Scott, and C. A. Araujo, "Ferroelectric Memories", *Science*, **246,** 1400-1405 (1989).

[5]B. H. Park, B. S. Kang, S. D. Bu, T.W. Noh, J. Lee, W. Jo, "Lanthanum-substituted Bismuth Titanate for Use in Non-volatile Memories," *Nature*, **401**, 682-684, (1999).

[6]M. Villegas, T. Jardiel, and G. Farias, "Sintering and Electrical Properties of $Bi_4Ti_{2.95}W_x O_{11.9+3x}$ Piezoelectric Ceramics" *J. Euro. Ceram. Soc.*, **24**, 1025-1029 (2004).

[7]R. Q. Chu, L. N. Zhang, and Z. J. Xu, "Effect of Niobium Doping on the Properties and Ferroelectric Domain Structures of $Bi_4Ti_3O_{12}$ Ceramics", *Phys. Stat. Sol. (a)*, **201**, R45-R48 (2003).

[8]M. Yokosuka, "Dielectric and Piezoelectric Properties of Mn-Modified $Bi_4CaTi_4O_{15}$ Based Ceramics", *Jpn. J. Appl. Phys.*, **41**, 7123-6 (2002).

[9]Y. Noguchi, M. Miyayama and Tetsuichi Kudo, "Direct Evidence of *A*-site-deficient Strontium Bismuth Tantalite and Its Enhanced Ferroelectric Properties", *Phys. Rev. B*, 2001, **63**, 214102-1-5 (2001).

[10]U. Chon, H. M. Jang, M. G. Kim, C. H. Chang, "Layered Perovskites with Giant Spontaneous Polarizations for Nonvolatile Memories", *Phys. Rev. Lett.*, **89**, 087601-1-4 (2002).

[11]T. Kojima, T. Watanabe, and H. Funakubo, "Ferroelectric Properties of Lanthanide-substituted $Bi_4Ti_3O_{12}$ Epitaxial Thin Films Grown by Metalorganic Chemical Vapor Deposition," *J. Appl. Phys.*, **93**, 1707-1712 (2003).

[12]H. Irie, M. Miyayama and T. Kudo, "Structure Dependence of Ferroelectric Properties of Bismuth Layer-structured Ferroelectric Single Crystals", *J. Appl. Phys.*, **90**, 4089-4094 (2001).

[13]M. Miyayama, I. Yi, "Electrical Anisotropy in Single Crystals of Bi-layer Structured Ferroelectrics", *Ceram. Int.*, **26**, 529-533 (2000).

[14]T. Takenaka, and K. Sakata, "Grain Orientation and Electrical Properties of Hot-forged $Bi_4Ti_4O_{12}$ Ceramics", *Jpn. J. Appl. Phys.*, **19**, 31-39 (1980).

[15]T. Kimura, T. Yoshimoto, and N. Lida, "Mechanism of Grain Orientation During Hot-pressing of Bismuth Titanate" *J. Am. Ceram. Soc.*, **72**, 85-89 (1989).

[16]J. A. Horn, S. C. Zhang, U. Selvaraj, "Templated Grain Growth of Textured Bismuth Titanate", *J. Am. Ceram. Soc.*, **82**, 921-926 (1999).

[17]E. Sabolsky, S. Kwon and E. Suvaci, "Dielectric and Electromechanical Properties of (001)-textured PMNPT", *12th IEEE International Symposium on the Applications of Ferroelectrics*, 393-396 (2000).

[18]T. Takeuchi, T. Tani, Y. Saito, "Piezoelectric Properties of Bismuth Layer-structure Ferroelectric Ceramics With a Preferred Orientation Processed by the Reactive Templated Grain Growth Method", *Jpn. J. Appl. Phys.*, **38**, 5553-5556 (1999).

[19]T. Takeuchi, T. Tani, Y. Saito, "Unidirectionally Textured $CaBi_4Ti_4O_{15}$ Ceramics by the Reactive Templated Grain Growth With an Extrusion", *Jpn. J. Appl. Phys.*, **39**, 5577-5580 (2000).

[20]J.T. Zeng, Y. X. Li, Q. B. Yang, "Fabrication and Properties of Grain Oriented $CaBi_4Ti_4O_{15}$ Piezoelectric Ceramics, " *J. of Euro. Ceram. Soc.*, 25(12), 2727-2730 (2005).

[21]J.T. Zeng, Y.X. Li, D. Wang, Q.R. Yin, "Electrical Properties of Neodymium Doped $CaBi_4Ti_4O_{15}$ Ceramics," *Solid State Communication*, **133**, 553-557 (2005).

[22]J.T. Zeng, Y.X. Li, Q.B. Yang, Q.R. Yin, "Ferroelectric and Piezoelectric Properties of Vanadium Doped $CaBi_4Ti_4O_{15}$ Ceramics," *Materials Science and Engineering*, **B117**, 241-245 (2005).

[23]M. Osada, M. Tada, M. Kakihana, T. Watanabe and H. Funakubo, "Cation Distribution and Structural Instability in $Bi_{4-x}La_xTi_3O_{12}$, *Jpn. J. Appl. Phys.*, **40,** 5572-5575 (2001).

[24]H. Irie, and M. Miyayama, "Dielectric and Ferroelectric Properties of $SrBi_4Ti_4O_{15}$ Single Crystals" *Appl. Phys. Lett.*, **79**, 251-253 (2001).

Electroceramic Materials and Applications
Edited by R. W. Schwartz

ELECTRICAL PROPERTIES OF $BaBi_4Ti_4O_{15}$ CERAMICS MODIFIED BY Bi_2SiO_5

Tomoaki Inoue, Masafumi Kobune, Jiro Abe and Tetsuo Yazawa
Graduate School of Engineering, University of Hyogo. Hyogo 671-2201, Japan

ABSTRACT

The effects of the addition of 0-16.7 mol% Bi_2SiO_5 (BSO) to high-density $BaBi_4Ti_4O_{15}$ (BBT) ceramics sintered in a powder bed are discussed based on analyses of their dielectric, ferroelectric and piezoelectric properties. The Curie temperature of BBT samples increased gradually from 391°C to 478°C as the BSO content increased. The electromechanical coupling factor, k_t, for the thickness vibration mode and the mechanical quality factor, Q_m, for the radial vibration mode of the sample containing 9.1 mol% BSO were 29% and 1476, respectively. The temperature coefficients of the fundamental resonant frequency, $|TCF_{fr}|$, and the fundamental antiresonant frequency, $|TCF_{fa}|$, were 58-67 ppm/°C and 50-66 ppm/°C, respectively, indicating excellent thermal properties suitable for practical applications. BBT ceramics containing 9.1 mol% BSO are shown to be potential piezoelectric element materials for piezoelectric resonators.

INTRODUCTION

Ferroelectrics are promising materials for data storage in digital memory system applications, and as such, have attracted much attention recently for piezoelectric and pyroelectric applications such as piezoelectric resonators and pyroelectric infrared sensors. In the case of $Pb(Zr, Ti)O_3$ (PZT)[1-3] and relaxor-PZT systems,[4-6] which have a typical perovskite structure, the research and development of element materials for important electronic components such as ferroelectric random access memories (FeRAMs) and piezoelectric actuators has received extensive coverage in the literature. However, the following issues that have come up recently demand immediate attention: (1) reduction of the strain on the environment due to these lead-based materials, (2) improvement of the polarization fatigue degradation caused by repetitive read/write operation in ferroelectric applications and (3) development of new materials with high Curie temperature ($T_c \geq 300$°C) and high piezoelectric coefficients for piezoelectric applications. Bismuth-layer-structured ferroelectrics (BLSF) such as $SrBi_2Ta_2O_9$ (SBT) and $BaBi_4Ti_4O_{15}$ (BBT) have received attention as promising candidates for addressing these issues. BLSF ceramics afford fatigue-free FeRAMs capable of operating under low voltage, though their relatively small remanent polarization (P_r=7.0-12.0 $\mu C/cm^2$)[7] is a shortcoming. Thus, it has been difficult to fabricate high-density FeRAMs using BLSF ceramics as element materials.[7-10] So far, there have been a few reports on the improvement of the physical properties by adding a small amount of Bi_2SiO_5 (BSO) to BLSF. Kijima et al.[11, 12] reported that the addition of 40 mol% BSO to $Bi_4Ti_3O_{12}$ (BIT) is effective in decreasing the crystallization temperature and improving the surface morphology of the ferroelectric thin films. Idemoto et al.[13, 14] found that BSO addition to SBT and $(Bi, La)_4Ti_3O_{12}$ (BLT) induced substitution of Si for the B site of the pseudo-perovskite block in BLSF, resulting in an improvement in P_r. It is well known from the several findings described above that BSO is an effective additive for increasing the sinterablity of ferroelectrics and improving their electrical properties. On the other hand, although BSO addition to BLSF such as BIT, SBT and BLT has been investigated, there have been no studies on BBT as a ferroelectric with the same bismuth layer structure.

In the present study, we attempt to fabricate high-density BBT ceramics through the addition of 0-16.7 mol% BSO. The electrical properties (dielectric, piezoelectric and ferroelectric properties) of the BBT ceramics fabricated are investigated in detail. Thus, the effect of BSO addition on electrical properties is revealed and the materials' piezoelectric properties are discussed in connection with applications such as piezoelectric resonators.

EXPERIMENTAL PROCEDURE

The BSO powders used as a sintering aid were synthesized by the chemical solution reaction. The starting material $BiCl_3$ was dissolved in methyl alcohol and mixed for 24 h. The resultant $BiCl_3$ solution was hydrolyzed by adding the appropriate amounts of tetraethoxysilane (TEOS) and HCl, whereby the BSO wet gel was formed. After drying the BSO wet gel at 65°C for 48 h, BSO powders were produced by firing the dried gel at 400°C for 3 h in air. BBT ceramics were fabricated by the solid-state reaction of the conventional mixed-oxide-carbonate method. Reagent-grade Bi_2O_3, TiO_2 and $BaCO_3$ were used as starting materials. First, the mixture of raw materials was ground for 1 h in ethanol in a planetary zirconia ball mill and then calcined BBT powders were obtained by firing the ground powders at 900°C for 2 h in air. After the addition of 0-16.7 mol% BSO powder and 15 wt. % organic binder to the calcined BBT powders, the mixture was ground for 12 h in the same zirconia wet ball mill. The mixture was dried at 105°C for 1.5 h and sieved using a 120-mesh (opening: 125 μm) screen. The granules were pressed at around 196 MPa into discs, 12.0 mm in diameter and roughly 1.0 mm in thickness. The green compacts were sintered at 1080°-1120°C for 3 h in air in a powder bed[15] of $Bi_4Ti_3O_{12}$ (BIT). The sintered bodies obtained were polished to a thickness of 0.3-0.6 mm and silver electrodes were attached to them by firing the silver paste coated on their surfaces at 750°C for 10 min in air. The samples were poled under a DC field of 3.0 MV/m for 30 min at 230°C in a silicone oil. The crystalline phases were analyzed using an X-ray diffractometer (XRD; Rigaku, RINT2200). The bulk density was measured by Archimedes' method using automatic density measuring equipment (Mettler-Toledo, AG204). The dielectric properties were measured under an applied field of 25 V/cm at 100 kHz in vacuo over the temperature range 0°-450°C using a multi-frequency LCR meter (Yokogawa-Hewlett-Packard, YHP4274A). Piezoelectric properties were measured by the resonant-antiresonant frequency method on the basis of EMAS-6100[16] using an impedance/gain-phase analyzer (Hewlett-Packard, HP4194A). Ferroelectric properties were evaluated from the measurements of P-E hysteresis loops carried out at room temperature, under an applied field of 100 kV/cm using a ferroelectric film test system (Radiant Tech., RT-66A). The temperature dependence of resonant and antiresonant frequencies was investigated between -20 and 150°C.

RESULTS AND DISCUSSION

(1) Physical properties of BSO

The BSO powder synthesized by the chemical solution reaction process was white. Fluorescent X-ray analysis results suggest that the synthesized BSO powder is composed of Bi_2O_3:SiO_2 = 1:1. It was confirmed based on these XRD results that the synthesized BSO powder is amorphous. Figure 1 shows the thermogravimetric-differential thermal analysis (TG-DTA) curves for BSO powders. A significant decrease in TG is observed at around 700°C; the DTA curve also exhibits three small endothermic peaks, indicated by arrows. While the endothermic peak labeled A is due to H_2O evaporation, peaks represented B (around 810°C) and C (around 900°C) correspond well to the melting temperatures of Bi_2O_3 and $Bi_{12}SiO_{20}$, respectively, as shown in the phase diagram.[17] It appears based on the above findings that BSO is effective as a sintering aid for fabricating high-density ceramics at the relatively low temperature range of 900°-1100°C.

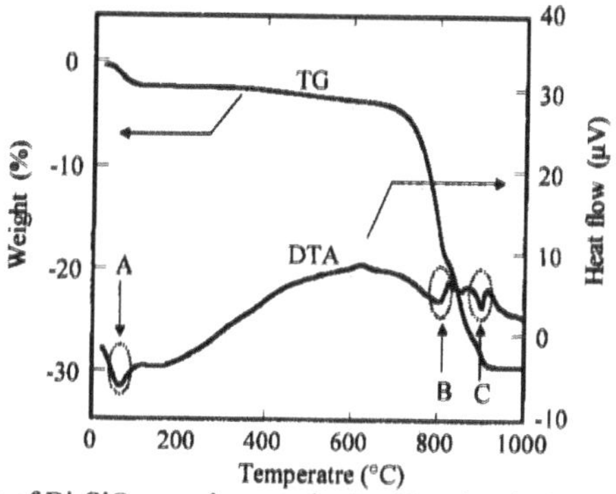

Fig. 1. TG-DTA curves of Bi_2SiO_5 powders synthesized by chemical solution reaction process.

(2) Crystal structure and solubility limit of BSO

The relative density of BSO-free BBT samples was estimated to be approximately 95% of the theoretical density (7.473×10^3 kg/m^3).[18] The bulk density of the present BBT ceramics containing 0-16.7 mol% BSO additives was in the range 7.1-7.3×10^3 kg/m^3. The XRD patterns of BBT ceramics with 0-16.7 mol% BSO are shown in Fig. 2. The data suggest that all BBT samples were almost completely composed of a single-phase bismuth layer structure. In the case of BBT samples containing 4.8-16.7 mol% BSO, however, a weak BSO reflection was observed at around 29° in addition to the strong BBT peak. On the basis of these results, we speculate that the solubility limit of BSO under the present experimental conditions is in the vicinity of 4 mol%, i.e., less than 4.8 mol%.

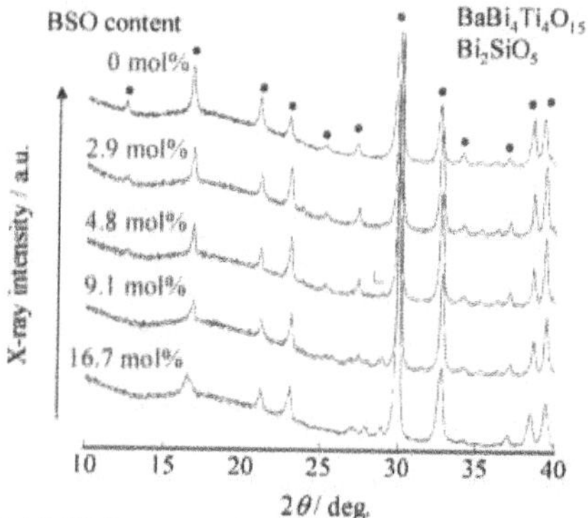

Fig. 2. XRD patterns of $BaBi_4Ti_4O_{15}$ ceramics with 0-16.7 mol% Bi_2SiO_5.

(3) Dielectric properties

Figure 3 plots the temperature dependence of the relative permittivity, ε_r, of BBT ceramics containing 0-16.7 mol% BSO. The variation in Curie temperature, T_c, of BBT ceramics containing 0-16.7 mol% BSO is shown in Fig. 4. These results indicate that T_c increases almost linearly with BSO content. The Curie temperatures of BSO-free BBT ceramics and BBT ceramics containing 16.7 mol% BSO reached 391°C and 478°C, respectively. These results presumably suggest that a small amount of BSO (< 4.8 mol%) is soluble in BBT, whereas in the range 4.8-16.7 mol%, BSO partly precipitates as a secondary phase, as revealed by the previous XRD results.

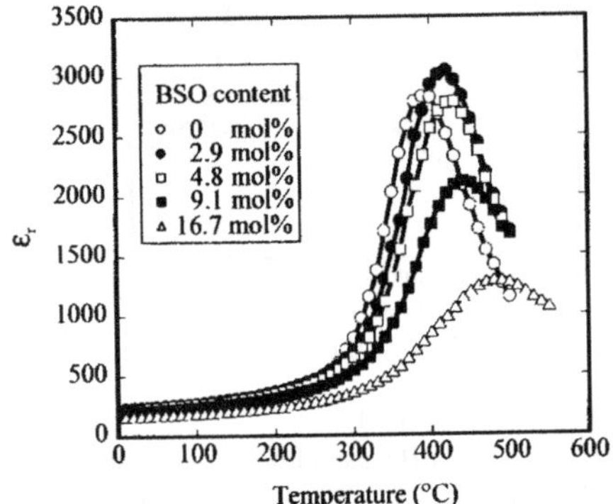

Fig. 3. Temperature dependence of relative permittivity for $BaBi_4Ti_4O_{15}$ ceramics with 0-16.7 mol% Bi_2SiO_5.

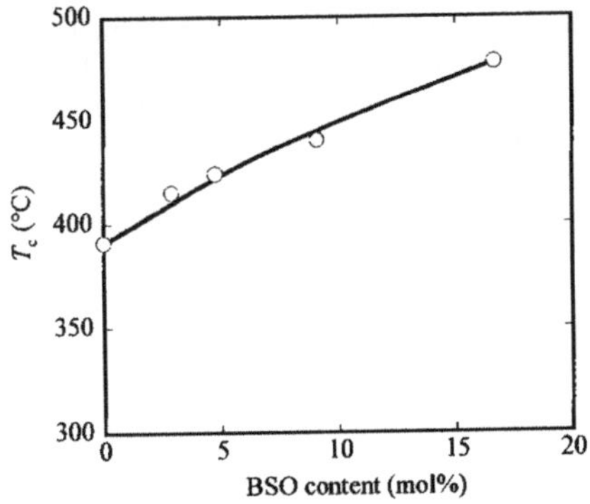

Fig. 4. Variation in Curie temperature of $BaBi_4Ti_4O_{15}$ ceramics with 0-16.7 mol% Bi_2SiO_5.

Palka et al.[19] investigated the physical properties of a Si-doped $PbTiO_3$ ceramic by neutron diffraction, revealing the partial substitution of Si into the $PbTiO_3$ lattice and the concomitant precipitation of a $PbSiO_3$ phase along the grain boundaries. On the basis of these findings, it is thought that in the case of BSO with a structure similar to $PbSiO_3$, Si is incorporated with $BaBi_4Ti_4O_{15}$ and simultaneously Si partly precipitates along the grain boundaries. Fig. 5 plots the relative permittivity, $\varepsilon_{33}^T/\varepsilon_0$, of BBT ceramics containing 0-16.7 mol% BSO as a function of

BSO content. As can be seen, $\varepsilon_{33}{}^{T}/\varepsilon_0$ of BSO-containing BBT ceramics exhibited a maximum (= 223) at 2.9 mol%. On the other hand, $\varepsilon_{33}{}^{T}/\varepsilon_0$ showed a tendency to decrease almost linearly with increasing BSO content in the range 4.8-16.7 mol%. The above XRD data suggest that this decrease is closely related to the precipitation of BSO as a secondary phase at around 4.8 mol%.

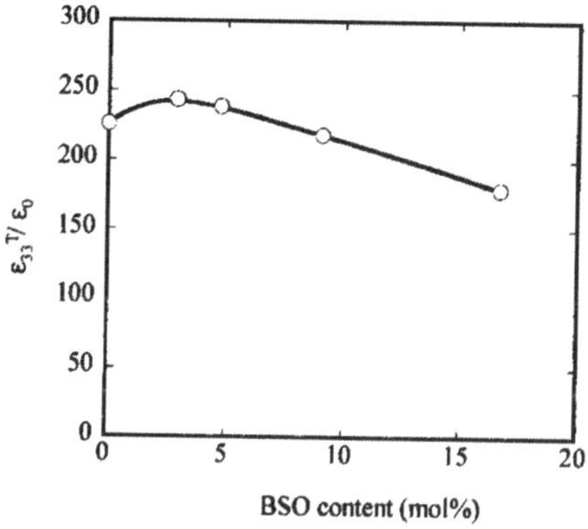

Fig. 5. Relative permittivity of $BaBi_4Ti_4O_{15}$ ceramics with 0-16.7 mol% Bi_2SiO_5 as a function of Bi_2SiO_5 content.

(4) Piezoelectric properties

Figure 6 shows the electromechanical coupling factors, k_p (radial vibration mode) and k_t (thickness vibration mode), and the piezoelectric coefficient, d_{31}, of BBT ceramics containing 0-16.7 mol% BSO as a function of BSO content. Here, k_p and k_t represent the efficiency of transforming electrical energy into mechanical energy in, respectively, the radial and thickness vibration modes for a thin circular plate, while d_{31} denotes the piezoelectric strain coefficient in the length vibration mode for a thin rectangular plate poled along the thickness direction. The BSO-free BBT samples had the following piezoelectric parameters: k_p=7.0%, d_{31}=-5.9 pC/N and k_t=22%. In particular, the k_t value almost coincided with that of a BBT single crystal reported by Sawase et al.[20] As easily seen from the formula for d_{31}, the $\varepsilon_{33}{}^{T}/\varepsilon_0$ behavior is reflected in this d_{31} behavior and consequently, d_{31} of the sample containing 2.9 mol% BSO exhibited a maximum (d_{31}=-7.0 pC/N). Meanwhile, k_p also indicated a maximum (k_p=8.8%). However, k_t exhibited a different behavior because of its different vibration mode. Thus, k_t of BSO-containing BBT ceramics attained a maximum (k_t=28%) at 9.1 mol%, producing an improvement in piezoelectric properties. The mechanical quality factor, Q_m, of BBT ceramics containing 0-16.7 mol% BSO as a function of BSO content is plotted in Fig. 7. After increasing almost linearly with BSO content up to 9.1 mol%, Q_m reached a maximum (1476), and then decreased significantly at 16.7 mol%. We believe that Q_m increased because Si was partially substituted for the B site of the pseudo-perovskite block in BBT, resulting in the formation of $BaBi_4(Si, Ti)_4O_{15}$ in the grains.[13, 14] As predicted from the above TG-DTA results of BSO, these results indicate that BSO addition is effective in promoting the sinterability of BBT ceramics, enabling the fabrication of high-density sintered bodies and improving piezoelectric properties. By contrast, a marked deterioration in piezoelectric properties was confirmed for the BBT samples containing 16.7 mol% BSO. This is presumably attributable to the increase in volume fraction of BSO (paraelectrics) precipitating as a secondary phase along grain boundaries upon the addition of excess BSO.

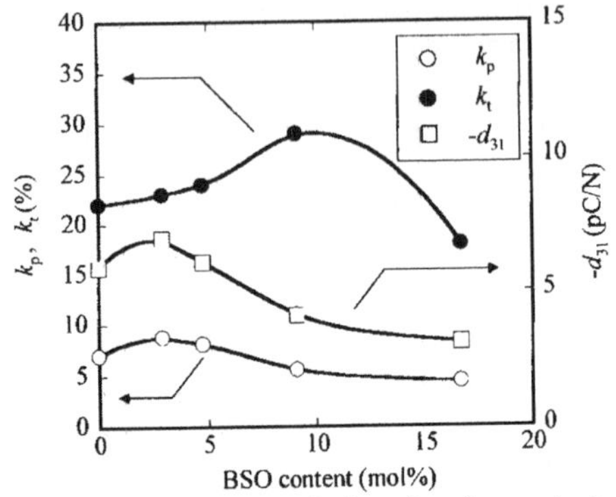

Fig. 6. Electromechanical coupling factors and piezoelectric constant of $BaBi_4Ti_4O_{15}$ ceramics with 0-16.7 mol% Bi_2SiO_5 as a function of Bi_2SiO_5 content.

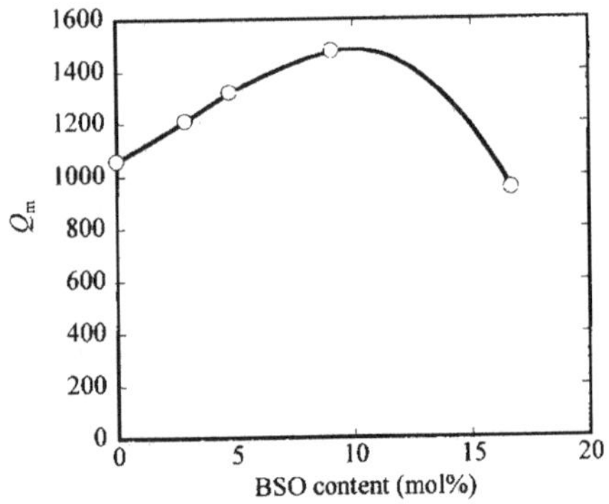

Fig. 7. Mechanical quality factor of $BaBi_4Ti_4O_{15}$ ceramics with 0-16.7 mol% Bi_2SiO_5 as a function of Bi_2SiO_5 content.

(5) Ferroelectric properties

Figure 8 shows the (a) P-E hysteresis loops and (b) BSO content dependence of the remanent polarization and the coercive field for BBT ceramics containing 0-16.7 mol% BSO. The remanent polarization, $2P_r$, exhibited a maximum ($2P_r$=11.5 μC/cm^2) at 2.9 mol%, while the coercive field, $2E_c$, remained almost constant ($2E_c$=89-100 kV/cm), regardless of the BSO content. The remanent polarization exhibited a maximum at the composition (BSO=2.9 mol%) where both k_p and d_{31} attained their maxima, as is frequently seen in lead-based piezoelectric materials. The fact that $2P_r$ of the present material is slightly lower than those of BBT ceramics with a similar composition described in the literature[7, 14] is due to low maximum applied voltage of the present measuring instrument. Thus, we believe that $2P_r$ of the present sample increases further if its polarization is saturated under an applied electric field of around 30 MV/m.

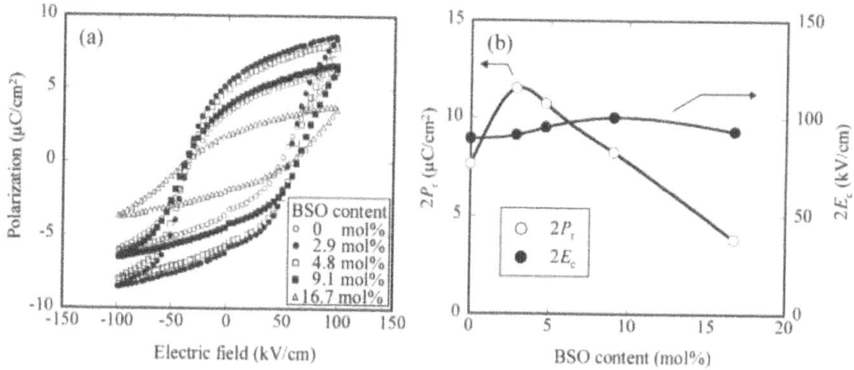

Fig. 8. (a) P-E hysteresis loops and (b) BSO content dependence of the remanent polarization and the coercive field for BBT ceramics with 0-16.7 mol% BSO.

(6) Thermal properties of resonant-antiresonant frequency

Figure 9 shows the temperature dependence of the resonant and antiresonant frequencies in radial vibration mode for BBT ceramics containing 9.1 mol% BSO in the temperature range -20°-150°C. As can be seen, the resonant and antiresonant frequencies both showed a tendency to decrease linearly with increasing temperature.

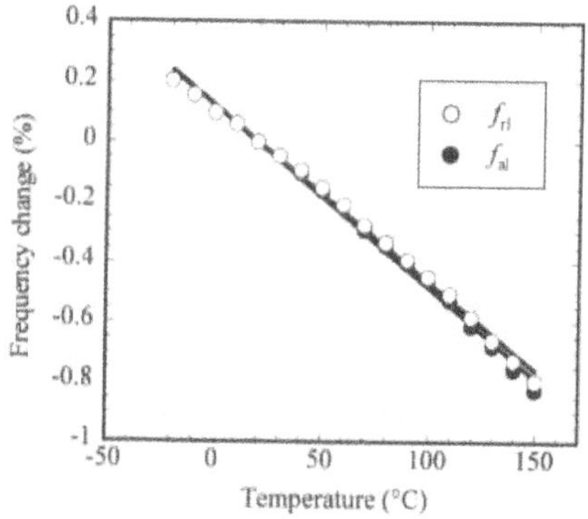

Fig. 9. Temperature dependence of resonant and antiresonant frequencies in radial vibration mode of $BaBi_4Ti_4O_{15}$ ceramics with 9.1 mol% Bi_2SiO_5. Here, f_{r1} and f_{a1} denote the resonant frequency and the antiresonant frequency, respectively.

The temperature coefficients of resonant and antiresonant frequencies (TCF)[21-24] are derived from

$$TCF = (f_{max} - f_{min})/(170 \cdot f_{rt}) \cdot 10^6 \ (ppm/°C) \qquad (1)$$

where f_{max} and f_{min} denote the maxima and minima of the resonant and antiresonant frequencies in the temperature range -20°-150°C, respectively. f_{rt} represents the resonant and antiresonant frequencies at 20°C. Figure 10 shows the change in TCF of BBT ceramics containing 0-16.7 mol% BSO. It was found that the resonant and antiresonant frequencies of all samples shifted slightly toward lower frequencies. The temperature coefficients of the fundamental resonant frequency, $|TCF_{fr1}|$, and the fundamental antiresonant, $|TCF_{fa1}|$, were 58-67 ppm/°C and 50-66 ppm/°C, respectively, indicating excellent thermal properties suitable for practical applications. On the basis of the above results, we investigated the relationship between the temperature characteristics of resonant-antiresonant frequencies and the BSO content of BBT ceramics containing 9.1 mol%, whose Q_m exhibits a maximum (= 1476). As a result, we obtained relatively small $|TCF_{fr1}|$ and $|TCF_{fa1}|$ values that remained almost constant over the BSO content range studied. Thus, we have demonstrated that BBT ceramics containing 9.1 mol% BSO show great promise as element materials for piezoelectric resonators.

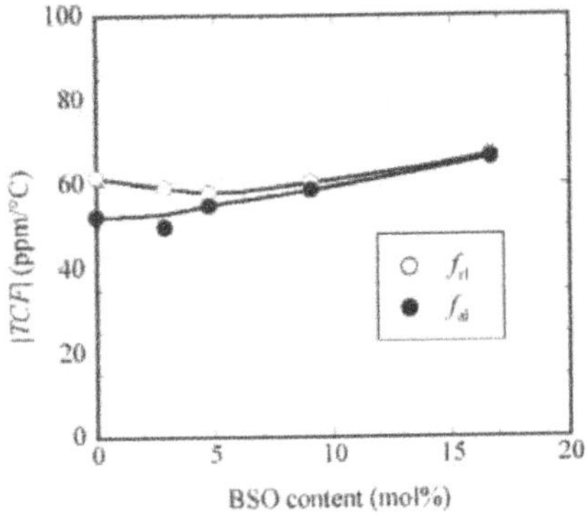

Fig. 10. Variation in temperature coefficients of resonant and antiresonant frequencies in radial vibration mode of $BaBi_4Ti_4O_{15}$ ceramics with 0-16.7 mol% Bi_2SiO_5. Here, f_{r1} and f_{a1} denote the resonant frequency and the antiresonant frequency, respectively.

CONCLUSIONS

High-density BBT ceramics with 0-16.7 mol% BSO were fabricated by powder-bed sintering. The physical properties of the fabricated BBT ceramics BSO were investigated in detail. The results of this study can be summarized as follows:

(1) The Curie temperature of BBT samples increases gradually from 391°C to 478°C as the BSO content increased.

(2) The electromechanical coupling factor, k_t, for the thickness vibration mode and the mechanical quality factor, Q_m for the radial vibration mode of BBT ceramics containing 9.1 mol% BSO were 29% and 1476, respectively.

(3) The temperature coefficients of the resonant frequency, $|TCF_{fr}|$, and the antiresonant frequency, $|TCF_{fa}|$, were 58-67 ppm/°C and 50-66 ppm/°C, respectively, indicating excellent thermal properties suitable for practical applications.

REFERENCES

[1]M. Kobune, Y. Nishioka, T. Inoue, and T. Yazawa, "Growth and Ferroelectric Properties of $Pb(Zr_{0.52}Ti_{0.48})O_3$ Thin Films Crystallized on MgO Single-Crystal Substrates by Hot Isostatic Pressing," *Journal of Crystal Growth*, **275** [1-2] e2421-25 (2004).

[2]S. Tashiro, T. Murata, K. Ishii, and H. Igarashi, "Grain Size Dependence of Third Nonlinear Piezoelectric Coefficient in Lead Zirconate Titanate Ceramics," *Jpn. J. Appl. Phys.*, **40** Part 1, No. 9B, 5679-82 (2001).

[3]T. Futakuchi, K. Tanino, H. Sawasaki, and M. Adachi, "Low-Temperature Mixed Sintering of $Pb(Zr_{1-x}Ti_x)O_3$-$Pb(Zr_{1/3}Nb_{2/3})O_3$ Ceramics and Their Pyroelectric Properties," *Jpn. J. Appl. Phys.*, **36** Part 1, No. 9B, 5981-83 (1997).

[4]H. Ouchi, K. Nagano, and S. Hayakawa, "Piezoelectric Properties of $Pb(Mg_{1/3}Nb_{2/3})O_3$-$PbTiO_3$-$PbZrO_3$ Solid Solution Ceramics," *J. Am. Ceram. Soc.*, **48** [12] 630-35 (1965).

[5]H. Ouchi, "Piezoelectric Properties and Phase Relations of $Pb(Mg_{1/3}Nb_{2/3})O_3$-$PbTiO_3$-$PbZrO_3$ Ceramics with Barium or Strontium Substitutions," *J. Am. Ceram. Soc.*, **51** [3] 169-76 (1968).

[6]M. Kobune, K. Muto, and Y. Akiyama, "Piezoelectric and Ferroelectric Properties of $Pb(Zn_{1/3}Nb_{2/3})O_3$- $PbTiO_3$-$PbZrO_3$ Ceramics," *J. Ceram. Soc. Jpn.*, **110** [1] 12-17 (2002).

[7]T. Kobayashi, Y. Noguchi, and M. Miyayama, "Polarization Properties of Superlattice-structured $Bi_4Ti_3O_{12}$-$BaBi_4Ti_4O_{15}$ Single Crystals and Ceramics: Comparison with $Bi_4Ti_3O_{12}$ and $BaBi_4Ti_4O_{15}$," *Jpn. J. Appl. Phys.*, **43** NO. 9B, 6653-57 (2004).

[8]C. A-Paz de Araujo, J. D. Cuchiaro, L. D. McMillan, M. C. Scott, and J. F. Scott, "Fatigue-free Ferroelectric Capacitors with platinum electrodes," *Nature*, **374** [13] 627-29 (1995).

[9]H. Irie, M. Miyayama, and T. Kudo, "Structure Dependence of Ferroelectric Properties of Bismuth Layer-structured Ferroelectric Single Crystals," *J. Appl. Phys.*, **90** [8] 4089-94 (2001).

[10]H. Irie, M. Miyayama, and T. Kudo, "Enhanced Ferroelectric Properties of V-Doped $BaBi_4Ti_4O_{15}$ Single Crystal," *J. Appl. Phys.*, **40** Part 1, [1] 239-43 (2001).

[11]T. Kijima, and H. Ishiwara, "Ulutra-Thin Ferroelectric Films Modified by Bi_2SiO_5," *Ferroelectrics*, **271**, 289-95 (2002).

[12]T. Kijima, and H. Ishiwara, "Si-Substituted Ultrathin Ferroelectric Films," *Jpn. J. Appl. Phys.*, **41** Part 2, No. 6B, L716-19 (2002).

[13]Y. Idemoto, T. Miyahara, N. Koura, T. Kijima, and H. Ishiwara, "Crystal Structure and Ferroelectric Properties of $(Bi,La)_4(Ti,Si)_3O_{12}$ as a Bulk Ferroelectric Material," *Solid State Communications*, **128** [6/7] 255-59 (2003).

[14]Y. Idemoto, "Ferroelectric Performances of Si-added Ferroelectric Oxides," Function and materials, **23** [8] 31-36 (2003) [in Japanese].

[15]M. Kobune, S. Fujii, K. Furuya, and R. Takayama, "Preparation of La- and Mg-Modified $PbTiO_3$ Ceramics and Their Pyroelectric Properties," *J. Ceram. Soc. Jpn.*, **102** [11] 999-1004 (1994).

[16]*Electrical Testing Method of Piezoelectric Ceramic Vibrators*, pp. 21-95 in EMAS-6100. Edited by Electronic Materials Manufacturers Association of Japan, 1993.

[17]Robert S. Roth, "Phase Equilibria Diagrams," **11**, pp.115 Edited and Published by The American Ceramic Society (1995).

[18]JCPDS CARD "$BaBi_4Ti_4O_{15}$" #35-0757

[19]V. R. Palkar, S. Chattopadhyay, P. Ayyub, M. Multani, S. K. Paranjape, and V. Siruguri, "A Study of The Structure and Composition of Si-doped $PbTiO_3$," *Materials Letters*, **32** [2/3] 171-74 (1997).

[20]H. Sawase, H. Kakemoto, S. Wada, and T. Tsurumi,"Growth of Bismuth Layer-structured Ferroelectrics $BaBi_4Ti_4O_{15}$ Single Crystals and Their Piezoelectric Properties," *Trans. Mater. Res. Soc. Jpn.*, **28** [1] 197-200 (2003).

[21]T. Sawada, A. Ando, Y. Sakabe, D. Damjanovic, and N. Setter, "Properties of the Elastic Anomaly in $SrBi_2Nb_2O_9$-based Ceramics," *Jpn. J. Appl. Phys.*, **42** Part 1, No. 9B, 6094-98 (2003).

[22]H. Ogawa, M. Kimura, A. Ando, and Y. Sakabe, "Temperature Dependence of Piezoelectric Properties of Grain-Oriented $CaBi_4Ti_4O_{15}$ Ceramics," *Jpn. J. Appl. Phys.*, **40** Part 1, No. 9B, 5715-18 (2001).

[23]M. Suzuki, H. Nagata, J. Ohara, H. Funakubo, and T. Takenaka, "$Bi_{3-x}M_xTiTaO_9$ (M = La or Nd) Ceramics with High Mechanical Quality Factor" *Jpn. J. Appl. Phys.*, **42** Part 1, No. 9B, 6090-93 (2003).

[24]M. Kimura, T. Sawada, A. Ando, and Y. Sakabe, "Energy Trapping Characteristics of Bismuth Layer Structured Compound $CaBi_4Ti_4O_{15}$." *Jpn. J. Appl. Phys.* **38** Part 1, No. 9B, 5557-60 (1999).

Electroceramic Materials and Applications
Edited by R. W. Schwartz

PREPARATION AND CHARACTERIZATION OF LEAD-FREE PIEZOELECTRIC $(K_{0.5}Bi_{0.5})TiO_3$ CERAMICS

Takashi Obayashi and Takahiro Wada
Department of Materials Chemistry, Ryukoku University
Seta Otsu 520-2194 Japan

ABSTRACT

High-density potassium bismuth titanate $(K_{0.5}Bi_{0.5})TiO_3$ (KBT) ceramics with a relative density of about 95% were obtained by a conventional solid-state reaction and sintering method with the addition of 2 mol% of MnO_2 or SiO_2. The crystal structure, microstructure, dielectric and piezoelectric properties were characterized. The dielectric and piezoelectric properties of the fabricated KBT ceramics were comparable to those of pure KBT ceramics produced by the polymerized complex method. KBT ceramic with the addition of MnO_2 had electromechanical coupling factors of k_p = 14.8%, k_{31} = 11.0% and k_t=33.8% while the corresponding values of KBT ceramic with the addition of SiO_2 were k_p=19.2%, k_{31}=13.0% and k_t = 33.1%. The effects of the addition of MnO_2 and SiO_2 on the sintering of KBT are discussed.

INTRODUCTION

Lead zirconate titanate (PZT) ceramics have been widely used as piezoelectric materials for various applications such as filters, oscillators and actuators. Recently, lead-free piezoelectric ceramic materials have attracted attention from the viewpoint of environmental awareness. Sodium bismuth titanate $(Na_{0.5}Bi_{0.5})TiO_3$ (NBT)-based solid solution have been extensively studied by some groups,[1,2] because NBT is a ferroelectric material with the high Curie temperature of 320°C.

Potassium bismuth titanate, $(K_{0.5}Bi_{0.5})TiO_3$ (KBT), is also a ferroelectric material with the high Curie temperature of 380°C. However, research on KBT has rarely been reported,[3,4] possibly because it is difficult to fabricate high-density KBT ceramics due to the high volatility of the potassium component at sintering temperatures. Previously, we reported on the creation of high quality KBT fine powder prepared by a polymerized complex method (PCM). KBT ceramic with a relative density of about 95% was then fabricated from the KBT fine powder.[5] Recently, KBT ceramics with a relative density of 97% were obtained by a hot-pressing method.[6]

In this study, we attempted to fabricate high-density KBT ceramic using a combination of a conventional solid-state reaction and sintering. First, we searched for the appropriate sintering additive of KBT. It was found that a small amount of MnO_2 or SiO_2 worked effectively as a sintering additive for KBT. High-density KBT ceramics were fabricated using a conventional solid-state reaction and sintering with an addition of small amount of MnO_2 or SiO_2. The phases in the obtained high density KBT ceramics were characterized by X-ray diffraction analysis, the microstructure was observed using a scanning electron microscope (SEM), and the dielectric and piezoelectric properties were measured. The measured characteristics were compared with those of non-doped KBT ceramics fabricated by the polymerized complex method.

EXPERIMENTAL PROCEDURE

$KHCO_3$ (99.5%), Bi_2O_3 (99.99%), and rutile-type TiO_2 (99.9%) were used as raw materials. These raw materials and the sintering additive (MnO_2 or SiO_2) were weighed according to the compositional formulas 1.00KBT-0.02MnO_2 [KBT-0.02MnO_2] or 1.00KBT-0.02SiO_2 [KBT-0.02SiO_2]. The weighed powders were dry-mixed for 1.5 h using an agate mortar and pestle. The mixed powders were calcined in the powder-form at 850°C for 2 h and then re-calcined in the pressed pellet-form at 900°C for 2 h. The re-calicined powders were pulverized and mixed with polyvinyl alcohol (PVA), pressed into disks of 10 mm diameter under a pressure of 700 kgf/cm^2 and sintered at 1040°C for 2 h in air.

The densities for the obtained ceramics were measured by the Archimedes method. The phases in the samples were identified by powder X-ray diffraction (XRD) using Cu-K_α radiation. The microstructure of fractured surface of the sample was observed using a scanning electron microscope (SEM).

The upper and lower surfaces of the obtained disc samples were polished for the characterization of the dielectric and piezoelectric properties. The polished surfaces of the disks were coated with silver paste and then fired at 700°C for 15 min. The relative dielectric constants, ε_r, and loss tangents, tanδ, for unpoled samples were measured at room temperature using an inductance capacitance-resistance (LCR) meter (Agilent 4263B). The temperature dependences of dielectric constants, ε_r, and loss tangents, tanδ, were measured in the temperature range from room temperature to 600°C and for frequencies of 10 kHz, 100 kHz and 1 MHz using an automated dielectric measurement system with a multifrequency LCR meter. The P-E hysteresis loops for both unpoled and poled specimens were observed at room temperature using a ferroelectric tester, TF2000FE (aixACCT).

Specimens for the piezoelectric measurements were poled in a stirred silicone bath at 150°C by applying a direct (dc) electric field of 4 kV/mm for 30 min. The electromechanical coupling factors, k_p, k_{31} and k_t were measured using the resonance-antiresonance method based upon the EMAS-6002 standard using an impedance analyzer (YHP4192A).

RESULTS

Crystal structure

KBT ceramic with a relative density of about 95% was obtained using a conventional solid-state reaction and sintering with the addition of 2 mol% of MnO_2 or SiO_2. The XRD patterns of KBT-0.02MnO_2 and KBT-0.02SiO_2 ceramics are shown in Fig. 1. The XRD pattern of non-doped KBT ceramics fabricated by the polymerized complex method (KBT-PCM) is also shown for reference. All the peaks in KBT-0.02MnO_2 can be indexed on the basis of tetragonal perovskite structure, but for the spectra of KBT-0.02SiO_2, some small peaks corresponding to a secondary phase such as $K_2Ti_6O_{13}$ were observed.

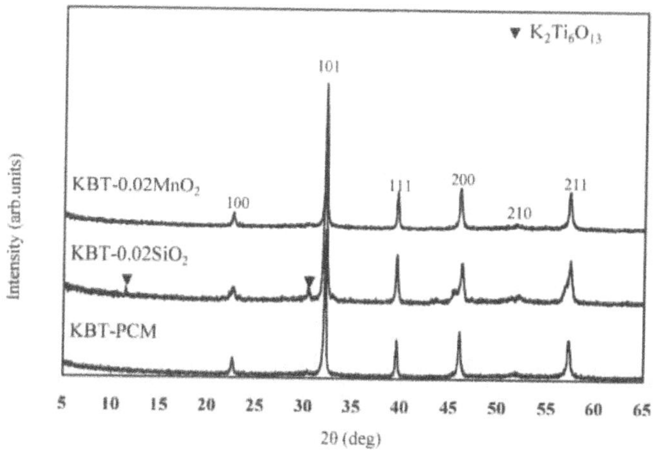

Fig. 1. Powder X-ray diffraction patterns of KBT-0.02MnO_2, KBT-0.02SiO_2 and KBT-PCM ceramics. All the peaks in the XRD pattern in KBT-0.02MnO_2

Microstructure

SEM micrographs of the fracture surface of KBT-0.02MnO_2 and KBT-0.02SiO_2 are shown in Figs. 2(a) and 2(b), respectively. Figure 2(c) and 2(d) shows those of non-doped KBT ceramics fabricated by the polymerized complex method (KBT-PCM) and the conventional solid-state reaction and sintering approach (KBT-SSRM), respectively. The densities of KBT-0.02MnO_2, KBT-0.02SiO_2 and KBT-PCM were about 95% while that of KBT-SSRM was 70%. Inhomogeneous grains and a high density of pores were observed in the microstructure of the

KBT-SSRM sample as can be seen in Fig. 2(d). The KBT-PCM sample, in contrast, had a homogeneous microstructure with small grains of about 0.26 μm without pores. Both the KBT-0.02MnO_2 and KBT-0.02SiO_2 samples also exhibited homogeneous microstructures without pores. The grain size of the KBT-0.02MnO_2 sample was about 0.5 μm while that of the KBT-0.02SiO_2 sample was about 0.3 μm. The grain size of the KBT-0.02SiO_2 sample was as large as that of the KBT-PCM sample. The grain size of the KBT-0.02MnO_2 sample was about 0.2 μm larger than that of both the KBT-PCM and KBT-0.02SiO_2 samples.

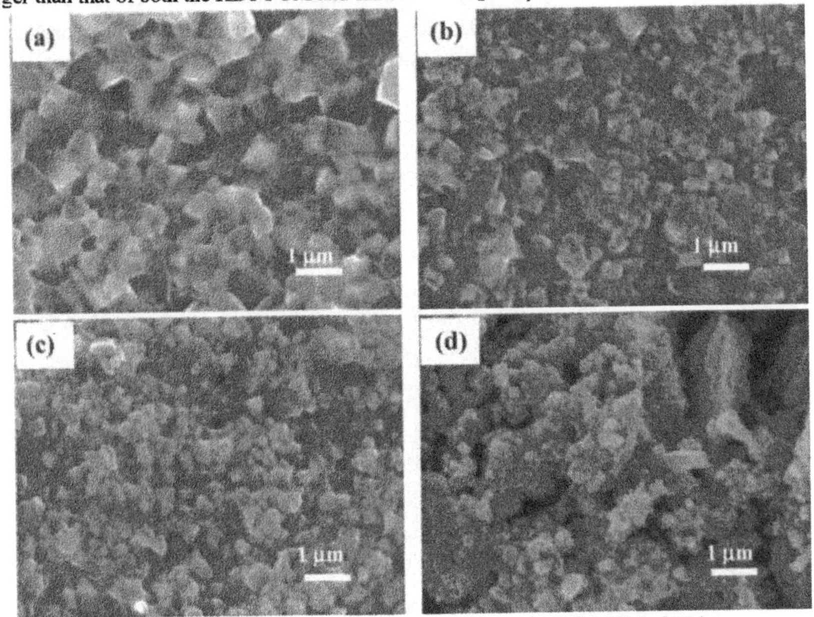

Fig. 2. SEM micrographs of the fracture surface of KBT-0.02MnO_2 (a), KBT-0.02SiO_2 (b), KBT-PCM (c) and KBT-SSRM (d) ceramics.

Table 1. Relative density and dielectric properties of KBT-0.02MnO_2, KBT-0.02SiO_2, KBT-PCM and KBT-SSRM ceramics.

Specimens	Relative densities (%)	Dielectric properties			
		1 kHz		100 kHz	
		ε_r	tanδ	ε_r	tanδ
KBT-0.02MnO_2	94.8	508	0.02	512	0.03
KBT-0.02□□O_2	95.2	531	0.05	528	0.05
KBT-PCM	94.3	632	0.06	621	0.05
KBT-SSRM	70.2	-	-	-	-

Dielectric Properties

The dielectric properties (ε_r and tanδ) for KBT-0.02MnO_2 and KBT-0.02SiO_2 ceramics are summarized in Table I. Those of the KBT-PCM are shown for reference. The ε_r and tanδ values at 1 kHz for KBT-0.02MnO_2 were 508 and 0.02, respectively. The corresponding values for the KBT-0.02SiO_2 sample were 531 and 0.05, respectively. These ε_r are smaller than that of KBT-PCM and the tanδ values are smaller than that for KBT-PCM.

The temperature dependences of the dielectric constant, ε_r, and the loss tangent, tanδ, for KBT-0.02MnO_2, KBT-0.02SiO_2 and KBT-PCM are shown in Figs. 3(a), 3(b) and 3(c), respectively. The temperature dependence of the dielectric constants was found to correspond to a single broad feature from which the temperatures of the maximum dielectric constant, T_{max}, are determined. The temperatures of the dielectric constant extrema correspond to the phase transition temperature between the paraelectric and ferroelectric phases at about 380°C; the measured values of T_{max} are summarized in Table II. For KBT-0.02MnO_2, the values of T_{max} at 10 kHz, 100 kHz and 1 MHz were 369, 374 and 382°C, respectively, while for KBT-0.02SiO_2 they were 367, 373 and 382°C, respectively, and for KBT-PCM, these are 368, 374 and 384°C, respectively. The frequency dispersion of T_{max} suggests that the KBT ceramics are typical relaxor-structure ferroelectrics. No low temperature phase transition at about 330°C could be inferred from the temperature dependences of the loss tangents shown in Fig. 3.[6]

The P-E hysteresis loops for unpoled and poled KBT-0.02MnO_2, KBT-0.02SiO_2 and KBT-PCM ceramics are shown in Figs. 4(a), 4(b) and 4(c) respectively. Relatively saturated hysteresis loops were observed for the poled specimens. The remnant polarization, P_r, and coercive field, E_c, for the poled KBT-0.02MnO_2 were 10.8 $\mu C/cm^2$ and 22 kV/cm, respectively while those for poled KBT-0.02SiO_2 are 9.80 $\mu C/cm^2$ and 33 kV/cm, respectively. The P_r for these samples was larger than that of KBT-PCM, 7.0 $\mu C/cm^2$. The E_c of KBT-0.02MnO_2 was smaller than that of KBT-0.02SiO_2. The reason for this is thought to be the difference in grain size between these samples. The P_r of our samples was smaller than that of KBT ceramics fabricated by the hot-pressing technique.[6]

Piezoelectric properties

The typical frequency dependence of the impedance Z and the phase of the planar extension mode for KBT-0.02MnO_2 are shown in Fig. 5. The electromechanical coupling factors of the other modes such as longitudinal vibration through the transverse mode and thickness longitudinal mode were measured for these samples. The determined piezoelectric properties of KBT-0.02MnO_2, KBT-0.02SiO_2 and KBT-PCM are summarized in Table III. The planar electromechanical coupling factors, k_p, for KBT-0.02MnO_2, KBT-0.02SiO_2 and KBT-PCM were 14.8%, 19.2% and 15.9%, respectively. The k_p of KBT-0.02SiO_2 was larger than those for KBT-0.02MnO_2 and KBT-PCM.

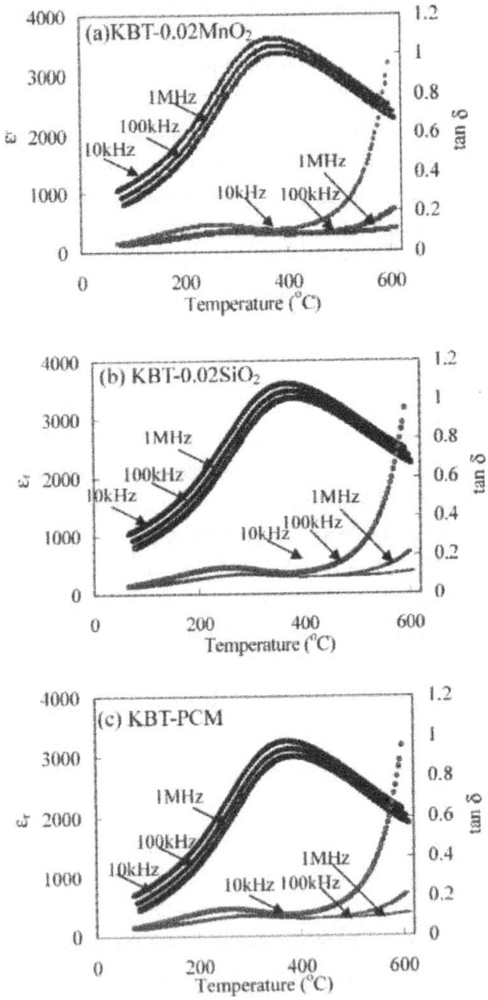

Fig. 3. Temperature dependences of the dielectric constants, ε_r and loss tangents, tanδ, for KBT-0.02MnO_2 (a), KBT-0.02SiO_2 (b) and KBT-PCM (c).

Table II. Temperatures of the maximum dielectric constant (T_{max}) for KBT-0.02MnO$_2$, KBT-0.02SiO$_2$ and KBT-PCM ceramics, observed in Fig. 3.

Specimens	T_{max} (°C)		
	10 kHz	100 kHz	1 MHz
KBT-0.02MnO$_2$	369	374	382
KBT-0.02SiO$_2$	367	373	380
KBT-PCM	368	374	384

The electromechanical coupling factors, k_{31}, of KBT-0.02MnO$_2$, KBT-0.02SiO$_2$ and KBT-PCM were 11.0%, 13.0% and 10.8%, respectively. The k_{31} for KBT-0.02SiO$_2$ was a little larger than the other samples. The thickness longitudinal electromechanical coupling factors, k_t, of KBT-0.02MnO$_2$ and KBT-0.02SiO$_2$ were 33.8% and 33.1%, respectively. These k_t values were smaller than that of KBT-PCM, 39.5%.

DISCUSSION

The relative density of KBT ceramics sintered with an addition of 2 mole% of MnO_2 or 2 mole% of SiO_2 was as high as 95%, while that of non-doped KBT (KBT-SSRM) was 70%. High-density KBT ceramics without any sintering additive cannot be prepared by the conventional sintering process, because the KBT tends to decompose and the K component vaporizes during the sintering process. From these results, it is clear that these oxides work effectively as sintering additives for KBT fabrication.

The effects of the addition of SiO_2 and MnO_2 on the sintering of KBT are discussed on the basis of these results. No diffraction peaks from impurity phases were observed in the X-ray diffraction pattern of KBT-0.02MnO$_2$. For KBT-0.02MnO$_2$, the added Mn dissolves in the Ti site (B site of perovskite structure) in $(K_{0.5}Bi_{0.5})TiO_3$. Therefore, a vacancy is formed at the $(K_{0.5}Bi_{0.5})$ site (The A site of the perovskite structure) or $(K_{0.5}Bi_{0.5}V_{0.02})(Ti_{1.00}Mn_{0.02})O_3$. Generally, vacancies at the A site of the perovskite structure promote the diffusion of ions during the sintering process. KBT-0.02MnO$_2$ powder tends to sinter at lower temperatures than the decomposition temperature of KBT. High-density KBT ceramics can be fabricated by solid phase sintering. In the case of KBT-0.02MnO$_2$, the movement of ions in the grains is assisted by the presence of vacancies at the A site of the KBT. Therefore, grains in the KBT-0.02MnO$_2$ can grow easily. This is confirmed by the observation that the grain size of KBT-0.02MnO$_2$ is larger than that of KBT-0.02SiO$_2$ and non-doped KBT (KBT-PCM).

In the X-ray diffraction pattern of KBT-0.02SiO$_2$, small peaks belonging to a secondary impurity phase such as $K_2Ti_6O_{13}$ were observed. This is a consequence of part of the potassium component in KBT dissolving in the additive, SiO_2 during the sintering process. It can be understood from the binary phase diagram of SiO_2 and K_2O that a liquid phase is formed at the sintering temperature, 1040°C.

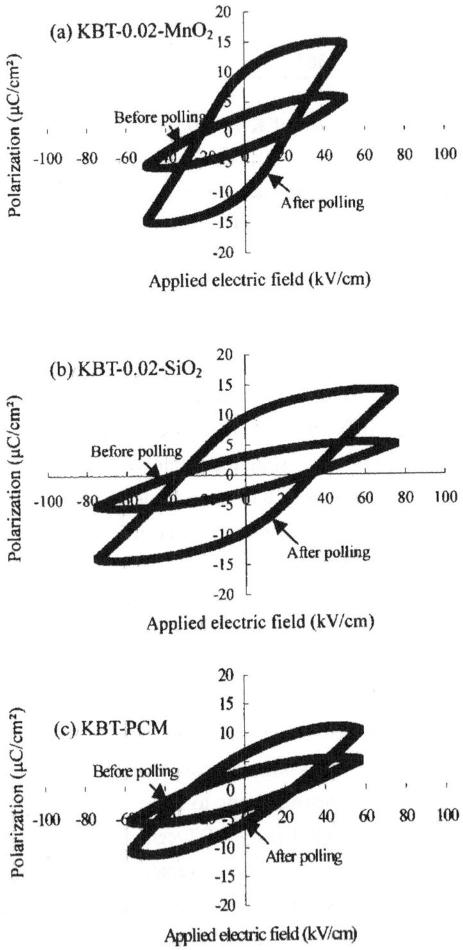

Fig. 4. P-E hysteresis loops for KBT-0.02MnO_2(a), KBT-0.02SiO_2(b) and KBT-PCM(c) ceramics.

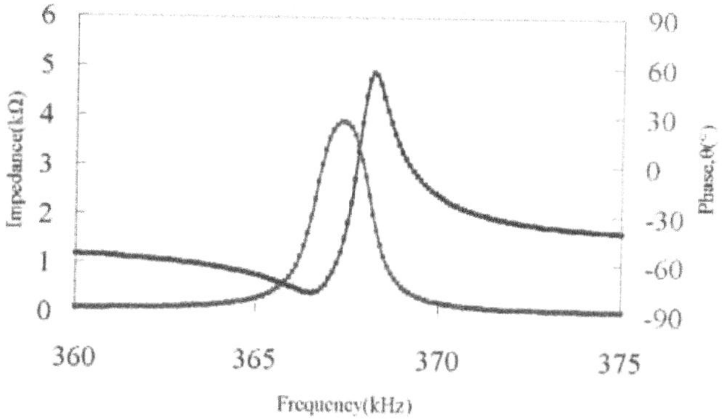

Fig. 5. Frequency dependence of the impedance Z of the planar extension mode for KBT-$0.02MnO_2$.

Table III. Piezoelectric properties of KBT-$0.02MnO_2$, KBT-$0.02SiO_2$ and KBT-PCM ceramics.

Specimens	KBT-$0.02MnO_2$	KBT-$0.02SiO_2$	KBT-PCM
k_p (%)	14.8	19.2	15.9
k_{31} (%)	11.0	13.0	10.8
k_t (%)	33.8	33.1	39.5

The movement of ions is assisted by the presence of a small amount of liquid in which the crystalline phase has limited solubility. Such a sintering process is referred to as "liquid phase sintering". The liquid changes to a glass during the cooling process, because the liquid contains SiO_2 as its main constituent. The small amount of glass formed in KBT can not be identified by X-ray diffraction. High-density KBT-$0.02SiO_2$ ceramics can be fabricated by the liquid phase sintering.

CONCLUSION

High-density KBT ceramics were obtained by a conventional solid-state reaction and sintering method with an addition of 2 mol% of MnO_2 or SiO_2. The sintering of KBT is promoted by the deferent mechanism caused by the addition of MnO_2 and SiO_2. The dielectric and piezoelectric properties of these KBT ceramics were comparable to those of the pure KBT ceramics fabricated by the polymerized complex method.

ACKNOWLEDGMENT

This work was supported in part by a grant under the High-Tech Research Center Program for private universities from the Ministry of Education, Culuture, Sports, Science and Technology of Japan. The authors are also grateful to Dr. Y. Matsuo, T. Saito and T. Maeda for meaningful discussion.

REFERENCES

[1]T. Takenaka, K. Maruyama and K. Sakata, "$(Bi_{1/2}Na_{1/2})TiO_3$-$BaTiO_3$ System for Lead-Free Piezoelectric Ceramics", Jpn. J. Appl. Phys., **30**, 2236-2239 (1991).

[2]T.Takenaka, T. Okuda and K. Takegahara, "Lead-free piezoelectric ceramics based on $(Bi_{1/2}Na_{1/2})TiO_3$-$NaNbO_3$", Feroelectrics, **196**, 175-178 (1997).

[3]C. F. Buhrer, "Some Properties of Bismuth Perovskites", J. Chem. Phys., **36**, 798-803 (1962).

[4]T. Wada, K. Toyoike, Y. Imanaka and Y. Matsuo, "Dielectric and Piezoelectric Properties of $(A_{0.5}Bi_{0.5})TiO_3$-$ANbO_3$ (A = Na, K) Systems", Jpn. J. Appl. Phys., **40**, 5703-5705 (2001).

[5]T. Wada, A. Fukui and Y. Matsuo, "Preparation of $(K_{0.5}Bi_{0.5})TiO_3$ Ceramics by Polymerized Complex Method and their Properties", Jpn. J. Appl. Phys., **41**, 7025-7028 (2002).

[6]Y. Hiruma, R. Aoyagi, H. Nagata and T. Takenaka, "Ferroelectric and Piezoelectric Properties of $(Bi_{1/2}K_{1/2})TiO_3$ Ceramics", Jpn. J. Appl. Phys., **44**, 5040-5044 (2005).

Electroceramic Materials and Applications
Edited by R. W. Schwartz

ULTRA-WIDE BANDWIDTH, THIN FILM ELECTRO-OPTIC MODULATORS

Pingsheng Tang, A. L. Meier, D. J. Towner and B. W. Wessels
Department of Material Science and Engineering and Materials Research Center
Northwestern University
2220 Campus Drive
Evanston, Illinois, 60208, USA

ABSTRACT

Electro-optic (eo) modulators for ultra-wide bandwidth optical communication systems are being developed using $BaTiO_3$ ferroelectric thin films. These modulators have low voltage-length products at 1-1.6 micron wavelengths. Realization of high-speed traveling-wave $BaTiO_3$ thin film electro-optic waveguide modulators with 3-dB modulation bandwidth of 15 GHz at 1.55 μm wavelength is reported. Quasi-velocity matched electrodes for $BaTiO_3$ thin film modulators are achieved by addition of a SiO_2 buffer layer. Its microwave and electro-optic response are characterized.

INTRODUCTION

High-speed electro-optic waveguide modulators are important components for high-bit-rate fiber optic communication networks and information processing systems. Ferroelectric barium titanium electro-optic modulators offer the potential for low drive voltages, small size and integration on a silicon substrate. A $BaTiO_3$ thin film electro-optic waveguide modulator with a low voltage-length product has been recently demonstrated at near-infrared wavelengths of 1-1.6 μm.[1] Voltage-length products as low as 0.25 V·cm and 0.5 V·cm have been measured for a 5-mm long device at wavelengths of 973 nm and 1561 nm, respectively. The effective electro-optic coefficients are calculated as 420 pm/V at 973 nm and 360 pm/V at 1561 nm. These coefficients, however, are still lower than bulk values. Further improvements of the effective electro-optic coefficients are anticipated through optimizing the $BaTiO_3$ thin film ferroelectric domain structure and poling the multidomain $BaTiO_3$ thin film.

While large electro-optic coefficients are observed, $BaTiO_3$ also has a very large dielectric constant (ε_x=2200, ε_z=56). The large dielectric constant limits its high-speed and wideband operation due to the large phase velocity mismatch between the optical and microwave signals. To increase the microwave phase velocity, thin film electro-optic modulators on low microwave index substrates have been proposed[1]. By using a composite structure of $BaTiO_3$ on MgO, a lower effective dielectric constant can be obtained[2-4]. However, the optical wave still propagates at a higher phase velocity than that of the microwave. To obtain phase-matched electrodes for the high-speed, wideband operation of these modulators, traveling-wave electrodes and a SiO_2 buffer layer are required to further increase the phase velocity of the microwave signals[5]. The high frequency operation of a low-voltage, electrooptic modulator based on a strip-loaded $BaTiO_3$ thin film waveguide structure has been demonstrated. The epitaxial $BaTiO_3$ thin film on an MgO substrate forms a composite structure with a low effective dielectric constant of 20.8 at 40 GHz. A 3.9 V half-wave voltage with a 3.7 GHz 3-dB bandwidth and a 150 pm/V effective electrooptic coefficient was obtained for the 3.2mm-long modulator at 1.55 μm[5]. Broadband modulation up to 40 GHz was measured with a calibrated detection system. Previously reported numerical simulation and experimental results of the modulator response indicate that the $BaTiO_3$ thin film modulator has the potential of operation in excess of 40 GHz 3-dB bandwidth through optimized design[5].

Here we present experimental results for $BaTiO_3$ thin film waveguide modulators on a traveling-wave coplanar strip (CPS) electrode with a SiO_2 buffer layer that is designed to match the effective indices of optical waves and microwaves. We demonstrate that, by employing a low index SiO_2 buffer layer of optimal thickness between coplanar strip electrode and $BaTiO_3$ thin film, the phase velocity of the optical wave and the microwave can be nearly matched. High-speed and wideband thin film $BaTiO_3$ waveguide modulators are achieved.

FABRICATION

A strip-loaded waveguide designed for 1550 nm wavelength was formed in a ~600 nm thick $BaTiO_3$ thin film on a MgO substrate as previously described[5]. The cross-section of a strip-loaded $BaTiO_3$ thin film waveguide modulator with the coplanar strip electrode is shown in Fig. 1. The $BaTiO_3$ thin film was epitaxially grown on the (100) oriented MgO substrate using a low pressure metalorganic chemical vapor deposition process[6]. The waveguide region was defined using photolithography with the propagation direction aligned to the <110> crystallographic direction of the MgO substrate. The Si_3N_4 strip-loaded layer was 4 µm wide and 120 nm thick, fabricated by standard plasma enhanced chemical vapor deposition and reactive ion etching processes. SiO_2 buffer layer was deposited over the waveguide by e-beam evaporation to modify the effective microwave index. Metal electrode layers of 10 nm Cr and 800 nm Au were deposited by e-beam evaporation, followed by a standard lift-off process. The width of the CPS electrodes is 90 µm, and the electrode gap is 15 µm. The CPS electrodes have a 5 mm long interaction length.

Fig. 1. Schematic cross-section of a $BaTiO_3$ thin film waveguide modulator with a CPS electrode.

CHARACTERIZATION AND ANALYSIS

The microwave velocity characteristics of the traveling-wave electro-optic modulators were measured with a vector network analyzer (Agilent-8722ES) and a microwave probe station.

The microwave group velocity of the coplanar strip electrodes was evaluated from time-domain measurements of their transmission response. Because the wavelength (from several millimeters to several meters) of the microwaves is much larger than the dimension of the CPS electrodes, the microwave field in the structure is considered as quasi-stationary. Thus the microwave group velocity of V_{mg} is approximately equal to the microwave phase velocity of V_{mp} in this structure at frequencies of 50 MHz-40.05 GHz[7]. Fig. 2 shows the measured microwave velocity as a function of the SiO_2 buffer layer thickness. As the SiO_2 buffer layer thickness increases, the microwave velocity also increases. When the SiO_2 buffer layer thickness is 1050 nm thick, a microwave velocity of 0.136 m/ns is achieved, which is very close to the optical wave velocity of 0.139 m/ns (The effective optical refractive index is 2.16 at 1550 nm). The velocity difference between microwave and optical wave is only 2% for 1050 nm SiO_2 buffer layer on the CPS electrodes. The solid line in Fig. 2 is a simulated result for SiO_2 film on 600 nm $BaTiO_3$ thin film on a MgO substrate. The simulated result is in excellent agreement with the measured data. The difference is due to $BaTiO_3$ thickness variation (from 570 nm to 620 nm). The dashed line in Fig.2 is the optical wave velocity of the $BaTiO_3$ thin film waveguide. From the data in Fig. 2, the effective microwave index, N_m, is calculated to be 3.1 for the $BaTiO_3$/MgO composite, and the corresponding calculated effective dielectric constant is 18.3. It should be noted that the value of the effective composite $BaTiO_3$/MgO dielectric constant is smaller than that of z-cut $LiNbO_3$ ($\varepsilon_r = \sqrt{\varepsilon_x \cdot \varepsilon_z} = 35.1$), and much smaller than that of z-cut $BaTiO_3$ ($\varepsilon_r = \sqrt{\varepsilon_z \cdot \varepsilon_x} \sim 351$). Furthermore, the effective microwave index of 3.1 for the thin film composite structure is much smaller than that of bulk $BaTiO_3$ ($N_m = \sqrt{0.5 + 0.5 \cdot \varepsilon_r} = 13.3$), but it is much larger than the effective optical index of 2.16 for $BaTiO_3$ thin film waveguide modulators[4]. When a 1050nm thick SiO_2 buffer layer is employed, the effective microwave index decreased to 2.20, which is nearly matched to the effective optical index.

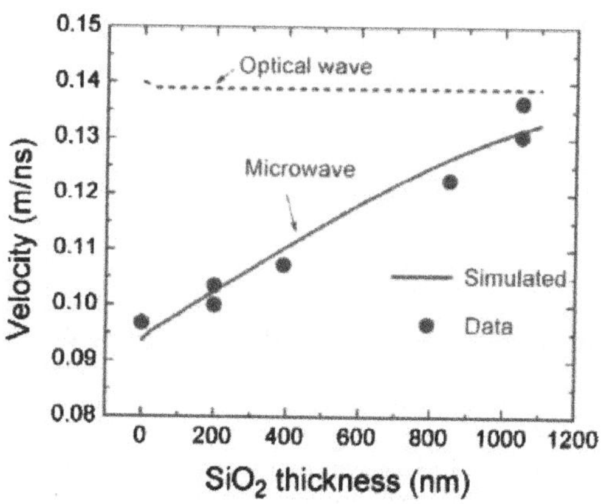

Fig. 2. Velocity versus SiO_2 buffer layer thickness. Solid line is a simulated microwave velocity for a 600 nm thick $BaTiO_3$ film. The dashed line is the optical velocity of the $BaTiO_3$ thin film waveguide.

The high frequency response of $BaTiO_3$ thin film modulators with 850 nm SiO_2 buffer layer was tested up to 40 GHz, as shown in Fig. 3. The modulator was biased at 4 V DC level and had a half-wave voltage of 5.8 V at 1561 nm wavelength. The –3 dB bandwidth of the modulator was measured to be 15 GHz, primarily limited by velocity mismatch (V_{mp} = 0.121 m/ns. V_{op}=0.139 m/ns at 1.55 μm), impedance mismatch and microwave loss of the electrode. From the measured S-parameters, the modulator electrode loss coefficient and impedance were calculated. The loss coefficient is 10.2 dB/cm at 40 GHz. The loss coefficient is attributed to the thin 800nm Au electrode, and presumably can be reduced by electroplating a thicker Au electrode. The impedance of the electrode structure is about 40 Ω at 40 GHz.

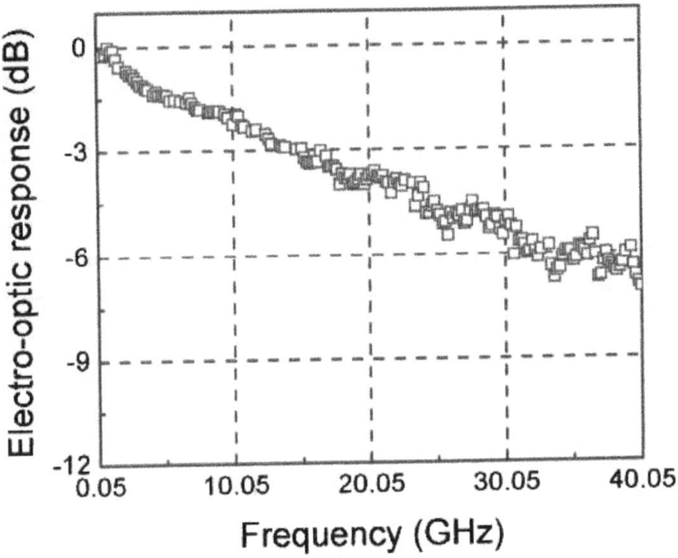

Fig. 3. Measured electro-optic modulation response.

These results indicate that it is possible to design and fabricate a $BaTiO_3$ thin film modulator with 3-dB bandwidth greater than 40 GHz by optimizing the buffer layer thickness and impedance.

CONCLUSION

High-speed traveling-wave $BaTiO_3$ thin film waveguide modulator have been designed and fabricated. A nearly velocity-matched traveling-wave coplanar strip electrode using a SiO_2 buffer layer in $BaTiO_3$ thin film waveguide modulators was developed. A 15 GHz modulation bandwidth was achieved for the first time in thin film $BaTiO_3$ waveguide modulators. These results indicate the potential of high-speed, wideband $BaTiO_3$ thin film electro-optic modulators in optical communications and integrated optics.

ACKNOWLEDGEMENTS

This work was supported by the US Air Force through contract AFRL-33615-02-C-5053, the NSF through grants ECS-0123469 and DMR-0076977.

REFERENCES

[1]P. Tang, D. Towner, A. Meier, and B. Wessels, "$BaTiO_3$ Thin Film Waveguide Modulator with Low Voltage Length Product at Near-Infrared Wavelengths of 0.98 μm and 1.5 μm," *Optics Letters*, **30**, 254 (2005).

[2]K. Atsuki and E. Yamashita, "Transmission Line Aspects of the Design of Broadband Electrooptic Traveling-Wave Modulators", *IEEE J. Lightwave Tech.*, **5**, 316-319 (1987).

[3]D. Gill, B. Block, C. Conrad, B. Wessels, and S. Ho, "Thin Film Channel Waveguides Fabricated in Metalorganic Chemical Vapor Deposition Grown $BaTiO_3$ on MgO", *Appl. Phys. Lett.*, **69**, 2968-2970 (1996).

[4]A. Petraru, J. Schubert, M. Schmid, and C. Buchal, "Ferroelectric $BaTiO_3$ Thin Film Optical Waveguide Modulators," *Appl. Phys. Lett.*, **81**, 1375-1377 (2002).

[5]P. Tang, D. Towner, T. Hamano, A. Meier, and B. Wessels, "Electrooptic Modulation Up to 40 GHz in a Barium Titanate Thin Film Waveguide Modulator", *Opt. Express*, **12**, 5962-5967 (2004).

[6]D. Towner, J. Ni, T. Marks, and B. Wessels, "Effects of Two-Stage Deposition on the Structure and Properties of Heteroepitaxial $BaTiO_3$ Thin Films", *J. Cryst. Growth*, **255**, 107-113 (2003).

[7]E. Yamashita, K. Atsuki, and T. Ueda, "An Approximate Dispersion Formula of Microstrip Lines for Computer-Aided Design of Microwave Integrated Circuits', *IEEE Trans. Microwave Theory and Tech.*, **27**, 1036-1038 (1979).

Electroceramic Materials and Applications
Edited by R. W. Schwartz

REFINED POSITION OF THE MORPHOTROPIC PHASE BOUNDARY FOR $Pb(Mg_{1/3}Nb_{2/3})O_3$-$PbZrO_3$-$PbTiO_3$ CERAMICS

Masafumi Kobune, Yoko Maekawa, Atsushi Mineshige, Tetsuo Yazawa and Hiroshi Nishioka
Graduate School of Engineering, University of Hyogo, Hyogo 671-2201, Japan

ABSTRACT

The morphotropic phase boundary (MPB) for the PMNZT series of high-density ceramics ($xPb(Mg_{1/3}Nb_{2/3})O_3$-$yPbZrO_3$-$zPbTiO_3$; $x = 0.1$–0.2, $y = 0.400$–0.486, $z = 0.395$–0.477) is relocated based on analyses of the dielectric, piezoelectric and ferroelectric properties of incremental compositions. The PMNZT ceramics, fabricated by the columbite route, exhibit a single-phase perovskite structure. The refined MPB line is shifted toward the $PbTiO_3$ side in the ternary field by approximately 3.6 mol% from the position reported in the literature. The PMNZT 14.2/41.6/44.2 (100x/100y/100z) composition on the MPB line is shown to have properties suitable for practical applications.

INTRODUCTION

$Pb(Mg_{1/3}Nb_{2/3})O_3$ (PMN) has a Curie temperature (T_c) of −15°C, relative permittivity (ε_r) of 12600 at T_c, and spontaneous polarization (P_s) of 6 $\mu C/cm^2$. This material, first discovered by Smolenskii et al.,[1] is a typical relaxor-ferroelectric substance. Since 1965, PMN-Pb(Zr, Ti)O_3 (PMN-PZT) ceramics[2] have been investigated extensively for potential piezoelectric applications such as piezoelectric transducers,[3] piezoelectric resonators[4] and acoustic applications.[5] In the development of piezoelectric materials,[6-9] the compositions of the ternary form, $xPb(Mg_{1/3}Nb_{2/3})O_3$-$yPbZrO_3$-$zPbTiO_3$ (denoted by PMNZT 100x/100y/100z), subject to the most intensive research are PMNZT 12.5/44.0/43.5[3] and PMNZT 37.5/25.0/37.5.[4] The former is a soft-mode piezoelectric material with $\varepsilon_r = 1246$, an electromechanical coupling factor (k_p) of 49.2% (radial vibration mode), and a mechanical quality factor (Q_m) of 101, whereas the latter is an excellent hard-mode piezoelectric material with high Q_m (= 2051). The mechanical quality factor of this hard material is increased by the addition of small amounts of MnO_2 and NiO, without causing deterioration in k_p. Ouchi has also succeeded in fabricating piezoelectric materials in which Pb^{2+} at A-site ions in the perovskite structure of PMNZT 37.5/26.5/36.0 (similar to the composition in Ref. 4) are partially substituted by a small amount of Ba^{2+} or Sr^{2+}. Electrical measurements have revealed that this partial substitution of A-site ions affords a high ε_r of 3185–3527 and a high k_p of 63–65%.[5] Tashiro et al. recently reported the nonlinear coefficients on the morphotropic phase boundary (MPB) line to elucidate the nonlinear phenomena frequently observed when PMNZT ceramics are driven at high power near the resonance frequency.[10] However, the pinpoint composition of the ceramic studied was not reported in even their study in spite of the importance of practical use for piezoelectric applications. Although the PMNZT 37.5/25.0/37.5 and similar compositions achieve physical properties suitable for use of the materials as piezoelectric resonators and in acoustic applications, the low T_c (< 300°C) of this composition has prevented its use in practical applications.[5] For piezoelectric actuators and ultrasonic sensors, a high T_c of >300°C is required, in addition to high k_p (>65%) and large piezoelectric coefficient (d_{31}, radial vibration mode, > −130 pC/N). The PMNZT 12.5/44.0/43.5 composition also fails to satisfy these requirements. However, there have as yet been no reports on the relevant ternary PMN-$PbZrO_3$-$PbTiO_3$ (PMN-PZ-PT) compositions applicable to the above applications.

In the present study, the compositions near PMNZT 12.5/44.0/43.5 are investigated in detail in order to identify compositions that satisfy the above requirements and improve the MPB composition reported in previous research.[3, 5] The optimal pinpoint composition and relevant piezoelectric properties for piezoelectric applications such as piezoelectric actuators and ultrasonic sensors are discussed based on this compositional analysis.

EXPERIMENTAL PROCEDURE

Reagent-grade PbO, Nb_2O_5, $MgCO_3$, ZrO_2 and TiO_2 were used as starting materials. PMN-PZ-PT sintered bodies were fabricated by a two-step solid-state reaction using a columbite precursor, as follows.[11] A mixture of 50 mol% Nb_2O_5 and 50 mol% $MgCO_3$ was first mixed for 30 min in ethanol using a planetary wet zirconia ball mill and then calcined in air at 1000°C for 4 h. The powders obtained were used as the $MgNb_2O_6$ precursor for preparation of $Pb(Mg_{1/3}Nb_{2/3})O_3$. PbO, ZrO_2 and TiO_2 were added to this $MgNb_2O_6$ and ground for 30 min in ethanol using the ball mill described above. The mixture was then dried and calcined in air at 850 °C for 3 h. To the calcined powder was added 2 mass% polyvinyl alcohol (PVA) solution followed by further grinding for 12 h in the same ball mill. The wet powders were dried at 105°C for 1.5 h and sieved using a 100 mesh (opening, 150 μm) screen. The granules were pressed at 128 MPa into discs of 12.0 mm diameter by approximately 1.0 mm in thickness. The green compacts were then placed in a high-purity MgO crucible and sintered at 1220°C for 3 h in air by the powder-bed method[12] using calcined powders with the same composition as the compacts. The samples were subsequently polished to a thickness of 0.6 mm. After drying, silver electrodes were attached to the surfaces of the discs by coating the surfaces with silver paste followed by firing at 750°C for 10 min in air. The samples were poled under a dc field of 3.5 MV/m for 30 min at room temperature in a silicone oil bath.

The crystalline phase and lattice constants were analyzed by X-ray diffractometry (XRD; RINT2200, Rigaku). The bulk density was measured by the Archimedean method using an automatic density determination instrument (AG204, Mettler-Toledo). A fluorescent X-ray analyzer (RIX2000, Rigaku) was employed for chemical analysis by the fundamental parameter method. The grain size and microstructure were observed under a digital high-definition microscope (VH-7000, Keyence) and by scanning electron microscopy (SEM; JSM-5310, Jeol). The dielectric properties were measured under an applied field of 25 V/cm at 1 kHz in vacuo over a temperature range of 0–400°C using a multi-frequency LCR meter (YHP4274A, Yokogawa-Hewlett-Packard). Piezoelectric properties were measured by the resonant-antiresonant frequency method on the basis of EMAS-6100[13] using an impedance/gain-phase analyzer (HP4194A, Hewlett-Packard). Ferroelectric properties were evaluated from measurements of the P–E hysteresis loops at room temperature under applied fields of 10–40 kV/cm, which were carried out using a ferroelectric film test system (RT-66A, Radiant Tech.).

RESULTS AND DISCUSSION

(1) PMNZT 10–20/40.0–48.6/39.1–47.7 ceramics

The ternary phase diagram for the PMNZT system is shown in Fig. 1. All the PMNZT ceramics prepared in this study were confirmed to have a single-phase perovskite structure with no pyrochlore structure (paraelectric phase). Although the MPB line of the ternary system has been reported previously,[3] the location of the line may not be sufficiently precise as the measurements used to determine the line were based on PMN content intervals of 5.3–

12.5 mol%. The MPB line found by Ouchi et al.[3] is shown in the figure. It is apparent from the figure that the MPB line, which indicates the structural phase transition between the tetragonal and rhombohedral phases, lies further to the $PbTiO_3$ (PT) side than suggested by Ouchi et al. The relative densities of the present PMNZT ceramics are in the range 94.4–98.4%. This high relative density is attributed to good adjustment of the PbO evaporation rate from discs with respect to the PbO supply from the embedded powders of the same composition in the powder-bed method, and appropriate choice of sintering conditions.

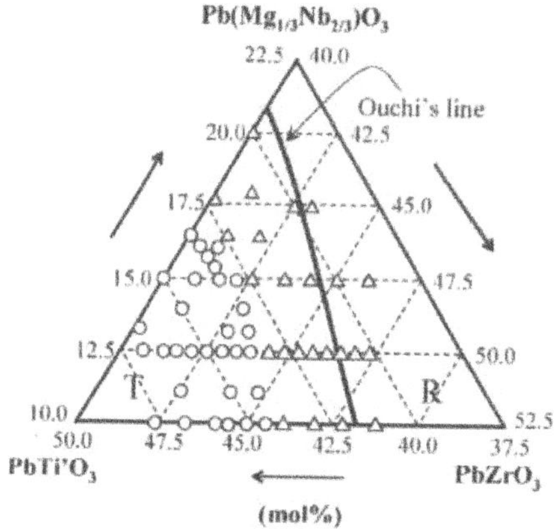

Fig. 1. Phase diagram of ternary $Pb(Mg_{1/3}Nb_{2/3})O_3$-$PbZrO_3$-$PbTiO_3$ system. Open circles and open triangles display the tetragonal phase and the rhombohedral phase, respectively. Ouchi's line: the MPB line reported by Ouchi et al.

The variation in the relative permittivity ($\varepsilon_{33}{}^T/\varepsilon_0$) at room temperature for the ternary PMNZT ceramics is shown in Fig. 2. The relative permittivities for PMNZT ceramics with compositions 12–16.2/40.4–42.9/43.4–45.5 are higher than 1600. This compositional region exists lies on the slightly Ti-rich side of the new MPB line. The dielectric loss (tanδ) for these compositions was in the range 1.1–3.4% regardless of the MPB composition, indicating a tendency to increase with increasing PZ content. This is attributable to deterioration of sinterability under the same sintering conditions with increasing PZ content, which is similar to the case for $Pb(Zn_{1/3}Nb_{2/3})O_3$-$PbZrO_3$-$PbTiO_3$ ceramics.[14] Figure 3 shows the variation in Curie temperature for these ternary PMNZT ceramics. It can easily be predicted from the Curie temperatures of pure PMN, PZ and PT (−15°C,[1] 230°C,[15] and 490°C,[16] respectively) that T_c decreases significantly with increasing PMN content and increases with increasing PT content. As can be seen from the figure, a contour map showing a gradual decrease in T_c from PT at the center toward PMN and PZ can be successfully drawn. The Curie temperatures of the present PMNZT samples were in the range 305–353°C, satisfying the objective T_c of greater than 300°C.

The variations in the electromechanical coupling factors k_p (radial vibration mode) and k_t (thickness vibration mode) for these ternary PMNZT ceramics are shown in Fig. 4. As predicted from the previous result for $\varepsilon_{33}{}^T/\varepsilon_0$, PMNZT ceramics with compositions of 13.5–15.0/42.0–42.2/

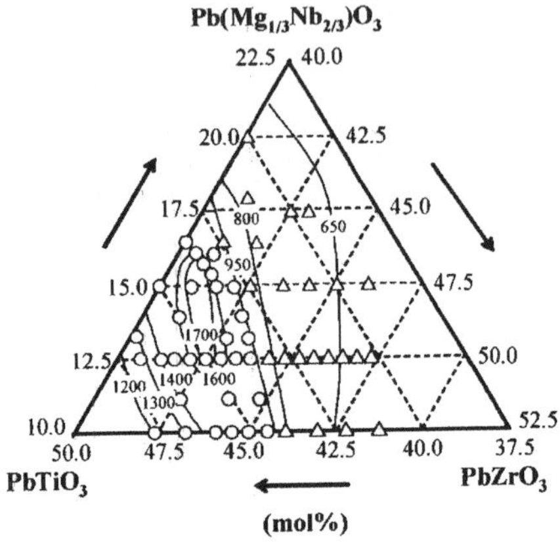

Fig. 2. Variation in relative permittivity, $\varepsilon_{33}{}^{T}/\varepsilon_0$ at room temperature of ternary $Pb(Mg_{1/3}Nb_{2/3})O_3$-$PbZrO_3$-$PbTiO_3$ ceramics. Open circles and open triangles display the tetragonal phase and the rhombohedral phase, respectively.

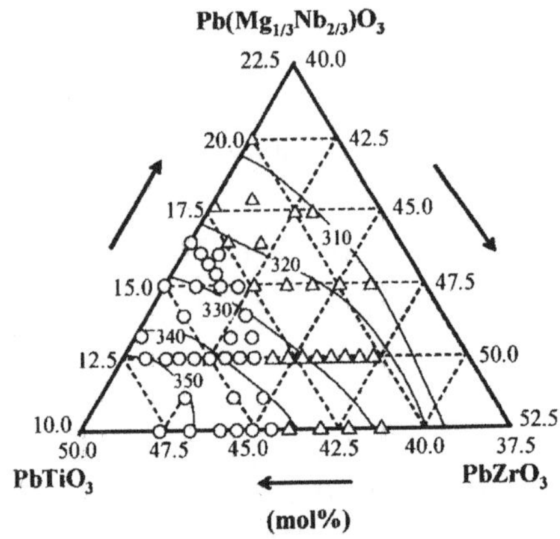

Fig. 3. Variation in Curie point, T_c of ternary $Pb(Mg_{1/3}Nb_{2/3})O_3$-$PbZrO_3$-$PbTiO_3$ ceramics. Open circles and Open triangles display the tetragonal phase and the rhombohedral phase, respectively.

42.5–44.1 exhibit k_p values exceeding 60%, reaching a maximum of 66%. This value is sufficiently high compared to the piezoelectric materials currently employed as ultrasonic sensors. Similarly, k_t for the PMNZT 13.5–16.5/41.7–43.4/41.6–43.1 ceramics is higher than 50%, reaching a maximum of 53%. These results demonstrate that the compositional region with high $\varepsilon_{33}{}^{T}/\varepsilon_0$ corresponds well to the region exhibiting the highest electromechanical coupling factors (k_p and k_t). Both compositional regions lie almost precisely on the new MPB line identified in this study. The variation in the piezoelectric coefficient d_{31} at room temperature is shown in Fig. 5. As similar tendency to that observed for $\varepsilon_{33}{}^{T}/\varepsilon_0$ and k_p is also seen for d_{31}, which is greater than 130 pC/N (maximum 180 pC/N) for PMNZT ceramics with compositions of 12.2–16.2/40.4–44.1/42.8–44.6. This result can be explained by the fact that d_{31} reaches a maximum when electromechanical coupling k_{31} and $\varepsilon_{33}{}^{T}$ reach maxima. The Young modulus $Y_{11}{}^{E}$ simultaneously takes a minimum value, as given by

$$d_{31} = k_{31}(\varepsilon_{33}{}^{T}/Y_{11}{}^{E})^{1/2} \quad (1)$$

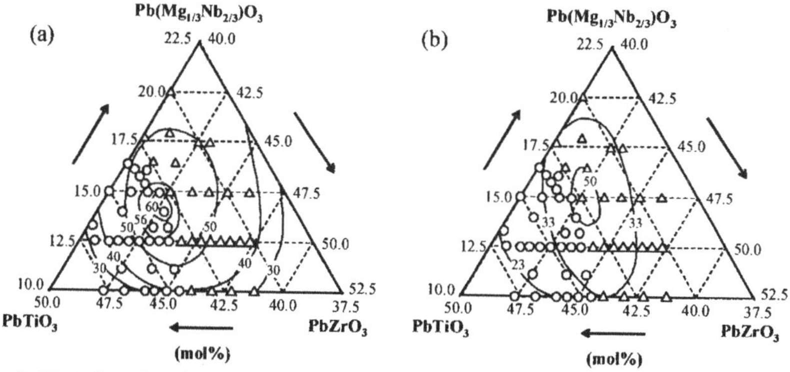

Fig. 4. Variations in electromechanical coupling factors, (a) k_p (%) and (b) k_t (%) at room temperature of ternary $Pb(Mg_{1/3}Nb_{2/3})O_3$-$PbZrO_3$-$PbTiO_3$ ceramics. Open circles and Open triangles display the tetragonal phase and the rhombohedral phase, respectively.

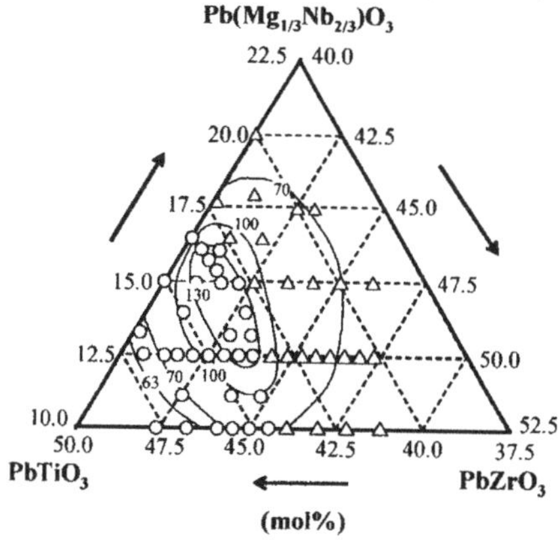

Fig. 5. Variation in piezoelectric coefficient, $-d_{31}$ (pC/N) at room temperature of ternary $Pb(Mg_{1/3}Nb_{2/3})O_3$-$PbZrO_3$-$PbTiO_3$ ceramics. Open circles and Open triangles display the tetragonal phase and the rhombohedral phase, respectively.

(2) PMNZT 14.2/100y/85.8-100y ceramics with $0.400 \leq y \leq 0.458$

It is evident from the physical properties of the PMNZT ceramics described above that PMNZT ceramics with x = 0.142 have excellent dielectric and piezoelectric properties. On the basis of this result, the dielectric, piezoelectric and ferroelectric properties of eight PMNZT 14.2/100y/(85.8 – 100y) ceramics with PZ contents (100y) of 40.0–45.8 were investigated in further detail. The additional PMNZT compositions considered here are plotted in Fig. 6 After revealing the optimal PZ and PT contents based on electrical properties, the precise composition of PMNZT for piezoelectric applications can be determined from this figure.

Figure 7 shows the dependence of $\varepsilon_{33}^T/\varepsilon_0$ and tanδ on composition at room temperature for these PMNZT ceramics. Although $\varepsilon_{33}^T/\varepsilon_0$ reaches 1400–2430 on the tetragonal side, it becomes significantly lower in the vicinity of the MPB with increasing PZ content, deteriorating to half that observed on the tetragonal side. The dielectric loss of the tetragonal side is in the range 1.2–1.4%, whereas that on the rhombohedral side reaches 1.6–2.2%, increasing from the vicinity of the MPB line. This behavior may be explained by the deterioration of sinterability caused by structural phase transition from the tetragonal to the rhombohedral phase. The dependence of T_c on composition is shown in Fig. 8, where T_c derived from the values for PMN, PZ and PT single crystals is also shown. The latter relationship is described by

$$T_{cPMNZT}\ x/y/z = xT_{cPMN} + yT_{cPZ} + zT_{cPT} \tag{2}$$

where T_{cPMN}, T_{cPZ} and T_{cPT} are –15°C,[1] 230°C,[15] and 490°C,[16] respectively. The observed T_c values are approximately 23°C higher than the calculated values, although both values exhibit a similar linear decrease with increasing PZ content. The rates of decrease estimated from the slope of the linear plot of T_c with PZ content are 3.0°C/mol% PZ (calculated) and 2.6°C/mol% PZ (observed). Thus, the maximum PZ content required to achieve T_c = 300°C is estimated to be

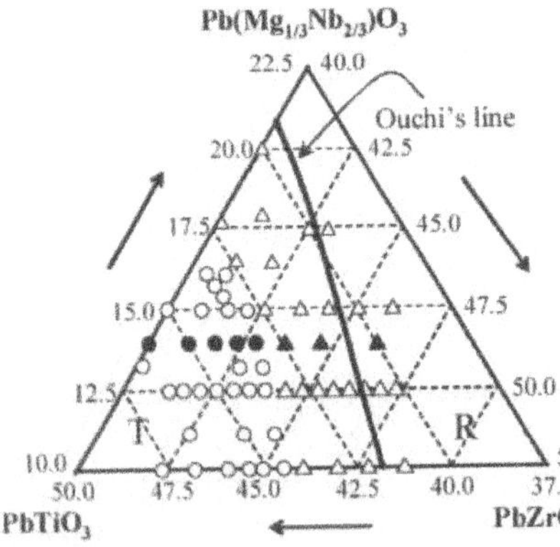

Fig. 6. A series of PMNZT 14.2/100y/85.8-100y compositions with y=0.400-0.458 further investigated on the basis of the results of dielectric and piezoelectric properties for the whole PMN-PZ-PT ceramics studied in this work. These compositions are plotted by the solid circles (tetragonal phase) and solid triangles (rhombohedral phase) in the figure.

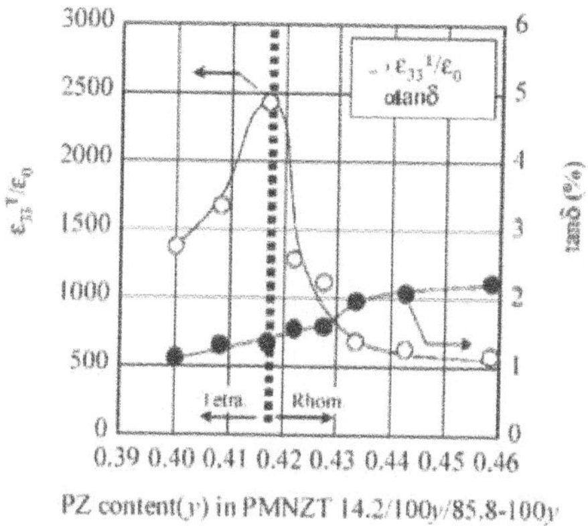

Fig. 7. Compositional dependence of relative permittivity, $\varepsilon_{33}^T/\varepsilon_0$ and dielectric loss, tanδ at room temperature for PMNZT 14.2/100y/85.8-100y ceramics.

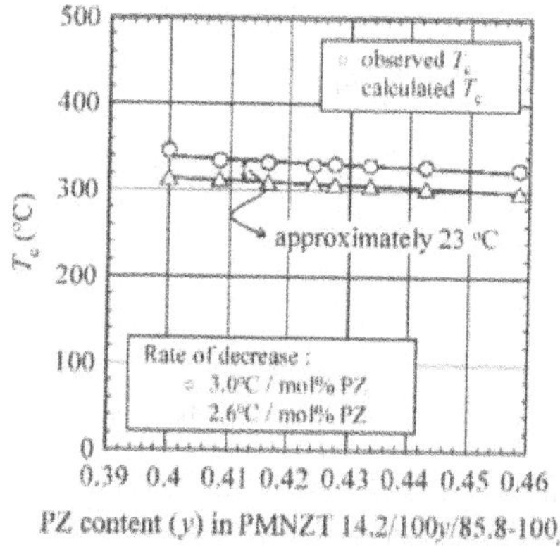

Fig. 8. Compositional dependence of Curie point, T_c for PMNZT 14.2/100y/85.8-100y ceramics.

53.4 mol% based on the observed values. The PZ content should therefore be limited to less than approximately 53.4 mol% for practical applications.

Figure 9 shows the dependence of k_p and d_{31} on composition at room temperature for these PMNZT ceramics. The values of k_p and d_{31} at y = 0.416 approach maxima near the MPB line, reached 70% and −243 pC/N, respectively. Figure 10 shows the dependence of remanent polarization (P_r) and coercive field (E_c) on composition, and the typical P–E hysteresis loop for

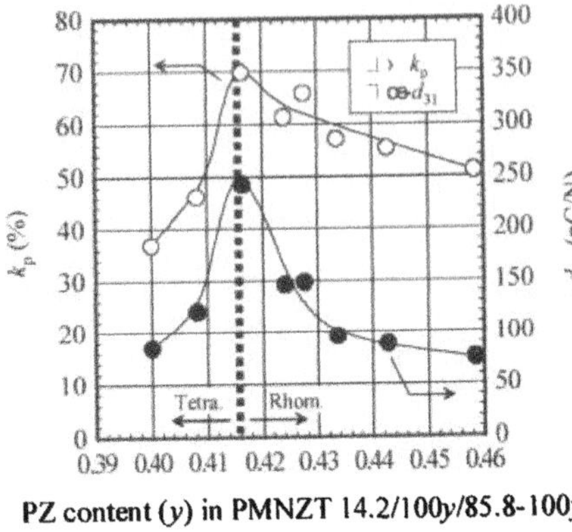

Fig. 9. Compositional dependence of electromechanical coupling factor, k_p and piezoelectric coefficient, $-d_{31}$ for PMNZT 14.2/100y/85.8-100y ceramics.

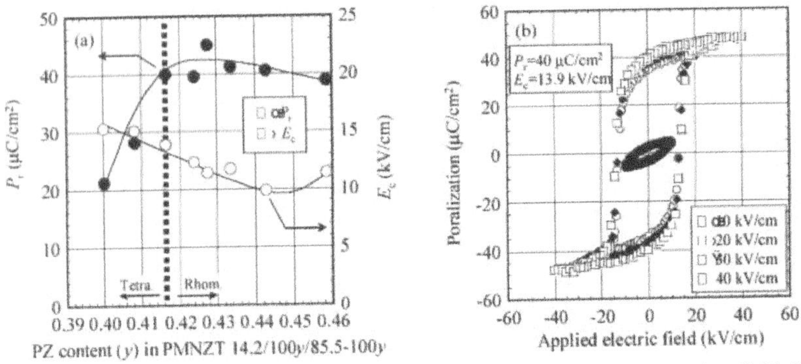

Fig. 10. (a) Compositional dependence of remanent polarization, P_r and coercive field, E_c for PMNZT 14.2/100y/85.8-100y ceramics and (b) typical P-E hysteresis loop of PMNZT 14.2/41.6/44.2 ceramic with a composition on the MPB line.

the PMNZT 14.2/41.6/44.2 ceramic (MPB composition). The compositions exhibiting the maximum values of P_r and k_p differ, $y = 0.427$ and 0.416, respectively. The reason for this disagreement is unclear. As confirmed in other literature,[14] however, E_c changes regardless of the MPB, decreasing slightly on the rhombohedral side. The typical PMNZT sample with the MPB composition exhibits a slim rectangular hysteresis loop with large remanent polarization

(P_r = 40 μC/cm²) and weak coercive field (E_c = 13.9 kV/cm). Figure 11 shows the MPB line determined from the piezoelectric properties and phase transition observations, with the lines of Ouchi et al.[3] and Tashiro et al.[10] shown for comparison. The refined position of the MPB line is shifted from Ouchi's line toward the PT side by approximately 3.6 mol%, but is in good agreement with Tashiro's line, which is plotted based on the compositional points exhibiting the maximum values of d_{31}.

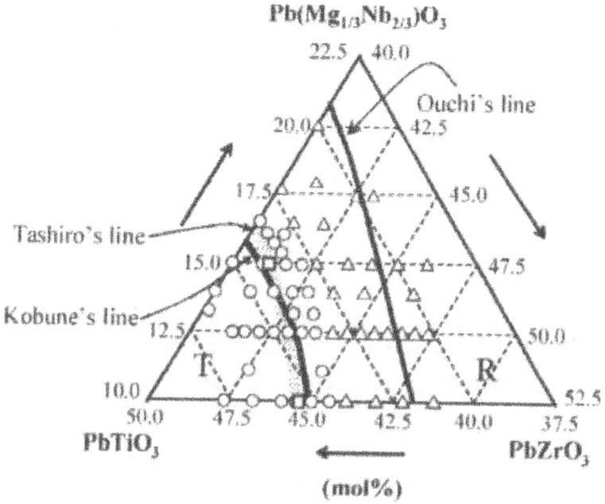

Fig. 11. Phase diagram of ternary $Pb(Mg_{1/3}Nb_{2/3})O_3$-$PbZrO_3$-$PbTiO_3$ system. Open circles and open triangles display the tetragonal phase and the rhombohedral phase, respectively. Ouchi's line: the MPB line reported by Ouchi et al., Kobune's line: the MPB line newly found by Kobune et al. and Tashiro's line: the MPB line recently reported by Tashiro et al., being broadly plotted using an open square.

A comparison of the analyzed and theoretical chemical constituents of the PMNZT 14.2/41.6/44.2 ceramic is shown in Table I. The analyzed values coincide well with the theoretical values within the accuracy of measurement, indicating that any compositional fluctuation of the sintered PMNZT bodies fabricated in the present study is negligible.

Table I. Comparison of Analyzed Values and Theoretical Values on Chemical Constituents of PMNZT14.2/41.6/44.2 Ceramics

Item	PbO (mol%)	MgO (mol%)	Nb_2O_5 (mol%)	ZrO_2 (mol%)	TiO_2 (mol%)
Analyzed values	49.838	3.963	3.568	20.700	21.942
Theoretical values	50.000	3.550	3.550	20.800	22.100

Table II shows a comparison of the piezoelectric properties for PMNZT 14.2/41.6/44.2, ceramic with similar compositions,[5, 9] and commercial PZT ceramics.[17] Thus, the present material having a high T_c of >300°C shows great promise as a practical piezoelectric material for piezoelectric actuators and ultrasonic sensors, on the basis of exhibiting excellent piezoelectric properties such as $\varepsilon_{33}^T/\varepsilon_0 = 2430$, $\tan\delta = 1.4\%$, $k_p = 70\%$, $k_t = 50\%$ and $d_{31} = -243$ pC/N as compared with those of other PMNZT and PZT ceramics.

Table II. Comparison of Piezoelectric Properties for PMNZT Ceramic with the Pinpoint Composition, the PMNZT Ceramics with Similar Compositions and Commercial PZT Ceramics

Ceramics	PMNZT composition	$\varepsilon_{33}^T/\varepsilon_0$	$\tan\delta$ (%)	T_c (℃)	k_p (%)	$-d_{31}$ (pC/N)	Q_m
Present study	14.2/41.6/44.2	2434	1.4	328	70	243	71
Ouchi *et al.* [5]	37.5/25.5/37	2137	1.9	-	64	198	65
Koval *et al.* [9]	12.5/44/43.5	-	-	-	40	106	71
P-5E[17]	-	1490	0.4	280	56	131	970
P-7[17]	-	1930	1.4	300	65	207	80
P-7B[17]	-	3200	2.2	180	65	303	180

CONCLUSIONS

High-density ceramics with compositions of $xPb(Mg_{1/3}Nb_{2/3})O_3$-$yPbZrO_3$-$zPbTiO_3$ (PMNZT 100x/100y/100z) having x=0.1-0.2, y=0.400-0.486 and z=0.395-0.477 were fabricated by columbite route. The dielectric, piezoelectric and ferroelectric properties of the fabricated PMNZT ceramics were investigated in detail. The results of this study can be summarized as follows:

(1) All the obtained PMNZT ceramics had a single-phase perovskite structure.

(2) The refined morphotropic phase boundary (MPB) line was shifted toward the $PbTiO_3$ (PT) side in the ternary field by approximately 3.6 mol% from the position reported in the literature.

(3) From measurements of dielectric, piezoelectric and ferroelectric properties, the PMNZT 14.2/41.6/44.2 composition on the MPB line had excellent physical properties consisting of $\varepsilon_{33}^T/\varepsilon_0=2430$, $\tan\delta=1.4\%$, $T_c=331°C$, $k_p=70\%$, $k_t=50\%$, $d_{31}=-243$ pC/N, $P_r=40$ μC/cm^2 and $E_c=13.9$ kV/cm.

(4) The results of physical properties demonstrate that the PMNZT 14.2/41.6/44.2 ceramic is a potential element material for piezoelectric actuators and ultrasonic sensors.

REFERENCES

[1]G. A. Smolenskii and A. I. Agranovskaya, "Dielectric Polarization of a Number of Complex Compounds," *Fizika Tverdogo Tela,* **1** [10] 1562-72 (1959).

[2]B. Jaffe, R. S. Roth and S. Marzullo, "Properties of Piezoelectric Ceramics in The Solid-Solution Series Lead Titanate Zirconate-Lead Oxide: Tin Oxide and Lead Titanate-Lead Hafnate," *J. Res. Natl. Bur. Std.*, **55**, 239-54 (1955).

[3]H. Ouchi, K. Nagano and S. Hayakawa, "Piezoelectric Properties of $Pb(Mg_{1/3}Nb_{2/3})O_3$-$PbTiO_3$-$PbZrO_3$ Solid Solution Ceramics," *J. Am. Ceram. Soc.*, **48** [12] 630-35 (1965).

[4]H. Ouchi, M. Nishida and S. Hayakawa, "Piezoelectric Properties of $Pb(Mg_{1/3}Nb_{2/3})O_3$-$PbTiO_3$-$PbZrO_3$ Ceramics Modified with Certain Additives," *J. Am. Ceram. Soc.*, **49** [11] 577-82 (1966).

[5]H. Ouchi, "Piezoelectric Properties and Phase Relations of $Pb(Mg_{1/3}Nb_{2/3})O_3$-$PbTiO_3$-$PbZrO_3$ Ceramics with Barium or Strontium Substitutions," *J. Am. Ceram. Soc.*, **51** [3] 169-76 (1968).

[6]G. Zhilum, L. Longtu, G. Suhua and Z. Xiaowen, "Low Temperature Sintering of Lead-Based Piezoelectric Ceramics with Transient Liquid Phase"; Proceedings of the 4th International Symposium on Science and Technology of Sintering, Tokyo, **2**, 920-25 (1987).

[7]H. Yamaguchi, "Behavior of Electric-Field-Induced Strain in PT-PZ-PMN Ceramics," *J. Am. Ceram. Soc.*, **82** [6] 1459-62 (1999).

[8]K. H. Yoon and H. R. Lee, "Effect of Ba^{2+} Substitution on Dielectric and Electric-Field-Induced Strain Properies of PMN-PZ-PT Ceramics," *J. Am. Ceram. Soc.*, **83** [11] 2693-98 (2000).

[9]V. Koval, C. Alemany, J. Briancin and H. Brunckova, "Dielectric Properties and Phase Transition Behavior of xPMN-(1-x)PZT Ceramic Systems," *J. Electroceramics*, **10**, 19-29 (2003).

[10]S. Tashiro, W. Tokunaga, K. Ishii and K. Nagata, "Nonlinear Coefficients in Compositions near Morphotropic Phase Boundary of $PbZrO_3$-$PbTiO_3$-$Pb(Mg_{1/3}Nb_{2/3})O_3$ Piezoeletcric Ceramics," Abstracts of The 22nd Meeting on Ferroelectric Materials and Their Applications, 25-26 (2005) [in Japanese].

[11]S. L. Swartz and T. R. Shrout, "Fabrication of Perovskite Lead Magnesium Niobate, "*Mater. Res. Bull.*, **17** [10] 1245-50 (1982).

[12]M. Kobune, S. Fujii, K. Furuya and R. Takayama, "Preparation of La- and Mg-Modified $PbTiO_3$ Ceramics and Their Pyroelectric Properties," *J. Ceram. Soc. Jpn.*, **102** [11] 999-1004 (1994).

[13]*Electrical Testing Method of Piezoelectric Ceramic Vibrators*, pp. 21-95 in EMAS-6100. Edited by Electronic Materials Manufacturers Association of Japan, 1993.

[14]M. Kobune, K. Muto and Y. Akiyama, "Piezoelectric and Ferroelectric Properties of $Pb(Zn_{1/3}Nb_{2/3})O_3$- $PbTiO_3$-$PbZrO_3$ Ceramics," *J. Ceram. Soc. Jpn.*, **110** [1] 12-17 (2002).

[15]S. Roberts, "Dielectric Properties of Lead Zirconate and Barium-Lead Zirconate," *J. Am. Ceram. Soc.*, **33**, [2] 63-66 (1950).

[16]J. P. Remeika and A. M. Glass, "The Growth and Ferroelectric Properties of High Resistivity Single Crystals of Lead Titanate ," *Mater. Res. Bull.*, **5** [1] 37-46 (1970).

[17]Murata Manufacturing's Catalog, *Piezoelectric Ceramics (Piezotite)*, Cat. No. P19-7, 9-10 [in Japanese].

Electroceramic Materials and Applications
Edited by R. W. Schwartz

Thermoelectric and Energy Harvesting Materials for Solid State Power Conversion

Electroceramic Materials and Applications
Edited by R. W. Schwartz

FABRICATION OF GRAIN-ALIGNED THICK FILMS OF THERMOELECTRIC OXIDES BY AN ELECTROPHORETIC DEPOSITION METHOD UNDER HIGH MAGNETIC FIELDS

Taichi Okamoto, Shigeru Horii, Toshiaki Kumagai, Jun-ichi Shimoyama, Kohji Kishio
Department of Superconductivity, University of Tokyo
7-3-1 Hongo, Bunkyo-ku
Tokyo 113-8586, Japan

Tetsuo Uchikoshi, Tohru Suzuki, Yoshio Sakka
National Institute for Materials Science (NIMS)
1-2-1 Sengen, Tsukuba
Ibaraki 305-0047, Japan

ABSTRACT

We report the preparation of grain-aligned $[Ca_2CoO_{3-\delta}]_{0.62}CoO_2$ (Ca349) thick films by a simultaneous use of electrophoretic deposition (EPD) and magneto-science methods, and multi-layered thick films composed of Ca349, and $Ca_{0.9}La_{0.1}MnO_{3-\delta}$(Mn113) and Al_2O_3 using the EPD method. The thick films with the orders from 10^{-6} to 10^{-3} m in thickness were fabricated at high deposition rates of the order of 10^{-5} m/min by optimization of the suspension component and the control of applied electric field for the EPD process. The EPD thick films processed through appropriate heat-treatment for densification showed almost same thermoelectric properties with their bulk samples. The Ca349 thick films deposited under high magnetic fields exhibited high degree of *c*-axis grain-orientation, and remarkable reduction of resistivity perpendicular to the applied magnetic field was observed. The Al_2O_3/Mn113/Al_2O_3/Ca349/Al_2O_3 multi-layered thick films have been successfully prepared by the optimization of the EPD process. Our present results indicate the effectiveness of the EPD process (under high magnetic fields) as an integration technique of thermoelectric modules for high temperature power generation.

INTRODUCTION

Discovery of the excellent thermoelectric (TE) properties of Na_xCoO_2 ($x \sim 0.5$)[1] has triggered developments of TE oxides for production of practical TE modules for the power generation using high-temperature waste heat. In addition to Na_xCoO_2, misfit layered cobaltites, such as $[Ca_2CoO_{3-\delta}]_{0.62}CoO_2$[2, 3] and $[BiSrO]_{1.10}CoO_2$[4], were reported as *p*-type TE compounds, while $CaMnO_3$[5], $LaNiO_3$[6] and $(ZnO)_mIn_2O_3$[7] were candidates as *n*-type TE oxides. In order to improve TE properties of their polycrystals, grain-alignment[8-10] and densification methods[9] have been proposed.

Recently, production of TE modules using the oxide bulk materials and their properties of power generation at high temperatures have been reported.[11, 12] These TE modules are produced by using grain-aligned and/or densified polycrystalline bulks which were formed into cubic shape. In practice, from viewpoints of downsizing and high integration degree in TE modules, one can recognize that packaging using films rather than bulks is more adequate. However, in the case of standard vapor growth method to prepare thin films, such as the pulsed laser deposition, there are disadvantageous points of slow growth rate and difficulty of fabrication of films with a high orientation degree without using single crystalline substrates. In this context, introduction of a deposition process by which thicker films can be grown at high deposition rates is required for production of highly integrated TE modules.

Based on the above background, we focused on the electrophoretic deposition (EPD) method,[13-15] which is a colloidal process that ceramic particles are shaped on the electrode directly from a stable colloid suspension by a DC electric field. Our group has already reported the preparation of *c*-axis aligned $[Ca_2CoO_{3-\delta}]_{0.62}CoO_2$ bulks, which have their easy axes of magnetization parallel to the *c*-axis, by the combination of the colloidal process and the magnetic alignment.[9] Therefore the EPD is an extension of our preparation technique of the grain-aligned bulks. Furthermore, this method enables the deposition of thick films with orders of 10^{-6}-10^{-3} m in thickness at deposition rates of 1-100 μm/min and is available under high magnetic fields[16, 17], which is one of the most desirable processes in the sense of the above purpose. Namely, by introduction of simultaneous use of the EPD and magnetic alignment processes, it is expected that grain-oriented thick films are obtained at high deposition rates.

In this study, we have prepared the thick films of $[Ca_2CoO_{3-\delta}]_{0.62}CoO_2$ (Ca349) and $Ca_{0.9}La_{0.1}MnO_{3-\delta}$ (Mn113) by the EPD method and evaluated these TE properties. For the Ca349 compound, we also attempted to fabricate the grain-aligned thick films by EPD under magnetic fields (M-EPD). Furthermore, for the demonstration of an integration technique of TE modules, multi-layered thick films composed of Ca349, Mn113 and insulating α-Al_2O_3 have been fabricated.

EXPERIMENTAL

$[Ca_2CoO_{3-\delta}]_{0.62}CoO_2$ (Ca349) and $Ca_{0.9}La_{0.1}MnO_{3-\delta}$ (Mn113) powders were synthesized by standard solid-state reaction technique using $CaCO_3$, Co_3O_4, Mn_3O_4 and La_2O_3 as starting materials. Each calcined powder with single phase characterized by the X-ray diffraction (XRD) method was fined by ball-milling. The average particle sizes of the resulting powders analyzed by a precipitation method were ~ 0.9 μm and ~ 0.7 μm in diameter for Ca349 and Mn113, respectively. We chose a commercial α-Al_2O_3 fine powder (Sumitomo Chem. Co., Ltd., AKP-50) with the average particle size of ~ 0.3 μm in diameter as an insulating material between *p*- and *n*-type of TE layers for the fabrication of multi-layered TE thick films.

In order to prepare the suspension of each compound, each powder was dispersed in ethanol. Optimized pH values controlled by addition levels of dispersant were ~5, ~6 and ~4 for Ca349, Mn113 and α-Al_2O_3, respectively. Note that volume fraction of each compound was 10 vol.% to the volume of ethanol. Zeta potentials of the particles of Ca349, Mn113 and α-Al_2O_3 in the suspension were +10 mV, +15 mV and +40 mV, respectively.

The experimental configuration of EPD without a magnetic field is shown in Fig. 1(a). A pair of glass electrodes coated by carbon, 10×30 or 20×30 mm^2 in area with 20 mm spacing, were held on conducting metal poles, and were dipped into the suspension. The EPD was examined in an electric field of 10.0 kV/m for Ca349 and 2.5-20.0 kV/m for Mn113 for 5-60 min at room temperature, respectively. The deposited films were heated at 500°C for 12 h in air to remove the solvent, dispersant and carbon substrate. Then Ca349 films were densified by four times repetition of intermediate uniaxial pressing under 392 MPa and sintering at 900°C for 12 h in air. This densification process is expressed as "IP", and films and bulks without the IP process are expressed "as-deposited" films and "as-sintered" bulks, respectively. Mn113 films were densified only by sintering at 1350°C for 24 h in air. Both Ca349 and Mn113 films were furnace-cooled after sintering in order to eliminate oxygen deficiencies, i.e. $\delta = 0$.

The experimental configuration of M-EPD is shown in Fig. 1(b). The Ca349 suspension was placed at the center of a magnetic field in the superconducting magnet, and then a magnetic field ($\mu_0 H_a$) of 1 T or 3 T was applied for alignment of the Ca349 particles. Incidentally, palladium plates with the size of 20×20 mm^2 in area were used as electrodes, and electrical field was applied parallel to the magnetic field. The M-EPD process was performed at 20.0 kV/m for 20 min. The deposited films were densified by cold isostatic pressing (CIP) under 392 MPa instead of the IP process. The films processed through CIP were heated at 500°C for 12 h in air to remove the organic components, then were furnace-cooled after sintering at 900°C for 12 h in air to control δ to be zero.

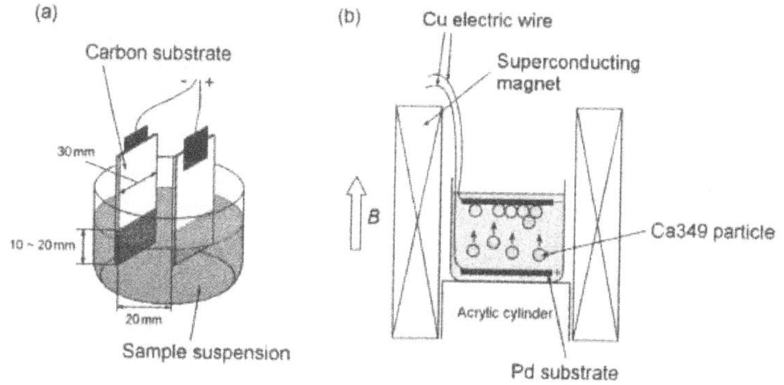

Fig. 1. Schematic diagram of the experimental configurations of the EPD (a) and the M-EPD (b).

Electrical resistivity measurement was performed by a four-probe method from 5 K to 300 K. Seebeck coefficient was examined also by a four-probe method from 90 K to 300 K. These measurements for the grain-aligned Ca349 films were done for a parallel direction to the bottom surface ($\perp H_a$). The XRD measurement was performed for the bottom surface of the films. Microstructures of cross-sectional surfaces of thus obtained thick films were observed by a scanning electron microscope (SEM).

RESULTS AND DISCUSSION

Fabrication and TE Properties of Ca349 and Mn113 Thick Films

Figure 2 shows changes of thickness of the Ca349 and Mn113 thick films fabricated by EPD in 0 T at various electric fields (E) as a function of deposition time (t). The deposition rates were 58 μm/min at 20.0 kV/m for Mn113 and 66 μm/min at 10.0 kV/m for Ca349, respectively, which were sufficiently high and were more than 100 times higher than those of the vapor growth methods. Obviously, the thickness was linearly increased with the deposition time for all the Mn113 films. Moreover, the deposition rate was found to be almost proportional to the electric field.

From the principle of conservation of mass, Hamaker *et al.* proposed the following equation as the basis for the kinetics of EPD.[18]

$$\frac{dw}{dt} = f\mu Ec \,, \tag{1}$$

where w, μ, c and f are the mass of deposition per unit area, the electrophoretic mobility, the particle concentration and a factor taking into account the absorption rate, respectively. Ignoring the reduction of particle concentration, which means c is constant, dw/dt is proportional to E. Thickness is represented as w/ρ, where ρ is the mass density. Since the ρ of the Mn113 film was found to be constant from the microstructure observation by SEM, thickness of the films can be considered to be proportional to w. That is, thickness is proportional to t and E. From the comparison between our results and the above theory, one can recognize that the reduction of particle concentration in the suspension with t is ignored within the deposition time we performed.

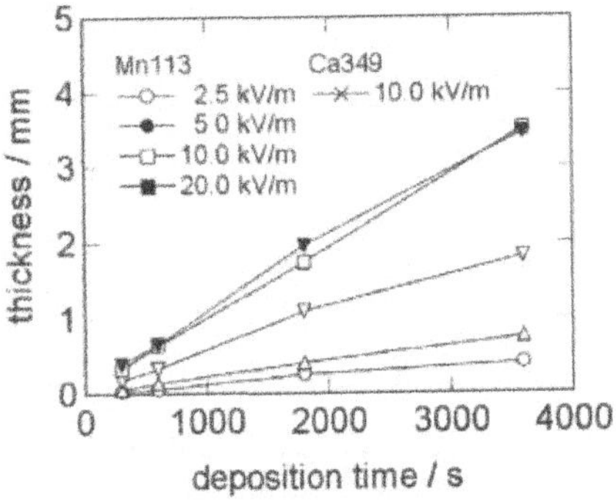

Fig. 2. t-dependence of thickness at different electric fields in 0 T for the suspensions of Mn113 and Ca349 in the EPD process.

Figures 3(a) and 3(b) show temperature dependence of electrical resistivities of the thick films of Ca349 and Mn113 measured under applying current perpendicular to the deposition direction, respectively, also with the resistivities of the bulk samples of both compounds for comparison. Note that the Mn113 thick film for the resistivity measurement were fabricated in a condition of E = 10.0 kV/m. As shown in Fig. 3(a), electrical resistivity of the as-deposited Ca349 was one order of magnitude higher than that of the as-sintered bulk. SEM images did not show the grain-aligned microstructure for both film and bulk samples, therefore this remarkable difference can be understood only in terms of difference of relative densities between the as-deposited film (~ 40 %) and as-sintered bulk (~ 60 %). On the other hand, the resistivities of both the Ca349 film and bulk were drastically decreased by the introduction of the IP process and, interestingly, they showed almost same values each other, which is due to improvement of relative densities by IP (~ 80 %). In the case of Mn113 (Fig. 3(b)), a behavior of resistivity for the EPD film without IP was coincident with that of the as-sintered bulk. Both samples showed almost same relative densities (~ 97 %), and densification was achieved only through the sintering process without IP. This result indicates that the resistivities of thick films fabricated by the EPD process are intrinsically depending only on the relative density and are very close to the behaviors of bulks rather than epitaxial thin films[19].

Figures 4(a) and 4(b) show temperature dependence of Seebeck coefficients of the films of Ca349 and Mn113 prepared in 0 T, respectively. For comparison, data of the bulk samples of both compounds were also shown in Fig. 4. In both cases, one can obviously recognize that Seebeck coefficient is insensitive to the fabrication process and densification process (or relative density). From these results in Fig. 3 and Fig. 4, it was found that the EPD process enabled fabrication of the thick films at high deposition rates without deterioration of TE properties.

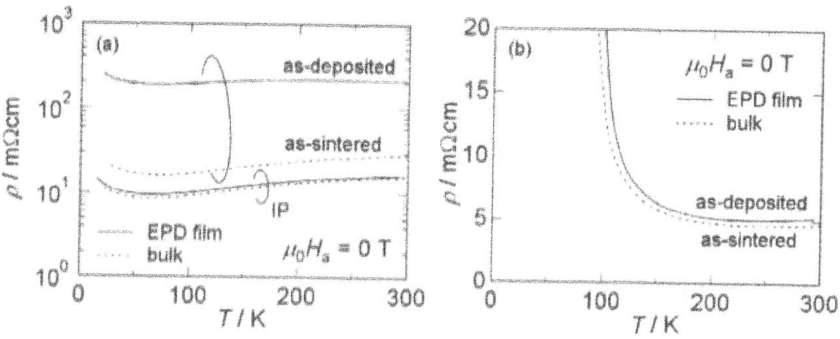

Fig. 3. Temperature dependence of electrical resistivities for the EPD films and bulks of Ca349 with and without the IP process (a) and of Mn113 (b). The EPD process was performed under 0 T.

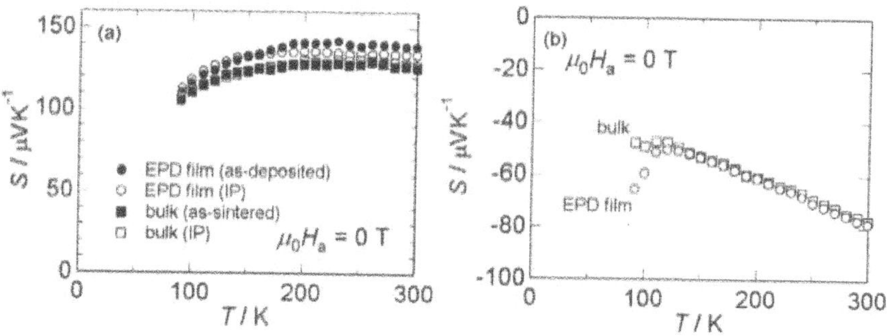

Fig. 4. Temperature dependence of Seebeck coefficients for the EPD films and bulks of Ca349 with and without the IP process (a) and of Mn113 (b).

Fabrication of Grain-Aligned Ca349 Thick Film

In the previous section, effectiveness of the EPD method was shown for the preparation of thick films. We have attempted to fabricate grain-oriented thick films by the M-EPD method for the Ca349 compound which possesses layered crystal structure and its easy axis of magnetization parallel to the *c*-axis. Figure 5 shows the XRD patterns of the bottom surface ($\perp H_a$) of Ca349 thick films deposited under various magnetic fields ($\mu_0 H_a$ = 0, 1, and 3 T), also with a XRD pattern of the Ca349 powder for comparison. For all the deposited Ca349 films, diffraction peaks except for the (0 0 *l* 0) peaks were suppressed in the M-EPD films, while the (0 0 *l* 0) peaks were dramatically enhanced. For the evaluation of degrees of *c*-axis grain-alignment, Lotgering factors (F)[20] for these M-EPD films were estimated from these XRD patterns in Fig. 5. F is expressed as $(P - P_0) / (1 - P_0)$, where $P = \Sigma I(0\ 0\ l\ 0) / \Sigma I(h\ k_1\ l\ k_2)$ and $P_0 = \Sigma I_0(0\ 0\ l\ 0) / \Sigma I_0(h\ k_1\ l\ k_2)$. ΣI and ΣI_0 are the sums of peak intensities for the aligned samples and the randomly oriented samples, respectively. The estimated F's were 0.76, 0.95 and 0.97 for the M-EPD films deposited under $\mu_0 H_a$ = 0, 1, and 3 T, respectively. Clear difference between the EPD films with and without the applied magnetic field emerged from microstructure observation by SEM. Figures 6(a) and 6(b) show SEM images of the cross-sectional surface for the Ca349 thick films deposited in 0 T and 1 T, respectively. Platelet grains, which are spread along *ab* direction, were aligned in a perpendicular direction to H_a for the film deposited under $\mu_0 H_a$ = 1 T (Fig. 6(b)), while the EPD film deposited under $\mu_0 H_a$ = 0 T showed randomly-oriented microstructure (Fig. 6(a)). Therefore, the estimated F in the EPD deposited under $\mu_0 H_a$ = 0 T is reflection of the degree of grain-orientation only at the surface, and these results indicate that sufficiently high magnetic fields are required for *c*-axis grain-orientation at a whole part of the films.

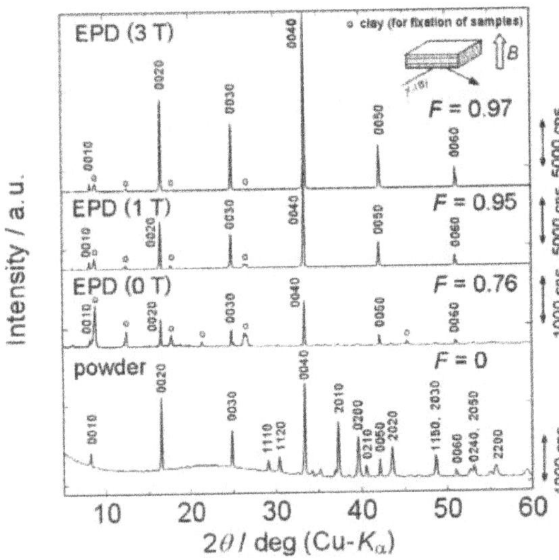

Fig. 5. X-ray diffraction patterns of the bottom surface ($\perp H_a$) for the M-EPD Ca349 films deposited under $\mu_0 H_a$ = 0, 1 and 3 T. The powder pattern is also show for comparison. F is the degree of *c*-axis orientation calculated from the Lotgering's method[20].

In our previous study[9], the minimum magnetic field (H_m) for grain-alignment in Ca349 was estimated from the magnetic anisotropy at 300 K using the grain-oriented and densified bulk.

In this sense, the $\mu_0 H_m$ values were estimated to be ~ 3 T and ~ 1 T for the Ca349 particles with 0.1 and 0.2 μm in diameter, respectively. From particle size distribution of the Ca349 fine powders used for M-EPD, the particles over 0.1 and 0.2 μm in diameter accounted for 95 % and 85 % of the Ca349 particles, respectively. Therefore, one can understand that the magnetic fields of 1 T and 3 T applied in the M-EPD experiment are sufficient for the grain-alignment of almost all the Ca349 particles.

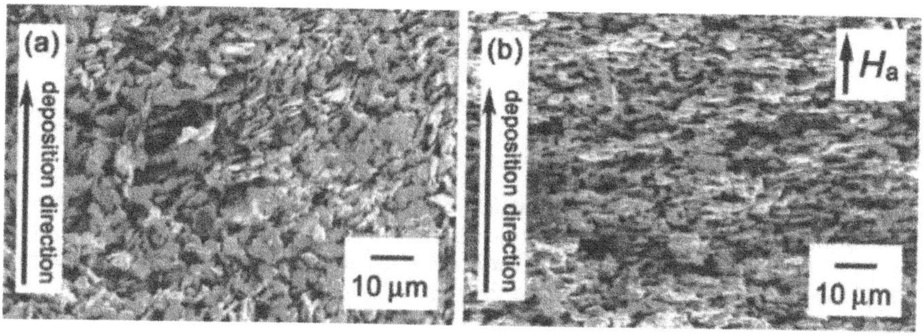

Fig. 6. Secondary electron images of cross-sectional surface of the Ca349 thick films deposited under the magnetic fields of 0 T (a) and 1 T (b). Deposition proceeded from bottom to top on these images. Relative densities were ~ 40 % for both films.

Figure 7 shows temperature dependence of electrical resistivity perpendicular to H_a (or the deposition direction) for the Ca349 films which were fabricated under $\mu_0 H_a$ = 0 and 1 T, also with data for the grain-aligned and densified Ca349 bulk reported by Sano *et al.*[21]. Electrical resistivities of the both Ca349 thick films without CIP showed similar values and were apparently independent of H_a. Taking relative densities for both the films (~ 40 %) into account, this result indicates in the case of porous film that electrical resistivity is dominantly affected by grain boundaries rather than grain-oriented microstructure. On the other hand, for the samples through the CIP process, the Ca349 thick film deposited in 1 T showed lower electrical resistivity (12 mΩcm at 300 K) compared with that deposited in 0 T (17 mΩcm at 300 K). The relative densities of both the thick films with CIP were improved into ~ 60 %, which is reflection of the grain-orientation effect. Seebeck coefficients of both films were ~ 157 μV/K at 300 K, that is, the power factor of the Ca349 thick film deposited in 1 T was enlarged to 2.1×10^{-6} W/cmK at 300 K. These results indicate the effectiveness of the M-EPD process for fabrication of the Ca349 thick films with high TE performance. However, the reduction of electrical resistivity in the thick film is insufficient compared with resistivity of the Ca349 bulk, and introduction of other densification processes, such as hot-forging and SPS, is required for achieving higher relative density exceeding 90 %.

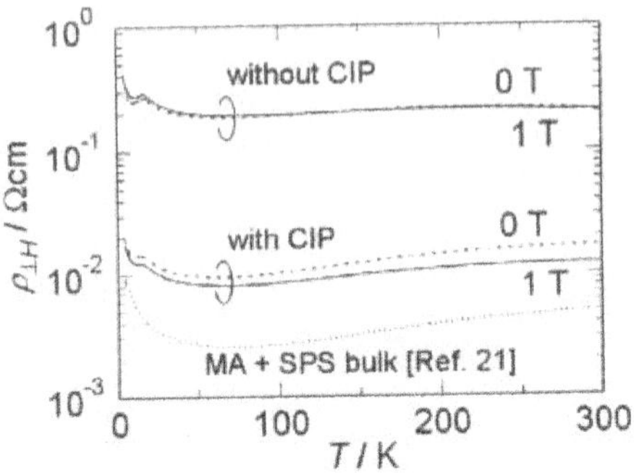

Fig. 7. Temperature dependence of electrical resistivity for the Ca349 thick films deposited under $\mu_0 H_a$ = 0 T and 1 T. The data for the films processed through the CIP were also shown. MA+SPS bulk[21] represents the grain-aligned and densified Ca349 bulk prepared by magnetic alignment (MA) process and spark plasma sintering (SPS) method.

Fabrication of Ca349, Mn113 And α-Al_2O_3 Multi-Layered Thick Film

For the purpose of development of an integration technique of TE modules, a multi-layered thick film composed of Ca349, Mn113, and α-Al_2O_3 was prepared by the EPD method, three of which are *p*- and *n*-type of TE compounds and an insulation compound, respectively. The EPD process was performed under E = 10.0 kV/m for 2 h in 0 T, for fabrication of each layer. The photo image of thus obtained thick film is shown in Fig. 8. The stacking microstructure of α-Al_2O_3/Mn113/α-Al_2O_3/Ca349/α-Al_2O_3 on the Ni substrate was clearly observed. Thickness of the multi-layered film was ~3 mm and thickness of each layer is smaller than 1 mm. For practical use, further investigations are required in terms of improvement of undulated interface, suppression of cracks and densification of Ca349 and Mn113 films.

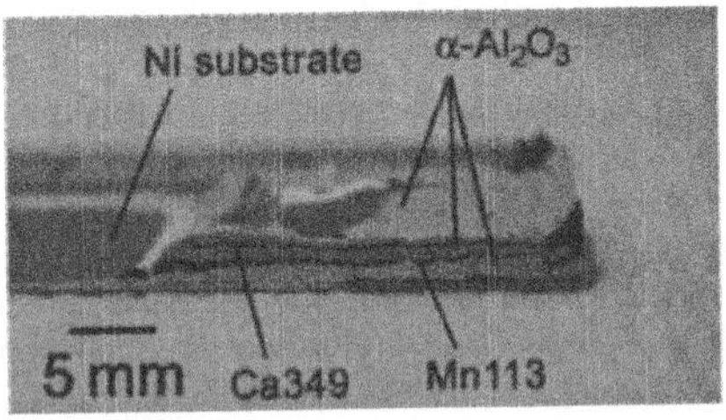

Fig. 8. The photo image of α-Al_2O_3/Mn113/α-Al_2O_3/Ca349/α-Al_2O_3 multi-layered thick film deposited on the Ni substrate.

CONCLUSION

We have developed a novel fabrication technique for TE modules: the EPD method. The thick films of Ca349 and Mn113 have been successfully prepared at high deposition rates of ~ 60 μm/min without deterioration of their original TE properties. TE properties of Ca349 thick films have been improved by grain-alignment of the M-EPD process. Multi-layered thick film using Ca349, Mn113 and α-Al_2O_3 has been prepared. It is suggested that the combination of colloidal and magneto-scientific processes is practical as an integration technique of TE modules.

REFERENCES

[1]I. Terasaki, Y. Sasago, and K. Uchinokura, "Large thermoelectric power in $NaCo_2O_4$ single crystals," *Phys. Rev. B*, **56**, R12685 (1997).

[2]Y. Miyazaki, K. Kudo, M. Akoshima, Y. Ono, Y. Koike, and T. Kajitani, "Low-Temperature Thermoelectric Properties of the Composite Crystal $[Ca_2CoO_{3.34}]_{0.614}[CoO_2]$," *Jpn. J. Appl. Phys.*, **39**, L531-33 (2000).

[3]A. C. Masset, C. Michel, A. Maignan, M. Hervieu, O. Toulemonde, F. Studer, B. Raveau, and J. Hejtmanek, "Misfit-layered cobaltite with an anisotropic giant magnetoresistance: $Ca_3Co_4O_9$," *Phys. Rev. B*, **62**, 166-175 (2000).

[4]R. Funahashi, I. Matsubara, and S. Sodeoka, "Thermoelectric properties of $Bi_2Sr_2Co_2O_x$ polycrystalline materials," *Appl. Phys. Lett.*, **76**, 2385-87 (2000).

[5]M. Ohtaki, H. Koga, T. Tokunaga, K. Eguchi, and H. Arai, "Electrical Transport Properties and High-Temperature Thermoelectric Performance of $(Ca_{0.9}M_{0.1})MnO_3$ (M = Y, La, Ce, Sm, In, Sn, Sb, Pb, Bi)," *J. Solid State Chem.*, **120**, 105-111 (1995).

[6]R. Funahashi, S. Urata, and M. Kitawaki, "Exploration of *n*-type oxides by high throughput screening," *Appl. Surf. Sci.*, **223**, 44-48 (2004).

[7]Y. Masuda, M. Ohta, W.-S. Seo, W. Pitschke, and K. Koumoto, "Structure and Thermoelectric Transport Properties of Isoelectronically Substituted $(ZnO)_5In_2O_3$," *J. Solid State Chem.*, **150**, 221-227 (2000).

[8]S. Tajima, T. Tani, S. Isobe, and K. Koumoto, "Thermoelectric properties of highly textured $NaCo_2O_4$ ceramics processed by the reactive templated grain growth (RTGG) method," *Mater. Sci. Eng. B*, **86**, 20-25 (2001).

[9]S. Horii, I. Matsubara, M. Sano, K. Fujie, M. Suzuki, R. Funahashi, M. Shikano, W. Shin, N. Murayama, J. Shimoyama, and K. Kishio, "Thermoelectric Performance of Magnetically *c*-Axis Aligned Ca-based Cobaltites," *Jpn. J. Appl. Phys.*, **42**, 7018-22 (2003).

[10]R. Funahashi, S. Urata, T. Sano, and M. Kitawaki, "Enhancement of thermoelectric figure of merit by incorporation of large single crystals in $Ca_3Co_4O_9$ bulk materials," *J. Mater. Res.*, **18**, 1646-51 (2003).

[11]I. Matsubara, R. Funahashi, T. Takeuchi, S. Sodeoka, T. Shimizu, and K. Ueno, "Fabrication of an all-oxide thermoelectric power generator," *Appl. Phys. Lett.*, **78**, 3627-29 (2001).

[12]R. Funahashi, S. Urata, K. Mizuno, T. Kouuchi, and M. Mikami, "$Ca_{2.7}Bi_{0.3}Co_4O_9$ / $La_{0.9}Bi_{0.1}NiO_3$ thermoelectric devices with high output power density," *Appl. Phys. Lett.*, **85**, 1036-38 (2004).

[13]P. Sarker and P. S. Nicholson, "Electrophoretic Deposition (EPD): Mechanisms, Kinetics, and Application to Ceramics," *J. Amer. Ceram. Soc.*, **79**, 1987-2002 (1996).

[14]I. Zhitomirsky, "Cathodic Electrodeposition of Ceramic and Organoceramic Materials. Fundamental Aspects," *Adv. Coll. Interf. Sci.*, **97**, 279-317 (2002).

[15]A. R. Boccaccini and I. Zhitomirsky, "Application of Electrophoretic Deposition Techniques in Ceramic Processing," *Curr. Opin. Solid State Mater. Sci.*, **6**, 251-260 (2002).

[16]T. Uchikoshi, T. S. Suzuki, H. Okuyama, Y. Sakka, and P. S. Nicholson, "Electrophoretic deposition of alumina suspension in a strong magnetic field," *Europ. Ceram. Soc.*, **24**, 225-229 (2004).

[17]T. Uchikoshi, T. S. Suzuki, H. Okuyama, and Y. Sakka, "Fabrication of textured alumina by electrophoretic deposition in a strong magnetic field," *J. Mater. Sci.*, **39**, 861-865 (2004).

[18]H. C. Hamaker and E. J. W. Verway, "The Role of the Forces between the Particles in Electrodeposition and Other Phenomena," *Trans. Faraday Soc.*, **36**, 180-185 (1940).

[19]H. W. Eng, W. Prellier, S. Hébert, D. Grebille, L. Méchin, and B. Mercey, "Influence of pulsed laser deposition growth conditions on the thermoelectric properties of $Ca_3Co_4O_9$ thin films," *J. Appl. Phys.*, **97**, 013706 (2005).

[20]F. K. Lotgering, "Topotactical reactions with ferrimagnetic oxides having hexagonal crystal structures—I," *J. Inorg. Nucl. Chem.*, **9**, 113-123 (1959).

[21]M. Sano, S. Horii, I. Matsubara, R. Funahashi, M. Shikano, J. Shimoyama, and K. Kishio, "Synthesis and Thermoelectric Properties of Magnetically *c*-Axis-Oriented $[Ca_2CoO_{3-\delta}]_{0.62}CoO_2$ Bulk with Various Oxygen Contents," *Jpn. J. Appl. Phys.*, **42**, L198-L200 (2003).

Electroceramic Materials and Applications
Edited by R. W. Schwartz

MICROSTRUCTURE STUDIES OF $Ca_3Co_4O_9$ THIN FILMS ON GLASS SUBSTRATES

Y. F. Hu and Qiang Li
Materials Science Department
Brookhaven National Laboratory
Upton, NY 11973

E. Sutter
Center for Functional Nanomaterials
Brookhaven National Laboratory
Upton, NY 11973

W. D. Si
Physics Department
Brookhaven National Laboratory
Upton, NY 11973

ABSTRACT

High-quality *c*-axis oriented $Ca_3Co_4O_9$ thin films have been grown on glass substrate by pulsed-laser deposition. Cross-sectional transmission electron microscopy (TEM) shows stacking faults abundant throughout the films. Detailed microstructure analysis showed that the film growth starts amorphous. However, once the amorphous material turns entirely into crystalline, areas/domains with layer sequencing can be seen. These are precursors after which the film begins to grow in a layer-by-layer fashion. The misalignment of the small grains that form originally are associated with the introduction of stacking faults.

INTRODUCTION

Thermoelectric (TE) materials can convert heat flow into electric energy directly without producing carbon dioxide gas, radioactive substances, or other emissions, thus are considered to play a key role in clean energy generation. The quality of a TE material is quantified by the thermoelectric figure of merit $ZT=S^2T/(\rho\kappa)$, where S, T, ρ and κ are the Seebeck coefficient, absolute temperature, electric resistivity and thermal conductivity, respectively. To obtain high thermoelectric efficiency, therefore, large S values and low ρ and κ values are simultaneously required. Thermoelectric oxide materials have been widely studied in the past few years due to their high chemical stability at high temperatures, and are thus likely to have profound impact in the thermoelectric applications such as power generation and Peltier cooler. Among them, layered cobaltates have been extensively investigated since the discovery of a high thermoelectric performance in $NaCo_2O_4$.[1] After this discovery, many other layered cobaltates which have misfit structure have been found and most of them show reasonable TE properties.[2-6] The misfit layered cobaltates $Ca_3Co_4O_9$ and $Ca_2Co_2O_5$ are especially more promising in the thermoelectric applications due to their high thermoelectric power and stability in air,[7-11] and thus have been extensively studied in single crystalline and bulk polycrystalline forms.[1, 8-11]

On the other hand, the studies of TE properties of cobaltate films are somewhat limited.[12-15] In many TE applications, such as thermochemistry-on-a-chip, bio-thermoelectric chip, and active cooling for microelectronic processor, film devices are required that allow localized cooling/heating at points of interest. High quality TE thin films, having the crystal structure of single crystals, can have low resistivity and high Seebeck coefficient intrinsic to their electronic band structures. At the same time, thin films are expected to have thermal conductivity lower than that of the single crystals due to strong phonon scatterings at both surfaces and film/substrate interfaces.[16, 17]

Glass is an amorphous material and has very low thermal conductivity, therefore, cobaltate films on glass substrates will have additional advantages in practical TE device applications by lowering the overall thermal conductivity of the device. We have prepared high-quality *c*-axis oriented $Ca_3Co_4O_9$ thin films directly on glass substrates by pulsed laser deposition.[18] Cross-sectional transmission electron microscopy revealed good overall *c*-axis alignment of these $Ca_3Co_4O_9$ films, with a high density of stacking faults throughout the films. The structural defects in the films are desirable to enhance the thermoelectric performance since they may result in lower thermal conductivity without affecting the resistivity and thermal power of the films at higher temperatures. In this paper, we report detailed microstructure analysis of the $Ca_3Co_4O_9$ films on glass substrates. It suggests that the misalignment of the initially grown small crystalline grains may cause the stacking faults in the films.

EXPERIMENT

Thin films of $Ca_3Co_4O_9$ were fabricated using pulsed laser deposition (PLD). The $Ca_3Co_4O_9$ target was prepared from high purity $CaCO_3$ and Co_3O_4 powders. The stoichiometrically mixed powders were calcined two times at 880-890°C for 24 hours in flowing air with intermediate grinding, and then pressed into a disk for final sintering at 900°C for 24 hours in flowing O_2 gas. Commercial glass (fused silica) wafer (from Sydor Optics, Inc.) was used as substrate. The substrates were cleaned in acetone and methanol prior to the deposition. The films (~ 160 nm thick) were deposited at a substrate temperature of 700°C, oxygen pressure of 300 mTorr, and laser energy density of ~1.5 J/cm^2. After deposition, the films were cooled to room temperature in ~ 1 atmosphere in oxygen. The microstructure of the films was then investigated using x-ray diffraction (XRD) and cross-sectional transmission electron microscopy (XTEM). The XTEM specimens were prepared by tripod polishing, a method particularly suitable for obtaining large, homogeneously thin electron transparent regions. Polishing was followed by brief ion-milling (<10 min) at a sputtering angle of 4° to minimize any structural modifications due to specimen preparation. The samples were examined in JEOL JEM 3000F microscope operating at 300 kV.

RESULTS AND DISCUSSTION

Figure 1 shows a typical XRD pattern for the $Ca_3Co_4O_9$ film grown on glass substrate. The XRD patterns exhibit nearly perfect *c*-axis alignment for the thin film (note: the log-scale was used for the intensity counts).

A typical cross-sectional TEM overview image of the $Ca_3Co_4O_9$ film and the adjacent glass substrate is shown in Fig. 2. Two interfaces can be seen in Fig. 2, indicated by the arrows 1 and 2. Between the crystalline $Ca_3Co_4O_9$ film (indicated by arrow 2) and glass substrate, there is an amorphous layer ~ 5 - 10 nm thick with composition related to that of $Ca_3Co_4O_9$. The overview TEM image shows that the film is crystalline and exhibits the well-ordered layered structure typical of calcium cobaltate. Some structural defects are present in the films and they are mostly stacking faults, as indicated by arrows 3 and 4 in Fig. 2, that can have different lengths.

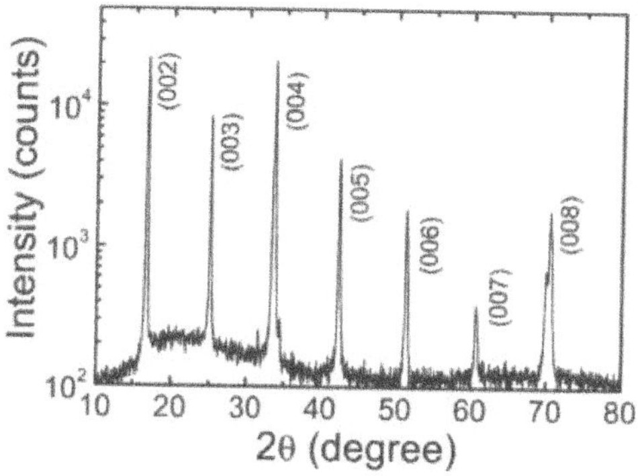

Figure 1. X-ray diffraction (XRD) patterns for a 160 nm thick $Ca_3Co_4O_9$ film grown on glass substrate.

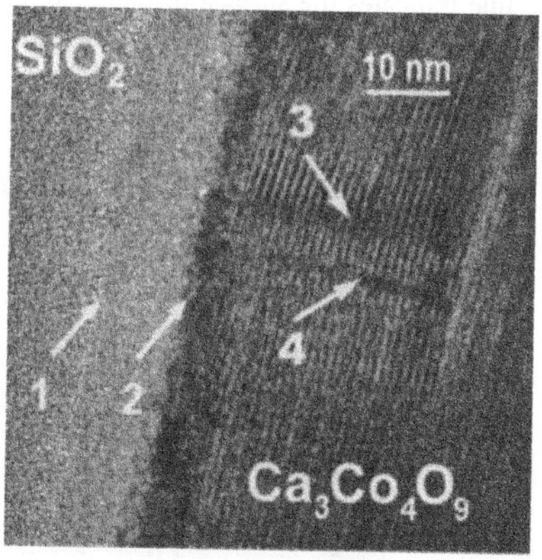

Figure 2. The cross-sectional transmission electron microscopy (TEM) overview image of the $Ca_3Co_4O_9$/glass interface region for a $Ca_3Co_4O_9$ film. Between the glass substrate (indicated by Arrow 1) and well ordered $Ca_3Co_4O_9$ layered film (indicated by Arrow 2) there is an amorphous region, with composition related to that of $Ca_3Co_4O_9$. Arrows 3 and 4 indicate the stacking faults.

A typical high-resolution electron microscopy (HREM) image of the layered $Ca_3Co_4O_9$ film near the interface is shown in Fig. 3. The white line indicates the surface of the glass substrate. It can be seen that interface between the glass and the initially growing amorphous cobaltate layer is relatively sharp despite of the amorphous structure of both. This initial layer is truly amorphous over a thickness of ~ 2-3 nm. Further formation of small randomly oriented crystalline grains can be observed embedded in the amorphous material, as indicated by the arrows in Fig. 3. These small crystalline grains later merge together and become larger grains.

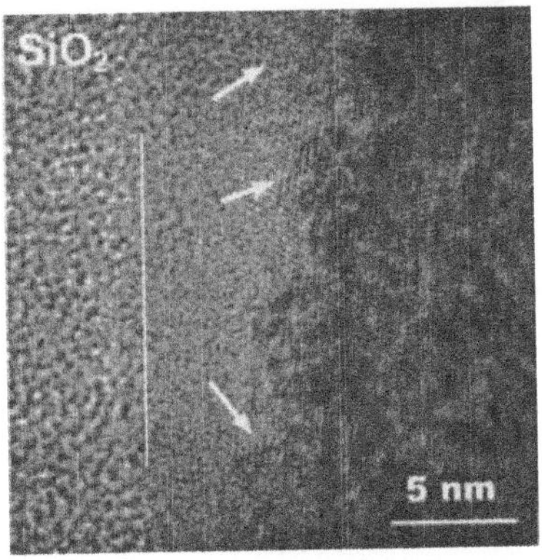

Figure 3. High-resolution electron microscopy (HREM) image of the $Ca_3Co_4O_9$ film near the interface. The white line indicates the surface of the glass substrate. The arrows indicate the small crystalline grains embedded inside the amorphous material that is related to $Ca_3Co_4O_9$.

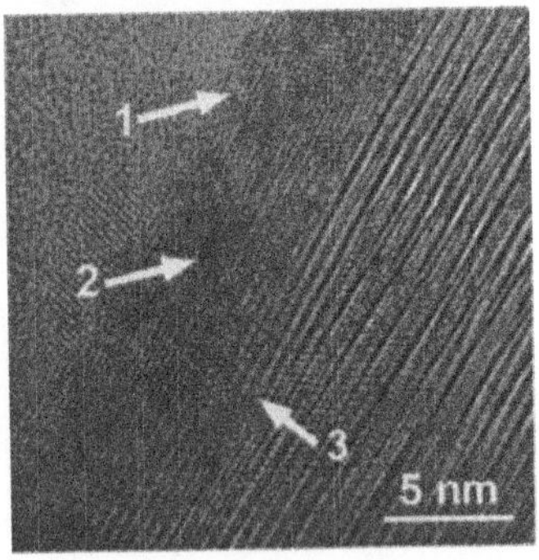

Figure 4. HREM image of the layered $Ca_3Co_4O_9$ film near the interface with the stacking faults. Arrows 1 and 2 indicate two random crystalline grains, and arrow 3 indicate the stacking faults.

Although the initially grown crystalline grains are randomly oriented, once the amorphous material turns entirely into crystalline, areas/domains with layer sequencing can be seen. These are precursors after which the film begins to grow in a layer-by-layer fashion. Figure 4 shows a HREM image of the layered $Ca_3Co_4O_9$ film near the interface with the stacking faults. Arrow 1 indicates random crystalline grains while arrow 2 points to a grain that shows locally some layer stacking. Thus within a few nano-meter thickness, the film assumes the layered structure typical for the cobaltate system. This growth behavior suggests the self-assembly may play a role in the film growth mechanism. The stacking faults can also be seen in the Fig. 4, as indicated by the arrow 3. The stacking faults origins are associated with grain boundaries between small randomly oriented crystalline grains. In areas where the initial layer has large number of misoriented grains sequence of stacking faults get introduced. Such a sequence of closely spaced stacking faults can be seen in Fig. 5. Once a stacking fault is introduced the strain due to it can be seen to persist at different lengths sometimes even through the entire film in direction always perpendicular to the film substrate interface.

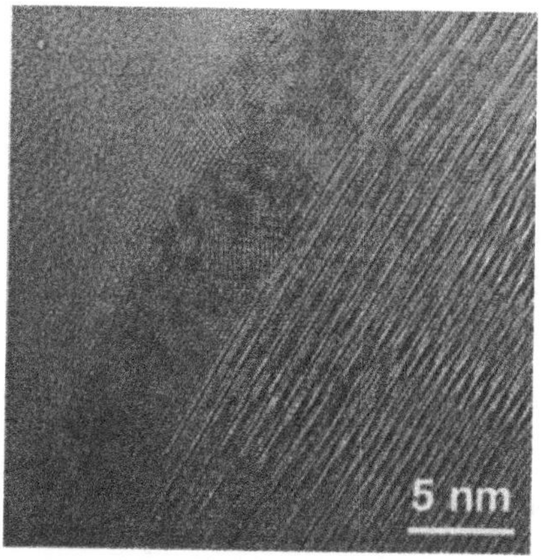

Figure 5. HREM image of the layered $Ca_3Co_4O_9$ film near the interface with abundant stacking faults.

As reported earlier, the stacking faults do not seem to affect the electrical transport of the film very much at higher temperatures. The Seebeck coefficient (~ 130 μV/K) and resistivity (~ 4.3 mΩ•cm) of these films on glass substrate at room temperature were found comparable to the single crystal samples.[18] However, the presence of these structural defects would have substantially lower κ than that of the single crystal samples, therefore, the ZT value of these films is expected to be higher than the single crystal samples. The exact growth mechanism of the stacking faults and why they do not affect the electrical transport at higher temperatures are certainly worth of further investigation, but is beyond the scope of this paper. By understanding this, it would help to engineer structural defects in the TE films, and enhance the thermoelectric performance by significantly reducing the thermal conductivity caused by the structural defects.

CONCLUSIONS

In summary, we have investigated the microstructure of $Ca_3Co_4O_9$ films grown on glass substrates by PLD. The XRD and TEM analysis show that the films are the highly oriented along the *c*-axis. The film structure is initially amorphous followed by formation of small randomly oriented crystalline grains that upon merging become precursors of the layered structure. The misalignment of the small grains that form originally are associated with the introduction of stacking faults. Those stacking faults would reduce the thermal conductivity of the film without affecting the resistivity and thermal power at higher temperatures, therefore, help to enhance the thermoelectric performance of the $Ca_3Co_4O_9$ films.

ACKNOWLEDGMENT

This work was supported by the U. S. Dept. of Energy, Office of Basic Energy Science, under contract No. DE-AC-02-98CH10886.

REFERENCES

[1]I. Terasaki, Y. Sasago, and K. Uchokura, "Large Thermoelectric Power in $NaCo_2O_4$ single crystals," *Phys. Rev. B*, **56**, 12685-87 (1997).

[2]S. Li, R. Funahashi, I. Matsubara, K. Ueno, and H. Yamada, "High Temperature Thermoelectric Properties of Oxide $Ca_9Co_{12}O_{28}$," *J. Mater. Chem.*, **9**, 1659-60 (1999).

[3]R. Funahashi, I. Matsubara, and S. Sodeoka, "Thermoelectric Properties of $Bi_2Sr_2Co_2O_x$ Polycrystalline Materials," *Appl. Phys. Lett.*, **76** 2385-87 (2000).

[4]S. Hebert, S. Lambert, D. Pelloquin, and A. Maignan, "Large Thermopower in a metallic Cobaltite: The Layered Tl-Sr-Co-O Misfit," *Phys. Rev. B*, **64**, 172101 (2001).

[5]D. Pelloquin, A. Maignan, S. Hebert, C. Martin, M. Hervieu, C. Michel, L. B. Wang, and B. Raveau, "New Misfit Cobaltites $[Pb_{0.7}A_{0.4}Sr_{1.9}O_3][CoO_2]_{1.8}$ (A = Hg, Co) with Large Thermopower," *Chem. Mater.*, **14**, 3100-05 (2002).

[6]Y. Miyazaki, T. Miura, Y. Ono, and T. Kajitani, "Preparation and Low-Temperature Thermoelectric Properties of the Composite Crystal $[Ca_2(Co_{0.65}Cu_{0.35})_2O_4]_{0.624}CoO_2$," *Jpn. J. Appl. Phys.*, **41**, L849-51 (2002).

[7]S. Li, R. Funahashi, I. Matsubara, K. Ueno, and H. Yamada, "Synthesis and Thermoelectric Properties of the New Oxide Materials $Ca_{3-x}Bi_xCo_4O_{9+\delta}$ $(0.0 < x < 0.75)$," *Chem. Mater.*, **12**, 2424-27 (2000).

[8]R. Funahashi, I. Matsubara, S. Sodeoka, H. Ikuta, T. Takeuchi, and U. Mizutani, "An Oxide Single Crystal with High Thermoelectric Performance in Air," *Jpn. J. Appl. Phys., Part 2*, **39**, L1127-29 (2000).

[9]Y. Miyazaki, K. Kudo, M. Akoshima, Y. Ono, Y. Koike, and T. Kajitani, "Low-Temperature Thermoelectric Properties of the Composite $[Ca_2CoO_{3.34}]_{0.614}[CoO_2]$," *Jpn. J. Appl. Phys., Part 2*, **39**, L531-33 (2000).

[10]A. Masset, C. Michel, A. Maignan, M. Hervieu, O. Toulemonde, F. Studer, B. Raveau, and J. Hejtmanek, "Misfit-layered Cobaltite with an Anisotropic Giant Magnetoresistance: $Ca_3Co_4O_9$," *Phys. Rev. B*, **62**, 166-75 (2000).

[11]M. Shikano and R. Funahashi, "Electric and Thermal Properties of Single-Crystalline $(Ca_2CoO_3)_{0.7}CoO_2$ with $Ca_3Co_4O_9$ Structure," *Appl. Phys. Lett.*, **82**, 1851-53 (2003).

[12]Y. Yoshida, T. Kawai, Y. Takai, and M. Yamaguchi, "Thermoelectrical Properties and the Microstructure of $[Ca_2CoO_3]_xCoO_2$ Thin Films Fabricated on the Various Substrates Using the Pulsed Laser Deposition," *J. Ceram. Soc. Japan*, **110**, 1080 (2002).

[13]I. Matsubara, R. Funahashi, and M. Shikano, "Cation Substituted $(Ca_2CoO_3)_xCoO_2$ Films and Their Thermoelectric Properties," *Appl. Phys. Lett.*, **80**, 4729-31 (2002).

[14]H. Minami, K. Itaka, H. Kawaji, Q. J. Wang, H. Koinuma, M. Lippmaa, "Rapid Synthesis and Characterization of $(Ca_{1-x}Ba_x)_3Co_4O_9$ Thin Films Using Combinatorial Methods," *Appl. Surf. Sci.*, **197-198**, 442-47 (2002).

[15]H. W. Eng, W. Prellier, S. Hebert, D. Grebille, L. Mechin, and B. Mercey, "Influence of Pulsed Laser Deposition Growth Conditions on the Thermoelectric Properties of $Ca_3Co_4O_9$ Thin Films," *J. Appl. Phys.*, **97**, 013706 (2005).

[16]L. D. Hicks and M. D. Dresselhaus, "Effect of Quantum-well Structures on the Thermoelectric Figure of Merit," *Phys. Rev. B*, **47**, 12727 (1993).

[17]R. Venkatasubramanian, E. Siivola, T. Colpitts, and B. O'Quinn, "Thin-film Thermoelectric Devices with High Room-temperature Figures of Merit," *Nature*, **413**, 597-603 (2001).

[18]Y. F. Hu, E. Sutter, W. D. Si, and Qiang Li, "Thermoelectric properties and microstructure of *c*-axis oriented $Ca_3Co_4O_9$ thin films on glass substrates," *Appl. Phys. Lett.* **87**, 171912 (2005).

Electroceramic Materials and Applications
Edited by R. W. Schwartz

IMPROVING THERMOELECTRIC DEVICE PERFORMANCE AND DURABILITY THROUGH THE INTEGRATION OF ADVANCED, AEROGEL-BASED CERAMICS

Jeff Sakamoto*, Thierry Caillat, Jean-Pierre Fleurial, Steve Jones and Jong-Ah Paik
Jet Propulsion Laboratory, California Institute of Technology
Pasadena, CA 91109

Winny Dong
Cal Poly, Department of Materials Engineering
Pomona, CA

*Email: Jeff.S.Sakamoto@jpl.nasa.gov

ABSTRACT

The recent interest in thermoelectric technology has spurred the development of advanced materials and device technology. Specifically, current work at the Jet Propulsion Laboratory has identified aerogel as an effective sublimation barrier for a wide range of thermoelectric technologies such as Skutterudites, SiGe and PbTe/TAGS. Aerogel is typically known as highly porous (>90-99% porous) silicon dioxide, which has very low thermal and electrical conductivity. The porous network consists of a relatively homogeneous network of pores ranging from angstroms to hundreds of nanometers and as such creates a tortuous path for vapor to permeate. As a result, when thermoelectric materials are coated with aerogel, sublimation rates are dramatically reduced. To date, coupons and devices based on Skutterudite, and PbTe/TAGS have been encapsulated with aerogel and the sublimation rates (g/cm^2h) have been reduced by several orders of magnitude compared to unprotected elements. For example, unprotected or uncoated Skutterudite (n-$CoTe_{0.05}Pd_{0.05}Sb_3$ and p-$CeRu_3FeSb_{12}$) lose Sb through sublimation at approximately 2×10^{-2} (g/cm^2h) at 700C in high vacuum while aerogel encapsulated coupons lose Sb at approximately 3×10^{-5} (g/cm^2h) under the same conditions. This is a **700-fold** decrease in sublimation rate. The details regarding the integration of aerogel into thermoelectric technology as a sublimation barrier are presented. Additionally, this report presents thermal stability analyses of aerogel for extended periods of time up to 1000C. Thermal stability is critical if aerogel is to be used in thermoelectric generator technology, which commonly requires operation at elevated temperatures. Linear shrinkage was used as the metric to evaluate resistance to sintering as a function of temperature in air and in vacuum. Interestingly, aerogel sinters considerably more when heated in air versus in high vacuum (10^{-6} torr), especially above 500C. Moreover, linear shrinkage at 700C in high vacuum is negligible and as such indicates that silica-based aerogel is compatible with current Skutterudite technology.

INTRODUCTION

The Jet Propulsion Laboratory (JPL) is currently developing a new class of thermoelectric generators capable of delivering up to 7 W/kg. This work involves the development of Skutterudite compounds that have increased ZT and operate over larger ΔTs compared to SiGe and PbTe/TAGS. Employing the new materials is expected to increase thermal-to-electric efficiency. The development program also involves in the integration of aerogel into thermoelectric technology. Aerogel is expected to improve both performance and durability. The basic concept involves casting aerogel into or around a thermoelectric device such that intimate contact is made between the thermoelectric elements and the aerogel **(Figure 1).** Since aerogel is synthesized using sol-gel chemistry it can be cast in place, thus filling all free void space in a device. Once in place, the inherently low thermal conductivity of aerogel (a few mW/m-K) channels heat through the thermoelectric elements, thus minimizing parasitic heat loss. A more intriguing effect was, however, recently discovered and involves the use of aerogel as a sublimation suppression barrier **(Figure 2).** The inherently micro-meso porous structure of aerogel combined with the cm-scale mean free path of heavy metal vapors significantly slows the sublimation rates of practically all thermoelectric materials relevant to thermoelectric power generation (Skutterudite, SiGe, and PbTe/TAGS). Altogether, the integration of aerogel into thermoelectric technology could have a significant impact on how thermoelectric generators, and to a lesser extent coolers, are designed and fabricated. Intense efforts supported by NASA's Science and Mission Directorate, Project Prometheus and a DOE sponsored project to improve fuel efficiency of vehicles using thermoelectrics are considering aerogel for use in the next generation thermoelectric generator technology.

Efficient thermal-to-electric conversion efficiency using thermoelectric materials is primarily determined by the ZT of the thermoelectric materials and the maximum hot-side (T_{hot}) operating temperature. The latter parameter, T_{hot}, is often limited by the sublimation of volatile metal species, which is considered one of the primary modes of degradation in thermoelectric generators. Current efforts at JPL are focusing on Skutterudite-based thermoelectric technology, thus Antimony (Sb) is the volatile, subliming species. Essentially, sublimation reduces the effective cross-section of the thermoelectric elements resulting in increased electrical resistance and thermal impedance **(Figure 3).** As Sb leaves n-type Skutterudite, the reaction (not in mass balanced form since it occurs in stages) is as follows: $CoSb_3 \longrightarrow CoSb_2 + CoSb + Sb$ (vapor). Simultaneous decomposition into lower cobalt antimonides occurs such that $CoSb_2$ forms first followed by the eventual formation of CoSb as shown in an n-type Skutterudite coupon heated in vacuum (10^{-6} torr) for 72 hours at 700C **(Figure 4).** The p-type Skutterudite ($CeFe_3Ru_1Sb_{12}$) decomposes into Ce, Fe and Ru diantimonides as Sb sublimes and a similar depletion band forms along the perimeter of coupons or legs.

Thermoelectric technology is known for reliability resulting from simplicity of operation, i.e. no moving parts, static thermal environment, etc. Assuring long term reliability will involve suppressing sublimation. In the past, Radioisotope Thermoelectric Generators (RTG) employed SiO_2 and Si_3N_4 coatings or a cover gas of Argon to suppress sublimation for SiGe and PbTe/TAGS technology, respectively. During the course of development these schemes were characterized using simple mass loss experiments of coupons and power output of prototypical devices. In the end, values for acceptable sublimation rates were determined and correspond to a projected power reduction based on the abovementioned increase in electrical and thermal resistance associated with sublimation (Figure 3). Using the procedures developed under previous programs, several goals have been established for effectively suppressing sublimation in Skutterudite-based technology: 1) Sublimation of Sb must not result in more than a 5% reduction in cross-section over 10 years of operation, which will require a Sb sublimation (flux) rate of $\sim 1 \times 10^{-7}$ g/cm^2 hours or depletion band formation $\sim 10^{-8}$ cm/hour, 2) The method for suppressing sublimation should also prevent sublimed species from condensing on the cold-side circuitry, which will likely result in short circuiting and 3) The method for suppressing sublimation should not have a negative impact on the system performance, e.g. if a coating is considered it must have low thermal and electrical conductivity to prevent thermal and electrical shorting.

Recent investigations have determined that silica-based aerogel could potentially meet all three requirements. Aerogel is made through Sol-Gel chemistry in which liquid precursors are combined to form a gel, which is then supercritically dried in an autoclave[1] **(Figure 5).** The resulting structure is comprised of an interconnected network of pores ranging from the micro to meso scale and porosity ranging from 90 to 99% (density from 30 to 300 mg/cc). The nanometer scale porosity combined with the homogeneous dispersion of pores creates a highly tortuous path for vapor to permeate. It is for this reason that aerogel is currently being considered as a sublimation barrier **(Figure 2).** The mean free path of most heavy metal vapors, like Sb, are in the range of centimeters, thus it is expected that the permeation rate of Sb vapor through aerogel should be low. Additionally, since aerogel has low thermal and electrical conductivity it should not have a negative impact on the device efficiency and it is likely that the device performance can be improved. For example, in SiGe technology significant parasitic heat loss resulted from gaps between the SiGe unicouples and the multi-foil insulation used to prevent the hot-side components from heating the cold-side components. Since aerogel is cast in place, intimate contact can be made between the thermoelectric elements and the thermal insulation, thus reducing parasitic heat loss. Although the benefits of using aerogel as thermal insulation are clear, the emphasis of this report is on the use of aerogel for sublimation suppression. In this report, detailed analyses of aerogel sublimation suppression coatings for Skutterudite-based technology are presented. Specifically, the sublimation rates of aerogel-coated coupons in various environments, temperatures and time periods are covered in detail. Additionally, detailed analyses of aerogel stability at elevated temperature will be presented.

EXPERIMENTAL

The standard two-step sol-gel process was used to prepare the aerogel in this work. The first step involves combining and refluxing requisite amounts of tetraethylorthosilicate (TEOS) or tetramethylorthosilicate (TMOS), ethanol and nitric acid. Water, ammonia hydroxide base, and acetonitrile are then added. This solution or "sol" was then poured into the appropriate molds containing sublimation coupons, and allowed to solidify or "gel". After gelation, the samples were transferred into a critical point dryer pre-filled with acetonitrile. The apparatus was sealed, heated to 295C and pressurized to 5.5 MPa. These are the conditions required to establish supercritical conditions for acetonitrile

The n-type and p-type Skutterudite compositions were $CoSb_3$ and $CeFe_3Ru_1Sb_{12}$, respectively. Both types were prepared by combining the appropriate pure metals, melting in a quartz ampoule, followed by heating to 1200C for 24 hours, cooling to room temperature and re-heating at 700C for 48 hours. The ampoules were then transferred to a glove box where they were broken to extract the ingots. The ingots were then milled in a vibratory mill to produce a fine powder and hot pressed at 120 MPA in a graphite die at 700C under flowing argon. The hot pressed slugs were then diced into the appropriate geometry.

Thermal stability was characterized using linear shrinkage as the metric. Aerogel coupons (cubes approximately 2cm by 2.5cm by 4cm) were heated to various temperatures in high vacuum (10^{-6} torr) and ambient air for various times. The longest dimension was measured before and after heating to determine the percent shrinkage. FTIR analysis was performed to investigate the change of aerogel before and after heating. Aerogel as made, heated at 1000 °C in vacuum, and heated at 1000°C in air were separately prepared and investigated with FTIR in the range of 400 ~ 4000 cm^{-1}.

RESULTS AND DISCUSSION

Initially, sublimation barrier studies of Skutterudite coupons were conducted in accordance with protocol established under the SiGe and PbTe/TAGS RTG development programs in which uncoated coupons and coupons coated with various thin films were heated in their respective projected operating environment and periodically weighed to determine mass loss. SiGe-based, RTG technology employed 25 micron thick, Chemical Vapor Deposited (CVD) composite films of SiO_2 and Si_3N_4 to suppress sublimation. PbTe/TAGS-based technology employed a cover gas consisting of approximately 1 atm of argon to suppress sublimation. Both approaches were attempted to suppress sublimation in Skutterudite-based technology and neither provided adequate results. Various sputtered films were rapidly consumed by the highly corrosive Sb vapor and 1 atm of argon reduced sublimation, but only by a factor of 10 where a factor 10^5 is required. One somewhat promising approach involved the hot pressing of 12 micron Ti foil to the surface of Skutterudite elements. Significant reduction in Sb sublimation rates was achieved, but with time the Ti was consumed by Sb to form $TiSb_2$, which resulted in coating embrittlement.

To achieve significant reductions in sublimation rates an alternative approach using aerogel was pursued. To date, three iterations of coupon design have been investigated in which Skutterudite n and p-type coupons were encapsulated by various forms of aerogel **(Table I).** The first type consisted of cube-shaped Skutterudite coupons **(Figure 6),** which were placed in an alumina, TGA crucible and then filled with a silica sol. The silica sol is composed of TEOS, acetonitrile, water, ammonia hydroxide and an approximately 20 wt% alumina powder to improve thermal stability (discussed in detail below). The amount of each component varied depending on the density of aerogel. The sol gelled around the coupon and the gel was allowed to age for 1~2 days and was supercritically dried in acetonitrile at 5.5 MPa and 295C. The sublimation rates of most coupons were measured using a custom designed ThermoGravimetric Analyzer (TGA) purchased from Netzsch Instruments. The TGA (STA 449 C Jupiter ®) was outfitted with all alumina components and as such could withstand exposure to the corrosive metal vapors such as Sb. A turbo molecular pump was integrated into the apparatus to allow testing under high vacuum (10^{-6} torr). The TGA profile was increased at a rate of 10°C/minute up to predetermined temperatures for isotherms at 400, 500, 600 and 700C. Each isotherm was held for one hour to achieve a linear weight loss profile used to measure the sublimation rate. The slope of the weight versus time plot provides the weight loss in terms of a flux normalized by surface area of the exposed coupon. The units used are g/cm^2h and are consistent with units used in the past for SiGe sublimation evaluation. In some cases the isothermal time interval of one hour was not sufficient and the interval was extended to 10 hours to obtain a greater degree of accuracy. As a rough approximation a flux rate of $\sim 10^{-5}$ g/cm^2h over a one hour interval for a 2 gram sample is within the detection limit of the apparatus.

Table I. Coupon Configurations

Generation	Coupon Configuration	Sol	Supercritically Drying
1st	Square-cube	TEOS/Acetonitrile/Water/NH_4OH/Alumina powder	Acetonitrile, 5.5MPa 295C
2nd	Press fit in graphite cup	TEOS/Acetonitrile/Granular oxide/Water/NH_4OH	Acetonitrile, 5.5MPa 295C
3rd	Press fit in graphite cup	TMOS/Acetonitrile/Fumed silica/Granular oxide/Water/NH_4OH	Acetonitrile, 5.5MPa 295C

First Generation Coupon Testing

The first generation coupon experiments were conducted for uncoated Skutterudite coupons and coupons encapsulated with 50 and 100 mg/cc aerogel **(Figures 7)**. The uncoated sample is an n-type Skutterudite coupon in which the sublimation rates were undetectable ($<10^{-5}$), 5.3×10^{-4}, 5.2×10^{-3}, and 2.2×10^{-2} g/cm^{2}h for the isothermal plateaus at 400, 500, 600 and 700C, respectively. [Interestingly, measurements in high vacuum produced an unusual buoyancy phenomenon in which the samples appeared to gain wait with increasing time, but this was later mitigated by longer isotherms. Also, at 400C the weight loss response was too small to accurately measure a loss rate. Incidentally, the p-type sublimation rates (not shown) are similar and not shown for brevity.] When the Skutterudite cubes are encapsulated in the 50 mg/cc and 100 g/cc aerogel the weight loss is suppressed by a factor of 7 and 20 at 700C, respectively. Although this was the first attempt at suppressing sublimation with aerogel the results do not meet NASA requirements for performance duration, but the results clearly indicate that the aerogel did indeed slow sublimation, especially as the density of the aerogel increases. Interestingly, all post-test analyses indicated that cracks propagated through the aerogel coatings and appeared to originate from the tops of the Skutterudite cubes **(Figure 8)**. It is believed that these cracks as well as leakage from the bottom of the cubes were the primary paths for Sb vapor to escape. Likewise, the aerogel processing and coupon design were re-engineered to assure that the only Sb loss occurred as a result of permeation through the aerogel. The second generation coupon testing addressed these issues.

Second Generation Coupon Testing

It was believed that the cracks in the aerogel were caused by two factors. First, it is common for the aerogel to shrink between 1 and 5 % when heated to 700C. Secondly, the orthogonal edges of the Skutterudite cubes acted as stress intensifiers. Thus, as the aerogel shrinks around the cube, the edges of the cube likely initiated cracks, which propagated through the thickness of the aerogel coating. To mitigate cracking and simultaneously eliminate Sb vapor from escaping from the bottom of the coupon, a second generation coupon was designed and implemented. Basically, a graphite cup was precision machined such that a 6.4mm diameter cylindrical coupon could be pressed or forced into it **(Figure 9a)**. The press fit occluded all surfaces, but the horizontal surface facing upward. Additionally, all the edges of the graphite cup were chamfered to eliminate orthogonal edges and the entire cup/coupon assembly was fully encapsulated in aerogel. Another improvement was to further increase the density of the aerogel since the first generation experiments indicated that the higher the density the lower the sublimation rate. However, the particular Sol-Gel process used typically results in greater aerogel shrinkage, associated with supercritical drying, as the density of the gel increases. The primary modification to the previous Sol-Gel procedure was the addition of oxide powders (Titania) into the silica sol before casting. The titania oxide powder was added, because it is known to opacify infrared radiation[2]. With these modifications, the aerogel was considered a composite with overall density of 120 mg/cc comprised of a 40 mg/cc silica aerogel network "gluing" together a homogeneous distribution of >80 mg/cc titania. Finally, to further eliminate the potential effects of shrinkage associated with high temperate exposure, pure antimony pellets were also tested, which required heating to 500C instead of 700C to achieve substantial sublimation rates.

Since Sb was known to be the primary element lost from sublimation, a fundamental study was conducted on pure Sb pellets encapsulated with second generation coupon design **(Figure 9a)**. Uncoated Sb pellets, 40 mg/cc pure silica aerogel coatings and the improved composite aerogel coatings (120 mg/cc) were compared using the TGA apparatus. The uncoated, antimony sublimation rate was approximately 1 g/cm^2h at 500C, while the 40 mg/cc and composite aerogel coatings were 1.5×10^{-3} and non-detectable for the aerogel composite over the 1 hour isotherm **(Figure 9c)**. To increase signal sensitivity, the aerogel composite encapsulated Sb sample was held for 10 hours at 500C and a sublimation rate of 7×10^{-5} g/cm^2h was measured **(Figure 9d)**. Essentially, combining the second generation coupon testing with the composite aerogel resulted in a 14000 fold decrease in sublimation rate of pure Sb. After testing, discolored regions in the aerogel appeared to indicate where the Sb vapor permeated and condensed upon cooling **(Figure 9b)**. A noticeable "cloud" of Sb appeared to condense directly above the Sb pellet, but did not permeate entirely through the aerogel (similar post-test analyses confirmed condensation of Sb in aerogel using SEM/EDS). More interesting was the barely noticeable condensation that appeared to condense along the vertical, graphite cup-to-aerogel interface. It is believed that negligible condensation along this interface indicates that a good seal is made and significantly reduced Sb vapor loss associated with leakage as was suggested was the primary mode of Sb vapor loss in the first generation coupon design. Although these results were encouraging for pure Sb coupons, more meaningful results could only be achieved testing Skutterudite coupons. This would require testing at 700C, thus requiring further modifications to the Sol-Gel chemistry to decrease shrinkage associated with heating to higher temperatures. Hence a third generation coupon test group was conducted in which the second generation coupon design was used and dramatic changes to the aerogel formulation were made.

Third Generation Coupon Testing

The promising results achieved with the second generation coupon design, prompted characterization of n and p-type Skutterudite coupons. Essentially, the same tests were conducted for the Skutterudite coupons as was done for the Sb coupons except that the temperature range was increased to 700C, which would require greater thermal stability to reduce shrinkage. As previously mentioned, severe cracking resulted when the aerogel coatings shrank greater than 5% around the coupons. It is important to note that there are several steps in aerogel processing that can cause aerogel to shrink, and likely crack, when used as an encapsulant. First, shortly after the liquid sol transforms into free standing gel, the gel is often aged for > 24 hours. Ageing can cause some shrinkage although typically only a few %. Secondly, as supercritical fluid leaves the gel, the gels often expand ~ 10% the original volume and contract or shrink as the autoclave is cooled and all the supercritical fluid is expelled. Significant shrinkage can be caused by this step especially as the density of the aerogel network is increased above ~75 mg/cc using the TEOS-based sol. Controlling this shrinkage is extremely important if aerogel is to be used as a cast-in-place encapsulant. The third mechanism for causing shrinkage, as previously mentioned, is that associated with heating, especially above 500C, and will be discussed in more detail below. From what has been said, the second step is the most challenging step since this work seeks to increase aerogel density, while minimizing shrinkage. Thus, significant changes to the Sol-Gel chemistry were made to minimize shrinkage associated with supercritical drying. First, Tetramethylorthosilicate (TMOS) was used in place of TEOS. Secondly, 325 mesh fumed silica powder was added along with granular (dense) silica powder. The procedure was nearly identical as was used for the TEOS sol and the specific concentrations of each constituent in the aerogel were as follows: 75 mg Silica aerogel, 20 mg/cc fumed silica and 55 mg/cc granular silica powder. The modifications to the Sol gel chemistry dramatically improved the reproducibility and eliminated cracking especially when working with higher density aerogel. The details as to why these modifications were made are not within the scope of this report and will be reported in future publications.

The combined effects of optimized Sol-gel formulation and coupon design are striking. The sublimation rates for n and p type Skutterudite coupons at the beginning-of-life (BOL) measured in the TGA are in the 10^{-4} g/cm^{2}h range **(Figure 10).** To improve accuracy and evaluate long term sublimation suppression a new test protocol was established. Extended period tests were implemented with out the use of the TGA apparatus. Samples requiring greater than 10 hour isotherms were impractical, thus a new protocol involving heating coupons in a quartz ampoule under high vacuum (10^{-6} torr) for week long intervals was implemented. The ampoules had an inner diameter of 2.54cm and were 200cm long. The ampoules extended out of the furnace such that half of the ampoule was heated in the furnace and half was not. As Sb vapor sublimed and escaped it condensed on the colder walls of the ampoule. Between intervals, the encapsulated coupons were cooled and weighed to measure mass loss rates **(Figure 11).** With the exception of one data set, which could have had a partially cracked aerogel coating, it is apparent that the sublimation rates for all samples significantly decreased during the first few days of heating. Most samples initially lost weight at rates ranging from the high 10^{-5} to low 10^{-4} g/cm^{2}h, but after approximately 3 days practically all samples were in the low 10^{-5} g/cm^{2}h range **(Figure 12).** This is a **700-fold** decrease in sublimation rate. Incidentally, the relatively high rates in the beginning of testing are typical and similar observations were made when testing SiGe sublimation rates. This could also be a result of slight oxidation of the Skutterudite coupons during the supercritical extraction process. The oxidation film burns off initially until the Skutterudite surface is exposed and the weight loss becomes constant. Regardless, the data clearly shows that aerogel can slow the sublimation rate of Skutterudite at 700C by approximately 3 orders of magnitude compared to the uncoated samples. Although, the suppression rates are approximately off by two orders of magnitude, significant improvements can be made. First, current efforts are underway to increase the aerogel network density by a factor of 5 and likely result in significant reductions in sublimation rate. Secondly, alternative, thin film coatings can be easily integrated with aerogel coatings to provide additional sublimation suppression. Additionally, cover gas, alternative device designs and improved Skutterudite materials chemistry can all be considered if necessary to reduce the sublimation rates to the target rate in the 10^{-7} g/cm^{2}h range.

Thermal Stability

RTGs typically operate in a static thermal environment for extended (up to decades) periods of time. Thus, all components are expected to maintain their thermal and mechanical integrity for the life of the generator. A study was conducted to characterize the thermal stability of silica-based aerogel. The primary emphasis was on linear shrinkage associated with heating. As mentioned above, linear shrinkage in the range of <10% is considered acceptable. Cubic coupons were prepared and heated to various temperatures in air and in high vacuum. One group of experiments involved heating a coupon for one hour at a certain temperature, cooling, measuring linear shrinkage and heating to the next higher temperature in air and under high vacuum (10^{-6} torr). Two types of aerogel were compared. Pure silica was evaluated and a composite of 20 wt% colloidal alumina powder and silica aerogel made up 80 wt%. The addition of alumina has been shown to increase viscosity of aerogel, thus improving thermal stability[2]. The data for these are shown in **(Figures 13 and 14)** from which it is apparent that shrinkage for both aerogel was negligible up to 600C. The shrinkage appears to increase exponentially with increasing temperature and clearly is more pronounced in the samples heated in air above 700C. Post heating FTIR analysis indicates that the primary difference between samples heated in air versus vacuum was the loss of Si-OH bonds, which caused accelerated sintering. This observation agrees with work done by T. Woignier and J. Phaliippou[3]. This will be explained in more detail in a later publication. Thermal stability as a function of time at 1000C was compared between a pure silica aerogel and silica/alumina composite aerogel (20wt% alumina/80 wt% silica) **(Figure 15)**. Both samples were heated in high vacuum at 1000C. In both cases, the linear shrinkage appeared to level off near the 10% shrinkage mark.

CONCLUSION

A new application for aerogel has been proposed and evaluated for use in thermoelectric technology. The inherently tortuous pore network of aerogel significantly slows the rate of sublimation of volatile metal vapors from the surface of Skutterudite materials and is likely to exhibit the same behavior in most thermoelectric materials such as PbTe, SiGe or TAGS. At 700C under high vacuum, the sublimation of uncoated Skutterudite coupons was determined to be 2.15 x 10^{-2} g/cm^{2}h while coupons encapsulated with aerogel were reduced by a factor of more than 700 (~3 x 10^{-5} g/cm^{2}h). Although preliminary calculations indicate that a sublimation rate in the 10^{-7} g/cm^{2}h is required, it is believed that further improvements to aerogel processing and the composition may approach the suppression goal. Additionally, it is not unreasonable to think that conventional coatings can be combined with aerogel to further enhance sublimation suppression. The thermal stability of aerogel was also characterized and no significant degradation was observed up to 700C. Despite the high surface area, the aerogel resisted significant sintering up to 1000C in high vacuum. Altogether, the prospect of integrating aerogel into thermoelectric technology is appealing in that durability is improved without sacrificing performance and it is likely that aerogel may increase RTG performance by providing improved thermal insulation.

ACKNOWLEDGEMENTS

This work was performed by the California Institute of Technology's Jet Propulsion Laboratory and funded by NASA's Science and Mission Directorate and Project Prometheus under contract No. NAS7-03001.

REFERENCES

[1]J. Brinker, G. Sherrer, Sol Gel Science and Technology, Academic Press (1990).

[2]R. Saliger, T. Heinrich, T. Gleissenger and J. Fricke, "Sintering Behavior of alumina-modified silica aerogels," *J. Non-Crystal. Solids*, **186**, 113-117 (1995).

[3]T. Woignier and J. Phalippou, *J. of Mater. Sci.*, **25**, 3118-3126 (1990).

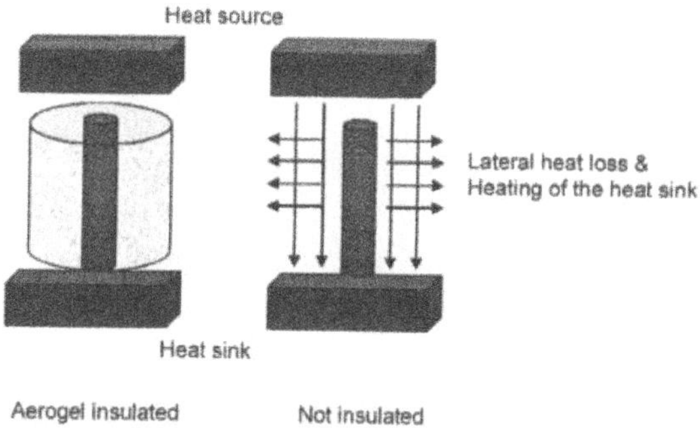

Figure 1. A typical thermoelectric device consists of n and p-type thermoelectric legs, a heat source and heat sink. The single leg on the left is encapsulated in aerogel, which channels heat through the thermoelectric element and prevents parasitic heat loss.

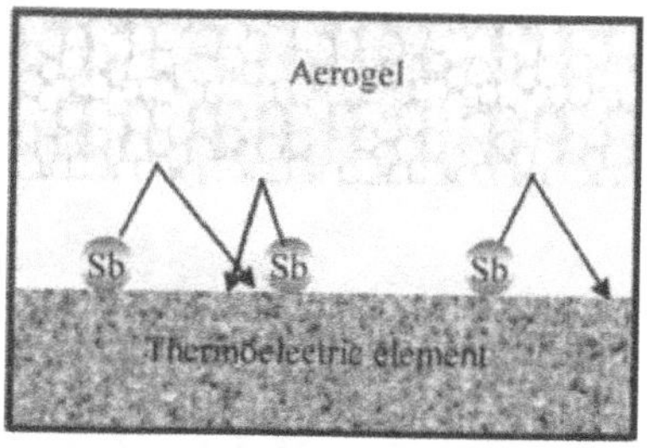

Figure 2. Under high vacuum (10^{-6} torr), the mean free path of most heavy metal vapors, like Sb, are far too long to rapidly penetrate the highly-tortuous, porous, aerogel network (not to scale). The Sb atoms are "reflected" back to the SKD surface, thus establishing a highly-localized, equilibrium environment.

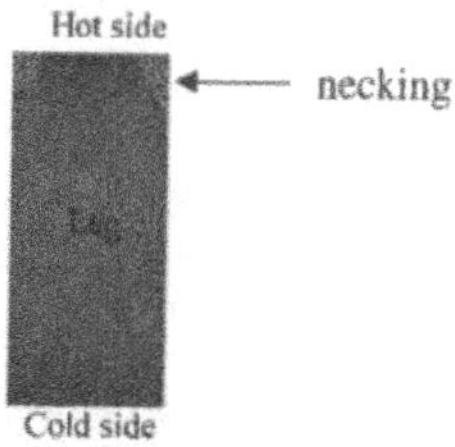

Figure 3. Practically all thermoelectric materials considered for use in RTGs are susceptible to sublimation. Typically, sublimation causes a reduction in effective cross-section causing a necking effect.

Figure 4. Depletion bands are formed as Sb sublimes from Skutterudite.

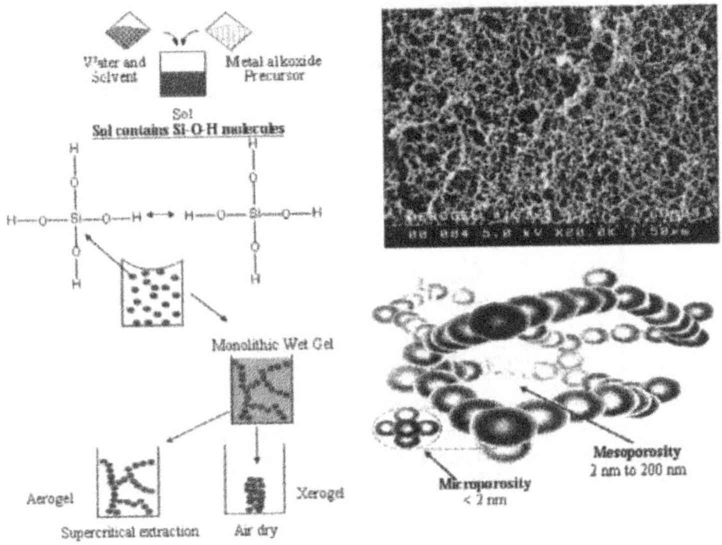

Figure 5. Silica aerogel is synthesized using Sol-Gel synthesis, which involves the combining of liquid precursors. The slight addition of water and acid causes hydrolysis of silica precursors such as TEOS or TMOS. Typically, base is added to cause hydrolyzed silica units to condense forming an interconnected network of silica. The majority of the volume is occupied by a diluting solvent. If the diluting solvent, such as acetonitrile, ethanol or methanol, is allowed to evaporate under ambient conditions, the capillary pressure associated with evaporation causes severe shrinkage and produces a xerogel (~50% porous). To eliminate the capillary pressure, the expanding solvent is extracted under supercritical conditions in an autoclave, which preserves the pore volume and produces aerogel. A typical microstructure imaged in an SEM reveals a homogeneous distribution of fine pores mesopores (40 mg/cc silica). Meso and micropores are typically present in aerogel.

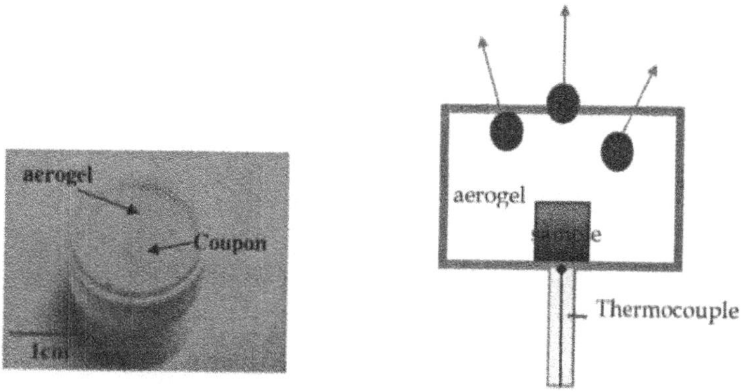

Figure 6. The first generation coupons consisted of a cube of Skutterudite encapsulated by aerogel, which was contained in an alumina TGA crucible.

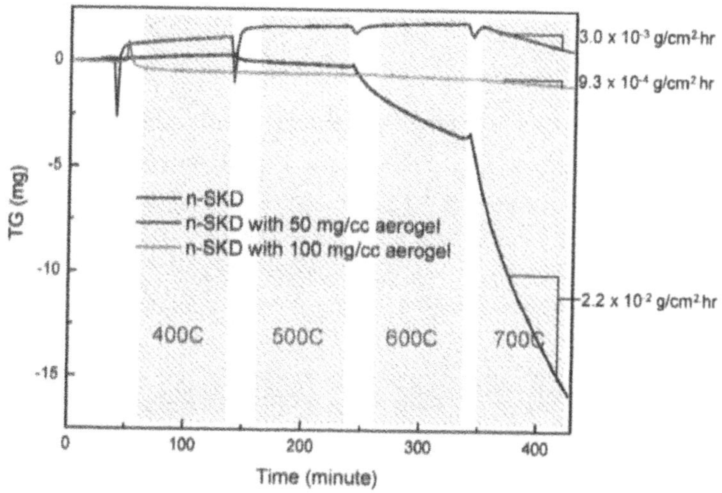

Figure 7. TGA, weight loss vs. time at various isotherms for n-type Skutterudite coupons. At 700C, sublimation rates are significantly reduced for samples encapsulated in aerogel.

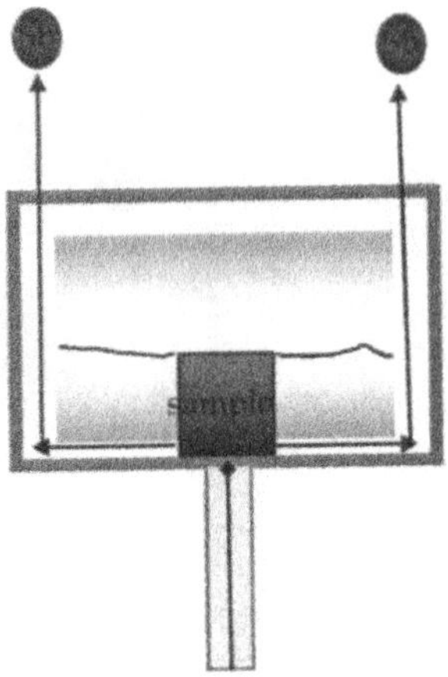

Figure 8. Cracks radiating from the edges of the coupon and shrinkage exposing the bottom of the coupon were likely the cause of the higher than expected sublimation rates.

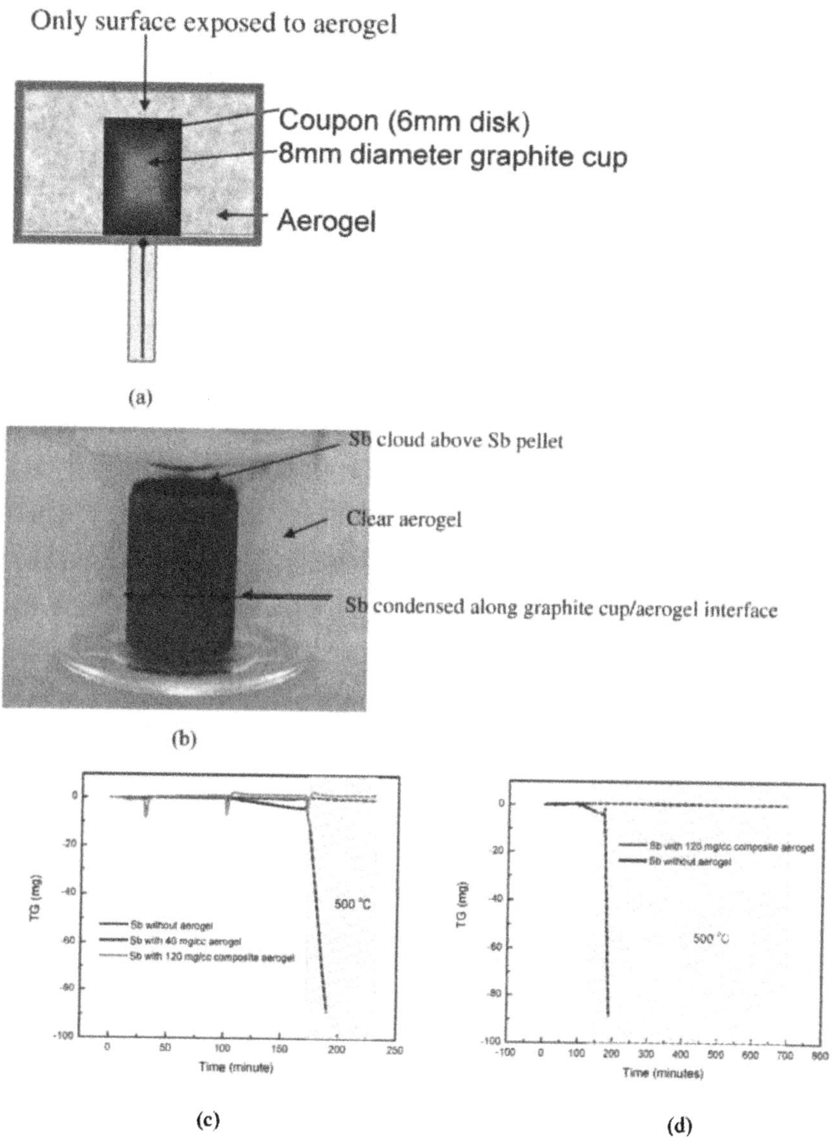

Figure 9. The second generation coupon design (a). Sb vapor appeared to condense above the Sb pellet and along the edges of the graphite cup (b). TGA plots for the uncoated, 40 mg/cc and 120mg/cc aerogel encapsulated coupons (c) and the extended test (d).

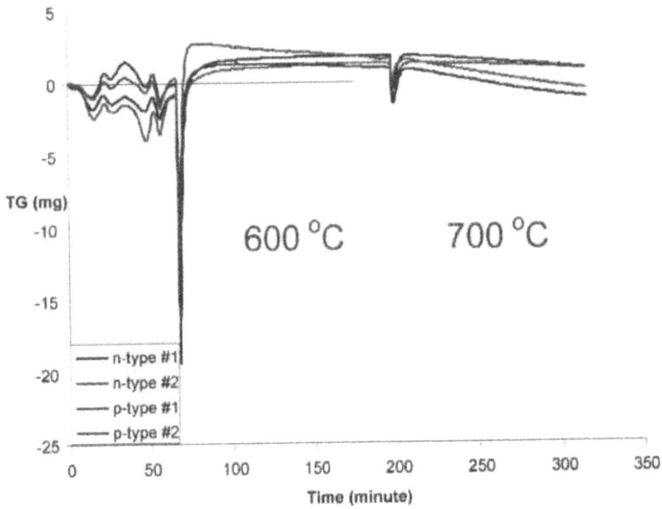

Figure 10. TGA data for two identical sets of coupons of n and p type Skutterudite encapsulated using the third generation scheme.

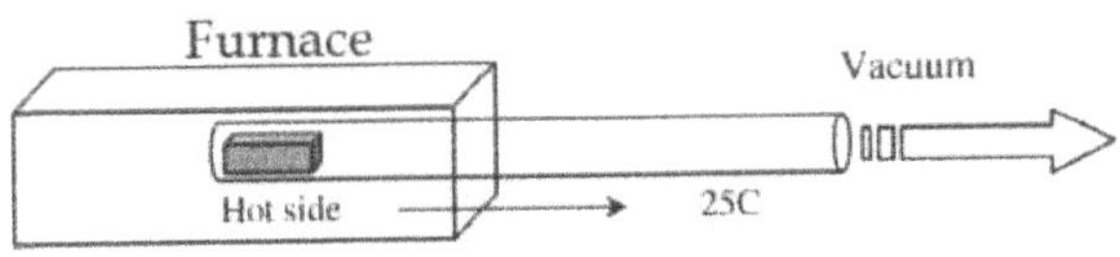

Figure 11. Coupons were placed in long ampoules, heated and evacuated.

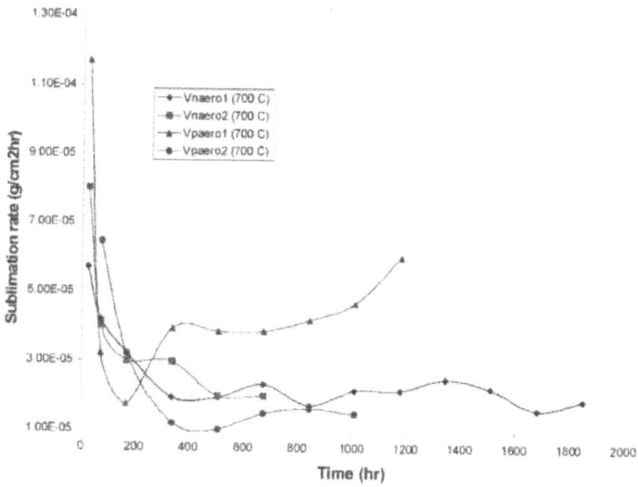

Figure 12. Extended life test of aerogel encapsulated n and p-type samples (same samples used to acquire data in Figure 10).

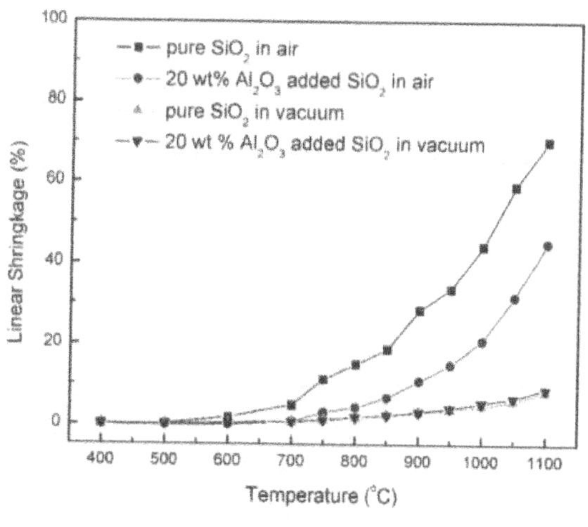

Figure 13. Linear shrinkage as a function of temperature for pure silica and alumina doped silica aerogel in air and in high vacuum.

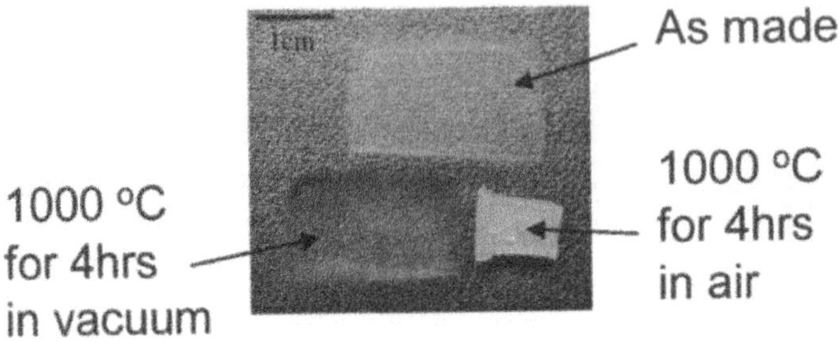

Figure 14. Aerogel cubes heated in vacuum shrank considerably less than those heated in air.

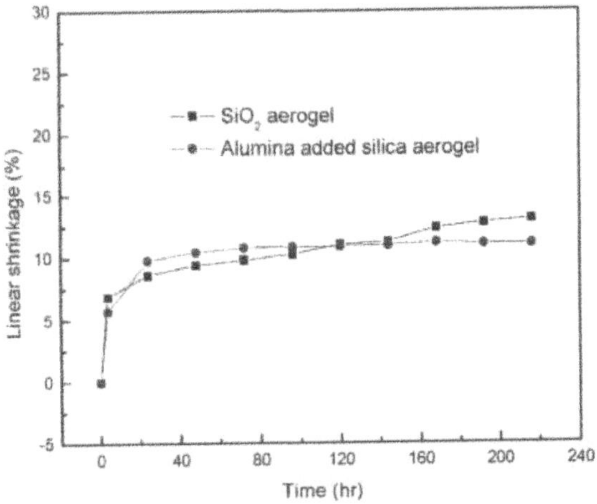

Figure 15. Linear shrinkage as a function of time at 1000C in high vacuum.

Electroceramic Materials and Applications
Edited by R. W. Schwartz

CONTROL OF MAGNETIC ANISOTROPY IN $(Bi_{1-x}Pb_x)_2(AE_{1-y}RE_y)_2Co_2O_z$ AND FABRICATION OF MAGNETICALLY ALIGNED BULKS

T. Kumagai, S. Horii, T. Okamoto, J. Shimoyama, and K. Kishio
Department of Applied Chemistry, University of Tokyo
7-3-1 Hongo, Bunkyo-ku, Tokyo, Japan

T. Uchikoshi, T. S. Suzuki, and Y. Sakka
National Institute of Materials and Science
1-2-1 Sengen, Tsukuba, Ibaraki, Japan

M. Mikami, and R. Funahashi
National Institute of Advanced Industrial Science and Technology
1-8-31 Midorigaoka, Ikeda, Osaka, Japan

ABSTRACT

We have successfully fabricated two different types of grain-aligned bulks of thermoelectric Bismuth-based cobalt oxides using a magnetic grain-alignment process. The Pb-doping in $(Bi_2Sr_2O_4)_{0.55}CoO_2$, which shows an easy axis of magnetization parallel to the *a*-axis, was effective for enhancement of magnetic anisotropy along the *a*-axis, while the Ca-doping in $[(Bi_{0.5}Pb_{0.5})_2Sr_2O_4]_{0.55}CoO_2$ converted the easy axis along the *a*-axis to that along the *c*-axis. Moreover, the magnetic anisotropy along the *c*-axis was enhanced by the partial Pr-doping for the Ca sites. The changes of their crystal structures revealed that distortion of the block layer was arisen from the substitution of the Ca-doping or the co-doping of Ca and Pr for the Sr sites in $[(Bi_{0.5}Pb_{0.5})_2Sr_2O_4]_{0.55}CoO_2$, leading to the changes of their magnetic anisotropies through locally structural distortion of the CoO_2 layers. The *c*-axis and *a*-axis grain-aligned bulks were fabricated by the slip-casting method under a magnetic field of 10 T for $[(Bi_{0.5}Pb_{0.5})_2(Ca_{0.8}Pr_{0.2})_2O_4]_pCoO_2$ and $[(Bi_{0.75}Pb_{0.25})_2Sr_2O_4]_{0.55}CoO_2$, respectively. Especially for the *c*-axis grain-aligned bulk, microstructure observation showed highly grain-oriented, and large reduction of resistivity was observed in a perpendicular direction to the applied field. The control of the easy axis of magnetization by a crystallochemical process is useful for the application of a magnetic grain-alignment process.

INTRODUCTION

Thermoelectric (TE) power generation using oxides is a hopeful system as a retrieval technology of thermal energy from high temperature heat waste because oxides materials are chemically stable at high temperatures in air in general. Development of TE oxides having high TE performance directly leads to the increase of conversion efficiency from heat to electricity. The performance of TE materials is expressed as the figure of merit, $Z = S^2/\rho\kappa$, and the power factor, $P = S^2/\rho$, where S, ρ and κ are Seebeck coefficient, electrical resistivity and thermal conductivity, respectively. The dimensionless figure of merit, ZT, where T is the absolute temperature, over 1 is a general requirement level for practical use. A lot of researchers have made efforts to develop novel TE oxides simultaneously showing larger S and lower ρ and κ at high temperatures. Recently, it has been reported in single crystalline samples that layered cobaltites such as Na_xCoO_2[1], $(Ca_2CoO_{3-\delta})_{0.62}CoO_2$ (Ca349)[2] and $(Bi_2Sr_2O_4)_{0.55}CoO_2$ (BiSr222)[3] showed $ZT > 1$ above 800 K, all of which possess CdI_2-type of CoO_2 layers parallel to the *ab* plane and showing high electrical conductivity and Seebeck coefficient. Particularly, the Ca349 and BiSr222 compounds possess a rock-salt-type block layer, in which the valence of Co in the CoO_2 layers can be controlled, and the *b*-axis length the block layers (b_1) is different from that (b_2) of the CoO_2 layers. These incommensurate and anisotropic crystal structures generate large S and low κ. However, the usage of single crystalline bulks is not realistic from viewpoints of technical problems and high production cost of the crystal growth. Therefore, development of polycrystalline bulks with high TE performance has been promoted by through several approaches. Since Bi222 grains are tabular having large *ab* plane, which is parallel to the CoO_2 layers, mechanical grain-alignment[4-6] and densification[7] processes are effective for reducing resisitivity. Moreover, improvements in thermoelectric power were achieved by various substitutions for Bi222[8-10], which affects the Co valence. Especially, the grain-alignment along the *ab* plane is one of the most useful processes because anisotropic crystal structures of these cobaltites affect their anisotropies in resistivity (ρ_c / ρ_{ab}) rather than those in S and κ[11].

Our group has reported the magnetic grain-alignment method as another grain-alignment process for Ca349[12,13] and $[Ca_2(Co_{0.65}Cu_{0.35})_2O_4]_{0.624}CoO_2$ (CCCO)[13]. These compounds show the easy axes of magnetization parallel to the *c*-axis[14,15]. This magneto-scientific method has advantageous points that the degree of grain-orientation is high[16] and grain-oriented bulks with thick and large size can be fabricated. Therefore, the magnetic method enabled the formation of microstructures in which platelet grains spread along the *ab* plane were aligned perpendicular to the applied magnetic field. On the other hand, in the case of BiSr222 with the easy axis parallel to the *a*-axis, one can understand that the grain-alignment process is not enough to contribute to the improved conduction path along the *ab* plane if the magnetically grain-aligned bulk is prepared. This complicated requirement for the grain-alignment is quite contrast to the mechanical grain-alignment techniques[4-6]. In this study, for the expansion of effectiveness of the magneto-scientific method, we attempt to fabricate two different types of magnetically grain-aligned Bi-based cobaltites. One is the fabrication of the *c*-axis oriented bulk after the development of Bi-based cobaltites with the easy axis of magnetization along the *c*-axis by the crystallochemical method. The other is the fabrication of the *a*-axis grain-oriented bulk of BiSr222.

EXPERIMENTAL

BiSr222, $[(Bi_{1-x}Pb_x)_2Sr_2O_4]_{0.55}CoO_2$ ($(Bi_{1-x}Pb_x)$Sr222), $[(Bi_{0.5}Pb_{0.5})_2(Ca_{1-y}RE_y)_2O_4]_pCoO_2$ ($(Bi_{0.5}Pb_{0.5})(Ca_{1-y}RE_y)$222) powders were synthesized by the conventional solid-state reaction using Bi_2O_3, PbO, $SrCO_3$, $CaCO_3$, Co_3O_4, Pr_6O_{11}, Nd_2O_3, and Sm_2O_3 as starting powders. Powder mixtures with appropriate cationic ratios were calcined at 700°C for 6 h. After an intermediate grinding, $(Bi_{0.5}Pb_{0.5})(Ca_{1-y}RE_y)$222 and $(Bi_{1-x}Pb_x)$Sr222 powders were sintered at 750°C for 24h and at 800°C for 24 h, respectively, taking the difference of appropriate sintering conditions between them into account. All these heating procedures were made in air.

Obtained powders were fined through a ball-milling process to be ~ 1 μm in mean grain size. In order to determine their easy axes of magnetization, magnetically aligned powders in epoxy resin were fabricated under magnetic field of $\mu_0H_a = 8$ T. Magnetically aligned bulks were fabricated by slip-casting[16,17] in $\mu_0H_a = 10$ T using the fine powders. For comparison, the slip-

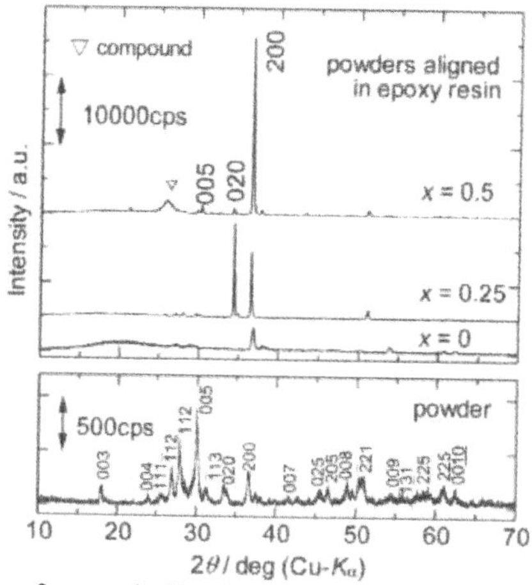

Fig. 1. XRD patterns of magnetically aligned $[(Bi_{1-x}Pb_x)_2Sr_2O_4]_{0.55}CoO_2$ powders (x = 0, 0.25, 0.5) of the surface perpendicular to the magnetic field, also with a powder pattern for $x = 0$ as a reference.

casted bulk under 0 T was also fabricated. For the fabrication of c-axis grain-oriented $(Bi_{0.5}Pb_{0.5})(Ca_{1-y}RE_y)$222 bulk, the applied direction of magnetic field (H_a) is parallel to the slip-casting direction, while for that of a-axis grain-oriented $(Bi_{1-x}Pb_x)$Sr222 bulk, the direction of H_a is perpendicular to the slip-casting. Incidentally, the casting direction is parallel to the vertical direction and dispersing medium of slurry is ethanol. These slip-casted samples were sintered for 6 h at 700°C in air for decarbonation of dispersant, then were pressed at 4 ton/cm^2 and sintered in air at 700°C for $(Bi_{0.5}Pb_{0.5})(Ca_{1-y}RE_y)$222 and at 800°C for $(Bi_{1-x}Pb_x)$Sr222. The intermediate pressing was repeated twice. Electrical resistivity from 5 K to 950 K and Seebeck coefficient from 90 K to 300 K were measured by a four-probe method, while Seebeck coefficient measurement above 300 K was performed by a two-probe method.

RESULTS AND DISCUSSIONS

Easy axes of magnetization and magnetic anisotropies in the Bi-based cobaltites

The changes of the easy axes in the Bi222 compounds doped by various cationic ions were investigated using the powders aligned under 8 T in epoxy resin. Figure 1 shows x-ray diffraction (XRD) patterns of the magnetically oriented $(Bi_{1-x}Pb_x)$Sr222 (x = 0, 0.25 and 0.5) powders. A powder XRD pattern of BiSr222 is also shown for comparison. These measurements were performed for the bottom surface of the solidified epoxy resin, which is perpendicular to H_a. For a sample with x = 0 (BiSr222), only the (200) peak was enhanced, which is clearly recognized from the comparison with the powder pattern. Furthermore, intensities of the (200) peaks are strengthened by the Pb-doping. At the composition of x = 0.25, the (020) peak also appeared. Obviously, the Pb-doping in the BiSr222 phase was effective for the enhancement of magnetic anisotropy with preserving the easy axis of magnetization parallel to the a-axis.

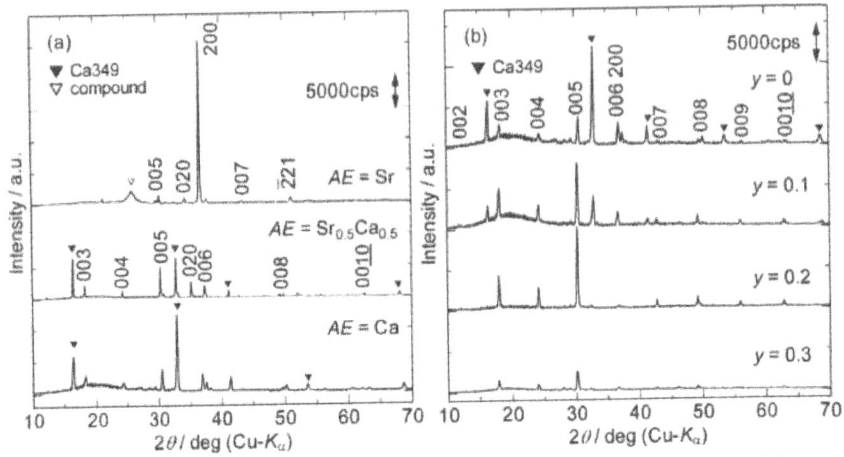

Fig. 2. Powder XRD patterns of various Bi-based cobaltites aligned under $\mu_0 H_a = 8$ T in epoxy resin; (a) $[(Bi_{0.5}Pb_{0.5})_2AE_2O_4]_pCoO_2$ (AE = Sr, $Sr_{0.5}Ca_{0.5}$, Ca). (b)$[(Bi_{0.5}Pb_{0.5})_2(Ca_{1-y}Pr_y)_2O_4]_pCoO_2$ (y = 0, 0.1, 0.2, and 0.3). The measurements were performed for the bottom surface ($\perp H_a$).

We focused on another cationic site in the block layer of BiSr222, Sr sites, and chose Ca as a substitution element. Figure 2 (a) show XRD patterns of magnetically oriented powders of $(Bi_{0.5}Pb_{0.5})(Sr_{1-y}Ca_y)222$ (y = 0, 0.5 and 1). The (00l) peaks for a sample with y = 0 were completely suppressed, while for samples with y = 0.5 and 1, intensities of the (00l) peaks were drastically enhanced. This means that the easy axis of magnetization converts the a-axis to the c-axis with an increase of the Ca-doping level. For more enhancement of the magnetic anisotropy ($\Delta\chi = \chi_c - \chi_{ab}$) and determination of the appropriate Pr-doping level showing the largest $\Delta\chi$ in $(Bi_{0.5}Pb_{0.5})(Ca_{1-y}Pr_y)222$ system, the easy axes of y = 0.1, 0.2, and 0.3 were clarified in the same method with the above experiment. As shown in Fig. 2 (b), one can easily see that intensities of the (00l) peaks were systematically increased up to y = 0.2, while those in y = 0.3 were suppressed. These results strongly suggest that the largest $\Delta\chi$ is found at $y \sim 0.2$ in the $(Bi_{0.5}Pb_{0.5})(Ca_{1-y}Pr_y)222$ system. Furthermore, in order to clarify the change of the easy axis in the $(Bi_{0.5}Pb_{0.5})(Ca_{1-y}RE_y)222$ system doped by various types of RE ions, the fine powders with y = 0.1 for RE = Nd and Sm were aligned in epoxy resin under $\mu_0 H_a = 8$ T. As can be seen in Fig. 3, the Nd-doping was found to be the most effective for enhancing the (00l) peaks among the three different RE-doped systems and this enhancement effect was almost same with that in the Pr-doped system. On the other hand, in the case of $(Bi_{0.5}Pb_{0.5})(Ca_{0.9}Sm_{0.1})222$, the (00$l$) peaks were suppressed and the (200) peak was grown. That is, the $\Delta\chi$ in the $(Bi_{0.5}Pb_{0.5})(Ca_{0.9}RE_{0.1})222$ systems depends on the type of RE.

Here, in order to clarify the conversion mechanism of the easy axis of magnetization in $(Bi_{0.5}Pb_{0.5})AE222$ (AE = Sr, Ca) and $(Bi_{0.5}Pb_{0.5})$(Ca, Pr)222, lattice parameters were analyzed from the XRD patterns. Figures 4 (a), 4 (b) and 4 (c) show the change of a-, b_1- and c-axis lengths of $(Bi_{0.5}Pb_{0.5})(AE_{1-y}Pr_y)222$ as a function of the mean ionic radius of (AE, Pr) sites surrounded by six-coordinate oxygen ions, respectively. Note that the b_1-axis corresponds to the

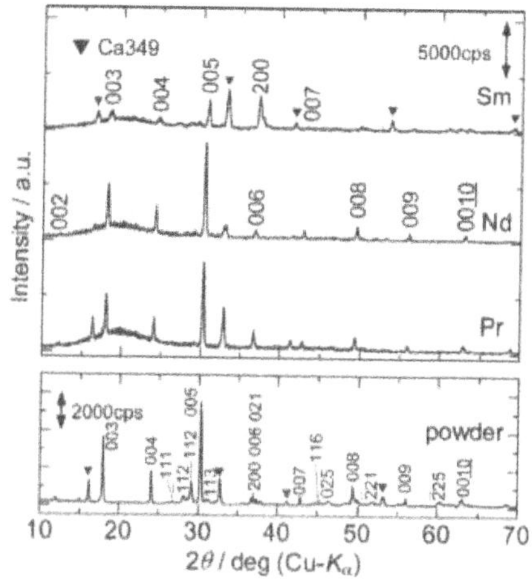

Fig. 3. XRD patterns of $[(Bi_{0.5}Pb_{0.5})_2(Ca_{0.9}RE_{0.1})_2O_4]_pCoO_2$ (RE = Pr, Nd and Sm) aligned under μ_0H_a = 8 T in epoxy resin. The measurements were performed for the bottom surface ($\perp H_a$).

b-axis of the block layer. The sizes of occupied ions in the (AE, Pr) sites are listed in the descending order: Sr^{2+}(1.16 Å), Pr^{3+}(1.01 Å), Ca^{2+}(1.00 Å). The a-axis length is almost independent of the mean ionic radii of (AE, Pr) sites of the $(Bi_{0.5}Pb_{0.5})$(AE, Pr)222 system. However, the b_1 and c-axis lengths are systematically decreased with the Ca-doping level in $(Bi_{0.5}Pb_{0.5})$Sr222. Interestingly, the Pr-doping induced contraction of the b_1- and c- axes despite the ionic radius of Pr^{3+} shows a similar value to that of Ca^{2+}. The b_2-axis length, which corresponds to the b-axis length of the CoO_2 layer, was unclear only from our XRD measurements. However, the transmission electron diffraction measurements[8,18] revealed that the b_2 length was almost equal to 2.8 Å and was not affected for the change of compositions of block layers. The values of the misfit ratio (b_2/b_1) in Fig. 3 (d) were obtained using the reported b_2 length (2.8 Å). Hence, the substitution of Ca for Sr in (Bi,Pb)Sr222 led to the increase of the misfit ratio, which is mainly due to the contraction of b_1 length. In $(Bi_{0.5}Pb_{0.5})$Sr222 and $(Bi_{0.5}Pb_{0.5})$Ca222, the easy axis of magnetization depends on magnetic state of Co ions in the CoO_2 layer, because the block layers of these compounds are composed of only non-magnetic ions. That is, the conversion of the easy axis in the $(Bi_{0.5}Pb_{0.5})$(AE, Pr)222 system is reflection of the magnetic states of Co ions in the CoO_2 layer.

Miyazaki *et al.* has reported that the modulation of Co-Co distance in the CoO_2 layers along the b-axis was reduced by the Bi-doping[19] and was enhanced by the Sr-doping[20] in Ca349[21], indicating that the substitution of ions with different ionic size into the block site affects local structure of the CoO_2 layers largely and indirectly. In $La_{2-2x}Sr_{1+2x}Mn_2O_7$[22], the easy axis of magnetization was dominated by the Jahn-Teller distortion of MnO_6 octahedron. Namely, magnetic states of transition metal ions are sensitive to the crystal field formed by six oxygen ions around the metal ions. Therefore, it is likely that the easy axis of magnetization and the value of $\Delta\chi$ in the CoO_2 layers of the misfit layered cobaltites depend on the modulated structure

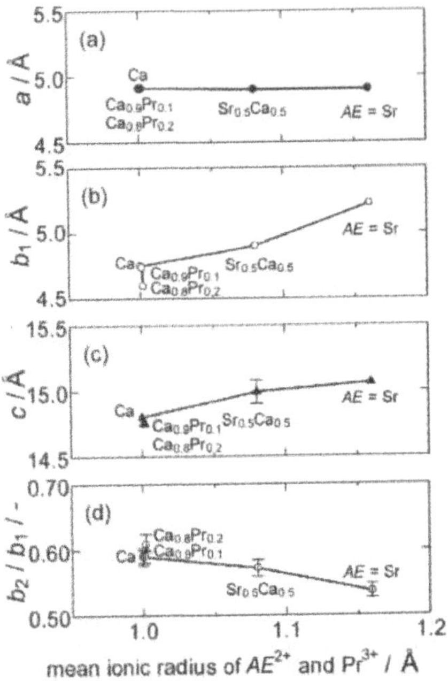

Fig. 4. Changes of lattice parameters (a, b_1, c) and misfit ratio (b_2/b_1) as a function of the mean ionic radius in (AE, Pr) sites in $(Bi_{0.5}Pb_{0.5})(AE_{1-y}Pr_y)222$.

of the CoO_2 layers. In the case of the Bi222, there are only non-magnetic ions in the block layer, therefore the changes of local structure of the CoO_2 layers lead directly to a direction of the easy axis of magnetization of the system.

On the other hand, the changes of magnetic anisotropies in the RE-doped $(Bi_{0.5}Pb_{0.5})Ca222$ system are complicated because $\Delta\chi$ is dominated by the magnetic anisotropy of the RE ions in the block layers in addition to the above structural effect. As mentioned above, the Pr- or Nd-doping showed a clear enhancement effect of $\Delta\chi$, while the Sm-doping did not show, apparently suggesting that $\Delta\chi$ depends on the effective Bohr magneton number (μ_{eff}) of the doped RE ions. However, the magnetic anisotropy of the doped RE ions need to take a crystal field formed by the surrounded oxygen ions into account. In order to elucidate the RE-doping effect systematically and quantitatively, the evaluations of χ and $\Delta\chi$ for the $(Bi_{0.5}Pb_{0.5})Ca222$ compounds doped by various RE ions using the grain-aligned bulks are required.

Fabrication of grain-aligned bulks in Bi-based cobaltite and their TE properties

Using the $(Bi_{0.5}Pb_{0.5})(Ca_{0.8}Pr_{0.2})222$ powder with the largest $\Delta\chi$ in all the prepared samples in this study, a c-axis grain-aligned bulk has been attempted to fabricate by the magnetic alignment method. The observation using a scanning electron microscope (SEM) was performed

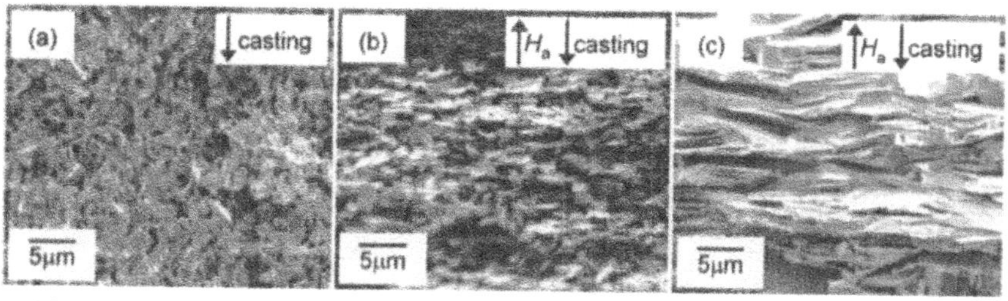

Fig. 5. SEM images of cross-sectional surface of the $[(Bi_{0.5}Pb_{0.5})_2(Ca_{0.8}Pr_{0.2})_2O_4]_zCoO_2$ bulks slip-casted under μ_0H_a = 0 T (a), μ_0H_a = 10 T (b), and processed through both slip-casting under μ_0H_a = 10 T and hot-forging (c).

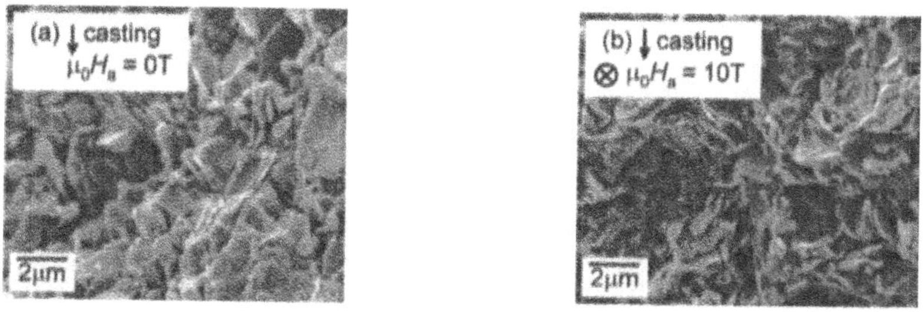

Fig. 6. SEM images of cross-sectional surface of the $[(Bi_{0.75}Pb_{0.25})_2Sr_2O_4]_{0.55}CoO_2$ bulks slip-casted under μ_0H_a = 0 T (a) and μ_0H_a = 10 T (b).

to elucidate the grain-alignment and densification effects on the internal microstructures. Figures 5 (a), 5 (b) and 5 (c) show SEM images of fractured surfaces of $(Bi_{0.5}Pb_{0.5})(Ca_{0.8}Pr_{0.2})$222 bulks slip-casted under 0 and 10 T and the hot-forged bulk after slip-casting under 10 T, respectively. Clearly, seen in Figs. 5 (b) and 5 (c), platelet grains which are spread along the *ab* direction, were aligned perpendicular to H_a, while the grains are randomly oriented in Fig. 4 (a). Moreover, the hot-forged and grain-aligned sample showed large grains with the size of ~ 10 μm in diameter and formed a densified microstructure, which is in a sharp contrast with the microstructure in Fig. 5 (a). The Lotgering factor[23], which is an index of the degree of grain orientation, was found to be 0.99 for the grain-aligned bulk and the grain-aligned and hot-forged bulk. Therefore, one can recognize that these high degrees of grain-orientation are clearly arisen from the magnetic alignment process and are preserved even for the introduction of the mechanical densification process.

Furthermore, using the $(Bi_{0.75}Pb_{0.25})$Sr222 compound with the easy axis of magnetization parallel to the *a*-axis, an *a*-axis grain-aligned bulk was fabricated by the same method. Note that the magnetic field was applied horizontally and was perpendicular to the casting direction. Figures 6 (a) and 6 (b) show SEM images of cross-sections of $(Bi_{0.75}Pb_{0.25})$Sr222 bulks fabricated under 0 and 10 T, respectively. Incidentally, the applied field in Fig. 6 (b) is perpendicular to the observed surface. As shown in Fig. 6 (a), one can see various aspects of platelet grains, which means these grains randomly oriented. On the other hand, from the microstructure in Fig. 6 (b), a lot of white lines can be seen and, however, each line runs in

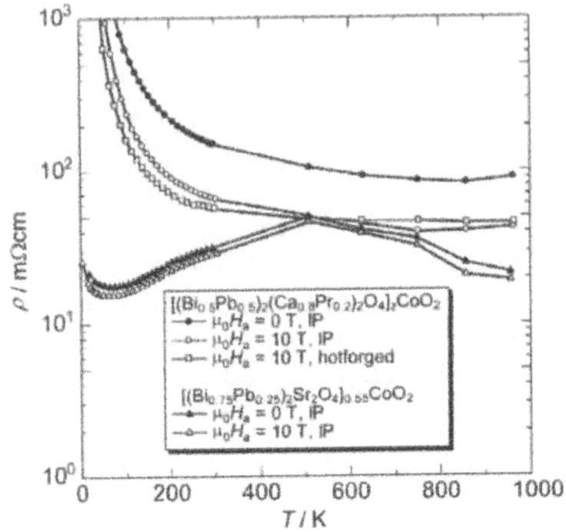

Fig. 7. Temperature dependence of resistivity of the slip-casted $[(Bi_{0.5}Pb_{0.5})_2(Ca_{0.8}Pr_{0.2})_2O_4]_zCoO_2$ and $[(Bi_{0.75}Pb_{0.25})_2Sr_2O_4]_{0.55}CoO_2$ bulks.

random directions. This suggests that the white line corresponds the edge of the platelet grain and the microstructure in Fig. 6 (b) is reflection of the a-axis grain orientation.

In order to clarify the relationship between microstructures and TE properties, the resistivity and Seebeck coefficient measurements were performed for the $(Bi_{0.5}Pb_{0.5})(Ca_{0.8}Pr_{0.2})222$ and $(Bi_{0.75}Pb_{0.25})Sr222$ bulks processed through both the magnetic grain-alignment and the intermediate-pressing (IP). Figure 7 shows temperature dependence of resistivity of the grain-aligned and densified $(Bi_{0.5}Pb_{0.5})(Ca_{0.8}Pr_{0.2})222$ and $(Bi_{0.75}Pb_{0.25})Sr222$ bulks. Note that the measurements of both $(Bi_{0.5}Pb_{0.5})(Ca_{0.8}Pr_{0.2})222$ and $(Bi_{0.75}Pb_{0.25})Sr222$ magnetic aligned bulks were performed for $I \perp H_a$ and $I \,//\, H_a$, respectively. For comparison, the bulks slip-casted under 0 T were also prepared and their resistivities and Seebeck coefficients were obtained for $I \perp$ slip-casting. Clearly, both bulks show lower resistivities compared to the bulks processed under 0 T. This is because conductive CoO_2 layers are aligned by magnetic grain-orientation process. However, resistivities of the grain-aligned and hot-forged $(Bi_{0.5}Pb_{0.5})(Ca_{0.8}Pr_{0.2})222$ bulk, which is more densified compared with the bulks processed IP, did not show significant difference from those of the IP bulks. At the current stage, we believe that this is due to the partial decomposition occurred during the hot-forging process at 800°C and/or the presence of micro-cracks in the bulk. In order to reduce the resistivity additionally, suppression of these phenomena by precise controls of the forging temperature and pressure are necessary. When we focus on the resistivity difference between $(Bi_{0.75}Pb_{0.25})Sr222$ and $(Bi_{0.5}Pb_{0.5})(Ca_{0.8}Pr_{0.2})222$, the compound containing Sr is more conductive than that containing Ca. The BiCa222 phase has reported to be crucially less conductive than Ca349 and BiSr222[4]. This is not favorable for a TE material so much, though this shows the easy axis of magnetization parallel to the c-axis.

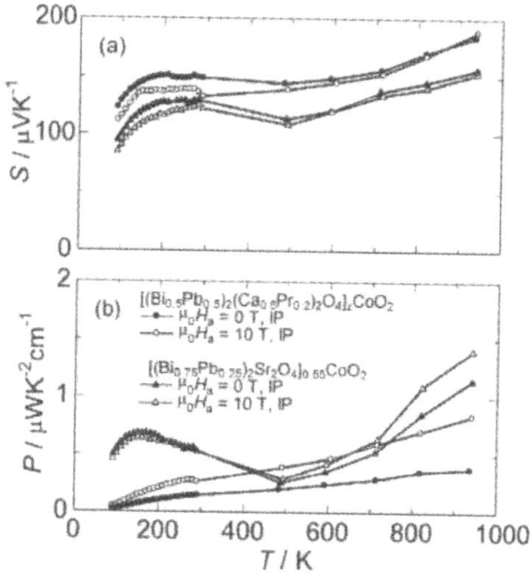

Fig. 8. Temperature dependence of Seebeck coefficients (a) and power factors (b) of the slip-casted $[(Bi_{0.5}Pb_{0.5})_2(Ca_{0.8}Pr_{0.2})_2O_4]_zCoO_2$ and $[(Bi_{0.75}Pb_{0.25})_2Sr_2O_4]_{0.55}CoO_2$ bulks.

Finally, Figs. 8 (a) and 8 (b) show temperature dependence of Seebeck coefficients and power factors of the $(Bi_{0.5}Pb_{0.5})(Ca_{0.8}Pr_{0.2})222$ and $(Bi_{0.75}Pb_{0.25})Sr222$ bulks, respectively. For both $(Bi_{0.5}Pb_{0.5})(Ca_{0.8}Pr_{0.2})222$ and $(Bi_{0.75}Pb_{0.25})Sr222$, Seebeck coefficients of the magnetically aligned and IP bulks did not show significant difference with the bulks fabricated under $\mu_0H_a = 0$ T. This is because their anisotropies of the Seebeck coefficient (S_c / S_{ab}) are smaller than that of the resistivity (ρ_c / ρ_{ab})[11]. Consequently, the power factor of $(Bi_{0.5}Pb_{0.5})(Ca_{0.8}Pr_{0.2})222$ at 950 K was improved almost twice larger by the magnetic alignment, which is due to the drastic decrease of the resistivity by the magnetic grain-orientation. However, $(Bi_{0.75}Pb_{0.25})Sr222$ did not show remarkable increase. For further improvement of TE properties in Bi-based cobaltites, optimization of the hot-forging process, precise control of initial compositions taking the misfit ratios into account, and development of the Sr-based (Bi, Pb)(Ca, Pr)222 with preserving $\Delta\chi > 0$ are required.

CONCLUSIONS

We have investigated the easy axes of the magnetization for the Bi-based cobaltites using the $(Bi_{1-x}Pb_x)Sr222$ (x = 0, 0.25, 0.5) and $(Bi_{0.5}Pb_{0.5})(Ca_{1-y}RE_y)222$ (RE = Pr, Nd, Sm, y = 0, 0.1, 0.2) powders. Their easy axes were drastically changed and were dependent on the type of doped cations. The Pb-doping in BiSr222 enhanced the magnetic anisotropy along the a-axis, and the Ca-doping converted the easy axis of magnetization along the a-axis to that along the c-axis. Moreover, the RE-doping (RE = Pr, Nd) showed the enhancement effect of magnetic anisotropy.

The conversion of the easy axis in $(Bi_{0.5}Pb_{0.5})(Sr_{1-y}Ca_y)222$ system was understood in terms of the change of magnetization states of Co ions in the CoO_2 layer, which was induced by the change of the misfit ratio. Using $(Bi_{0.5}Pb_{0.5})(Ca_{1-y}Pr_y)222$ with the easy axis along the c-axis, the *c*-axis grain-aligned bulk has been successfully fabricated by the slip-casting under the magnetic field of 10 T. The Lotgering factor, microstructure, and resistivity revealed that a high degree of the *c*-axis orientation, achieved in the whole bulk leading to the improvement of its power factor. The $(Bi_{0.75}Pb_{0.25})$Sr222 bulk was also fabricated. Our present works demonstrated that the development of materials suitable for the magneto-scientific process could be designed by the crystallochemical approach.

REFERENCES

[1] K. Fujita, T. Mochida, and K. Nakamura, "High-Temperature Thermoelectric Properties of $Na_xCoO_{2-\delta}$ Single Crystals," *Jpn. J. Appl. Phys.*, **40**, 4644-4647 (2001).

[2] R. Funahashi, I. Matsubara, H. Ikuta, T. Takeuchi, U. Mizutani, and S. Sodeoka, "An Oxide Single Crystal with High Thermoelectric Performance in Air," *Jpn. J. Appl. Phys.*, **39**, L1127-L1129 (2000).

[3] R. Funahashi, and M. Shikano, "$Bi_2Sr_2Co_2O_y$ whiskers with high thermoelectric figure of merit," *Appl. Phys. Lett.*, **81**, 1459-1461 (2002).

[4] H. Itahara, K. Fujita, J. Sugiyama, K. Nakamura, and T. Tani, "Highly Textured $Na_xCoO_{2-\delta}$ Ceramics Fabricated by Both Templated Grain Growth, and Reactive Templated Grain Growth Methods Using Single-Crystalline Particles as Templates," *J. Ceram. Soc. Jpn.*, **111** 227-231 (2003).

[5] H. Itahara, C. Xia, J. Sugiyama, and T. Tani, "Fabrication of textured thermoelectric layered cobaltites with various rock salt-type layers by using β-$Co(OH)_2$ platelets as reactive templates," *J. Mater. Chem.*, 61-66 **14** (2004).

[6] W. Shin, and N. Murayama, "Thermoelectric properties of (Bi,Pb)-Sr-Co-O oxide," *J. Mater. Res.*, **15**, 382-386 (2000).

[7] I. Matsubara, R. Funahashi, T. Takeuchi, and S. Sodeoka, "Thermoelectric properties of spark plasma sintered $Ca_{2.75}Gd_{0.25}Co_4O_9$ ceramics," *J. Appl. Phys.*, 462-465 **90** (2001).

[8] T. Fujii, I. Terasaki, T. Watanabe, and A. Matsuda, "Large In-Plane Anisotropy on Resistivity and Thermopower in the Misfit Layered Oxide $Bi_{2-x}Pb_xSr_2Co_2O_y$," *Jpn. J. Appl. Phys.*, **41**, L783 –L786 (2004).

[9] T. Itoh, and I. Terasaki, "Theremoelectric Properties of $Bi_{2.3-x}Pb_xSr_{2.6}Co_2O_y$ Single Crystals," *Jpn. J. Appl. Phys.*, **39**, 6658-6660 (2000).

[10] E. Iguchi, S. Katoh, H. Nakatsugawa, and F. Munakata, "Theremoelectric Properties (Resistivity and Thermopower) in $Bi_{1.5}Pb_{0.5}Ca_{2-x}M_xCo_2O_{8-\delta}$ (M = Sc^{3+}, Y^{3+}, or La^{3+})," *J. Solid State Chem.*, **167**, 472-479 (2002).

[11] T. Tani, and H. Itahara, "Topotactic Synthesis of Highly Textured Thermoelectric Oxide Ceramics," *Materials Integration*, **17**, 25-31 (2004) [in Japanese].

[12] M. Sano, S. Horii, I. Matsubara, R. Funahashi, M. Shikano, J. Shimoyama, and K. Kishio, "Synthesis and Thermoelectric Properties of Magnetically *c*-Axis-Oriented $[Ca_2CoO_{3-\delta}]_{0.62}CoO_2$ Bulk with Various Oxygen Contents," *Jpn. J. Appl. Phys.*, **42**, L198-L200 (2003).

[13] S. Horii, I. Matsubara, M. Sano, K. Fujie, M. Suzuki, R. Funahashi, M. Shikano, W. Shin, N. Murayama, J. Shimoyama, and K. Kishio, "Thermoelectric Performance of Magnetically *c*-Axis Aligned Ca-based Cobaltites," *Jpn. J. Appl. Phys.*, **42**, 7018-7022 (2003).

[14] A. C. Masset, C. Michel, A. Maignan, M. Hervieu, O. Toulemonde, F. Studer, B. Raveau, and J. Hejmanek, "Misfit-layered cobaltite with an anisotropic giant magnetoresistance: $Ca_3Co_4O_9$," *Phys. Rev. B*, **62**, 166-175 (2000).

[15] M. Suzuki, S. Horii, M. Sano, K. Fujie, K. Otzschi, J. Shimoyama, and K. Kishio, "Synthesis of $[Ca_2(Co_{1-x}Cu_x)_2O_4]_yCoO_2$ Single Crystals and Their Intrinsic Properties," *Mater. Res. Soc. Symp. Proc.*, **755**, 10.5.1-6 (2003).

[16] T. S. Suzuki, Y. Sakka, and K. Kitazawa, "Orientation Amplification of Alumina by Colloidal Filtration in a Strong Magnetic Field and Sintering," *Adv. Eng. Mat.*, **3**, 490-492 (2001).

[17] T. S. Suzuki, and Y. Sakka, "Fabrication of Textured Titania by Slip Casting in a High Magnetic Field Followed by Heating," *Jpn. J. Appl. Phys.*, **41**, L1272–L1274 (2004).

[18] M. Hervieu, P. Boully, C. Michael, A. Maignan, and B. Raveau, "A New Family of Misfit Layered Oxides with Double Rock Salt Layers $Bi_\alpha(A_{0.75\pm\varepsilon}Bi_{0.25\pm\varepsilon}O)MO_2$ (A = Ca, Sr and M = Co, Cr)," *J. Solid State Chem.*, **142**, 305-318 (1999).

[19] Y. Miyazaki, Y. Suzuki, M. Onoda, Y. Ishii, Y. Morii, and T. Kajitani: "Modulated Structure of Misfit Layered Oxide $[(Ca_{0.90}Bi_{0.10})_2(Co_{0.95}Bi_{0.05})O_3]_pCoO_2$," *Jpn. J. Appl. Phys.*, **43**, 6252-6258 (2004).

[20] Y. Miyazaki, T. Miura, Y. Ono, and T. Kajiwara, "Crystal Structure and Thermoelectric Properties of the Composite Crystal $[(Ca_{1-x}Sr_x)_2CoO_3]_pCoO_2$," *J. Jpn. Soc. Powder Powder Metall.*, **50**, 475-479 (2003) [in Japanese].

[21] Y. Miyazaki, M. Onoda, T. Oku, M. Kikuchi, Y. Ishii, Y. Ono, Y. Morii, and T. Kajitani, "Modulated Structure of Thermoelectric Compound $[Ca_2CoO_3]_{0.62}CoO_2$," *J. Phys. Soc. Jpn.*, **71**, 491-497 (2002).

[22] T. Kimura, Y. Tomioka, A. Asamitsu, and Y. Tokura, "Anisotropic Magnetoelastic Phenomena in Layered Manganite Crystals: Implication of Change in Orbital State," *Phys. Rev. Lett.*, **81**, 5920-5923 (1998).

[23] F. K. Lotgering, "Topotactical Reactions with Ferrimagnetic Oxides Having Hexagonal Crystal Structures," *J. Inorg. Nucl. Chem.*, **9**, 113-123 (1959).

Electroceramic Materials and Applications
Edited by R. W. Schwartz

HIGH-QUALITY EPITAXIAL FILM GROWTH OF SUPERCONDUCTING SODIUM-COBALT OXYHYDRATE, $Na_{0.3}CoO_2 \cdot 1.3H_2O$

Kenji Sugiura[1], Hiromichi Ohta[1,2], Kenji Nomura[3], Hiroshi Yanagi[4], Masahiro Hirano[3], Hideo Hosono[3,4,5], and Kunihito Koumoto[1,2]

[1]Nagoya University, Graduate School of Engineering, Furo-cho, Chikusa, Nagoya 464-8603, Japan
[2]CREST, JST, 4-1-8 Motomachi, Kawaguchi 332-0012, Japan
[3]ERATO-SORST, JST, in Frontier Collaborative Research Center, Tokyo Institute of Technology, Mail Box S2-13, 4259 Nagatsuta, Midori-ku, Yokohama 226-8503, Japan
[4]Materials and Structures Laboratory, Tokyo Institute of Technology, Mail Box R3-1, 4259 Nagatsuta, Midori-ku, Yokohama 226-8503, Japan
[5]Frontier Collaborative Research Center, Tokyo Institute of Technology, Mail Box S2-13, 4259 Nagatsuta, Midori-ku, Yokohama 226-8503, Japan

ABSTRACT
High-quality epitaxial films of superconducting sodium-cobalt oxyhydrate, $Na_{0.3}CoO_2 \cdot 1.3\ H_2O$ were successfully fabricated onto a (0001)-oriented α-Al_2O_3 substrate. The $Na_{0.8}CoO_2$ epitaxial film, which was grown by reactive solid-phase epitaxy (R-SPE), was converted into $Na_{0.3}CoO_2 \cdot 1.3\ H_2O$ epitaxial film via formation of $Na_{0.3}CoO_2$ epitaxial film by an oxidation - hydration process. The epitaxial relation was (0001) $[11\bar{2}0]$ $Na_{0.3}CoO_2 \cdot 1.3H_2O$ || (0001) $[1\bar{1}00]$ α-Al_2O_3. Presently available $Na_{0.3}CoO_2 \cdot 1.3H_2O$ epitaxial films exhibited similar electrical resistivity and superconducting transition behavior to those of bulk single crystalline $Na_{0.3}CoO_2 \cdot 1.3H_2O$.

INTRODUCTION
Since the discovery of superconducting cobalt oxyhydrate, $Na_{0.3}CoO_2 \cdot 1.3H_2O$ (superconducting transition temperature, $Tc \sim 4$ K)[1] in 2003, a number of researchers have attempted to clarify its complex superconducting transition mechanism[2-7]. Although a readily available supply of high-quality epitaxial film composed of $Na_{0.3}CoO_2 \cdot 1.3H_2O$ is considered to be essential to clarify the intrinsic physics of this material, powders and bulk single crystals have been used due to lack of the epitaxial film.

Since $Na_{0.3}CoO_2 \cdot 1.3H_2O$ has an extremely complicating layered crystal structure ($P6_3/mmc$; a = 0.2823 nm, c = 1.9621 nm) and contains volatile species of Na and H_2O (see the bottom-right of Fig. 1)[1,6], conventional high temperature vapor-phase-epitaxy (VPE) methods such as sputtering, molecular beam epitaxy (MBE) and pulsed-laser deposition (PLD) are inadequate for fabrication of $Na_{0.3}CoO_2 \cdot 1.3H_2O$ epitaxial films. The preservation of the chemical composition of $Na_{0.3}CoO_2 \cdot 1.3H_2O$ is almost impossible due to extremely large difference in vapor pressure between Na and/or H_2O and Co.

To overcome this difficulty, we explored a new approach, i.e., first high-quality epitaxial film of Na_xCoO_2 was formed on a substrate, and then the resulting films were subjected to oxidation and hydration with an expectation of preservation of the epitaxial relation in the converted phase.

In a recent publication, we reported that high-quality $Na_{0.8}CoO_2$ epitaxial films can be grown by reactive solid-phase epitaxy (R-SPE [9,10])[11]. Although the oxidation by Br_2 & hydration process explained in ref. 1 is preferable to convert Na_xCoO_2 powder into $Na_{0.3}CoO_2 \cdot 1.3H_2O$ powder, in our preliminary experiment, the R-SPE grown $Na_{0.8}CoO_2$ epitaxial films were observed to have decomposed into $Co(OH)_2$ during this process, due most likely to that the Br_2-oxidation process is too strong for thin (~ 200 nm) epitaxial films. In order to overcome this problem, we used an HNO_3 aqueous solution to oxidize $Na_{0.8}CoO_2$ epitaxial film by a milder process. As a result, we successfully fabricated a high quality epitaxial film of $Na_{0.3}CoO_2 \cdot 1.3H_2O$, which exhibited similar electrical resistivity and superconducting transitional behavior to those of bulk single crystals.

EXPERIMENTAL

Figure 1 illustrates the schematic procedure of fabricating superconducting $Na_{0.3}CoO_2 \cdot 1.3H_2O$ epitaxial films. First, high-quality epitaxial film of (111)-oriented CoO is deposited on a (0001) face of α-Al_2O_3 substrate (10 mm x 10 mm x 0.5 mm^t) at 700 °C by a pulsed-laser deposition (PLD) technique using a Co_3O_4 sintered disk as a target. The surface of the PLD-deposited CoO film is fully capped by a Y_2O_3-stabilized-zirconia (YSZ) single-crystalline plate (10 mm x 10 mm x 0.5 mm^t) in order to keep the surface clean. Then $NaHCO_3$ powder is put on the YSZ plate. The $Na_{0.8}CoO_2$ epitaxial film is grown by thermal annealing of the sandwiched specimen at 700°C for 1 h in air. Then the $Na_{0.8}CoO_2$ epitaxial film is immersed in an HNO_3 aqueous solution at 30°C for 5 min. Finally the film is immersed in NaCl aqueous solution for approximately 5h at 5°C.

Crystalline quality, orientation and thickness of the films were analyzed by high-resolution X-ray diffraction (HR-XRD, ATX-G, CuKα_1, Rigaku Co.) measurements. The geometric set-up used for our X-ray diffractmeter is described elsewhere [9-12]. The electrical resistivity was measured by a standard four-probe method using a physical property measurement system (PPMS, Quantum Design).

RESULTS AND DISCUSSION

Figure 2a (lower) shows out-of-plane XRD patterns of the R-SPE-grown-$Na_{0.8}CoO_2$ film on the (0001)-oriented α-Al_2O_3 substrate. Only intense Bragg diffraction peaks of 000l $Na_{0.8}CoO_2$ were seen together with 0006 α-Al_2O_3, indicating that the $Na_{0.8}CoO_2$ is highly c-axis oriented. The lattice parameter, c was calculated to be 1.091 nm, which corresponds well with the accepted value in other literature (c = 1.0811 nm [8]). Although an XRD pattern is observed in the resultant film after oxidation treatment by 1 mol L^{-1} HNO_3 aqueous solution (Fig. 2a (upper)), the peak positions were found to be shifted slightly towards the smaller q_z side. This indicates that the interlayer distance of the Na_xCoO_2 expanded slightly due to deintercalation of Na^+ ions from the $Na_{0.8}CoO_2$ lattice[5,6]. Figure 2b shows relationship between the c-axis lattice parameter of Na_xCoO_2 and the concentration of the HNO_3 aqueous solution. The lattice parameter increases gradually with increase of HNO_3-concentration, indicating that the Na^+ ion concentration in the Na_xCoO_2 lattice can be controlled by HNO_3-treatment [7]. The chemical composition of the film treated with 1 mol L^{-1} HNO_3 aqueous solution was evaluated to be $Na_{0.3}CoO_2$ by X-ray fluorescence (XRF) analysis. The lattice parameter, c of the $Na_{0.3}CoO_2$ film was 1.121 nm, which corresponds well with the accepted value in other literature (c = 1.1211 nm [6]). Intense Bragg diffraction peaks of $11\bar{2}0$ $Na_{0.3}CoO_2$ is seen together with $3\bar{3}00$ α-Al_2O_3 in the in-plane

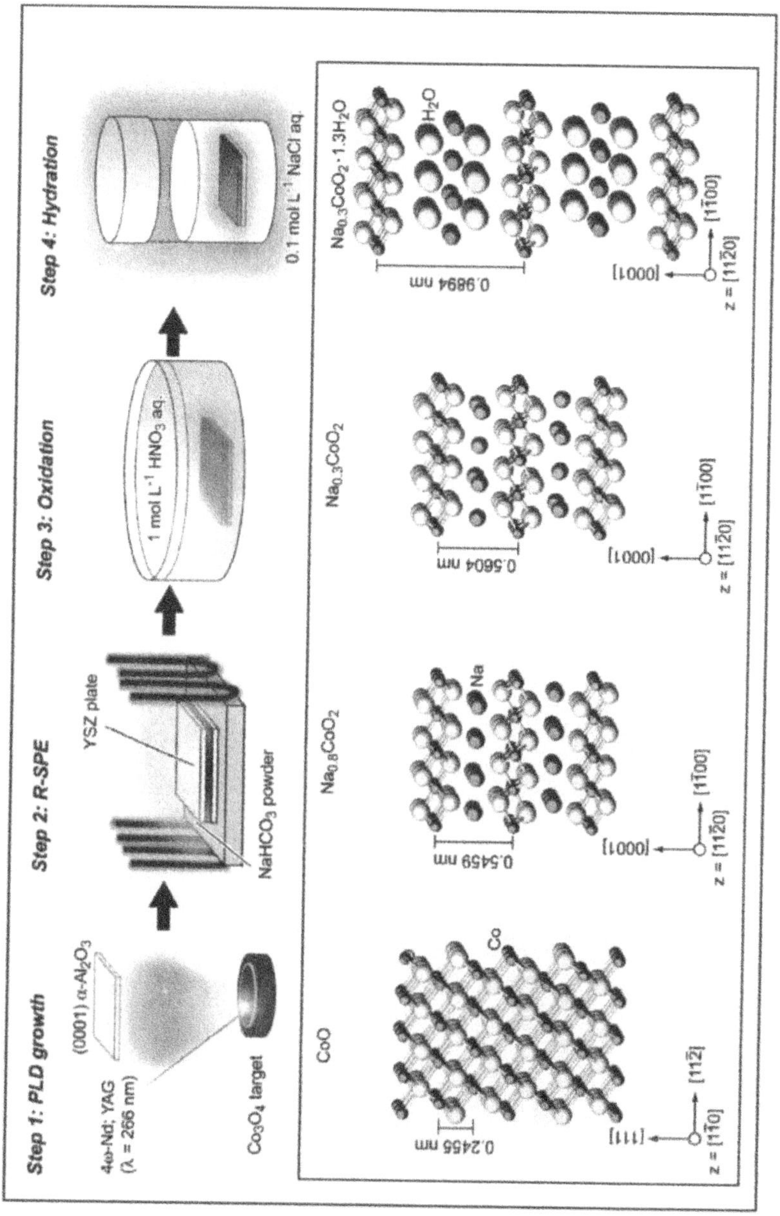
Step 1: PLD growth
Step 2: R-SPE
Step 3: Oxidation
Step 4: Hydration
4ω-Nd: YAG (λ = 266 nm)
(0001) α-Al2O3
Co3O4 target
YSZ plate
NaHCO3 powder
1 mol L-1 HNO3 aq.
0.1 mol L-1 NaCl aq.
CoO
Co
Na0.8CoO2
Na
Na0.3CoO2
Na0.3CoO2·1.3H2O
H2O
0.2455 nm
0.5459 nm
0.5604 nm
0.9894 nm
[111]
[112̄]
z = [11̄0]
[0001]
[11̄00]
z = [112̄0]

Figure 1

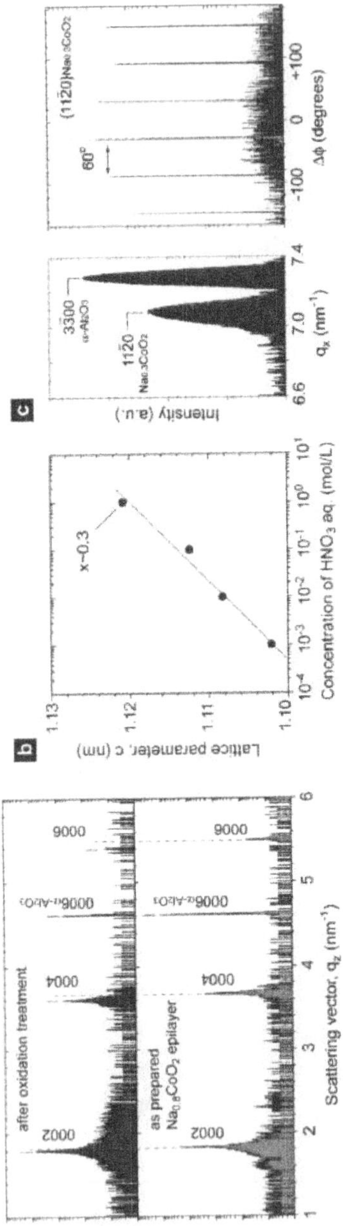

a
Intensity (a.u.)
after oxidation treatment
as prepared
$Na_{0.8}CoO_2$ epilayer
0002
0004
0006
Scattering vector, q_z (nm^{-1})
b
Lattice parameter, c (nm)
x~0.3
Concentration of HNO_3 aq. (mol/L)
c
Intensity (a.u.)
q_x (nm^{-1})
60°
Δφ (degrees)

Figure 2

XRD pattern (Fig. 2c (left)). Six-fold symmetry of {11$\bar{2}$0} $Na_{0.3}CoO_2$ was clearly seen in the in-plane rocking curve (Fig. 2c (right)). These results demonstrate that the $Na_{0.8}CoO_2$ epitaxial film was converted to the $Na_{0.3}CoO_2$ epitaxial film by the 1 mol L^{-1} HNO_3-treatment.

Then the $Na_{0.3}CoO_2$ epitaxial film was immersed in NaCl aqueous solution for 5 h at 5°C. Figure 3a shows out-of-plane XRD pattern of the resultant film which was taken at 5 °C. Intense Bragg diffraction peaks of 000*l* $Na_{0.3}CoO_2 \cdot 1.3H_2O$ are resolved at q_z = 1, 2, 3, 4, 5, and 6 nm^{-1} together with 0006 α-Al_2O_3. The in-plane XRD pattern (Fig. 3b) reveals that the $Na_{0.3}CoO_2 \cdot 1.3H_2O$ film was kept an epitaxial relation of (0001) [11$\bar{2}$0] $Na_{0.3}CoO_2 \cdot 1.3H_2O$ || (0001) [1$\bar{1}$00] α-Al_2O_3. These results demonstrate that the $Na_{0.3}CoO_2$ epitaxial film was successfully converted to an $Na_{0.3}CoO_2 \cdot 1.3H_2O$ epitaxial film by the hydration treatment.

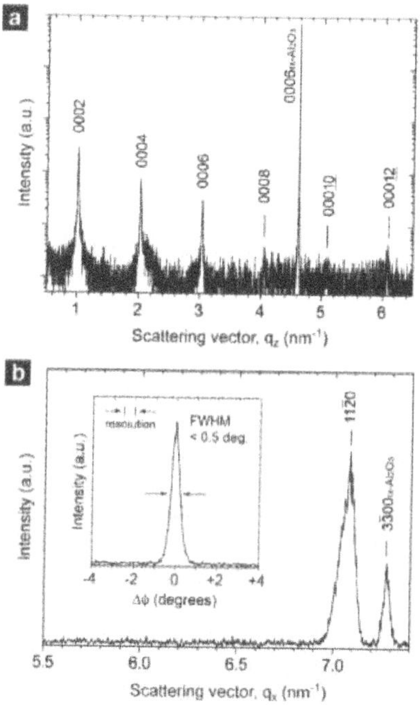

Figure 3

Figure 4 shows temperature dependence of the electrical resistivity of 275-nm-thick $Na_{0.3}CoO_2 \cdot 1.3H_2O$ epitaxial film, which was fabricated by the oxidation - hydration of $Na_{0.8}CoO_2$ epitaxial film. Superconducting transition behavior was observed at *Tc* ~ 4 K, the value of which is similar to that of bulk single crystals. The *Tc* value decreased with increasing magnetic field as shown in the inset. Resistivity of the $Na_{0.3}CoO_2 \cdot 1.3H_2O$ epitaxial film was ~ 500 μΩ·cm at 4 K, which value is comparable to that of bulk single crystal [3,4]. From these results, we concluded that a high quality $Na_{0.3}CoO_2 \cdot 1.3H_2O$ epitaxial film was successfully fabricated by the oxidation - hydration of $Na_{0.8}CoO_2$ epitaxial film.

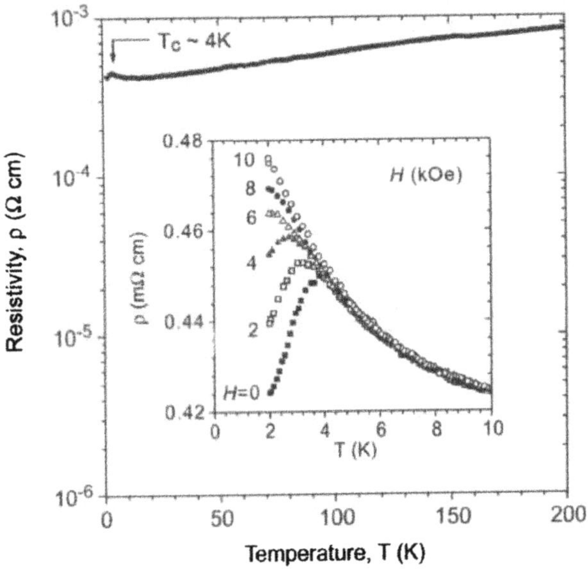

Figure 4

CONCLUSION

We have demonstrated herein that high-quality epitaxial films of superconducting sodium-cobalt oxyhydrate, $Na_{0.3}CoO_2 \cdot 1.3\ H_2O$ were successfully fabricated on a (0001)-oriented α-Al_2O_3 substrate by oxidation – hydration of $Na_{0.8}CoO_2$ the epitaxial film which was grown by reactive solid-phase epitaxy (R-SPE). The epitaxial relation was (0001) $[11\bar{2}0]$ $Na_{0.3}CoO_2 \cdot 1.3H_2O$ || (0001) $[1\bar{1}00]$ α-Al_2O_3. The resultant $Na_{0.3}CoO_2 \cdot 1.3H_2O$ epitaxial film exhibited similar electrical resistivity and superconducting transition behavior to those of bulk single crystals.

The present $Na_{0.3}CoO_2 \cdot 1.3H_2O$ epitaxial film has high enough crystal quality to clarify its intrinsic properties. We hope our achievement should accelerate future clarification of superconducting mechanism of $Na_{0.3}CoO_2 \cdot 1.3H_2O$.

REFERENCES

[1]K. Takada, H. Sakurai, E. Takayama-Muromachi, F. Izumi, R. A. Dilanian, and T. Sasaki, *Nature* 422, 53 (2003).

[2]R. E. Schaak, T. Klimczuk, M. L. Foo, R. J. Cava, *Nature* 424, 527 (2003).

[3]R. Jin, B. C. Sales, P. Khalifah, and D. Mandrus, *Phys. Rev. Lett.* 91, 217001 (2003).

[4]F. C. Chou, J. H. Cho, P. A. Lee, E. T. Abel, K. Matan, and Y. S. Lee, *Phys. Rev. Lett.* 92, 157004 (2004).

[5]M. L. Foo, R. E. Schaak, V. L. Miller, T. Klimczuk, N. S. Rogado, Y. Wang, G. C. Lau, C. Craley, H. W. Zandbergen, N. P. Ong, and R. J. Cava, *Solid State Commun.* 127, 33 (2003).

[6]K. Takada, K. Fukuda, M. Osada, I. Nakai, F. Izumi, R. A. Dilanian, K. Kato, M. Takata, H. Sakurai, E. Takayama-Muromachi, and T. Sasaki, *J. Mater. Chem.* 14, 1448 (2004).

[7]M. L. Foo, Y. Wang, S. Watauchi, H. W. Zandbergen, T. He, R. J. Cava, and N. P. Ong, *Phys. Rev. Lett.* 92, 247001 (2004).

[8]R. J. Balsys, and R. L. Davis, *Solid State Ionics* 93, 279 (1996).

[9]H. Ohta, M. Hirano, K. Nakahara, H. Maruta, T. Tanabe, M. Kamiya, T. Kamiya, and H. Hosono, *Appl. Phys. Lett.* 83, 1029 (2003).

[10]H. Ohta, K. Nomura, M. Orita, M. Hirano, K. Ueda, T. Suzuki, Y. Ikuhara, and H. Hosono, *Adv. Funct. Mater.* 13, 139 (2003).

[11]H. Ohta, S.-W. Kim, S. Ohta, K. Koumoto, M. Hirano, and H.Hosono, *Cryst. Growth Des.* 5, 25 (2005).

[12]H. Ohta, H. Mizoguchi, M. Hirano, S. Narushima, T. Kamiya, and H. Hosono, *Appl. Phys. Lett.* 82, 823 (2003).

Electroceramic Materials and Applications
Edited by R. W. Schwartz

EFFECT OF GRAIN SIZE OF PRECURSOR POWDER ON THERMOELECTRIC PROPERTIES OF TEXTURED $Ca_3Co_4O_9$ COMPOUNDS

M. Mikami, N. Ando, S. Urata, and R. Funahashi
National Institute of Advanced Industrial Science and Technology
2266-98 Anagahora, Shimoshidami, Moriyama,
Nagoya, Aichi 463-8560, Japan.

E. Guilmeau and D. Chateigner
CRISMAT-ENSICAEN Laboratory, UMR CNRS 6508,
6 boulevard Maréchal Juin,
14050 Cean Cedex, France.

ABSTRACT

Partially Bi substituted $[Ca_2CoO_3]_pCoO_2$ ceramics were synthesized by a hot-forging technique and the effect of the Bi-substitution on the microstructure and thermoelectric properties evaluated. The average grain size of the precursor powder prepared by a sold-state reaction increased with the increase of the Bi content, and the electrical resistivity (ρ) of the hot-forged sample was reduced accordingly. Since the Seebeck coefficient (S) was slightly increased by the Bi-substitution and less affected by the grain size, the power factor ($PF=S^2/\rho$) was improved. On the other hand, the ρ value of the hot-forged sample was reduced by using large grained precursor powder prepared by a solution growth method. Unlike in the case of the former samples made from as-sintered powder, the ρ value of the sample made from large grained powder was slightly increased with the increase of the Bi content. These results indicated that the main advantage of the Bi-substitution for the reduction of ρ value of hot-forged sample is the enhancement of the grain growth during the heat treatment processes.

INTRODUCTION

The recent discovery of high thermopower coexisting with low electrical resistivity in layered cobaltites has opened the route to the exploration of thermoelectric materials in oxides [1-5]. Due to their high thermal stability, high tolerance to oxidation, and low toxicity, these oxides are promising candidates for practical use in thermoelectric power generation in air at high temperatures. The conversion efficiency of a thermoelectric material is evaluated by the dimensionless figure-of-merit $ZT=(S^2/\rho\kappa)T$, where S is the Seebeck coefficient, ρ is the electrical resistivity, κ is the thermal conductivity, and T is the absolute temperature. ZT values of single crystals of Na_xCoO_2 and $[Ca_2CoO_3]_pCoO_2$ (CCO) along the ab-plane exceed 1 at 1000 K [2, 4, 5], the estimated criterion for practical application.

Some attempts to improve their thermoelectric efficiency by element partial-substitution have been reported [6-10]. Especially in the CCO system, it has been reported that the Bi-substitution for a part of the Ca site enhances its thermoelectric properties in sintered polycrystalline material [9]. Our recent investigation of the single-crystal CCO of mm size [11] indicated that the Bi-substitution had effect on the stability of the misfit CCO structure consisting of alternate stacking of a rocksalt-type [Ca_2CoO_3] layer and a CdI_2-type [CoO_2] layer [12-14]. More to the point, the Bi-substitution can moderate the misfit relation of these two subsystems and eliminate stacking faults and defects, resulting in the enhancement of electrical conductivity without degradation of *S*. However, it is still unclear that this Bi-substitution effect on the crystallographic structure has an influence on the polycrystalline bulk material, which is indispensable for the practical application as it is difficult to grow single crystals large enough for the construction of a thermoelectric device.

In this study, partially Bi-substituted $[Ca_2CoO_3]_pCoO_2$ polycrystalline materials were synthesized and the effect of the Bi-substitution on the microstructure and thermoelectric properties evaluated. Due to the layered structure of the CCO phase, ρ along the *c*-axis is considerably higher than in the *ab*-plane [12]. Therefore, the samples were fabricated by means of hot forging technique in order to obtain the highly textured CCO bulk material. Since the grain size of the precursor CCO powder has significant implication for the grain alignment and ρ parallel to the pressed plane of the hot-forged sample [15], precursor powders of various sizes were used in order to investigate the effect of the microstructure on thermoelectric properties.

EXPERIMENTAL

The $[(Ca_{1-2x}Bi_{2x})_2(Co_{1-x}Bi_x)O_3]_p[CoO_2]$ (x=0, 0.01, 0.02, 0.03 and 0.04, p=0.62 [16]) powders were prepared by solid-state reaction. Appropriate amounts of Co_3O_4 (99.9 %), $CaCO_3$ (99.5 %) and Bi_2O_3 (99.999 %) powders were mixed with an agate mortar. The mixture was calcined at 1073 K for 10 h in air, pulverized, and pressed into pellets, which were sintered at 1153 K for 20 h in an oxygen gas flow and ground into fine powder. In order to increase the grain size, the prepared powders were heated at 1123 K for 20 h in a solvent consisting of K_2CO_3 and KCl at a molar ratio of 4:1 [17]. The grains were grown at different weight ratios of solvent to CCO powder between 1wt% to 40wt%. The powder was collected from the solidified materials by dissolving the solvent in distilled water in an ultrasonic washing machine. The powders were cold-pressed into 20 thin-disks, which were finally stacked under uniaxial pressure of 190 MPa into a pellet of about 6 mm thickness and 20 mm diameter. The layered pellets were then hot-forged at 1153 K for 20 h in air under uniaxial pressure of 10 MPa. The obtained samples, with a diameter of about 22-25 mm and a thickness of about 3-4 mm, were cut into a bar shape with a typical size of $3\times3\times15$ mm^3 for measurement of thermoelectric properties.

Crystallographic structure and constituent analyses were performed with powder X-ray diffraction (XRD) using Cu $K\alpha$ radiation and an energy dispersive X-ray spectrometer (EDX), respectively. The microstructure was observed with a scanning electron microscope (SEM). The grain size of the CCO powders was measured using a HORIBA particle-size analyzer model LA-920. ρ was measured in air in the range of 300-1100 K using a conventional four-probe dc technique in the direction parallel to the pressed plane. S was calculated from a plot of thermoelectric voltage against temperature difference using an instrument designed by our laboratory. Two Pt-Pt/Rh (R-type) thermocouples were attached to both ends of the sample using silver paste and the Pt wires of the thermocouples used as voltage terminals. Measured S values were corrected by thermopower of the Pt wires to obtain the net S values of the samples.

RESULTS *and* DISCUSSION

The observation of the microstructure of prepared powders was conducted because our previous investigation has demonstrated that the size and the shape of the grains have an important effect on the gain alignment and the magnitude of ρ in the hot-forged CCO sample [15]. SEM images of the typical example of the prepared powders are shown in Fig. 1. Powders annealed in the mixture of K_2CO_3 and KCl consist of large and thick grains (Fig. 1(b) and 1(d)), while as-sintered CCO powders contain fine and thin particles (Fig. 1(a) and 1(c)). Thus, it is obvious that the annealing in the solvent is effective to grow CCO grains. The grains annealed in the solvent tend to form a platelet-shape derived from the layered crystallographic structure of CCO phase. Moreover, it can be seen that the grains of Bi-substituted CCO powder (Fig. 1(d)) are much larger than the non-substituted CCO (Fig. 1(b)) powder. Therefore, the grain size of CCO powder appeared to be significantly enhanced by the Bi-substitution.

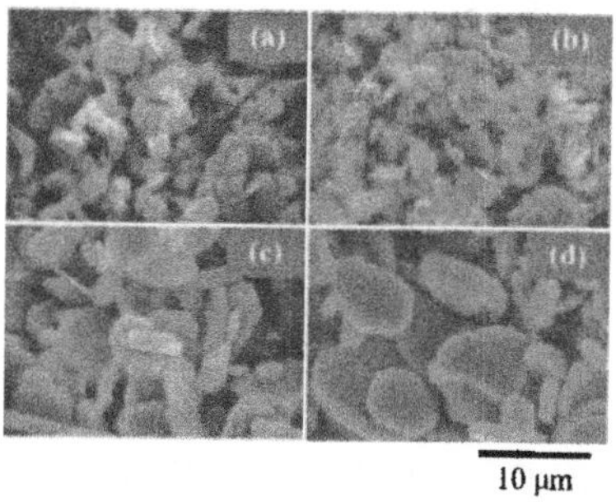

Fig. 1 SEM images of (a) as-sintered non and (b) Bi substituted CCO powder, (c) non and (d) Bi substituted powders heat-treated in a solvent mixture of K_2CO_3 and KCl with 10wt%.

The measurements of average grain size, represented in Fig. 2 and Fig. 3, confirmed this observation. As shown in Fig. 2, the average grain size increases with the increase of the amount of the solvent. This clearly proves the efficiency of solvent addition in improving the grain growth of the CCO phase. Since the solute of CCO was not entirely dissolved in the solvent even at 40wt%, the grains were grown by the recrystallization of the dissolved CCO on the residual solid particle. Therefore, the increase of the grain size is caused by the increase of the amount of the dissolved part of CCO powder. In addition, a small amount of Bi-substitution, about 1 mol% of sum total of Ca and Co cations, has a significant influence on the grain size.

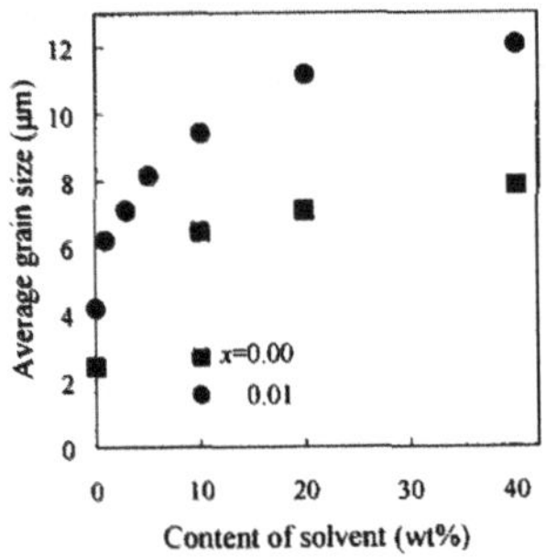

Fig. 2 Relationship between the weight ratio of solvent and average grain size of (■) CCO and (●) $[(Ca_{0.98}Bi_{0.02})_2(Co_{0.99}Bi_{0.01})O_3]_{0.62}[CoO_2]$ powders.

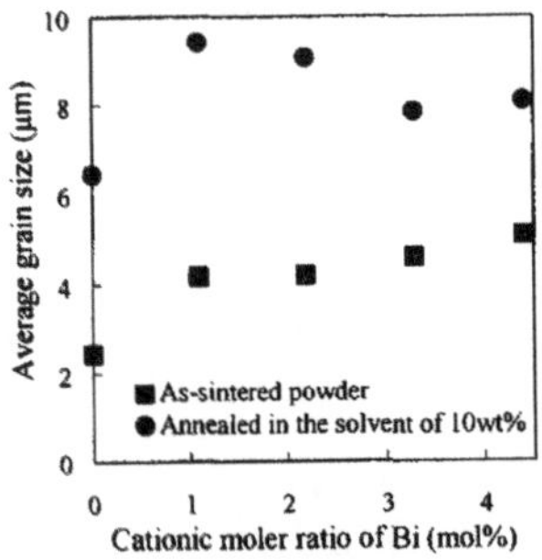

Fig. 3 Relationship between Bi composition and average grain size of $[(Ca_{1-2x}Bi_{2x})_2(Co_{1-x}Bi_x)O_3]_{0.62}[CoO_2]$ (x=0, 0.01, 0.02, 0.03 and 0.04) powders. (■) As-sintered and (●) annealed in a solvent mixture of K_2CO_3 and KCl with 10wt%.

The average grain sizes of CCO powder with the different Bi content were plotted in Fig. 3 in order to clarify the effect of the Bi-substitution on the grain growth. The grain size of as-sintered Bi substituted CCO powder is about twice as large as the non-substituted CCO and is increased with the increase of the Bi content. This result indicates that the Bi-addition can enhance the grain growth at the first solid-state reaction of raw powders. We propose that the liquid phase of Bi_2O_3, which melts at around 1100 K, at the sintering temperature of 1153 K facilitates the chemical reaction and the grain growth. In addition, since it has been reported that the Bi-substitution affects the modulation of misfit structural CCO phase [16], the alteration of crystallographic structure might affect the growth rate. On the other hand, the grain size of the powders annealed in the solvent of 10wt% decreases with the increase of the Bi content, although their gain size is much larger than the non-substituted powder. The increment ratio of grain size decreases with the increase with the Bi content. This result means that the Bi-substitution interferes with the growth of CCO grains in the solvent. Therefore, the Bi-substitution might reduce the solubility and the amount of the dissolved part of CCO.

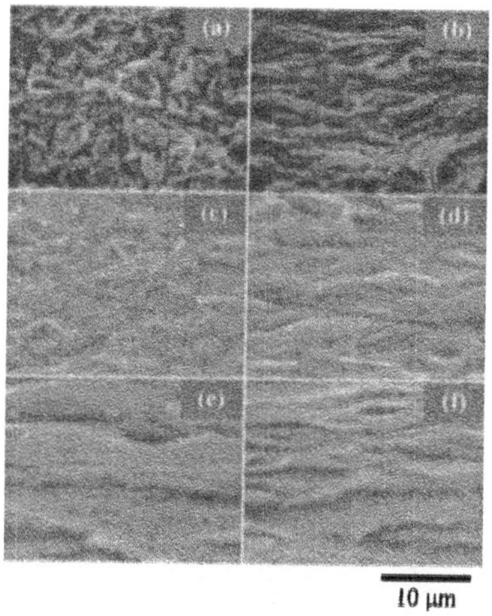

Fig. 4 SEM images of fracture transverse section of samples made from (a) as-sintered and (b) large grained CCO, (c) as-sintered and (d) large grained $[(Ca_{0.98}Bi_{0.02})_2(Co_{0.99}Bi_{0.01})O_3]_{0.62}[CoO_2]$. (e) as-sintered and (f) large grained $[(Ca_{0.92}Bi_{0.08})_2(Co_{0.96}Bi_{0.04})O_3]_{0.62}[CoO_2]$ powder.

XRD patterns of powders were in agreement with data from the literature on CCO phase [12]. There is no secondary phase. The diffraction peaks from the Bi-substituted samples tend to shift toward lower 2θ angle compared to the non-substituted one and the amount of the shift increases with the increase of the Bi content. This tendency indicates that the crystallographic structure is enlarged by the Bi-substitution. This result is consistent with the reported result of the precise structural analysis of the Bi substituted CCO [16]. The EDX analysis performed on the powders heated in the solvent showed the grain growth treatment causes no change of the CCO composition and no incorporation of K^+ or Cl^- ions contained in the solvent. These results indicate that the present grain-growth method has no contamination effects on CCO powders.

SEM images of the typical example of the fracture transverse section of the hot-forged samples are shown in Fig. 4. It can be seen that samples made from large grained powders (Fig. 4(b), 4(d) and 4(f)) are constructed from well-oriented grains with a size of several μm. The alignment of the grains is obviously improved when compared to samples made from as-sintered powders (Fig. 4(a) and 4(c)), where small misalignment grains can be observed. The effect of the grain size of the precursor powder on the grain alignment of hot-forged bulk has been demonstrated by means of the quantitative texture analysis in our previous report [15]. The grain size of the Bi-substituted hot-forged sample is obviously larger than the non-substituted one resulting from the size difference of the precursor powder as mentioned above. Interestingly, the grains in the sample made from as-sintered powders with a nominal composition of $[(Ca_{0.92}Bi_{0.08})_2(Co_{0.96}Bi_{0.04})O_3]_{0.62}[CoO_2]$ (Fig. 4(e)) seem to be well connected and aligned when compared to the non-substituted sample, although we could not evaluate the connectivity of the grains from a quantitative point of view. This change of the microstructure by the Bi-substitution might affect the magnitude of ρ described below.

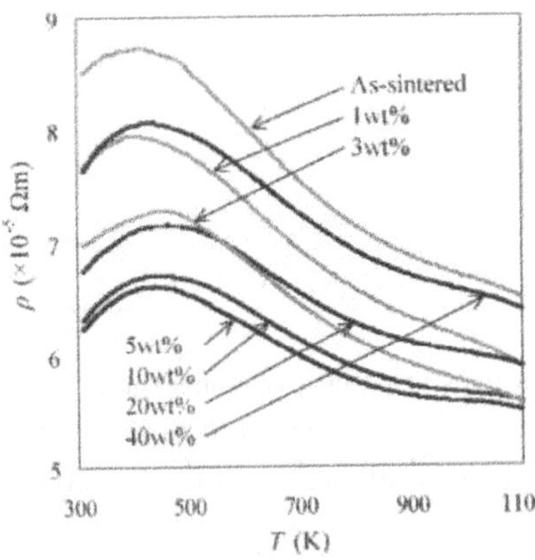

Fig. 5 T-dependence of ρ along the pressed plane of $[(Ca_{0.98}Bi_{0.02})_2(Co_{0.99}Bi_{0.01})O_3]_{0.62}[CoO_2]$ sample made from powders annealed in a mixture of K_2CO_3 and KCl with different weight ratio.

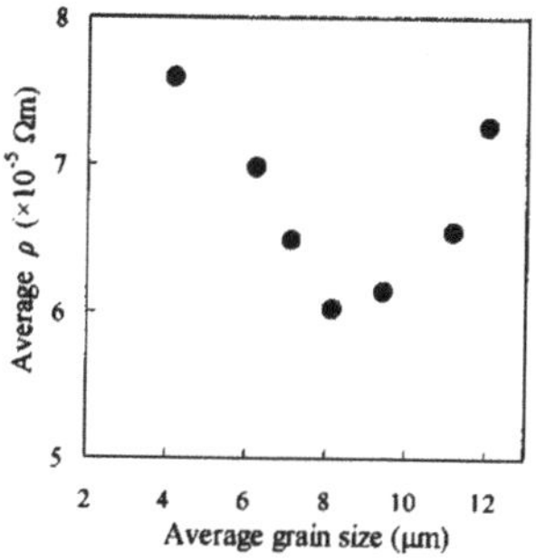

Fig. 6 Relationship between average ρ (from 373 K to 1073 K) and the average grain size of precursor powder of $[(Ca_{0.98}Bi_{0.02})_2(Co_{0.99}Bi_{0.01})O_3]_{0.62}[CoO_2]$.

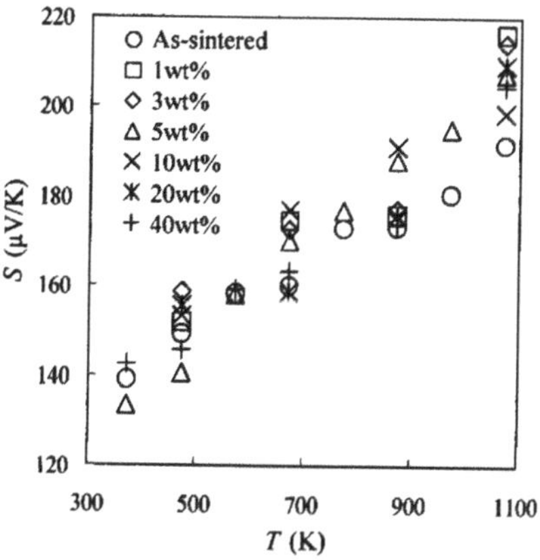

Fig. 7 T-dependence of S of $[(Ca_{0.98}Bi_{0.02})_2(Co_{0.99}Bi_{0.01})O_3]_{0.62}[CoO_2]$ sample made from powders annealed in a mixture of K_2CO_3 and KCl with different weight ratio.

In order to revalidate the influence of the grain size of precursor powder on the ρ value of the hot-forged sample [15], Bi-substituted samples were fabricated with the several powders of different grain sizes. The T-dependence of ρ along the direction parallel to the pressed plane of the samples with the nominal composition of $[(Ca_{0.98}Bi_{0.02})_2(Co_{0.99}Bi_{0.01})O_3]_{0.62}[CoO_2]$ is shown in Fig. 5. These samples were made from powders of different grain sizes as illustrated in Fig. 2. The inclination of ρ-T curve differs slightly between the sample made from powders heated in the solvent of 0wt% (as-sintered) to 3wt% and that in 5wt% to 40wt%, although the crystallographic structure and composition of these samples are almost the same evaluating by XRD and EDX. We could not elucidate the origin of this difference with our present data. The relationship between the average grain size of the precursor powder and the average ρ value (from 373 K to 1073 K) is shown in Fig. 6. The clear correlation between the grain size of precursor powder and ρ value is observed. The ρ value reaches minimum at the grain size of 8 μm. The similar grain size dependence of ρ value was observed in non-substituted CCO and we have demonstrated that the origin of the reduction of ρ is derived from the improvement of the grain alignment [15]. We suppose that the same effect is obtained in Bi-substituted CCO sample by analogy. As shown in Fig. 7, S was less affected by the grain size, although there is some variation. The clear correlation between the grain size of precursor powder and S value is not observed. Thus, the power factor ($PF=S^2/\rho$) was improved by using precursor powder of around 8-9 μm.

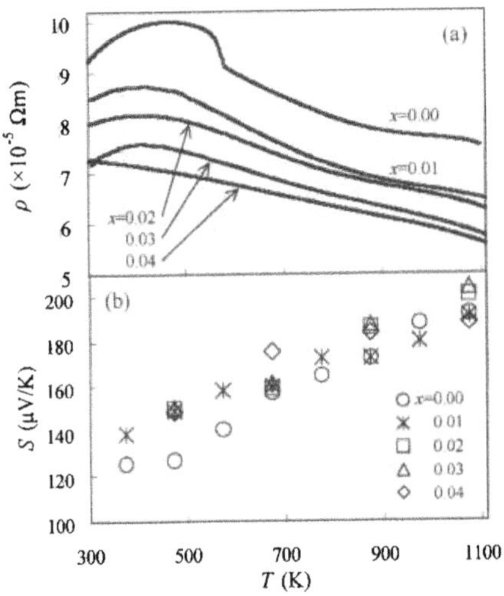

Fig. 8 T-dependence of (a) ρ and (b) S of hot-forged $[(Ca_{1-2x}Bi_{2x})_2(Co_{1-x}Bi_x)O_3]_{0.62}[CoO_2]$ (x=0, 0.01, 0.02, 0.03 and 0.04) sample made from as-sintered powders.

In order to examine the Bi-substitution effect, hot-forged samples with the different Bi content were synthesized and evaluated. The T-dependence of ρ along the direction parallel to the pressed plane of the samples made from as-sintered powder with the different Bi content is shown in Fig. 8(a). Above 600 K, ρ value in every sample decreases with the increase in T. On the other hand, a broad maximum around 450 K and a steep decrease around 570 K is clearly observed in non-substituted CCO sample. However, in the Bi-substituted samples, the sharp decline at 570 K is not observed and the hump around 450 K is reduced as increase the Bi content. Although we could not elucidate the origin of these changes of the *T*-dependence of ρ within our present data, the Bi-substitution might have a significant influence on the electronic state of CCO phase. The magnitude of ρ is decreased with the increase in the Bi composition. The reduction of ρ is caused by the decrease of the number of grain boundaries derived from the increase of grain size as illustrated in Fig. 3 and in Fig. 4. We propose that the improvement of connectivity and alignment of grains also has influence on the ρ value as illustrated in Fig. 4. Since the Bi-substitution slightly increases the magnitude of *S* especially below 600 K (Fig. 8(b)), the *PF* was improved.

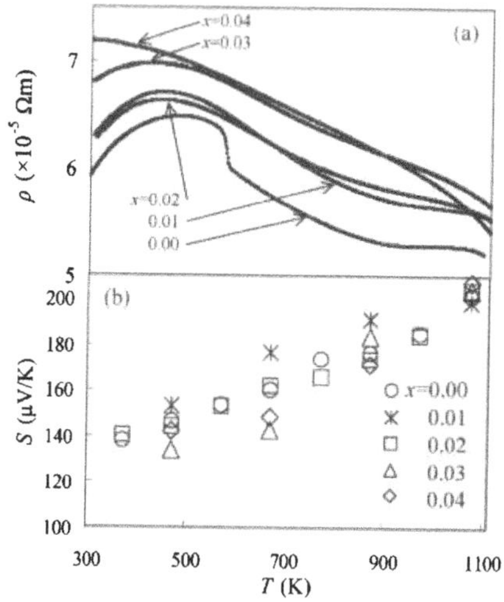

Fig. 9 *T*-dependence of (a) ρ and (b) *S* of hot-forged $[(Ca_{1-2x}Bi_{2x})_2(Co_{1-x}Bi_x)O_3]_{0.62}[CoO_2]$ (x=0, 0.01, 0.02, 0.03 and 0.04) samples made from powders annealed in the solvent mixture of K_2CO_3 and KCl with 10wt%.

As shown in Fig. 9(a), the effect of the Bi-substitution on the magnitude of ρ is completely changed when the large grained powders were used as a precursor. The ρ value is increased by the Bi-substitution despite the grain size is increased as mentioned above. This result indicates that the Bi-substitution intrinsically increases the ρ value of CCO phase. Since it has been suggested that

Bi^{3+} and Bi^{5+} substitute Ca^{2+} and Co^{3+} [16], the electrons would be injected. Thus, the Bi-substitution might reduce hole carrier concentration in the p-type CCO phase, resulting in the increase in ρ. However, a clear dependence of S on the Bi content was not observed as shown in Fig. 9(b). Additional measurements are required to elucidate the Bi-substitution effect on the electronic structure of CCO phase.

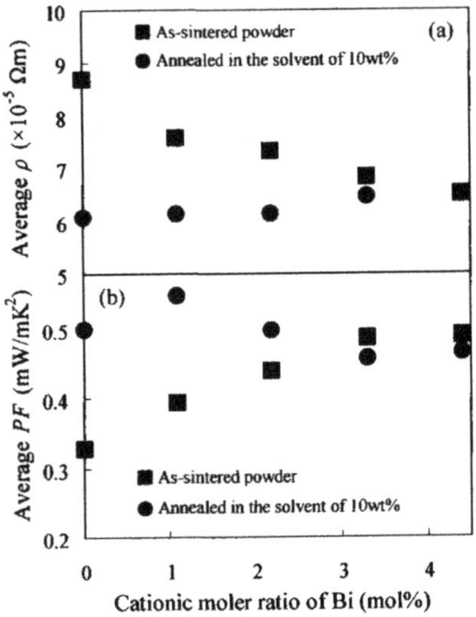

Fig. 10 The Bi-composition dependence of average ρ (from 373 K to 1073 K) and average PF (from 473 K to 1073 K). Hot-forged samples were made from (■) as-sintered and (●) large grained powders.

The relationship between the Bi content and the average ρ value (from 373 K to 1073 K) is shown in Fig. 10(a). This clearly shows the opposite influences of the Bi-substitution on the ρ value as mentioned above. Interestingly, ρ values of the sample made from the two kinds of precursor powder are almost the same at the nominal composition of $[(Ca_{0.92}Bi_{0.08})_2(Co_{0.96}Bi_{0.04})O_3]_{0.62}[CoO_2]$. This result indicates that the increase in grain size of precursor powder has little effect on the magnitude of ρ in the hot-forged sample. In other words, the grain size was sufficiently increased to fabricate the well-textured bulk at the first solid-state reaction of raw powders. This means that the Bi-substitution has almost the same contribution to the reduction of ρ as the increase in the grain size of the precursor powder, which has significant influence on the ρ value in the non-substituted CCO phase. Since the Bi-substitution seems to intrinsically increase the ρ value of CCO phase as mentioned above, the reduction of ρ by the Bi-substitution in the sample made from as-sintered powder as illustrated in Fig. 8 might be caused by the change of the microstructure of hot-forged bulk, such as the reduction of the number of grain boundaries and the improvement of grain alignment. Since S value is less affected by the change of the microstructure, the Bi-substitution can contribute to the improvement of PF. The average PF value (from 473 K to 1073 K) was summarized in Fig. 10(b). This indicates that the PF of non-substituted CCO bulk can be enhanced by two kinds of methods. One is the increase of the grain size of the precursor powder. Second is the Bi-substitution enough to enhance the chemical reaction and the grain growth, which is about 4.5 mol% of sum total of Ca and Co cations in this study. Consequently, the development of microstructure is significantly effective to improve the PF of CCO bulk material.

CONCLUSION

The effect of the Bi-substitution on the electrical properties of hot-forged CCO bulk material was evaluated from the viewpoint of microstructure. The reduction of grain boundaries and the enhancement of grain alignment were achieved by using large grained powder as a precursor in the case of CCO and a small amount of Bi substituted CCO phase. The optimum grain size for the reduction of ρ was around 7-8 μm. On the other hand, the sufficient amount of Bi-substitution, which was 4.5 mol% of sum total of Ca and Co cations in this study, facilitates the grain growth during the sintering processes. The Bi-substitution effect on the microstructure has significant influence on the reduction of ρ, resulting in the improvement of PF in the hot-forged sample.

REFERENCES

[1]I. Terasaki, Y. Sasago, and K. Uchinokura, "Large thermoelectric power in $NaCo_2O_4$ single crystals," *Phys. Rev. B*, **56**, 12685-87 (1997).

[2]K. Fujita, T. Mochida, and K. Nakamura, "High-temperature thermoelectric properties of $Na_xCoO_{2-\delta}$ single crystals," *Jpn. J. Appl. Phys.*, **40**, 4644-47 (2001).

[3]S. Li, R. Funahashi, I. Matsubara, K. Ueno, and H. Yamada "High temperature thermoelectric properties of oxide $Ca_9Co_{12}O_{28}$," *J. Mater. Chem.*, **9**, 1659-60 (1999).

[4]R. Funahashi, I. Matsubara, H. Ikuta, T. Takeuchi, U. Mizutani, and S. Sodeoka, "An oxide single crystal with high thermoelectric performance in air," *Jpn. J. Appl. Phys.*, **39**, L1127-29 (2000).

[5]M. Shikano and R. Funahashi, "Electrical and thermal properties of single-crystalline $[Ca_2CoO_3]_{0.7}CoO_2$ with a $Ca_3Co_4O_9$ structure," *Appl. Phys. Lett.*, **82**, 1851-53 (2003).

[6]K. Takahata, Y. Iguchi, D. Tanaka, T. Itoh, and I. Terasaki, "Low thermal conductivity of

the layered oxide $(Na,Ca)Co_2O_4$: Another example of a phonon glass and an electron crystal," *Phys. Rev. B*, **61**, 12551-55 (2000).

[7]T. Kawata, Y. Iguchi, T. Itoh, K. Takahata, and I. Terasaki, "Na-site substitution effects on the thermoelectric properties of $NaCo_2O_4$," *Phys. Rev. B*, **60**, 10584-87 (1999).

[8]I. Terasaki, I. Tsukada, and Y. Iguchi, "Impurity-induced transition and impurity-enhanced thermopower in the thermoelectric oxide $NaCo_{2-x}Cu_xO_4$," *Phys. Rev. B*, 65, 195106, (2002).

[9]S. Li, R. Funahashi, I. Matsubara, K. Ueno, S. Sodeoka, and H. Yamada, "Synthesis and thermoelectric properties of the new oxide materials $Ca_{3-x}Bi_xCo_4O_{9+\delta}$ ($0.0 < x < 0.75$)," *Chem. Mater.*, **12**, 2424-27 (2000).

[10]G. Xu, R. Funahashi, M. Shikano, I. Matsubara, and Y. Zhou, "Thermoelectric properties of the Bi- and Na- substituted $Ca_3Co_4O_9$ system," *Appl. Phys. Lett.*, **80**, 3760-62 (2002).

[11]M. Mikami, K. Chong, Y. Miyazaki, T. Kajitani, T. Inoue, S. Sodeoka, and R. Funahashi: in preparation for publication.

[12]A. C. Masset, C. Michel, A. Maignan, M. Hervieu, O. Toulemonde, F. Studer, and B. Raveau, "Misfit-layered cobaltite with an anisotropic giant magnetoresistance: $Ca_3Co_4O_9$," *Phys. Rev. B*, **62**, 166-75 (2000).

[13]S. Lambert, H. Leligny, and D. Grebille, "Three forms of the misfit layered cobaltite $[Ca_2CoO_3][CoO_2]_{1.62}$ - A 4D structural investigation," *J. Solid Chem.*, **160**, 322-31 (2001).

[14]Y. Miyazaki, M. Onoda, T. Oku, M. Kikuchi, Y. Ishii, Y. Ono, Y. Morii, and T. Kajitani, "Modulated structure of the thermoelectric compound $[Ca_2CoO_3]_{0.62}CoO_2$," *J. Phys. Soc. Jpn.*, **71**, 491-97 (2002).

[15]M. Mikami, E. Guilmeau, R. Funahashi, K. Chong, and D. Chateigner, "Enhancement of electrical properties of the thermoelectric compound $Ca_3Co_4O_9$ through use of large-grained powder," *J. Mater. Res.* **20**, 2491-97 (2005).

[16]Y. Miyazaki, Y. Suzuki, M. Onoda, Y. Ishii, Y. Morii, and T. Kajitani, "Modulated Structure of Misfit Layered Cobalt Oxide $[(Ca_{0.90}Bi_{0.10})_2(Co_{0.95}Bi_{0.05})O_3]_pCoO_2$," *Jpn. J. Appl. Phsy.* **43**, 6252-58 (2004).

[17]M. Mikami, S. Ohtsuka, M. Yoshimura, Y. Mori, T. Sasaki, R. Funahashi, and M. Shikano, "Effects of KCl addition on the K_2CO_3 flux growth of $Ca_3Co_4O_9$ crystals for a thermoelectric device," *Jpn. J. Appl. Phys.* **42**, 3549-51 (2003).

Electroceramic Materials and Applications
Edited by R. W. Schwartz

INVESTIGATION OF THERMOELECTRIC PROPERTIES OF METALLIC $NA_XCO_2O_4$ MATERIALS

Xiaofeng Tang
Department of Material Science & Engineering, Clemson University, Clemson, SC, 29634, USA

Ed Abbott, J. K. Kolis
Department of Chemistry, Clemson University, Clemson, SC, 29634, USA

Kelvin Aaron, Terry M. Tritt
Department of Physics and Astronomy, Clemson University, Clemson, SC, 29634, USA

In this paper, the transport properties, such as the resistivity, thermopower and thermal conductivity of Na-variant polycrystalline $Na_xCo_2O_4$ (x=1.0, 1.2, 1.44, 1.5) were systematically investigated. With the increase of Na concentration, the thermopower was found to monotonously increase from 75 μV/K to 104 μV/K, but the resistivity and thermal conductivity are not noticeably affected, indicating the carrier concentration n does not strongly depend on Na concentration and phonons are not markedly influenced by the increase of Na occupancy in the disordered Na layer.

The thermopower, resistivity and thermal conductivity of $NaCo_2O_4$ single crystals which were grown by typical NaCl flux method were also measured by our custom designed systems. Thermal conductivity of single crystals was measured by our custom-designed system "PTC" and is found to be ≈5W/m-K. A novel low temperature flux method where NaCl/NaOH was used as flux and metallic Co powders as Co source was developed to successfully synthesize Na-deficient $Na_xCo_2O_4$ crystals with size up to 5mm at low temperature of 550°C. The different behavior in resistivity of the crystals grown via low temperature flux method revealed there exist two different types of crystals, one is metallic and another is semiconducting.

I. INTRODUCTION

Thermoelectric (TE) materials can convert heat into electricity through Seebeck effect; also electricity can be converted into thermal energy via Peltier effect. Thermoelectric devices exhibit high stability and reliability, quietness (without moving parts) and small-scale localized cooling/heating[1]. The performance of thermoelectric materials, which is proportional to device efficiency, can be evaluated by the dimensionless figure-of-merit (FOM) which is defined as $ZT = \alpha^2 T / \rho(\kappa_L + \kappa_e)$, here α is the Seebeck coefficient, also called thermopower, ρ is the resistivity, κ_L and κ_e is lattice and electronic thermal conductivity, respectively. It's difficult to increase the efficiency of device since ZT is the combination of transport properties of materials and the three transport parameters α, ρ, κ are correlated.

Traditional TE theory considers oxides possessing low mobility as an ineffective TE material. However, in 1997 Terasaki[2] reported a new transition-metal oxide, $NaCo_2O_4$ with unexpected favorable TE properties. The $NaCo_2O_4$ single crystal was found to be potential as a good p-type TE material with low metallic in-plane resistivity ρ≈0.2 mΩ-cm, large thermopower α≈100 μV/K, although its mobility is pretty low, μ≈13 cm^2/V-s. The power factor ($\alpha^2 T/\rho$) value of this material was found to be ≈1.5 W/m-K, which is even larger than that of Bi_2Te_3 ≈1.2 W/m-K at 300K.

The $Na_xCo_2O_4$ oxide material has a hexagonal layered structure with edge-sharing 2D triangle CoO_2 sheets and Na layers alternately stacked along c-axis. The CoO_2 sheet is conducting and strongly electron correlated which appears to be the source of the corresponding small metallic electrical resistivity and large thermopower values. Wang et al. proposed the enhanced thermopower for this metallic oxide is attributed to the spin entropy in this correlated electron system, which is asserted by their experimental results, specifically the magnetic field suppression of in-plane thermopower. The intercalating Na layer, acting as the charge reservoir and donating electrons distributed among Co ions is insulating and highly disordered like an amorphous solid. Point defect scattering due to the partially and randomly occupancy of sodium atoms in oxygen environment leads to unexpected low thermal conductivity $\kappa \approx 2W/m\text{-}K$ for polycrystalline $Na_xCo_2O_4$. Fujita[4] reported that ZT of $Na_xCo_2O_4$ single crystal could reach 1.2 at 800K, comparable to traditional TE materials. This high ZT value suggests this oxide possessing high thermal/chemical stability and non-toxicity a promising p-type TE material for possible power generation application at high temperature. Recently, Takada[5] reported their discovery of surprising superconducting behavior under 5K in the hydrated form $Na_xCo_2O_4 \cdot yH_2O$, which may give insight into the mechanisms of superconductivity in the cuprates.

II. EXPERIMENTAL DETAILS

In order to precisely control the composition, a "rapid heat-up" technique[6] was applied to obtain polycrystalline samples with variant Na content. The starting materials, Na_2CO_3 and Co_3O_4 powders were mixed and ground, then calcined in the preheated furnace at 750°C for 12 hours. After being finely ground, the clacined powders were passed through 80 or 200 mesh sieves, then cold pressed into round pellets, and then were sintered as stacked pieces in air at 900-950°C for 12 hours. Single crystals $Na_xCo_2O_4$ were synthesized by NaCl flux method[2]. The mixture of starting powders, Na_2CO_3, Co_3O_4 and NaCl with the molar ratio of 1.2:2.0:11.0 were fired at 950°C and then cooled down to 850°C with very slow cool rate 0.3~0.6 °C/hr. In the novel low temperature NaCl/NaOH flux method, Ag tubes were employed to store the starting materials as the reaction container. Metallic Co powder instead of Co_3O_4 was used as Co source. The starting materials were fired to 550°C, and then were slowly cooled down to 420°C with cooling rate of 0.3-5.0 °C/hr[7].

The transport properties, including themopower, resistivity and thermal conductivity were measured using our custom-designed systems. Dupont silver paint 4922N and 4929N were applied for the electric/thermal contact. The details about the measuring systems were described in elsewhere[8,9]. Heat capacity measurements were performed using a Quantum-Design PPMS. SEM Hitachi-3500N with EDX was applied to investigate the microstructure and composition. X-ray diffraction with Cu K_α was used to study the crystalline phases. Measurements of the magnetic susceptibility were made by SQUID system with field 0.5T.

III. RESULTS AND DISCUSSIONS

A. Effect of Na content on the TE properties for Na-variant polycrystalline $Na_xCo_2O_4$

In $Na_xCo_2O_4$, ^{59}Co NMR studies[10] revealed there exist two distinct Co ions, Co^{4+} and Co^{3+}. The variety of Na concentration will change the Na occupancy in the intercalating Na layer and the ratio of Co^{4+}/Co^{3+} ions in the conducting CoO_2 layer; as a result, it may affect the TE properties. Because of the volatility nature of sodium at high temperature, it's a challenge to precisely control the composition through the tradition solid state reaction where excess sodium was typically added in the starting materials to compensate the Na loss. Based on DSC-TGA data, since there is no weight loss after 750°C and the $Na_xCo_2O_4$ phase is stable once it's formed, "rapid heat-up" techniques[6] were employed to prepare the precisely composition-controlled polycrystalline $Na_xCo_2O_4$ samples with four variant Na contents x=1.0, 1.2, 1.44 and 1.5 in our lab.

As shown in Fig. 1, all the four XRD powder patterns of Na-various $Na_xCo_2O_4$ samples resemble and the peaks are indexed as a single phase P2 (γ)-$NaCo_2O_4$., although four distinct phases depending on the Na content were reported[11], α phase ($1.8 \le x \le 2$), γ phase ($x<1$, $y \le 1$, $0.55 \le x/y \le 0.74$ for formula $Na_xCo_yO_2$), β phase ($1.1 \le x \le 1.2$) and α' phase (x=1.5) with different oxygen stacking sequences. For sample x=1.0, there are some traces of unreacted Co_3O_4, suggesting a small amount of Na was lost during the firing. As shown in table 1, the rough Na contents obtained by EDX are close to the nominal composition, which demonstrate, to some extent, this sample preparation technique is effective to prevent the Na evorapation.

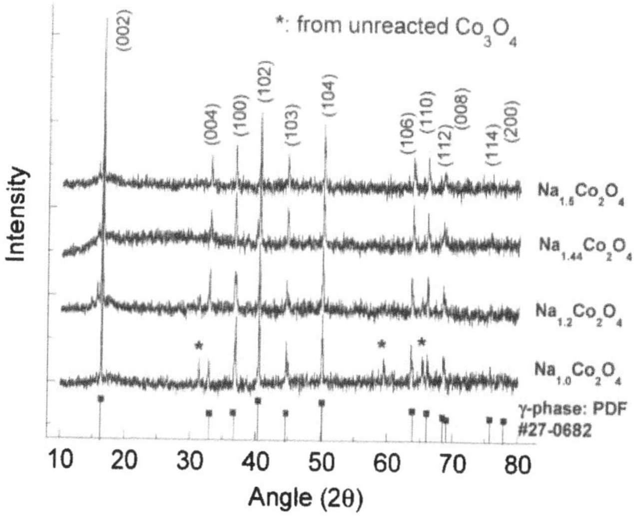

Fig. 1: Plot of X-Ray diffraction patterns for polycrystalline $Na_xCo_2O_4$.

Table 1: Polycrystalline $Na_xCo_2O_4$ samples made by "rapid heat-up" technique

Nominal composition $Na_xCo_2O_4$	Na content (by EDX)	Ratio of Co^{4+}/Co^{3+} ions	High T limit of α (μV/K)	Relative density (%)	Na occupancy
x = 1.0	0.96	1:1	154	76.1	50%
x = 1.2	1.13	2:3	189	74.3	60%
x = 1.44	1.45	7:18	235	76.4	72%
x = 1.5	1.57	1:3	249	77.8	75%

The temperature dependence of transport properties for Na-variant $Na_xCo_2O_4$ samples is displayed in Fig. 2. It is found that the thermopower monotonically increases with the increase of Na content, and for sample x=1.5 it reaches 104 μV/K. Although the obtained value is not as high as the reported one 120 μV/K in Ref. 6, the curve is smoother. Unlike thermopower, the resistivity is not noticeably affected by the Na content, which means the conducting paths are not evidently influenced. The $Na_{1.0}Co_2O_4$ sample has relatively higher resistivity probably due to the existence of unreacted Co_3O_4 indicated by XRD patterns and EDX. The other three samples have nearly the same resistivity magnitude within the experimental error, which is primarily from the measurements of sample dimensions. The elementary charge-transport process is believed to be the hopping of a hole from Co^{4+} ion to Co^{3+} ion. Apparently, the decrease of ratio of Co^{4+}/Co^{3+} as increase of x does not significantly affect the conducting path of hole hopping between the two types of Co ions.

As a 2D Fermi-liquid system, in $Na_xCo_2O_4$ the diffusive thermopower is believed to be proportional to m^*/n[12], where m^* is the effective mass of conducting carriers and n is the carrier concentration. Therefore the increase of thermopower with increasing Na concentration implies the increase of m^*, since n is almost unchanged which is indicated from the Na content independence of resistivity. Actually Terasaki estimated[13] the effective mass (m^*) in this oxide and found it is two orders of magnitude higher than the m^* of copper-based high temperature superconductors.

However, in the strongly correlated electron system $Na_xCo_2O_4$ where spin is not fixed but electrons are localized at the atomic sites, diffusive part α from the kinetic entropy transported by charge carriers does not produce large contribution to the total thermopower, instead the spin entropy is the dominant factor. When the hole hops from Co^{4+} (s=1/2) to Co^{3+} (s=0) in the charge transport processing, a spin -1/2 along with the hole carrying charge +1e is also transferred, which suggests a transfer of spin entropy $s = k_B \ln 2$. The moving spins carrying energy, call spin entropy are considered the likely source of the enhanced thermopower, which is asserted by the observed strong field dependence of in-plane α.[3] Meanwhile, based on the equation developed by Koshibae[14], the high temperature limit of thermopower α is found to monotonously increases with Na content due to the decrease of ratio of Co^{4+}/Co^{3+}, as shown in table 1.

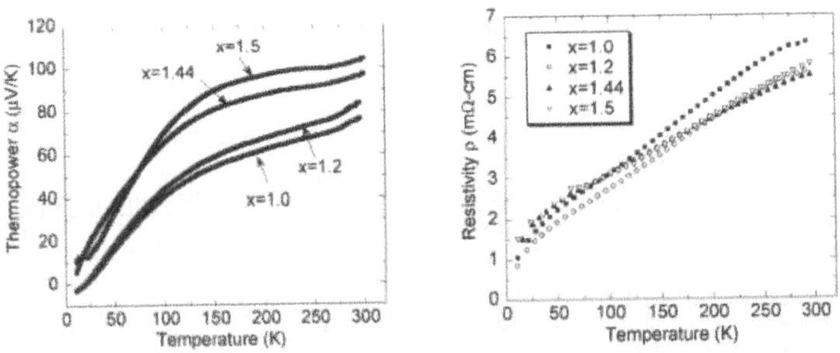

(a) Thermopower of $Na_xCo_2O_4$ (b) Resistivity of $Na_xCo_2O_4$

Fig. 2: Thermopower and resistivity of polycrystalline $Na_xCo_2O_4$ as function of temperature.

The temperature dependence of thermal conductivity κ for Na-variant $Na_xCo_2O_4$ is shown in Fig. 3(a). Though Na occupancy in partially occupied Na layer increase from 50% to 75% with increasing Na content from 1.0 to 1.5, κ is not remarkably changed, except for the sample $Na_{1.5}Co_2O_4$. Since the amorphous and insulating Na layer is believed to be the dominant factor scattering the phonons to reduce κ, and the randomness of Na layer is reduced due to the increase of Na occupancy from 50% to 75%, thermal conductivity is expected to increase when x increase. The fact that the measured κ does not clearly increase with Na content may suggest except the point defect scattering mechanisms due to partially occupied Na atoms, other phonon scattering mechanisms, like grain boundaries and impurities also could play important roles. The average grain size of $Na_{1.5}Co_2O_4$ sample is found to be relatively larger than x=1.0, 1.2, 1.44 samples, which implies less phonon-grain boundary scattering due to larger grains along with less randomness due to higher Na occupancy could be the reason behind the higher thermal conductivity of $Na_{1.5}Co_2O_4$.

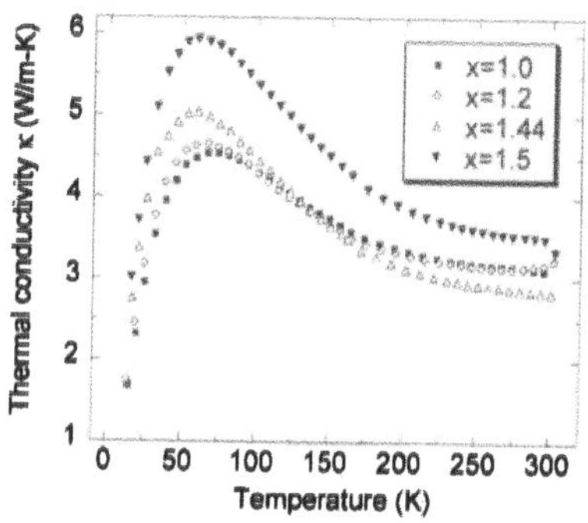

Fig. 3: Thermal conductivity of polycrystalline $Na_xCo_2O_4$ as function of temperature

B. Properties of $NaCo_2O_4$ single crystal synthesized by typical NaCl flux method

The typical size of single crystals that were synthesized by NaCl flux method with temperature up to 950°C is around 0.5-2 mm x 0.5-2 mm x 10-50 μm. The obtained hexagonal plate-like crystal, to some extent, reflects the hexagonal layered structure of this oxide. XRD powder pattern, as shown in Fig. 4, indicates the crystals are inclined toward c-axis and only have one single phase γ-$NaCo_2O_4$. The composition of Na in the crystals was found to be 1.0-1.2 via EDX. As displayed in Fig. 5(a), the thermopower α of single crystal was measured to be ≈100 μV/K and in-plane resistivity (with metallic behavior) ρ≈0.30 mΩ-cm. The linear temperature dependence of the in-plane resistivity at low temperature indicates the existence of electron correlation in this system. The value of power factor α^2T/ρ ~1.0W/m-K of single crystal at 300K is comparable to state-of-art thermoelectric Bi_2Te_3, almost 10 times larger than the polycrystalline samples due primarily to the high anisotropy in resistivity ($\rho_c >> \rho_a$) derived from the layered structure.

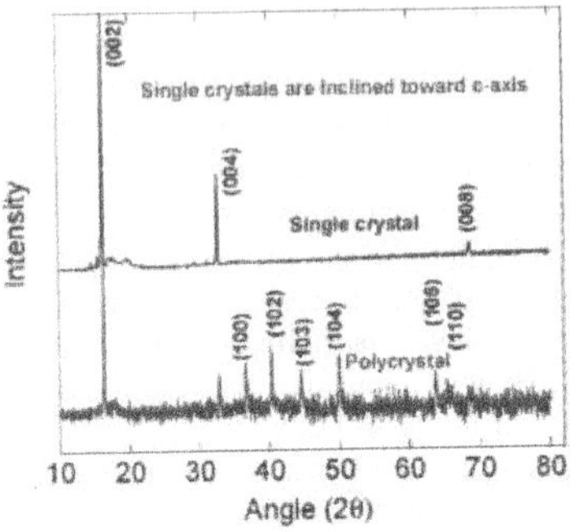

Fig. 4: XRD pattern of plate-like $NaCo_2O_4$ crystals

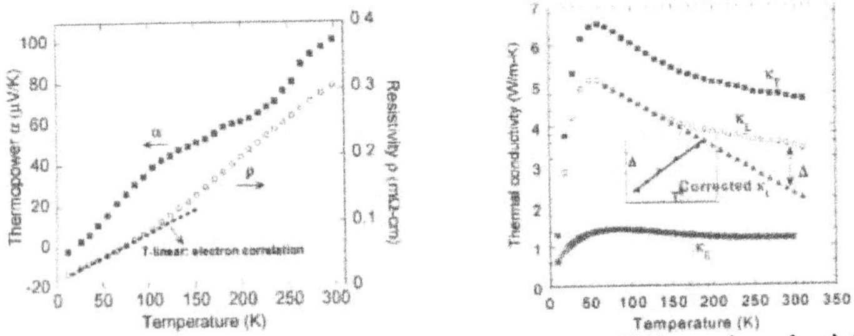

(a) Thermopower and in-plane resistivity (b) thermal conductivity

Fig. 5: Thermopower α and in-plane resistivity ρ of single crystals $NaCo_2O_4$.

The thermal conductivity κ measurement of the oxide crystals is always a challenge since the crystals are so tiny and brittle. The typical dimension of single crystals under investigation is (1-2) x (0.2-0.5) x (0.01-0.05) mm^3. In our lab, the "PTC" (parallel thermal conductance) system[14,15] was developed to perform the absolute steady state thermal conductivity measurement of single crystals. As presented in Fig. 5(b), the in-plane thermal conductivity was measured to be ≈5.0 W/m-K at 300K. The linear relation of $\Delta \sim T^3$ (inset of Fig. 5b) indicates the difference Δ between the measured and extrapolated lattice thermal conductivity is most likely due to the radiation loss.

The in-plane κ is close to the value reported by Satake et al.[16] using the Harman method where κ was obtained from the temperature difference induced by an applied current along with the thermopower and resistivity values. The in-plane κ of single crystals is larger than polycrystalline sample ≈2W/m-K, which can be understood since phonons are strongly scattered at the grain boundaries and along the c-axis. However, our measured value of thermal conductivity is much smaller than the datum (~19.0 W/m-K) reported by Fujita et al.[4], where in-plane thermal conductivity of $Na_xCo_2O_4$ crystals was calculated via measuring the in-plane thermal diffusivity by ac calorimetric method associated with optical pulse heating, the density by Archimedes method and the heat capacity by differential scanning calorimeter (DSC). Using our measured lattice thermal conductivity κ_L and calculated Debye temperature θ_D, we can the in-plane phonon mean free path l_{ph} via the equ. $\kappa_L = \frac{1}{3} c_v v l_{ph}$, where C_v is the volume heat capacity and v is the sound velocity. The l_{ph} was found to be ≈9.0Å, which is the same order of magnitude of the lattice constant (a=2.8 Å, c=10.8 Å) and is much smaller than the in-plane mean free path of conducting carriers l_a≈51Å at 300K[2].

C. Na-deficient $Na_xCo_2O_4$ single crystal grown by novel low temperature flux method

Because of the volatile nature of sodium, $Na_xCo_2O_4$ crystals are not easily grown via the typical NaCl flux method. Thus, finding a suitable flux medium with low melting point but without introducing impurities to reduce processing temperature is a feasible way to solve the sodium vaporization problem. Sodium hydroxide (NaOH) with melting point of 380°C is a suitable candidate. After several experiments, the mixture of NaCl and NaOH was selected to be the new flux. The plate-like $Na_xCo_2O_4$ crystals were successfully synthesized in our lab at low temperature 550°C through using the metallic Co powder as Co source and NaCl/NaOH as the flux.

Fig. 6 presents us some typical $Na_xCo_2O_4$ thin platelets with close-hexagonal shape and size up to 5mm. The typical dimension of $Na_xCo_2O_4$ crystals are 0.2~2.0mm x 0.2~2.0mm x 5~100μm. Most of large crystals do not possess a good hexagonal shape, though we still observe ~120° angles at the corner. Apparently there exist small cracks or holes on the large crystals and the surface with partial over-lap are not very flat. This is typical for layered-structure crystals grown by the flux method. Due to the impurities, thermal or composition fluctuation, the atoms deposition to form one layer at the melt-crystal interface was interrupted, and then another layer starts to form on the top. The synthesized single crystals were verified by XRD to be single phase γ-$NaCo_2O_4$, and Na composition was found to be deficient ~0.2-0.6 via EDX analysis.

The chemical reaction was believed to be:

$$2C_O + xNaOH + (2 - \frac{x}{4})O_2 \xrightarrow{NaCl} Na_xCo_2O_4 + \frac{x}{2}H_2O$$

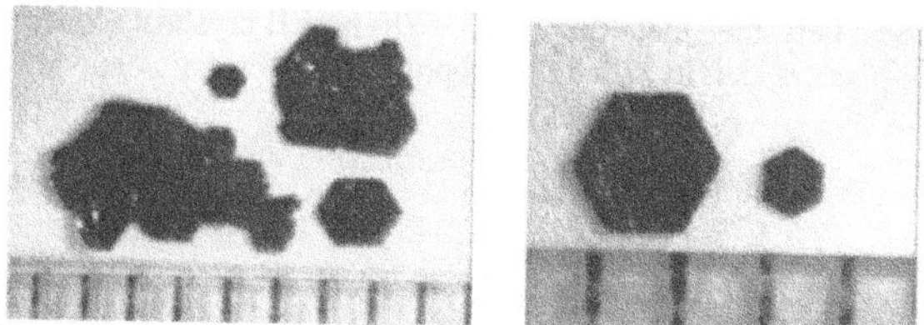

Fig. 6: Plate-like $Na_xCo_2O_4$ crystals by novel low temperature flux method

The resistivity and thermopower measurements revealed that there exist two types of single crystals synthesized via this new low temperature NaCl/NaOH flux method. For type I crystals, as shown in Fig. 7(a), the thermopower α (~50 μV/K at 300K) is smaller, but the resistivity ρ is higher than the crystals grown via typical high temperature NaCl flux method. Also the resistivity displays a metallic behavior, but it becomes semiconducting when the temperature decreased below ~50K. Fig. 7(b) exhibits another type crystal having small thermopower ~70μV/K at 300K and the resistivity behaving completely like a semiconductor. The band gap was found to be ~0.06 eV which was calculated from the resistivity at the high temperature region. For both types of crystals, the small thermopower can be explained by their low Na composition since thermopower was found to be associated with the ratio of Co^{3+}/Co^{4+} associated with Na composition[17].

Except individual single crystals, we also measured the properties of compact pellet of crystals which were pressed under pressure and temperature. All the grown Na-deficient $Na_xCo_2O_4$ crystals were passed through different size of funnels and separated into two groups, one group (group A) with larger crystal size >1mm and another group (group B) with crystals size of 0.5-1.0 mm. The platelets were placed in the mold and held at around 300°C for 20 minutes with pressure of 7000 pounds. Before the press, the mold was put in the ultrasonic bath to help lay down the crystals flatly layer by layer. Optical images of top surface and side view revealed most of the plate-like crystals are laid down flatly. As seen in Fig. 8, both groups behave like semiconductor as ascertained from the temperature dependence of the resistivity. There was little difference in thermopower and resistivity for group A and B though their crystals sizes are different. The values of thermopower of the pressed pellets are small and close to individual crystals, but resistivity is much higher, which is partially due to the boundaries between the crystals and out-of plane resistivity being involved. The band gap E_g was found to be 0.11eV and 0.09eV for group A and B, respectively.

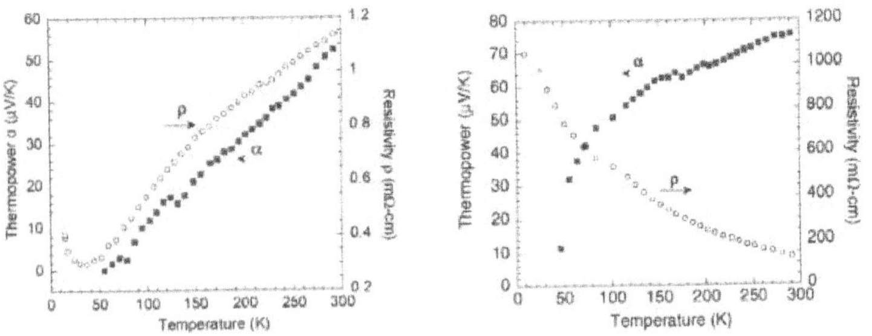

(a) type I crystals: metallic (b) type II crystals: semiconducting

Fig. 7: In-plane resistivity ρ and thermopower α of Na-deficient $Na_xCo_2O_4$ crystals

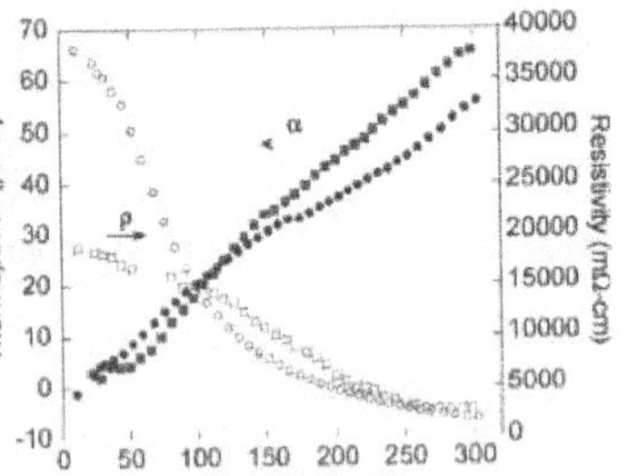

Fig. 8: Thermopower α and resistivity ρ for Na-deficient crystals pellets pressed under pressure and temperature (Solid and open square symbols represent α and ρ of group A, solid and open round symbols represent α and ρ of group B, respectively.)

IV. CONCLUSIONS

In summary, in polycrystalline $Na_xCo_2O_4$ samples, with the increases of Na concentration x, the thermopower was found to monotonously increase, but the resistivity and thermal conductivity were not strongly dependent.

The thermal conductivity of single crystals with size of (1-2) x (0.2-0.5) x (0.01-0.05) mm^3 grown via typical NaCl flux method was measured by our steady state "PTC" system and was found to be ≈5W/m-K. Using our measured lattice thermal conductivity κ_L and calculated Debye temperature θ_D, the in-plane phonon mean free path l_{ph} is estimated to be ≈8.9Å, which is the same order of magnitude of the lattice constant but is much smaller than the in-plane mean free path of conducting carriers l_a≈51Å at 300K.

Plate-like $Na_xCo_2O_4$ single crystals were successfully synthesized at low temperature 550°C via a novel low temperature NaOH/NaCl flux method and were found to be Na-deficient and possess two different behaviors in the temperature dependence of their resistivity.

ACKNOWLEDGEMENTS

We are grateful for the financial support for this research from the Office of Naval Research DOD-EPSCoR Grant #N00014-0310787 and SC EPSCoR/Clemson University Cost Share.

REFERENCES

[1] G. D. Mahan, Solid state physics, 51, 81 (1998).
[2] I. Terasaki, Y. Sasago and K. Uchinokura, Phys. Rev. B 56, 12685 (1997).
[3] Y. Wang, N. S. Rogado, R. J. Cava, and N. P. Ong, Nature 423, 425 (2003).
[4] K. Fujita, T. Mochida, and K. Nakamura, Jpn. J. Appl. Phys. part1 40, 4644 (2001).
[5] K. Takada, H. Sakurai, E. Takayama-Muromachi, F. Izumi, R. A. Dilanian and T. Sasaki, Nature 422, 53 (2003).
[6] T. Motohashi, E. Naujalis, R. Ueda, K. Isawa, M. Karppinen, and H. Yamauchi, Appl. Phys. Lett. 79, 1480 (2001).
[7] Xiaofeng Tang, Ed Abbott, Joe Kolis, Kelvin Aaron, ICT 2005 proceedings, Clemson (2005).
[8] A. L. Pope, R. T. Littleton and T. M. Tritt, Rev. Sci. Instr. 72, 3129 (2001).
[9] A. L. Pope, B. M. Zawilski and T. M. Tritt, Cryogenics 41, 725 (2001).
[10] R. Ray, A. Ghoshray, K. Ghoshray, and S. Nakamura, Phys. Rev. B, 59, 9454 (1999).
[11] C. Fouassier, G. Matejka, J-M. Reau, and P. Hagenmuller, J. Solid State Chem., 6, 532 (1973).
[12] T. Itoh, I. Terasaki. Jpn. J. Appl. Phys., 39, Pt. 1, 6658 (2000).
[13] I. Terasaki. Proceedings of 18th ICT, Baltimore, p 569(1999).
[14] B. M. Zawilski, R. T. Littleton, T. M. Tritt, Rev. Sci. Instr., 72, 1770 (2001).
[15] Kelvin Aaron. Master thesis. Clemson University (2005).
[16] Satake A., Tanaka H., Ohkawa T. et al., J. Appl. Phys., Vol. 96, 931 (2004).
[17] W. Koshibae, K. Tsutsui and S. Maekawa. Phys. Rev. B, 62(11), 6869 (2000).

Electroceramic Materials and Applications
Edited by R. W. Schwartz

PREPARATION AND THERMOELECTRIC PROPERTIES OF HIGHLY ORIENTED $Na_{1.5}Co_2O_4$ AND $Ca_3Co_4O_9$ CERAMICS BY THE SPARK PLASMA SINTERING METHOD

Keishi Nishio, Yasushi Kaneko, Kazuhiko Fukuda, Tsutomu Iida, Yoshihumi Takanashi and Atsuo Yasumori
Department of Materials Science and Technology, Tokyo University of Science
2641 Yamazaki, Noda-shi, Chiba 278-8510, Japan

Tsuyoshi Yoshimura, Kazuma Takahashi, Atsushi Torai, Kazuyasu Tokiwa and Tsuneo Watanabe
Department of Applied Electronics, Tokyo University of Science
2641 Yamazaki, Noda-shi, Chiba 278-8510, Japan

Thoru Kineri
Department of Materials Science & Environmental Engineering, Tokyo University of Science Yamaguchi
1-1-1 Daigaku-doori, Sanyo Onoda-shi, Yamaguchi 756-0884

ABSTRACT

We report on the preparation and thermoelectric properties of highly textured $Na_xCo_2O_4$ and $Ca_3Co_4O_9$ ceramics prepared by spark plasma sintering (SPS) combined with a modified Pechini's method. In Pechini's method, metal oxides can be obtained from a solution consisting of metal salts, organic solvents, and chelating agents for metal ions. We used a modified Pechini's method to prepare $Na_xCo_2O_4$ and $Ca_3Co_4O_9$ powders. Plate-like $Na_xCo_2O_4$ and $Ca_3Co_4O_9$ particles were synthesized from a solution consisting of metal acetic salts, citric acid, and ethylene glycol monomethyl ether. Plate-like phase $Na_xCo_2O_4$ powders (x=1.0-2.2) and $Ca_3Co_4O_9$ powder were obtained from the solution by heat treatment at 1173 K in air. The average diameter and thickness of the $Na_{1.5}Co_2O_4$ and $Ca_3Co_4O_9$ plate-like particles were 4 μm and 1 μm, respectively. Spark plasma sintering yielding $Na_xCo_2O_4$ and $Ca_3Co_4O_9$ ceramics is a new sintering method for such material. $Na_{1.5}Co_2O_4$ and $Ca_3Co_4O_9$ ceramics were obtained by sintering at 1073 K (SPS temperature) and 29.4 MPa (uni-axial pressure) for a few minutes in an Ar atmosphere (atmospheric pressure) and were of high density (relative density to single crystal 96%) and high orientation (Lotgering factor: $Na_{1.5}Co_2O_4$, 0.62; $Ca_3Co_4O_9$, 0.35). The plate-like grains were easily oriented by application of uniaxial pressure. The ceramics showed excellent thermoelectric properties.

INTRODUCTION

Thermoelectric materials have drawn interest because of their properties of large thermopower, low resistivity, and low thermoconductivity. Layered cobalt oxides, i.e., Na-Co-O and Ca-Co-O, have been studied by many researchers because they show low resistivity and low thermoconductivity[1-8]. $Na_xCo_2O_4$ consists of Na cations and CoO_2 blocks, and $Ca_3Co_4O_9$ consists of Ca_2CoO_3 blocks alternately stacked along the *c*-axis to form a layered structure. Thus, their physical properties are highly two-dimensional compared with *a* or *b*-axis direction. Although their carrier mobility is ten times smaller than in metal semiconductors, $Na_xCo_2O_4$ and $Ca_3Co_4O_9$ have large Seebeck coefficients that have been attributed to the strongly correlated system each ceramic possesses. However, mass production of $Na_xCo_2O_{4+\delta}$ and $Ca_3Co_4O_9$ single crystals has proven very expensive because large single crystals are difficult to obtain[7].

We used a modified Pechini's method to prepare $Na_xCo_2O_4$ (x=1.0-2.0) and $Ca_3Co_4O_9$ powders. Pechini's method[9] is a simple method for preparing metal oxide powders, whereby polymeric precursors are made from metal salts, ethylene glycol, and citric acid in a low-temperature heat treatment. Figure 1 shows Pechini's method of forming polymeric precursor with metal ions, ethylene glycol, and citric acid. This method, on which many studies have been reported[10-13], allows the metal cations to be mixed at the molecular level and stoichiometric compositions to be achieved by chelating the metal ions in solution with citric acid. Furthermore, it offers several advantages for fabrication of ceramic thin films; i.e., homogeneous compositions of high purity can be formed inexpensively by using a low-temperature heat treatment. The modified method of this study prepares a precursor solution from metal salts, citric acid, and ethylene glycol monomethyl ether.

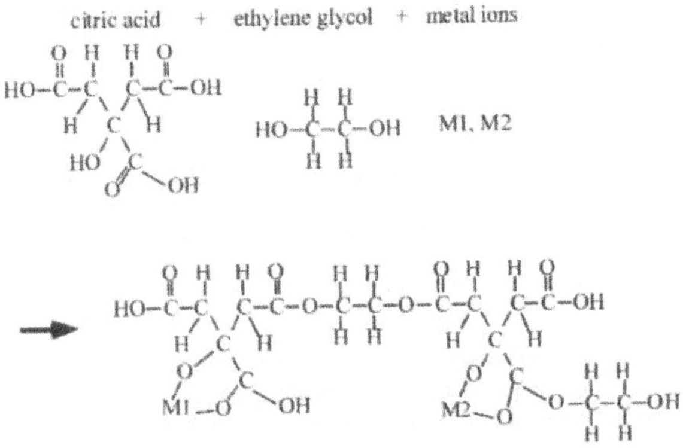

Figure 1 Pechini's method of forming polymeric precursor with metal ions, ethylene glycol and citric acid.

The spark plasma sintering (SPS) method is easy and cost effective, and it requires no previous sintering experience. SPS can be used to prepare metal compounds, ceramics, polymers, and thermoelectric semiconductors, and it an effective way of preparing ceramics with superior density and reduced grain growth[14-16]. Compared with conventional powder sintering, it has a lower temperature and shorter duration because it charges the intervals among the powder particles with electrical energy and momentarily applies highly energetic electrical spark discharges. The relatively low temperature and short sintering duration of the SPS process would be advantageous in keeping in check exaggerated grain growth.

We prepared plate-like precursor powder by using this modified Pechini's method and obtained highly oriented $Na_xCo_2O_4$ and $Ca_3Co_4O_9$ ceramics having excellent thermoelectric properties equivalent to those of a single crystal through application of uniaxial pressure.

EXPERIMENTAL PROCEDURE

The precursor solution was prepared with metal salts (CH_3COONa and $Co(NO_3)_2 \cdot 6H_2O$ for $Na_xCo_2O_4$, $Ca(NO_3)_2 \cdot 4H_2O$ and $Co(NO_3)_2 \cdot 6H_2O$ for $Ca_3Co_4O_9$), citric acid, and ethylene glycol monomethyl ether. These metal salts were dissolved in a solution consisting of citric acid and ethylene glycol monomethyl ether. The solution was refluxed at 333 K for 1 hr. The precursor solution was heated at 723 K for 2 hr after drying at 388 K. To obtain the precursor samples, the dried powder was heat treated at several temperatures from 973 to 1323 K in air for 1 hr.

Spark plasma sintering (SPS) was carried out with a spark plasma sintering system (DR. SINTER LAB 515S, Sumitomo Coal Mining Company, Ltd., Japan). The precursor powder was placed in a graphite die, and heat-treated to 1073 K at a rate of 100 K/min and kept at 1073 K for 5 minutes at 29.4 MPa in an Ar atmosphere (1013 hPa). The electric power was then turned off, and the sample was cooled to room temperature in the SPS chamber.

The SEM investigation was carried out using an SEM –5800LV system (JEOL Ltd., Japan). The XRD patterns of powders and ceramics were measured using a Rint-1000 system (Rigaku Co., Japan) and an X-ray diffractometer model CN4148 (Rigaku Co., Japan). The electrical properties were measured using a PPMS-6000 system (Quantum Design Japan).

The Seebeck coefficient and electrical resistivity were measured with the standard four-probe method (ULVAC-RIKO, ZEM-2) in a flowing He atmosphere in the temperature range from 300 to 1023 K.

RESULTS AND DISCUSSIONS

The modified Pechini's method using ethylene glycol monomethyl ether produced a homogeneous precursor solution. The precursor powder was prepared at a temperature above 400°C in order to remove organics from the powder. The solution (Na : Co = 1.5 : 2.0 atomic ratio) was heat treated at 400°C for 5 hr and then analyzed by TGA-DTA with heating rate of 5°C/min. Figure 2 shows DTA-DTG curves of the dried $NaCo_2O_4$ precursor powder.

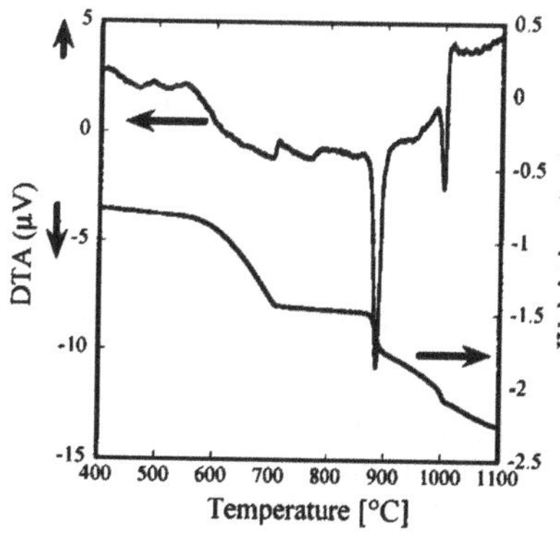

Figure 2 DTA-DTG curves of $NaCo_2O_4$ precursor powder dried at 400 °C.

The specimen exhibited several endothermic peaks and marked weight loss at 890°C and 1000°C. These endothermic peaks were probably due to decomposition or sublimate of the material because their appearance was accompanied by a weight loss in the specimen. Figure 3 shows XRD patterns of the $NaCoO_4$ powders prepared at (a) 400°C, (b) 800°C, (c) 950°C and (d) 1050°C for 1 hr.

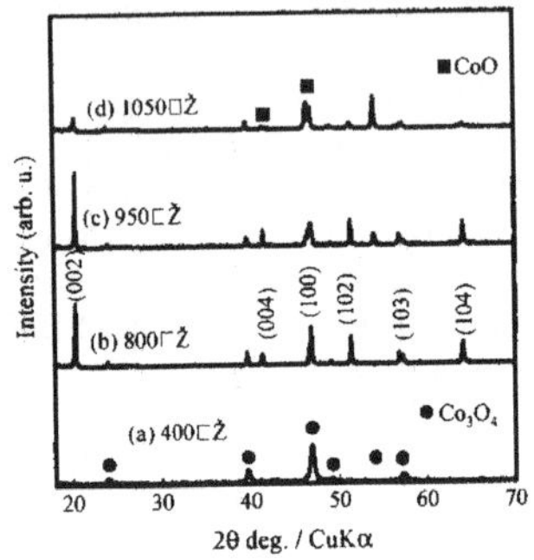

Figure 3 XRD patterns of $NaCoO_4$ powders prepared at (a) 400°C, (b) 800°C, (c) 950°C and (d) 1050°C for 1 hr.

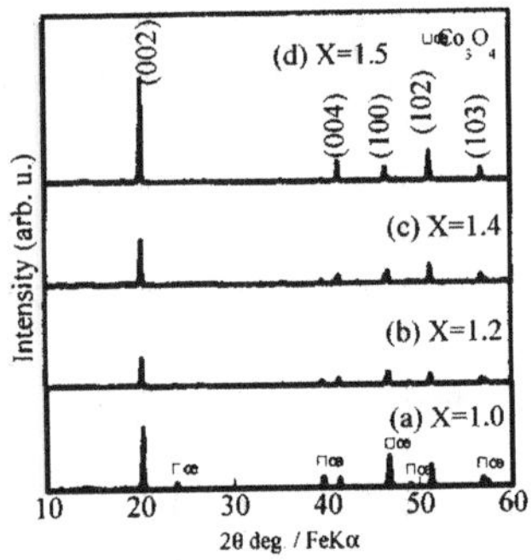

Figure 4 XRD pattern of $Na_xCo_2O_4$ powders prepared at 900°C for 1 hr: (a) x=1.0, (b) x=1.2, (c) x=1.4 and (d) x=1.5

Intense peaks of Co_3O_4 are evident in (a). On the other hand, the powders prepared above 700 °C showed the diffraction peaks of the $NaCoO_2$ phase, and the peak intensity of Co_3O_4 decreased with increasing heat treatment temperature. Diffraction peaks of CoO appeared in the pattern of the powders prepared above 950°C. The TGA-DTA and XRD analyses indicated that $NaCo_2O_4$ likely decomposed into Na_2O and CoO. Na_2O was sublimated in a temperature range above 950 °C. Figure 4 shows the XRD pattern of the $Na_xCo_2O_4$ powders prepared at 900°C for 1 hr: (a) x=1.0, (b) x=1.2, (c) x=1.4 and (d) x=1.5. In the patterns of the composition of up to x=1.4, Co_3O_4 and $NaCo_2O_4$ phases were investigated beside in the patterns of (d)[17-19]. The peak intensity of the Co_3O_4 phase decreased with increasing x. It disappeared from x=1.5 to x=2.0: a single phase of $NaCo_2O_4$ could be obtained. The localized Na ions would prevent thermal conduction and electrical conduction in the system. Figure 5 shows the XRD pattern of $Ca_3Co_4O_9$ powder prepared with the modified Pechini's method at 900°C. No impurity phase is evident.

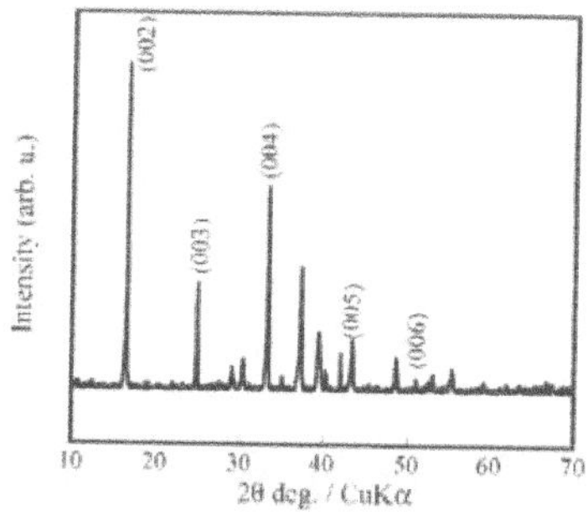

Figure 5 XRD pattern of $Ca_3Co_4O_9$ powder prepared with modified Pechini's method at 900°C.

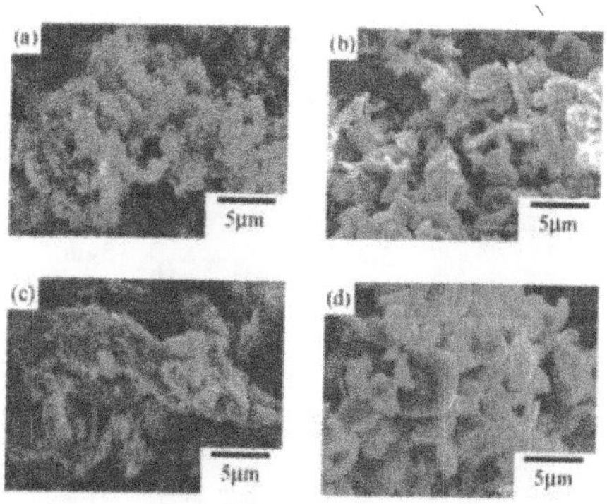

Figure 6 SEM images of the $Na_{1.5}Co_2O_4$ powders prepared at (a) 400°C, (b) 900°C, and (c) 1000°C and (d) $Ca_3Co_4O_9$ powder prepared at 900°C.

Figure 6 shows SEM images of the $Na_{1.5}Co_2O_4$ powders prepared at (a) 400°C, (b) 800°C, (c) 950°C and (d) $Ca_3Co_4O_9$ powder prepared at 900°C. The $Na_{1.5}Co_2O_4$ powder prepared at 400°C was Co_3O_4, and the amorphous phase consisted of small particles. The grain size increased with increasing heat treatment temperature. The shape of $Na_xCo_2O_4$ prepared with a solid-state reaction is spherical or cubic. However, the grain of the ceramic prepared with the modified Pechini's method was plate-like (see Figure 6 (b) and (d)). Average diameter and thickness of these plate-like particles were 4 μm and 1 μm, respectively. It is thought that the shape of crystal grains reflects the crystal structure. It seemed that the plate-like particles could easily be aligned by application of uniaxial pressure.

The $Na_xCo_2O_4$ and $Ca_3Co_4O_9$ ceramics were prepared under a reduction atmosphere made by placing a graphite die in a vacuum. The heat treatment in Ar at atmospheric pressure suppressed the sample reduction. Figure 7 shows the shrinkage of $Na_{1.5}Co_2O_4$ depending on the heat treatment condition. The uniaxial pressure (29.4 MPa) was applied to the sample while it was in the SPS chamber.

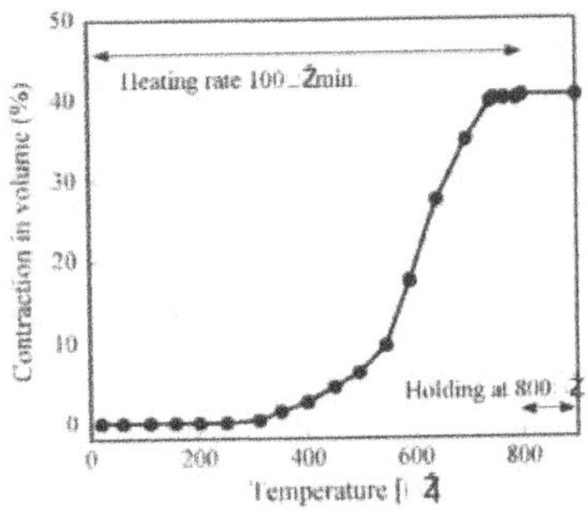

Figure 7 Shrinkage of $Na_{1.5}Co_2O_4$ under different heat treatment conditions.

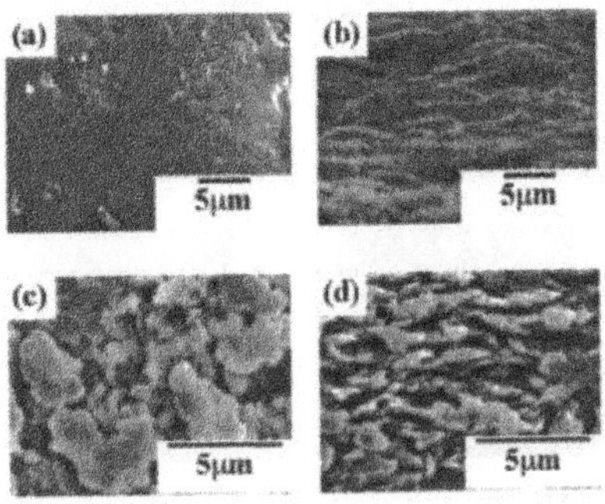

Figure 8 SEM images of $Na_{1.5}Co_2O_4$ ceramic ((a) surface and (b) cross section) and $Ca_3Co_4O_9$ ceramic ((c) surface and (d) cross section) after being etched in 1mol/L HCl aqueous solution.

The sample at low temperature shrank within a very short time. The obtained ceramic had a high bulk density, 95% larger than the theoretical X-ray density[17-20]. Figure 8 shows SEM images of $Na_{1.5}Co_2O_4$ ((a) surface and (b) cross section) and $Ca_3Co_4O_9$ ((c) surface and (d) cross section) after their surfaces were etched in a 1 mol/L HCl aqueous solution. The SEM images showed densification rather than grain growth. These results indicate that the shrinkage mechanism of the sample was not due to sintering, as in a heat treatment or Joule's heat treatment, but due to another mechanism. They also showed that the plate-like particles were aligned perpendicular to the uniaxial pressure direction.

Figure 9 shows XRD patterns of (a) $Na_{1.5}Co_2O_4$ and (b) $Ca_3Co_4O_9$ ceramics in the in-plane and out-of-plane directions, respectively.

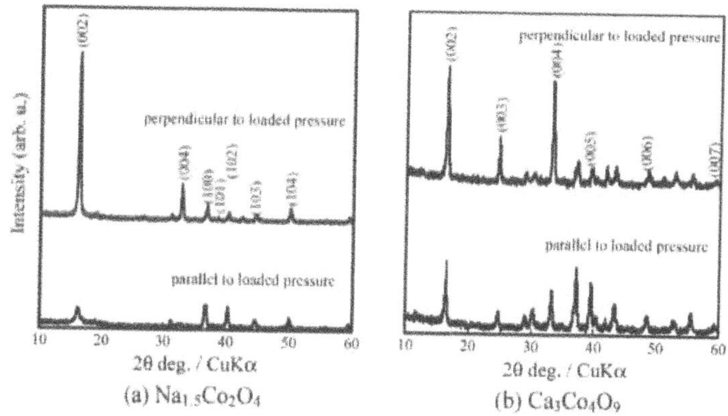

Figure 9 XRD patterns for (a) in-plane direction of $Na_{1.5}Co_2O_4$ and (b) out-of-plane direction of $Ca_3Co_4O_9$ ceramics prepared from plate-like powder synthesized with modified Pechini method.

The peaks of the *(00l)* phases are more intense than those of other phases in the in-plane patterns. The c-axis of the $Na_{1.5}Co_2O_4$ and $Ca_3Co_4O_9$ ceramics seemed to align perpendicularly to the uni-axial pressure direction. These results indicate that the plate-like grains were aligned in the c-axis direction by applying uniaxial pressure.

The degree of texture was evaluated in terms of the Lotgering factor[7],

$$f = (p - p_0)/(1 - p_0) \quad (1)$$

where

$$p = I_{(00l)} / I_{(hkl)} \quad (2)$$

for the diffraction pattern of the sample, and $p_0 = p$ for a randomly oriented reference powder diffraction pattern. The Lotgering factors of the $Na_{1.5}Co_2O_4$ and $Ca_3Co_4O_9$ ceramics were about 0.62 and 0.35, respectively[18-20]. These results confirm that the ceramic samples were oriented ceramic. $Na_xCo_2O_4$ consisted of Na cation and CoO_2 blocks, and $Ca_3Co_4O_9$ consisted of Ca_2CoO_3 and CoO_2 blocks alternately stacked along the *c*-axis to form a layered structure.

Table 1 shows the thermoelectrical properties of $Na_{1.5}Co_2O_4$ and $Ca_3Co_4O_9$. These ceramics had excellent anisotropic and thermoelectric properties. SPS sintering yielded dense ceramic after a short treatment, and the modified Pechini method formed highly textured ceramic from the plate-like particles. The $Na_{1.5}Co_2O_4$ ceramic had an in-plane direction resistivity of 2.40 mΩcm, a Seebeck coefficient of 121 μV/K, and a power factor of 6.20 $\mu W/cmK^2$ (at room temperature), as well as an out-of-plane direction resistivity of 6.11 mΩcm, a Seebeck coefficient of 77.2 μV/K, and a power factor of 0.98 $\mu W/cmK^2$ (at room temperature), The $Ca_3Co_4O_9$ ceramic had an in-plane direction resistivity of 9.68 mΩcm, a Seebeck coefficient of 138.3 μV/K, and a power factor of 1.98 $\mu W/cmK^2$ (at room temperature), as well as an out-of-plane direction resistivity of 22.5 mΩcm, a Seebeck coefficient of 132.9 μV/K, and a power factor of 0.78 $\mu W/cmK^2$ (at room temperature).

Table 1 Thermoelectric properties of highly textured $Na_{1.5}Co_2O_4$ and $Ca_3Co_4O_9$ ceramics

	ρ (mΩcm)	S (μV/K)	PF ($\mu W/cmK^2$),	relative density (%)
$Na_{1.5}Co_2O_4$ in-plane	2.40	121	6.20	95.5
out of plane	6.11	77.2	0.98	95.5
$Ca_3Co_4O_9$ in-plane	9.68	138	1.98	96.0
out of plane	22.5	133	0.78	96.0

CONCLUSIONS

Highly textured $Na_xCo_2O_4$ and $Ca_3Co_4O_9$ ceramics as thermoelectric materials were obtained by spark plasma sintering of a plate-like precursor powder. The plate-like particles were synthesized from a solution consisting of metal salts, citric acid, and organic solvents. The average diameter and thickness of the plate-like particle were 4 μm and 1 μm, respectively. The grains could be easily oriented by application of uniaxial pressure. Ceramics of high density (relative density to single crystal above 90%) and high orientation (Lotgering factor: $Na_{1.5}Co_2O_4$, 0.62; $Ca_3Co_4O_9$, 0.35) were obtained after sintering for 1 minute at 1073 K (SPS temperature) and 29.4 MPa (uniaxial pressure) in an Ar atmosphere (atmospheric pressure). These ceramics showed excellent thermoelectric properties.

REFERENCES

[1]I. Terasaki, Y. Sasago, K. Uchinokura, Phys. Rev. B56 (1997) R12685-R12687

[2]H. Yakabe, K. Kikuchi. I. Terasaki, Y. Sasago, K. Uchinokura, Proceedings of the 16th International Conference on Thermoelectrics, (1997) 523-527

[3]T. Itoh, T. Kawata, T. Kitajima, I. Terasaki, Proceedings of the 17th International Conference on Thermoelectrics, (1998) 567-569

[4]I.Terasaki. Proceedings of the 19th International Conference on Thermoelectrics, (2000) 20-24

[5]S. Li, R. Funahashi, I. Matsubara, K. Ueno, S. Sodeoka, H. Yamada, J. Mter. Chem., 9 (1999) 1659-1660

[6]S. Li, R. Funahashi, I. Matsubara, K. Ueno, S. Sodeoka, H. Yamada, Chem. Mater., 12 (2000) 2424-2427

[7]Shin Tajima, Toshihiko Tani, Shinya Isobe. Kunihito Koumoto, Mater. Sci. Eng., B86 (2001) 20-25

[8]Y. W. Gu, K. A. Khor, P. Cheang, Biomater., 25, (2004) 4127-4134
[9]M. P. Pechini, US Patent (1967) 3330697
[10]M. S. G..Baythoun, F. R. Sale, J. Materials Science 17 (1982) 2757-2769
[11]L. W. Tai, P. A. Lessing, J. Mater. Res., 7 (1992) 502-510
[12]S. Roy, W. Sigmund and F. Aldinger, J. Mater. Res. 14 (1999) 1524-1531
[13]Ch. L. Robert, F. Ansart, C..Deloget, M. Gaudon, A. Rousset, Mater. Res. Bull. 36 (2001) 2083-2101
[14]Y. W. Gu, K. A. Khor, P. Cheang, Biomater., 25, (2004) 4127-4134
[15]Dongli Wang, Lidong Chen, Qin Yao, Jianguo Li, Solid State Com., (2004) 615-618
[16]L. G. Yu, K. A. Khor, H. Li, P. Cheang, Biomater. 24 (2003) 2695-2705
[17]JCPDS 27-682
[18]JCPDS 9-418
[19]JCPDS 73-133
[20]JCPDS 48-1719

Electroceramic Materials and Applications
Edited by R. W. Schwartz

HIGH-TEMPERATURE THERMOELECTRIC PERFORMANCE OF STRONTIUM TITANATE DEGENERATE SEMICONDUCTORS

Hiromichi Ohta[1,2], S. Ohta[1] and K. Koumoto[1,2]

[1] Nagoya University, Graduate School of Engineering,
Furo-cho, Chikusa, Nagoya 464-8603, Japan

[2] CREST, Japan Science and Technology Agency,
4-1-8 Motomachi, Kawaguchi 332-0012, Japan

ABSTRACT

Nb-content dependence of the thermoelectric figure of merit, *ZT* of $SrTiO_3$ at high-temperature (1000 K) is clarified using heavily Nb-doped $SrTiO_3$ epitaxial films, which were grown on (100)-face of $LaAlO_3$ single crystal substrates by a pulsed-laser deposition. Carrier concentration, Hall mobility, Seebeck coefficient, and thermal conductivity of Nb-doped $SrTiO_3$ epitaxial films were experimentally evaluated at 1000 K with an aid of theoretical analysis. *ZT* reached 0.37 (20%-Nb-doped $SrTiO_3$ epitaxial film) at 1000 K, which is the largest value among *n*-type oxide semiconductors ever reported.

INTRODUCTION

High-temperature thermoelectric (HTE) materials exhibiting the figure of merit, *ZT* (= $S^2\sigma T/\kappa$, where *Z*, T, *S*, σ and κ are figure of merit, absolute temperature, Seebeck coefficient, electrical conductivity and thermal conductivity, respectively) exceeding that of conventional HTE materials such as SiGe alloy ($ZT \sim 0.7$ [1]) and β-$FeSi_2$ ($ZT \sim 0.3$ [2]) is intensely required for the future power generation. HTE materials may convert any heat sources which are thrown away from everywhere *e.g.* electric power plants, factories and automobiles to electric power by utilizing the Seebeck effect.

Recently, cobalt-oxide based layer-structured crystals including Na_xCoO_2 [3], $Ca_3Co_4O_9$ [4], and their derivative compounds have been developed as *p*-type materials having fairly high HTE performance, and the maximum *ZT* value generated from this compound group has reached 1 or larger up to date [5]. Additionally, modulated layered cobalt oxides have been found promising as *p*-type materials [6]. In contrast, *n*-type oxides so far proposed, such as Al-doped ZnO [7] and $In_2O_3(ZnO)_m$ (*m* = integer) [8, 9], only show rather low *ZT* and remain to be further improved, or otherwise novel oxide materials have to be explored.

We focused heavily carrier-doped $SrTiO_3$ as a promising candidate of *n*-type oxide semiconductor because it exhibits rather large $|S|$ due to that the large carrier effective mass (m^* = $6 \sim 10\ m_0$) [10, 11] leads to large $|S|$. Further, bulk single crystals of heavily La-doped $SrTiO_3$ have recently been found to have large power factor ($PF = S^2\sigma$) of 3.6×10^{-3} $Wm^{-1}K^{-2}$ at room temperature [12], which is comparable to that of practical Peltier material Bi_2Te_3 [13].

Very recently, we have examined carrier transport properties of Nb- and La-doped $SrTiO_3$ single crystals (carrier concentration, $n_e \sim 10^{20}$ cm^{-3}) at high temperatures (~1000 K) to clarify their thermoelectric response [14]. Although the experimental data suggested that fairly high *ZT* is expected in heavily Nb-doped $SrTiO_3$, the maximum *ZT* could not be clarified because the solubility of the Nb^{5+} ions in the $SrTiO_3$ lattice is substantially smaller than an optimal concentration. In order to overcome this problem, we fabricated $SrTiO_3$ epitaxial films having ~

10^{22} cm^{-3} Nb as dopant on (100)-face of $LaAlO_3$ substrate by pulsed-laser deposition (PLD). Here we report high-temperature (1000 K) thermoelectric performance of heavily Nb-doped $SrTiO_3$. The 20% (4×10^{21} cm^{-3}) Nb-doped $SrTiO_3$ epitaxial film exhibited $ZT \sim 0.37$ at 1000 K, which is the largest value among *n*-type oxide semiconductors ever reported.

EXPERIMENTAL

Epitaxial films of Nb-doped $SrTiO_3$ were grown on (100)-face of $LaAlO_3$ single crystal substrates at 700°C by PLD (4ω-Nd: YAG laser, $\lambda = 266$ nm, ~ 1 $Jcm^{-2}pulse^{-1}$, ~ 20 ns, 10 Hz) using $SrTiO_3$ ceramic targets containing up to 40% Nb as dopant. Crystalline quality and orientation of the films were analyzed by high-resolution X-ray diffraction (HR-XRD, ATX-G, $CuK\alpha_1$, Rigaku Co.). Electrical conductivity, carrier concentration and Hall mobility of the films were measured by d.c. four probe method (interval of the electrode: 6 mm) using van der Pauw configuration. Seebeck coefficient of the films was measured by a conventional steady state method.

RESULTS AND DISCUSSION

Figure 1A summarizes out-of-plane HR-XRD patterns of the Nb-doped $SrTiO_3$ films grown on (100)-face of $LaAlO_3$ single crystal substrates. Nb-contents named x_{Nb}, hereafter, of the Nb-doped $SrTiO_3$ films were (a) 0, (b) 0.05, (c) 0.1, (d) 0.2 and (e) 0.4, respectively. Only intense Bragg reflection peak of 200 $SrTiO_3$ was observed in all films together with 200 $LaAlO_3$, indicating highly crystallographic orientation of the films. Thickness fringes are clearly observed around the 200 $SrTiO_3$ peak indicating high-crystalline qualities of the films. Lattice constant of Nb-doped $SrTiO_3$ increases proportionally to Nb-concentration (Fig. 1B) indicating that Nb^{5+} (64.0 pm) is substituted at Ti^{4+} (60.5 pm) site [15]. This site selectivity of Nb is supported by the fact that Nb^{5+} ions act as donor. Smooth surface composed of square shaped domains with stepped & terraced structure are observed in the topographic AFM image of the Nb-doped $SrTiO_3$ film ($x_{Nb} = 0.2$, Fig. 1C). Figure 1D shows the reflection high energy electron diffraction (RHEED) pattern of the Nb-doped $SrTiO_3$ film ($x_{Nb} = 0.2$). Intense streak pattern indicates that the $SrTiO_3$ films were epitaxially grown.

The electrical conductivity (σ), carrier concentration (n_e), Hall mobility (μ_{Hall}), and Seebeck coefficient (S) for the Nb-doped $SrTiO_3$ epitaxial films were measured from 300 K to 1000 K. The n value was independent of temperature, while the μ_{Hall} value decreased proportionally to $T^{-1.5}$ above 700 K, indicating that the acoustic phonon plays a dominant role in the carrier scattering mechanism. On the other hand, the $|S|$ values gradually increased with temperature due to the monotonic decrease in chemical potential. As a result, the power factor ($PF = S^2\sigma$) value increased gradually with temperature, which is basically the same as that of Nb-doped $SrTiO_3$ bulk single crystal. In order to clarify the optimal x_{Nb}, we focused on high-temperature (1000 K) thermoelectric properties of the Nb-doped $SrTiO_3$ epitaxial films.

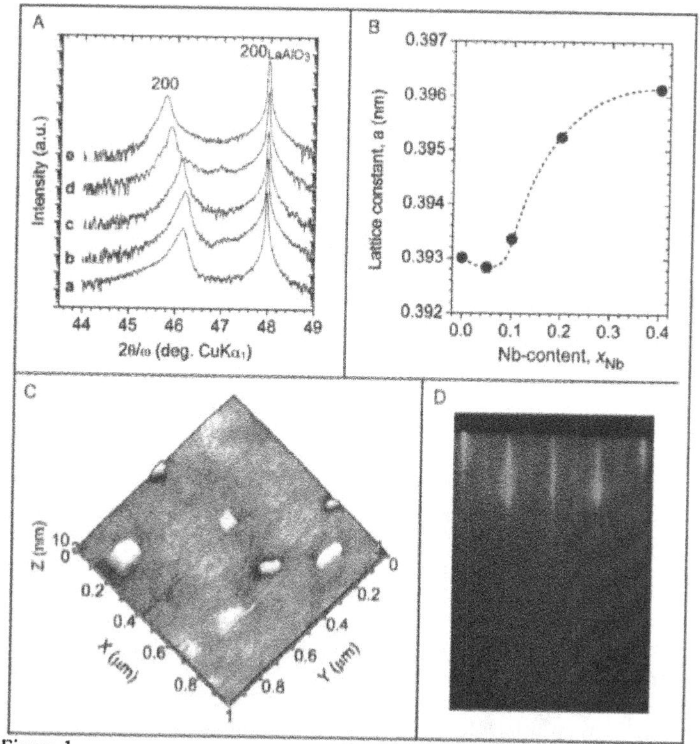

Figure 1

Figure 2A shows n_e and $|S|$ of the Nb-doped $SrTiO_3$ epitaxial films, which were measured at 1000 K under high-vacuum (~10^{-3} Pa) condition. The observed n_e-value of Nb-doped $SrTiO_3$ films increases direct proportionally with x_{Nb}. Further, these observed n_e-values correspond well with the ideal n_e-value (solid line), which was calculated based on the hypothesis that substituted Nb^{5+} at Ti^{4+} site fully generates carrier electron. Thus, n_e of Nb-doped $SrTiO_3$ films is simply controlled up to ~10^{22} cm^{-3} by means of Nb-doping. The observed S is entirely negative, indicating that the Nb-doped $SrTiO_3$ epitaxial films are n-type semiconductor, and $|S|$ decreases gradually with x_{Nb}. Although n_e of the 40%-Nb-doped $SrTiO_3$ film (x_{Nb} = 0.4) is extremely high (~7×10^{21} cm^{-3}), rather large $|S|$ (0.16 mVK^{-1}) is observed.

μ_{Hall} for the Nb-doped $SrTiO_3$ films decreased gradually with the x_{Nb} as shown in Fig. 2B. It is likely due to the fact that the carrier effective mass (m^*) for the Nb-doped $SrTiO_3$ films gradually increased with the x_{Nb} as discussed later. Thus, the mean free path of the carrier electron in the Nb-doped $SrTiO_3$ was almost independent of the x_{Nb}, which indicates that the contribution of the ionized impurity scattering to the carrier mobility is rather small.

In order to clarify the origin of the large |S| values in high carrier concentration films, we calculated m^* of the Nb-doped $SrTiO_3$ films at 1000 K by means of the following equations [16],

$$S=-\frac{k_B}{e}\left(\frac{(r+2)F_{r+1}(\xi)}{(r+1)F_r(\xi)}-\xi\right) \qquad (1)$$

where k_B, ξr, and F_r are the Boltzmann constant, chemical potential, scattering parameter of relaxation time ($r = 0$ [17]), and Fermi integral respectively. The F_r is given by

$$F_r(\xi)=\int_0^\infty \frac{x^r}{1+e^{x-r}}dx \qquad (2)$$

The n is given by

$$n_- = 4\pi\left(\frac{2mk_BT}{h^2}\right)^{3/2} F_{1/2}(\xi) \qquad (3)$$

where h, T, and m are the Planck constant, absolute temperature (T = 1000 K), and carrier effective mass, respectively. Here we neglect r since r becomes zero in the acoustic phonon scattering mechanism. The m^*/m_0 value, where m_0 is free electron mass, increases monotonically from ~7 ($x_{Nb} = 0$) to ~11 ($x_{Nb} = 0.4$) with x_{Nb} as shown in Fig. 2B. The μ_{Hall} values calculated using the m^* values (solid line) agree well with the observed m^* values, showing the validity of our analysis. From these results, we conclude that the large |S| values in high carrier concentration films result from the gradual increase in m^*/m_0 with x_{Nb}. The enhancement of m^* is partly due to an increase in the lattice constant or an increase in the distance between two neighboring Ti^{4+} ions, which leads to a decrease in the overlapping between Ti $3d$–t_{2g} orbitals.

PF values of Nb-doped $SrTiO_3$ at 1000 K are plotted in Fig. 2C. PF increases gradually with x_{Nb}. We estimated thermal conductivity (κ_{total}) at 1000 K according to the Wiedemann-Franz's law ($\kappa_{electron} = L\sigma T$, L: Lorentz number). Electronic contribution to the thermal conduction does not become negligible when the n_e exceeds 10^{21} cm^{-3}. κ_{total} increases with x_{Nb} when x_{Nb} exceeds 0.02 due to a significant increase in $\kappa_{electron}$ as shown in the Fig (dotted line). κ_{total} of dense ceramics (relative density: ~99 %) for Nb-doped $SrTiO_3$ was similar to calculated value as plotted in the Fig. Hence, we took contribution of $\kappa_{electron}$ into account to estimate ZT values of the epitaxial films.

ZT values of Nb-doped $SrTiO_3$ at 1000 K are plotted in Fig. 3. The ZT value increases with x_{Nb}, and it shows the maximum value of ~0.37 at $x_{Nb} = 0.2$ (n_e ~4.0×10^{21} cm^{-3}) as a result of the reduction of ZT due to the enhancement of the thermal conductivity with the carrier concentration. Temperature dependence of the ZT for the Nb-doped $SrTiO_3$ ($x_{Nb} = 0.2$) epitaxial film is also shown in the inset. The ZT value increases gradually with increasing temperature, demonstrating that the Nb-doped $SrTiO_3$ is good candidate for n-type thermoelectric oxide applicable at high temperatures.

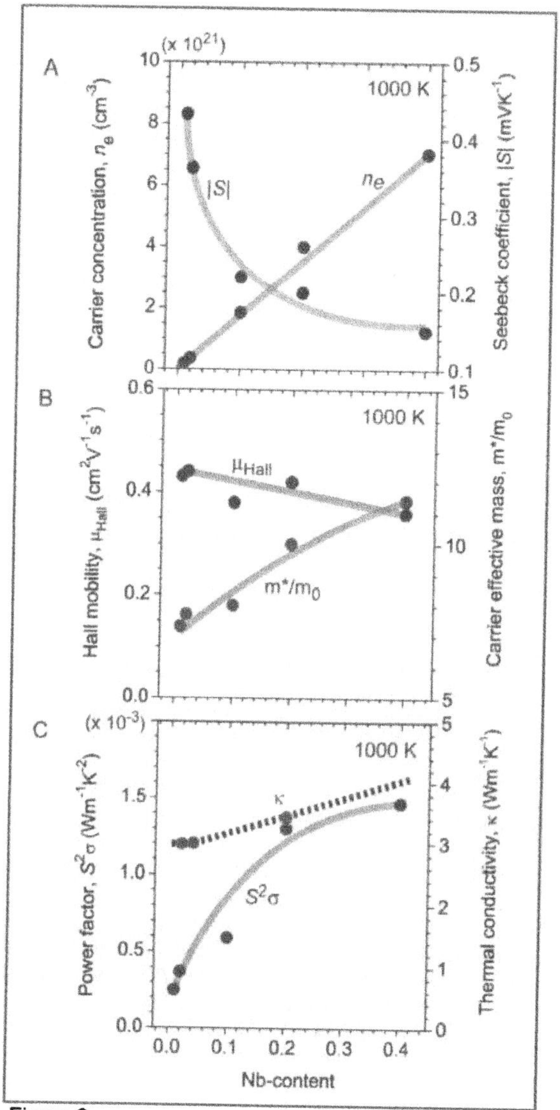
A
(x 10^21)
1000 K
|S|
ne
Carrier concentration, ne (cm-3)
Seebeck coefficient, |S| (mVK-1)
B
1000 K
μHall
m*/m0
Hall mobility, μHall (cm2V-1s-1)
Carrier effective mass, m*/m0
C
(x 10-3)
1000 K
κ
S2σ
Power factor, S2σ (Wm-1K-2)
Thermal conductivity, κ (Wm-1K-1)
Nb-content

Figure 2

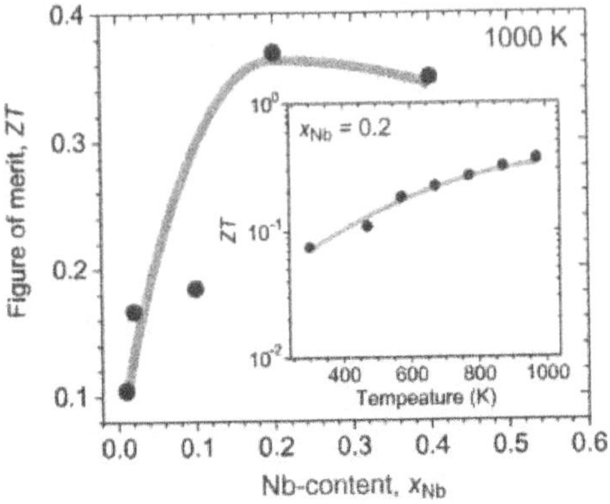

Figure 3

SUMMARY

We clarified experimentally the maximum ZT value for Nb-doped $SrTiO_3$ at 1000 K in terms of carrier concentration. ZT of Nb-doped $SrTiO_3$ increases with the Nb content (x_{Nb}) and it reaches ~0.37 at $x_{Nb} = 0.2$. Our results indicate that heavily Nb-doped $SrTiO_3$ is a good candidate for n-type thermoelectric oxide usable at high temperatures.

REFERENCES

[1]B. A. Cook, J. L. Harringa, S. H. Hann and C. B. Vining, *J. Appl. Phys.* **78**, 5474 (1995).

[2]I. Nishida, *Phys. Rev. B* **7**, 2710 (1973).

[3]I. Terasaki, Y. Sasago and K. Uchinokura, *Phys. Rev. B* **56**, 12685 (1997).

[4]R. Funahashi, I. Matsubara, H. Ikuta, T. Takeuchi, U, Mizutani and S. Sodeoka, *Jpn. J. Appl. Phys.* **39**, L1127 (2000).

[5]R. Funahashi and M. Shikano, *Appl. Phys. Lett.* **81**, 1459 (2002).

[6]G. Chen, M. S. Dresselhaus, G. Dresselhaus, J.-P. Fleurial and T. Caillat, *Inter. mater. Rev.* **48**, 45 (2003).

[7]M. Ohtaki, T. Tsubota, K. Eguchi and H Arai, *J. Appl. Phys.* **79**, 1816 (1996).

[8]H. Ohta, W-S. Seo and K. Koumoto, *J. Am. Ceram. Soc.* **79**, 2193 (1996).

[9]S. Isobe, T. Tani, Y. Masuda, W-S. Seo and K. Koumoto, *Jpn. J. Appl. Phys.* **41**, 731 (2002).

[10]T. Caillat, A. Borshchevsky and J. P. Fleurial, *J. Appl. Phys.* **80** 4442 (1996)

[11]H. P. R. Frederikse, W. R. Thurber and W. R. Hosler, *Phys. Rev.* **134**, A442 (1964).

[12]T. Okuda, K. Nakanishi, S. Miyasaka and Y. Tokura, *Phys. Rev.* **B 63**, 113104 (2001).

[13]Francis J. DiSalvo, *Science*, **285**, 703 (1999).

[14]S. Ohta, T. Nomura, H. Ohta and K. Koumoto, *J. Appl. Phys.* **97**, 34106 (2005).

[15]R. D. Shannon, *Acta Cryst. A* **32**, 751 (1976).

[16]C. B. Vining *J. Appl. Phys.* **69**, 331 (1991).

[17]K. Durczewski and M. Ausloos, *Phys. Rev.* **B 61**, 5303 (2000).

Electroceramic Materials and Applications
Edited by R. W. Schwartz

PREPARATION AND ANISOTROPIC THERMOELECTRIC PROPERTIES IN LAYERED COBALTITE THIN FILMS

Tsutomu Kanno, Akihiro Sakai, Satoshi Yotsuhashi, Akihiro Odagawa, and Hideaki Adachi

Advanced Technology Research Laboratories, Matsushita Electric Industrial Co., Ltd.,
3-4 Hikaridai, Seika-cho, Soraku-gun, Kyoto 619-0237, Japan

ABSTRACT

Orientation-controlled epitaxial thin films of layered cobaltites, A_xCoO_2 and $A_3Co_4O_9$ (A=Sr and Ca), were fabricated on α-Al_2O_3 single crystal substrates by means of rf-magnetron sputtering. Among these layered cobaltites, A_xCoO_2 (A=Sr and Ca) and $Sr_3Co_4O_9$ are metastable phases and fabrication was achieved by non-equilibrium process of sputtering. For the films with c axis aligned in film plane, anisotropic transport properties could be investigated along both parallel (||) and perpendicular (⊥) directions in respect to CoO_2 sheets. The present film configuration enabled the reliable measurement of Seebeck coefficient along c axis of layered cobaltites, which has not been precisely evaluated so far. Large anisotropy in resistivity was confirmed and perpendicular resistivity $\rho_\perp$ was 10-60 times larger than parallel resistivity $\rho_\parallel$ at room temperature. As for thermoelectric properties, relatively isotropic nature was observed in the Seebeck coefficient. Perpendicular Seebeck coefficient $S_\perp$ was approximately 1/3-1/2 of parallel $S_\parallel$.

INTRODUCTION

The family of layered cobaltites such as $Na_{0.5}CoO_2$, $Ca_3Co_4O_9$ and $Bi_2Sr_2Co_2O_y$ has been extensively studied due to its exceptionally large thermoelectricity among thermoelectric oxides.[1-4] The crystal structure of layered cobaltites is constructed by the alternate stacking of conducting CdI_2-type CoO_2 layers and insulating block layers along c axis. The triangular coordination of mixed-valent Co ions in the CoO_2 layer causes the characteristic degeneracy of $3d$ electrons, and this strongly correlated electron system is suggested to be the origin of large entropy transport (*i.e.* large Seebeck coefficient).[5] The crystal structure of Na_xCoO_2 is considered to be primitive because the insulating layer is Na monolayer. On the other hand, $Ca_3Co_4O_9$ has a rock-salt type insulating block layer consisting of three atomic layers. $Ca_3Co_4O_9$ has the lattice misfit along b axis between the square rock-salt layers and the triangular CoO_2 layers and this misfit cannot be expressed as a simple fraction.

Na_xCoO_2 shows excellent thermoelectric properties, however, it is unstable in humid air because of constituent reactive Na. In the device application, it is quite desirable for thermoelectric materials to have good stability in air. Although alkaline-free primitive cobaltites are not obtained by a standard process, ion-exchange methods are reported to be useful for the formation of alkaline-free primitive material using Na_xCoO_2 starting ceramics.[6,7] In addition, it is considerably worth studying in detail on anisotropic transport properties in layered cobaltites, in order to clarify the origin of their unusual thermoelectricity. Until now, few studies have been reported on the thermoelectricity along the perpendicular direction to the CoO_2 layers (*i.e.* along c axis) because available single crystals of most layered cobaltites are too thin (<500 mm) to measure physical properties along c axis.[1,8,9]

Meanwhile, thin-film processing is effective not only for device application but also for the growth of metastable phases.[10] Moreover, the growth orientation can be controlled by adjusting growth parameters, which enables aligned crystal with large dimension along the c axis.[11]

Here we report on the thin-film preparation of layered cobaltites with the primitive structure, A_xCoO_2 (A=Sr and Ca) as well as with the misfit structure, $A_3Co_4O_9$ (A=Sr and Ca), and succeeded in controlling the crystal orientation and measurement of anisotropic nature between parallel and perpendicular directions to CoO_2 layers.

EXPERIMENTAL PROCEDURE

Thin films were prepared by an rf-planar magnetron sputtering. We used composite A-Co-O targets with A/Co ratio 0.3-1.2. Sputtering deposition was carried out onto *c*-plane (0001), *a*-plane $(11\bar{2}0)$ and *m*-plane $(1\bar{1}00)$ of sapphire (a-Al_2O_3) substrates concurrently in a mixed atmosphere of Ar and O_2 (20-30%) with a total pressure of 5 Pa. The substrates were heated to 600-700°C during the deposition. Deposition rate varied with the growth orientation and film thickness, and it was approximately 800 Å for films on *a*-plane and *m*-plane, and 2000 Å for films on *c*-plane of sapphire. After the deposition, samples were slowly cooled down to room temperature in an oxidizing atmosphere at a rate of 5 °C/min. The color of the films was metallic black. The cationic composition (A/Co) of the films was measured by an energy dispersive x-ray spectroscopy (EDX). Crystal structure and growth orientation of all films were analyzed by means of θ-2θ scan of x-ray diffraction (XRD) employing Cu $K\alpha$ radiation, and in-plane alignment was examined by reciprocal space mappings. Electric resistivity was measured using the standard four-probe method with sputter-deposited Au electrodes. Measurement of Seebeck coefficient was performed by the steady-state method using a standard reference of constantan wire.

RESULTS AND DISCUSSION

Crystal Structure

Fig.1-Fig.4 show XRD patterns for Ca-Co-O (Ca/Co~0.4, Fig.1), Ca-Co-O (Ca/Co~0.7, Fig.2), Sr-Co-O (Sr/Co~0.3, Fig.3) and Sr-Co-O (Sr/Co~0.8, Fig.4), films grown on *c*-plane, *a*-plane and *m*-plane of sapphire substrates, respectively. Fig.1-4 (a) confirm the *c* axis-oriented epitaxial growth of layered cobaltites. The series of peaks Fig.1 (a) corresponds to the period of 5.41 Å and this suggests the successful creation of alkaline-free metastable primitive Ca_xCoO_2 structure isomorphous with ion-exchanged metastable Ca_xCoO_2 by the sputtering processing.[11] In Fig.2 (a), the longer periodicity 10.8 Å was observed. This relatively large lattice constant indicates the formation of layered cobaltite with rock-salt type insulating layer, namely, $Ca_3Co_4O_9$. In the same way, the period of 5.75 Å in Fig.3 (a) and 11.1 Å in Fig.4 confirm the successful fabrication of metastable Sr_xCoO_2 and $Sr_3Co_4O_9$, respectively. Examination of in-plane alignment by the reciprocal space mappings was performed and the heteroepitaxial growths of all Ca_xCoO_2, $Ca_3Co_4O_9$, Sr_xCoO_2 and $Sr_3Co_4O_9$ films on *c*-plane sapphire substrates were confirmed. The epitaxial relation of all films on *c*-plane sapphire is (001)[100]film||(0001)$[11\bar{2}0]Al_2O_3$. For the A_xCoO_2 films, the crystal phase was determined to be monoclinic β-phase by reciprocal space mappings.[12]

From Fig.1-Fig.4 (b) and (c), *a* axis orientation perpendicular to *a*-plane and *b* axis orientation perpendicular to *m*-plane of sapphire substrates were confirmed, respectively. For *b* axis-oriented films as shown in Fig.2 (c) and Fig.4 (c), two series of reflections were observed, due to the misfit structure along *b* axis, each of which corresponds to reflection from insulating rock-salt (RS) sublattice and from conducting CoO_2 sublattice.[2] As a consequence, both substrates (*a*-plane and *m*-plane of sapphire) resulted in the formation of layered cobaltites with layer structure perpendicular to the substrate surface.

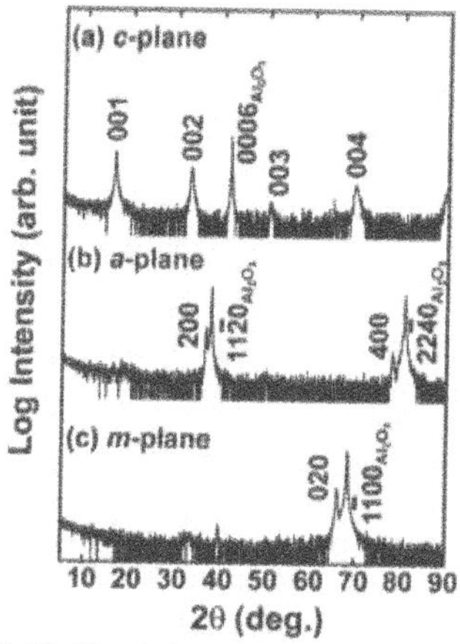

Fig.1: XRD patterns of Ca-Co-O (Ca/Co~0.4) thin films on (a) *c*-plane, (b) *a*-plane and (c) *m*-plane sapphire substrates.

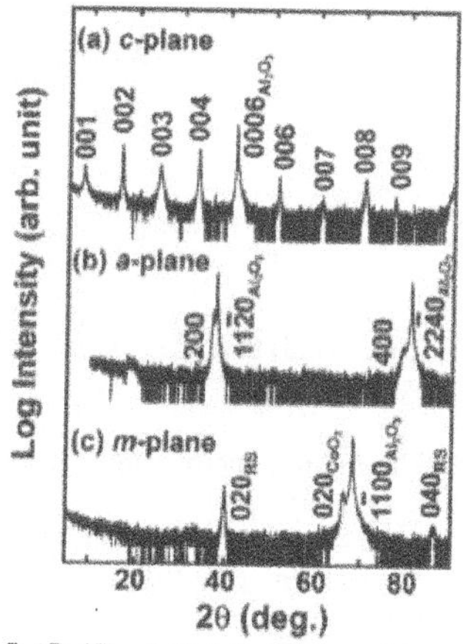

Fig.2: XRD patterns of Ca-Co-O (Ca/Co~0.7) thin films on (a) *c*-plane, (b) *a*-plane and (c) *m*-plane sapphire substrates.

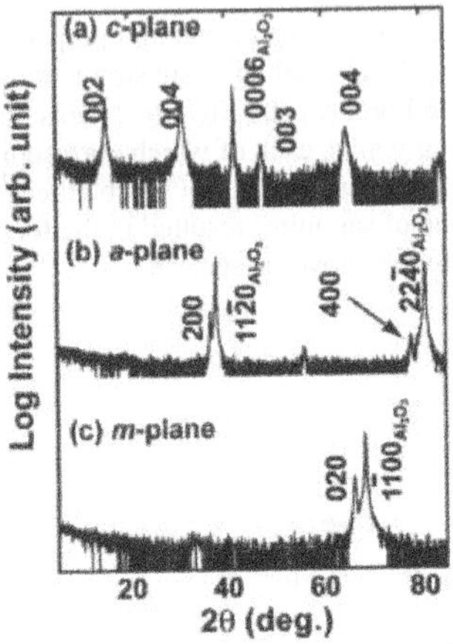

Fig.3: XRD patterns of Sr-Co-O (Sr/Co~0.3) thin films on (a) *c*-plane, (b) *a*-plane and (c) *m*-plane sapphire substrates.

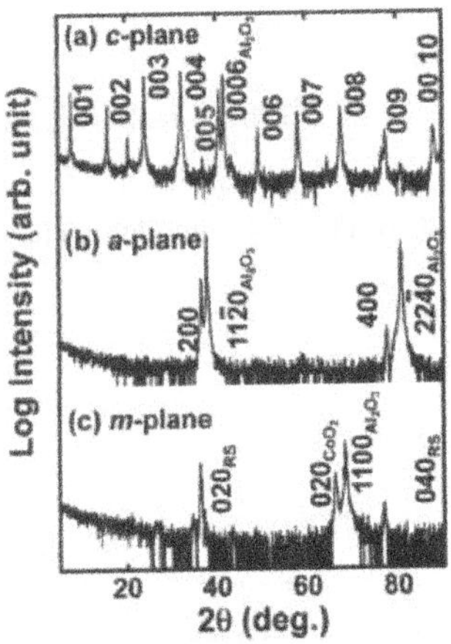

Fig.4: XRD patterns of Sr-Co-O (Sr/Co~0.8) thin films on (a) *c*-plane, (b) *a*-plane and (c) *m*-plane sapphire substrates.

Cross-sectional transmission electron microscopy (TEM) images were taken to examine microstructures of orientation-controlled epitaxial thin films. Fig.5 shows Ca_xCoO_2 film (Fig.5 (a)) and $Ca_3Co_4O_9$ film (Fig.5(b)) deposited on *m*-plane sapphire substrates, respectively. Well-ordered stacking sequences of layers standing perpendicular to the substrate surface were observed in both samples. Each periodicity of stacks corresponds to the interlayer spacing between CoO_2 sheets. Fig.6 shows $Sr_3Co_4O_9$ films on *a*-plane (Fig.6 (a) and (b)) and on *c*-plane

(Fig.6 (c) and (d)) sapphire substrates. Well-ordered stripes with the same periodicity, 11.08 Å, were observed in both films. These stripes correspond to the microstructure of layered $Sr_3Co_4O_9$, and the stacking direction of the stripes in respect to the substrate surface is consistent with the result of XRD analysis. For the film on *a*-plane, a small amount of intergrowth structure was observed, and from the interplanar spacing, this is considered to be Sr_xCoO_2 phase.

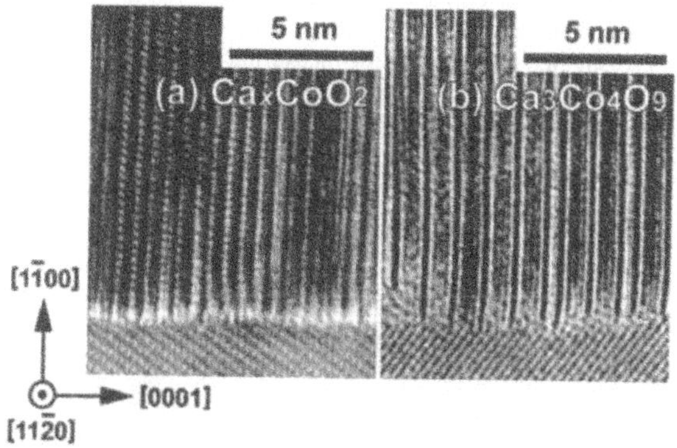

Fig.5: Cross-sectional TEM images of (a) CaxCoO2 and (b) Ca3Co4O9 thin films on m-plane sapphire substrates. Crystallographic directions of sapphire are shown besides the images.

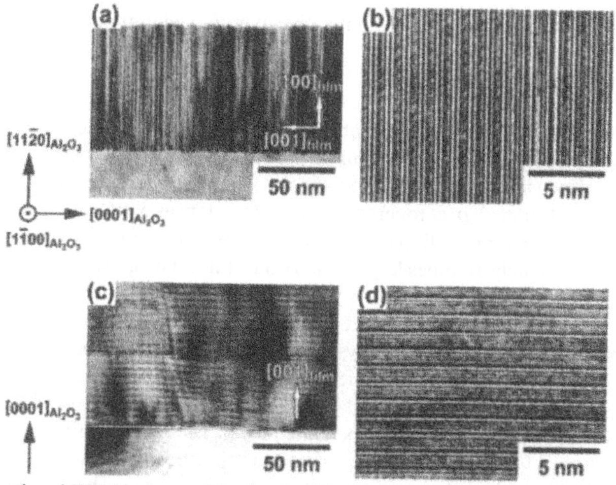

Fig.6: Cross-sectional TEM images of $Sr_3Co_4O_9$ thin films on *m*-plane ((a) and (b)) and *c*-plane ((c) and (d)) sapphire substrates. Crystallographic directions of sapphire are shown besides the images.

The epitaxial relation of all the films on *a*-plane of sapphire was determined to be [100]film||[11$\bar{2}$0]Al_2O_3 for out-of-plane and [001]*film||[0001]Al_2O_3 for in-plane geometry.

In the similar way, the epitaxial relation of all the films on *m*-plane in the out-of-plane direction is [010]film||[1$\bar{1}$00]Al_2O_3. As for in-plane relation, tilted *c* axis of monoclinic lattice for A_xCoO_2 and $A_3Co_4O_9$ (A=Sr and Ca) is not parallel to the c axis of sapphire but the normal axis of the CoO_2 layers is parallel, namely [001]*film||[0001]Al_2O_3.[12]

In the epitaxial layered cobaltite thin films on *a*-plane and *m*-plane sapphire, two distinct directions, parallel to CoO_2 layers (||) and perpendicular to CoO_2 layers (⊥), exist in the film plane as schematically visualized in Fig.7 for the case of A_xCoO_2. Thus, anisotropic transport properties can be investigated by choosing the measuring direction.

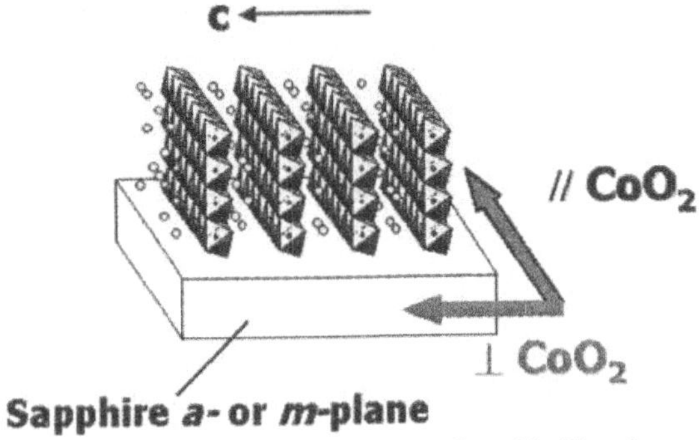

Fig.7: Schematic picture of the A_xCoO_2 film with *c* axis aligned in film plane.

Thermoelectric Properties

Electric resistivity was measured for the rectangular (approximately 5mm×1mm) samples along the direction parallel ($\rho_{\parallel}$) and perpendicular ($\rho_{\perp}$) to the CoO_2 sheets. Fig.8 shows the temperature dependence of the resistivity for Sr_xCoO_2 film on *m*-plane and Ca_xCoO_2 film on *m*-plane sapphire substrates. $\rho_{\parallel}$ at room temperature is as low as 5 mΩ cm for Ca_xCoO_2 and 2 mΩ cm for Sr_xCoO_2. $\rho_{\parallel}$ for both films shows the metallic behavior in the measured temperature range. $\rho_{\perp}$ is approximately 10 times larger than $\rho_{\parallel}$ but still has metallic behavior. Fig.9 shows the temperature dependence of the resistivity for $Sr_3Co_4O_9$ film on *a*-plane and $Ca_3Co_4O_9$ film on *m*-plane sapphire substrates. Below 300 K, parallel resistivity $\rho_{\parallel}$ of both $Sr_3Co_4O_9$ and $Ca_3Co_4O_9$ films shows metallic behavior until it shows the moderate metal-insulator transition around 50 K. Perpendicular resistivity $\rho_{\perp}$ is about 60 times larger than $\rho_{\parallel}$ for $Sr_3Co_4O_9$, and 40 times larger than $\rho_{\parallel}$ for $Sr_3Co_4O_9$. Temperature dependence of $\rho_{\perp}$ for both $Sr_3Co_4O_9$ and $Ca_3Co_4O_9$ is semiconductive through all measured temperature range. The result obtained using the present orientation-controlled $Ca_3Co_4O_9$ thin film is almost consistent with that in $Ca_3Co_4O_9$ single crystal, except for the slight difference in curvature of $\rho_{\perp}$-T curves.[2] The difference may be caused by the mixed contribution of $Ca_3Co_4O_9$ and the existing intergrowth of Ca_xCoO_2 phase.

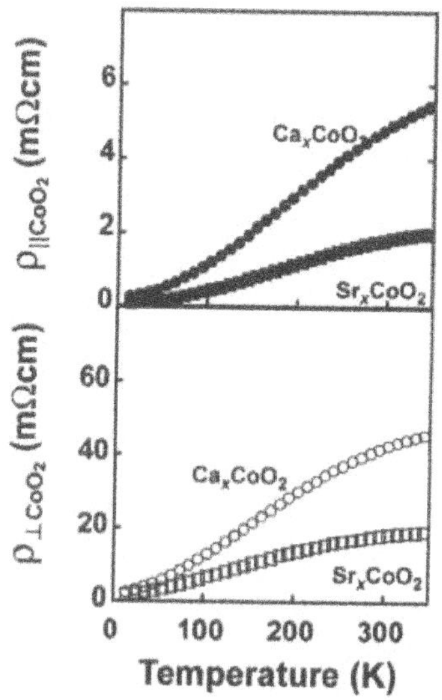

Fig.8: Temperature dependence of electric resistivity parallel to the CoO_2 layers (upper) and perpendicular to the CoO_2 layers (lower) in Sr_xCoO_2 and Ca_xCoO_2 thin films on *m*-plane sapphire substrates.

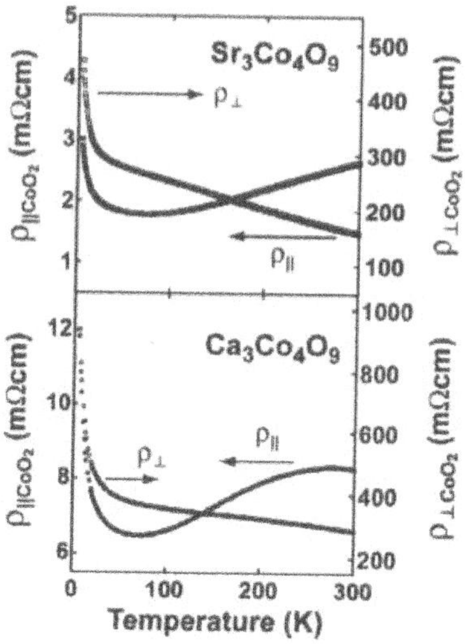

Fig.9: Temperature dependence of electric resistivity parallel to the CoO_2 layers (upper) and perpendicular to the CoO_2 layers (lower) in $Sr_3Co_4O_9$ and $Ca_3Co_4O_9$ thin films on *a*-plane sapphire substrates.

Seebeck coefficients for the rectangular-shaped samples (approximately 4 mm × 1 mm) were measured by a steady-state method using a standard reference of constantan wire. A sapphire substrate of the same rectangular dimension was attached just close to the reference wire in order to match the thermal conduction with the sample. The measurement was made possible for perpendicular Seebeck coefficients $S_{\perp}$ of layered cobaltites thin films, owing to the large *c* axis dimension in the *a* axis or *b* axis oriented epitaxial thin films. The temperature dependences of the Seebeck coefficients for Sr_xCoO_2 and Ca_xCoO_2 are shown in Fig.10. Parallel Seebeck coefficient S// for both Sr_xCoO_2 and Ca_xCoO_2 are around 60 μV/K at room temperature and decreased monotonically with the decrease of temperature. Perpendicular Seebeck coefficients $S_{\perp}$ at room temperature were approximately 25 μV/K, which is about a half of $S_{\parallel}$, and decreased as the temperature is lowered. The parallel Seebeck coefficient $S_{\parallel}$ for the present β-A_xCoO_2 film was comparable with that of the thermoelectric γ-Sr_xCoO_2 bulk made by ion-exchange method.[12] Seebeck coefficients for $A_3Co_4O_9$ (A=Sr and Ca) thin films on *a*-plane sapphire substrates are shown in Fig. 11. Large parallel Seebeck coefficients $S_{\parallel}$ more than 100 μV/K were observed at room temperature for both $Sr_3Co_4O_9$ and $Ca_3Co_4O_9$. Perpendicular Seebeck coefficients $S_{\perp}$ at room temperature were approximately 1/3 of $S_{\parallel}$.

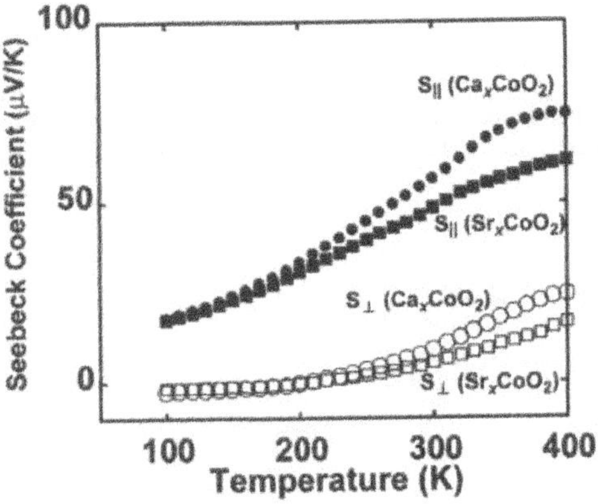

Fig.10: Temperature dependence of Seebeck coefficient in Sr_xCoO_2 and Ca_xCoO_2 thin films on *m*-plane sapphire substrates.

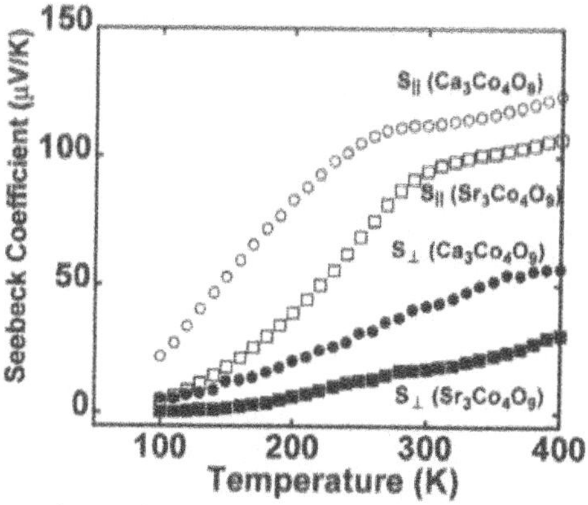

Fig.11: Temperature dependence of Seebeck coefficient in $Sr_3Co_4O_9$ and $Ca_3Co_4O_9$ thin films on *a*-plane sapphire substrates.

SUMMARY

We have prepared epitaxial thin films of primitive-type layered cobaltites A_xCoO_2 and misfit-type $A_3Co_4O_9$ (A=Sr and Ca). Sr_xCoO_2. Ca_xCoO_2 and $Sr_3Co_4O_9$ are metastable phases and the fabrication was made possible by non-equilibrium sputtering process. The crystal orientation of the film could be controlled by the surface plane of the sapphire substrates used. In the epitaxial films on *a*-plane and *m*-plane sapphire, CoO_2 layers grew perpendicularly to the substrate surface and two unique orientations parallel (||) and perpendicular (⊥) to the CoO_2 layers intersect in the film plane. Measurement of anisotropic properties was achieved by considering the transport direction and large anisotropy was confirmed in electric resistivity while more isotropic nature in the Seebeck coefficients. The present thin-film processing using sputtering, which is capable of fabricating metastable phase, has potential advantage in search for thermoelectric materials of higher performance and in basic research such as anisotropic transport properties.

ACKNOWLEDGMENTS

The authors would like to thank Prof. T. Kajitani, Dr. Y. Ono, Dr. Y. Miyazaki of Tohoku University, and Prof. I. Terasaki of Waseda University for their useful discussions. They also wish to thank Dr. K. Ohnaka for the continuous encouragements.

REFERENCES

[1]I. Terasaki, Y. Sasago, and K. Uchinokura, "Large Thermoelectric Power in $NaCo_2O_4$ Single Crystals"; *Phys. Rev. B*, **56**, R12685-7 (1997).
[2]Masset, A. C. et al., "Misfit-Layered Cobaltite with an Anisotropic Giant Magnetoresistance: $Ca_3Co_4O_9$", *Phys. Rev. B*, **62**, 166-75 (2000).

[3]Miyazaki, Y. et al., "Low-Temperature Thermoelectric Properties of the Composite Crystal $[Ca_2CoO_{3.34}]_{0.614}[CoO_2]$", *Jpn. J. Appl. Phys.*, **39**. L531-3 (2000).
[4]Funahashi, R. et al., "Thermoelectric Properties of $Bi_2Sr_2Co_2O_x$ Polycrystalline Materials", Appl. Phys. Lett., **76**, 2385-7 (2000).
[5]W. Koshibae, K. Tsutsui, and S. Maekawa, "Thermopower in Cobalt Oxides", *Phys. Rev. B*, **62**, R12685-7 (1997).
[6]Cushing, B. L. and Wiley, J. B., "Topotactic Routes to Layered Calcium Cobalt Oxides", *J. Solid State Chem.*, **141**, 385-91 (1998).
[7]Ishikawa, R. et al., "Low-Temperature Synthesis and Electric Properties of New Layered Cobaltite, Sr_xCoO_2", *Jpn. J. Appl. Phys.*, **41**, L337-9 (2002).
[8]Peleckis, G. et al., "Enhanced Thermoelectric Properties of Na_xCoO_2 Whisker Crystals", *Appl. Phys. Lett.*, **83**, 5416-8 (2003).
[9]Funahashi, R. et al., "An Oxide Single Crystal with High Thermoelectric Performance in Air", *Jpn. J. App. Phys.*, **39**, L1127-9 (2000).
[10]Minami, H. et al., "Rapid Synthesis and Characterization of $(Ca_{1-x}Ba_x)_3Co_4O_9$ Thin Films Using Combinatorial Methods", *Appl. Surf. Sci.*, **197-198**, 442-7 (2002).
[11]Matsuda, H. et al., "Preparation and Characteristics of 90° Rotated Biepitaxial Fe_3O_4 Thin Films", *J. Mater. Res.*, **17**, 1985 (2002).
[12]Kanno, T. et al., "Anisotropic Thermoelectric Properties in Layered Cobaltite A_xCoO_2 (A=Sr and Ca) Thin Films", *Appl. Phys. Lett.*, **85**, 739-41 (2004).

Electroceramic Materials and Applications
Edited by R. W. Schwartz

Chemical Methods of Preparation of Electroceramics

Electroceramic Materials and Applications
Edited by R. W. Schwartz

IMPACT OF SOLUTION CHEMISTRY ON SUCCESSFULLY DEPOSITING SOL-GEL PZT FILMS DIRECTLY ON COPPER SURFACES

Mark D. Losego and Jon-Paul Maria
The Electroceramic Thin Film Group, North Carolina State University
1001 Capability Drive, Research Building 1
Raleigh, NC 27695-7919

ABSTRACT

The integration of ferroelectric oxide thin films with base metal electrodes is both enticing for manufacturing development and intriguing for scientific investigation. Traditionally, the ferroelectric lead zirconate titanate (PZT) has been considered incompatible with base metal technology because its PbO volatility makes conventional thermodynamic processing impractical. However, recent advances in sol-gel processing science have shown that during film deposition, reactions at the PZT/Cu interface can be avoided if organic constituents are strategically removed. Strategic organic removal requires both careful design of process variables and solution chemistry. Thermal and atmospheric processing conditions are constrained to kinetically maintain an unoxidized copper substrate. Solutions processed within these confined conditions must form gels with sufficiently reduced organic content and properly consolidated gel networks such that phase-pure and crack-free ceramic films can be crystallized. The current work explores three solution chemistries that use different chelating ligands: alkanolamines, acetylactone, and acetic acid. It is found that the alkanolamine solution frustrates perovskite formation and is prone to cracking under processing conditions compatible with the copper substrate. The introduction of water vapor into the processing atmosphere is only moderately successful at resolving these issues. Using a more volatile chelating agent (acetylacetone) shifts the thermal process window nearer a copper compatible regime. Because of its weaker chelation strength, acetic acid solutions are observed to be the most compatible with the processing constraints imposed by the copper substrate.

INTRODUCTION

The development of ferroelectric thin film technology aims to broaden the functionality of microelectronic devices to include enhanced charge storage, electromechanical response, thermal sensitivity, and non-volatile memory. Historically ferroelectric thin films have been fabricated on noble metal or conductive oxide bottom electrodes to minimize substrate reactions/oxidation during high-temperature crystallization of the ferroelectric ceramic. Recently several groups have demonstrated ferroelectric thin films deposited directly on base metal bottom electrodes.[1-7] Commercially, base metals are attractive because of their low-cost and simple patternability; technologically, base metals offer lower resistivities and alluring functionality such as ferromagnetism and shape-memory response.

Unlike early reports of ferroelectric/base metal structures that include interfacial layers[8-11], recent reports have demonstrated ferroelectric/base metal interfaces free of reaction layers.[1, 2, 6, 7] Clean interfaces in the $(Ba,Sr)(Ti,Zr)O_3$ family of ferroelectrics are achieved by a thermodynamic equilibrium approach akin to the methodology of the multilayer capacitor industry.[1, 2, 4, 5, 7, 12] Without an interfacial layer, ferroelectric films on base metals exhibit properties comparable or superior to analogous films deposited on noble metal or conductive oxide electrodes.

Our recent report[3] of lead zirconate titanate (PZT) thin films deposited directly on copper foils deviates from the traditional thermodynamic process because of the difficulties introduced by the lead constituent (small thermodynamic window and PbO volatility). Using a chemical solution deposition (CSD) approach, we identified the specific process variables acting as sources for oxidation. A process flow, including careful solution chemistry selection and a novel composite gel architecture, were designed to circumvent these sources of oxidation. This paper explores the importance of adjusting PZT solution chemistry—particularly the chelating ligand—to achieve processing conditions compatible with copper substrates.

EXPERIMENTAL

Three chelated sol-gel solution chemistries with composition $Pb_{1.15}(Zr_{0.52}Ti_{0.48})O_3$ were prepared. Methanol acted as the solvent for all solutions. The metalorganic precursors consisted of lead acetate trihydrate powder, titanium (IV) isopropoxide, and zirconium (IV) propoxide (70 wt% in 1-propanol). The first solution used alkanolamines (diethanolamine and triethanolamine) to chelate the metal alkoxides and is based upon previous work by Kim *et al.*[13] The second solution used acetylacetone as a chelating agent and was prepared by simply mixing the precursors and acetylacetone in methanol. The final solution was chelated with acetic acid and is modeled after the inverted mixing order chemistry originally published by Assink and Schwartz.[14] All films were spun on 18 μm thick copper foils (Oak-Mitsui, PLSP 0.5 oz foil) at 3000 rpm for 30s. Films were dried on a hotplate at 250°C for 5 min, except for the acetylacetone films that were pyrolyzed directly at 300°C on the hotplate. Higher temperature pyrolysis/thermolysis was carried out in a 2 inch diameter controlled atmosphere tube furnace. Nitrogen or forming gas (99% N_2 + 1% H_2) were flowed at ~1 slm to control the furnace atmosphere. To achieve a "water vapor" atmosphere, nitrogen was bubbled through a heated (~75°C) water bath. Films consisted of 4-5 CSD layers and had a final film thickness ranging from 300-700 nm depending on the solution chemistry. Films were crystallized at 650°C for 30 min in a flowing nitrogen atmosphere. Although the ~10 ppm O_2 impurity in the nitrogen source is thermodynamically sufficient to oxidize copper at this temperature, the PZT film can kinetically limit this oxidation process.

PZT films were characterized with x-ray diffraction, dielectric measurements, and optical microscopy. X-ray diffraction data was collected on a Bruker AXS D-5000 x-ray diffractometer equipped with a GADDS area detector. The copper anode was operated at 40 kV / 30 mA and scans were collected for 15 min. Top electrodes (~200 μm diameter) were sputtered through a shadow mask for electrical measurement. Dielectric data was collected on a HP 4192A impedance analyzer. Dielectric tunability measurements used a sampling AC voltage of 0.05 V.

RESULTS AND DISCUSSION

PZT solutions prepared with alkanolamine chelating agents show less solution aging effects than other chelated sol-gel chemistries. However, to obtain high-quality ferroelectric films, alkanolamine chemistries require aggressive pyrolysis conditions to fully remove the alkanolamine ligands prior to crystallization. Intermediate pyrolysis steps of 400°C-450°C for 5-20 min in oxidizing atmospheres (air or O_2) are often used to achieve full pyrolysis and volatilization of alkanolamine ligands.[8, 15, 16] Previously, we demonstrated that thin film

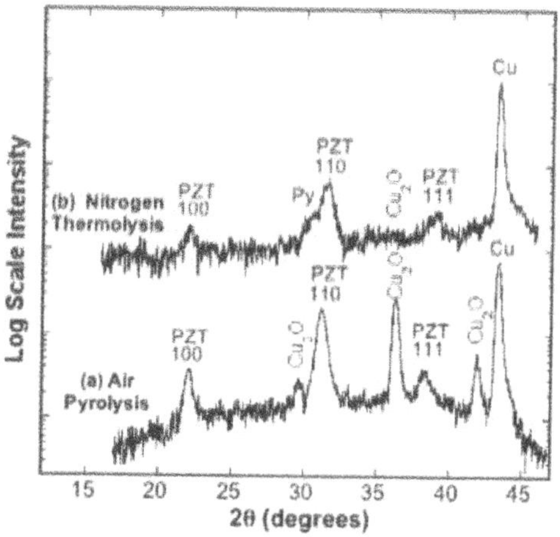

Fig. 1: *X-ray diffraction patterns comparing the phase assemblage of PZT films deposited on copper using an alkanolamine chelated solution; (a) pyrolyzed at 400°C in air and (b) thermolyzed at 400°C in nitrogen. Both samples were crystallized at 650°C in nitrogen.*

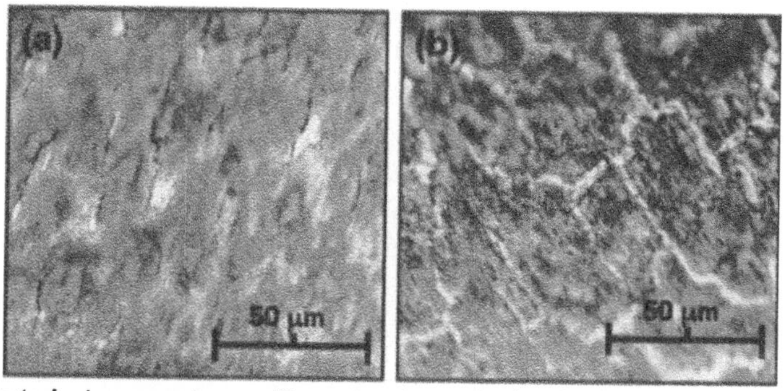

Fig. 2: *Optical microscopy images illustrating the cracks that develop when the alkanolamine chelated solution undergoes a 400°C thermolysis in nitrogen (a) after thermolysis; (b) after crystallization at 650°C in nitrogen.*

deposition processes conducted in air on copper substrates must be kept below ~250°C to kinetically avoid Cu_2O formation at the interface.[3] This constraint makes traditional processing of alkanolamine solutions incompatible with copper substrates; we demonstrate this in Fig. 1a.

Simply changing the pyrolysis atmosphere from air to nitrogen alleviates copper oxidation as illustrated in Fig. 1b. However, films processed in nitrogen are prone to severe cracking as shown in Fig. 2. Gels treated in nitrogen atmospheres are believed to primarily undergo a thermolysis process (heat-driven decomposition) rather than a pyrolysis process (oxidative combustion).[17] Thermolysis is less likely to combust the alkanolamine ligands into smaller, more volatile organic species. Consequently, the gels thermolyzed in nitrogen are expected to retain more organic content. This hypothesis is consistent with the frustrated perovskite formation evident in diffraction pattern (b) of Fig. 1. This incomplete organic removal may also inhibit sufficient M-O-M network formation. Without the rigidity of this network, the gel cannot maintain coherency upon densification. To improve organic ligand removal, two approaches are investigated: (1) the introduction of water vapor into the thermolysis atmosphere and (2) changing the chelating agent to a more volatile species.

Water vapor in the thermolysis atmosphere should increase the hydrolysis reaction rate, thereby accelerating the ligand removal process. Previous work in other sol-gel systems has demonstrated that water vapor atmospheres improve organic extraction, reduce process temperatures, and alleviate cracking.[18-20] Similar results are observed in this system. Fig. 3 demonstrates the reduction in film cracking when a nitrogen atmosphere containing water vapor is utilized. To completely suppress film cracking, water vapor was required during crystallization. Unfortunately, the water vapor atmosphere is too aggressive for the copper substrate. Copper oxidation is evident in the diffraction patterns shown in Fig. 4. Water vapor may directly enhance the copper oxidation rate, or by increasing the oxygen partial pressure via the reaction:

$$H_2O \rightarrow H_2 + 1/2O_2$$

it may indirectly create a more oxidizing atmosphere. If used only during thermolysis, water vapor fails to completely eliminate cracking but does improve perovskite crystallization (Fig. 4) relative to the dry nitrogen atmosphere (Fig. 1b). These results imply that although the water

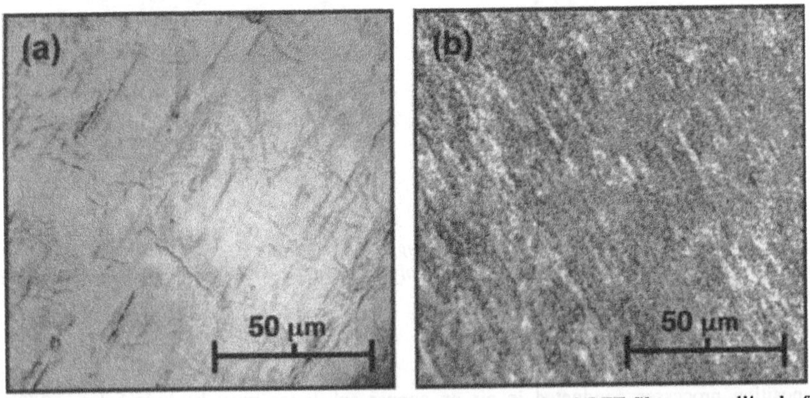

Fig. 3: *Optical microscopy images showing alkanolamine-chelated PZT films crystallized after undergoing thermolysis in a water vapor atmosphere. (a) Residual cracking occurs in water vapor thermolyzed sample crystallized in dry nitrogen; (b) No cracking is observed when the crystallization atmosphere contains water vapor. (Image (a) is taken after coating the PZT film with metal so that cracks are more visible.)*

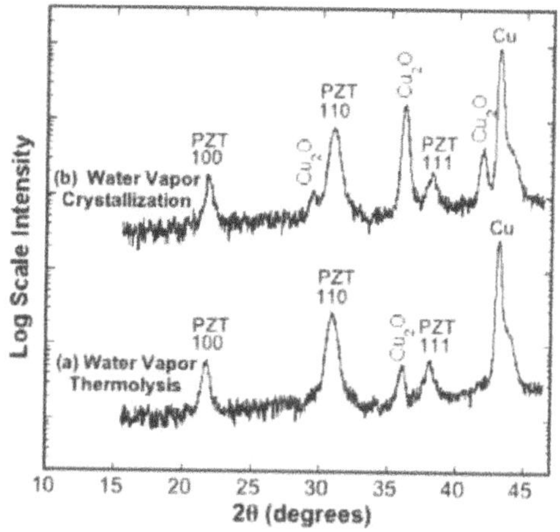

Fig. 4: *X-ray diffraction data illustrating how water vapor in the process atmosphere promotes copper oxidation; (a) PZT film thermolyzed in water vapor and crystallized in dry nitrogen; (b) PZT film thermolyzed and crystallized in water vapor.*

vapor improves the hydrolysis rate and gel network formation, it does not significantly enhance the volatility of the alkanolamines. Thus, many organics remain trapped in the gel's porosity, leading to the observed hairline cracks during crystallization.

To improve volatility of the organic constituents the alkanolamine chelating agent ($T_{boiling}$ > 200°C) is replaced with a lower boiling point β-diketonate, acetylacetone ($T_{boiling}$ ~136°C). This variation in solution chemistry immediately reduces the pyrolysis temperature necessary for crack-free films from ~400°C to ~300°C. Although this temperature remains incompatible with the copper substrate as illustrated in the diffraction pattern of Fig. 5, these uncracked films can be evaluated electrically. The electrical properties reported in Fig. 6 (relative permittivity < 350; dielectric loss > 0.10) illustrate why a Cu/PZT system containing a Cu_2O reaction layer cannot compete with current noble metal or conductive oxide bottom electrode technology. Attempts to lower the pyrolysis temperature (< 300°C) and reduce Cu_2O formation for this solution chemistry resulted in cracked films—consistent with the inadequate organic removal hypothesis.

As recently documented[3], an acetic acid chelation chemistry allows the deposition of PZT films on copper substrates without significant interfacial reactions. Because acetic acid is a weaker chelating agent that acetylacetone[21], this change in chemistry further assists organic extraction. As Fig. 7 demonstrates, this PZT/Cu system has dielectric properties comparable to conventional noble metal or conductive oxide electrodes. To manage residual cracking in acetic acid chelated films, a composite gel architecture was developed[3]; this scheme thermolyzes initial layers to maintain interface quality while subsequent layers are dried at elevated temperatures to

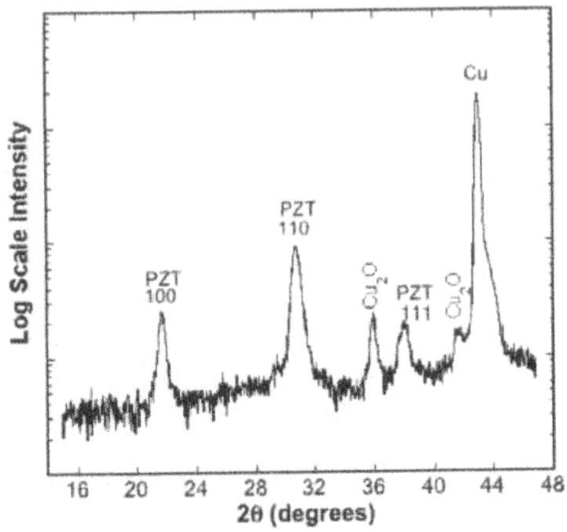

Fig. 5: *X-ray diffraction data for a PZT film prepared from an acetylacetone chelated solution on a copper substrate. The film was pyrolyzed at 300°C on a hotplate and then crystallized at 600°C in nitrogen.*

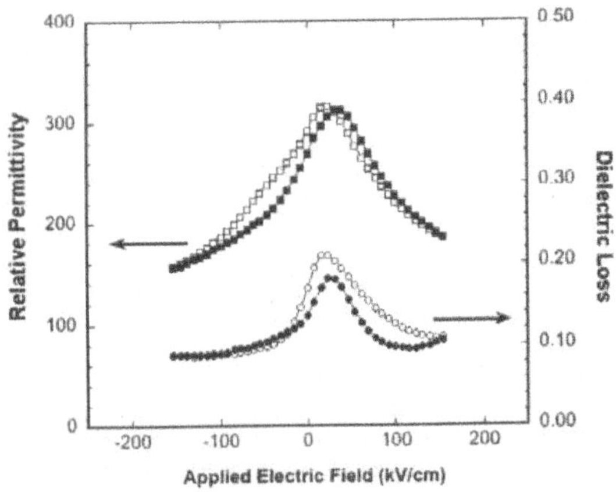

Fig. 6: *Dielectric tunability of the PZT film prepared on a copper foil using the acetylacetone chelated solution; this film contains a significant Cu_2O reaction layer at the PZT/Cu interface. Dielectric data was collected at 1 MHz and room temperature.*

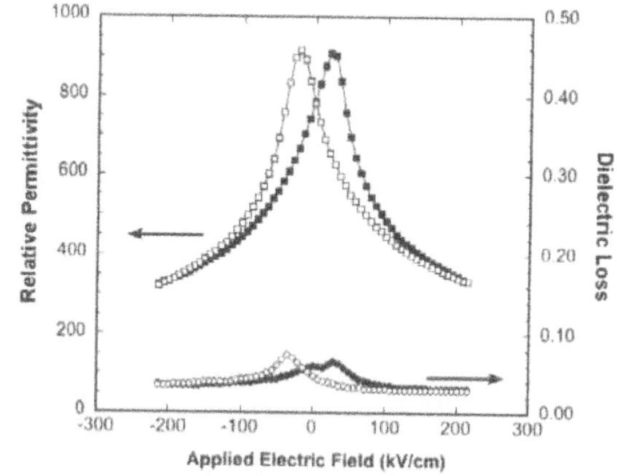

<u>Fig. 7:</u> *Dielectric tunability of a PZT film prepared on a copper foil with an acetic acid chelated solution using a composite gel architecture. In this sample a Cu_2O reaction layer is undetectable by x-ray diffraction. Dielectric data was collected at 10 kHz and room temperature.*

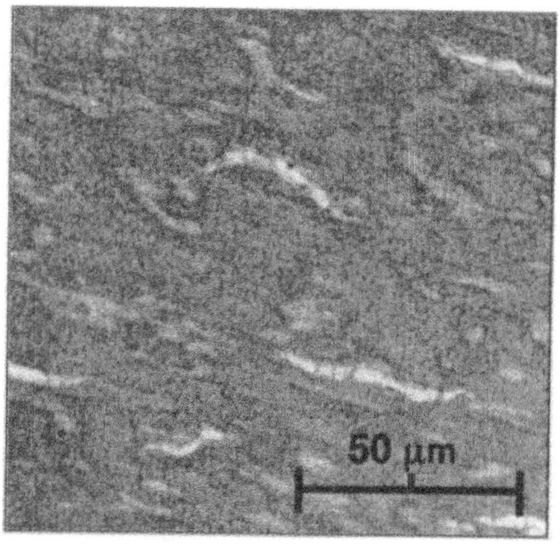

<u>Fig. 8:</u> *Optical microscopy image showing a cracked PZT film prepared from an alkanolamine chelated PZT solution using the composite gel architecture process.*

improve gel consolidation and prevent cracking. To verify that a combination of process design and solution chemistry is necessary to successfully synthesize PZT films on copper surfaces, a PZT film was prepared from the alkanolamine-chelated solution using the composite gel architecture. Fig. 8 shows that such a process fails to resolve the cracking problems for this solution chemistry. Thus, both a careful design of process variables and solution chemistry is necessary to achieve PZT films on copper surfaces that are crack-free and void of interfacial reactions.

CONCLUSIONS

Chemical solution deposited PZT thin films were prepared on copper substrates using different solution chemistries. When not applying equilibrium conditions to process oxide thin films on base metal substrates, process variables (temperature, atmosphere, etc.) are constrained to conditions that kinetically impede reactions at the film/substrate interface. The experiments in this work have demonstrated that the parameters consistent with the base metal substrate may not be optimal for processing sol-gel films of certain solution chemistries. Common consequences are film cracking and improper phase assemblage; these problems stem from inadequate organic removal and insufficient gel consolidation prior to crystallization. To resolve these problems, the solution chemistry can be tailored to provide proper organic removal and gel formation under the desired conditions. By shifting away from high boiling point, strongly bound chelating ligands (alkanolamines, acetylacetone), successful sol-gel films can be prepared at lower temperatures and in less aggressive/oxidizing atmospheres, resulting in a process that is compatible with copper substrates.

REFERENCES

[1]J. Ihlefeld, B. Laughlin, A. Hunt-Lowery, W. Borland, A. Kingon and J. P. Maria, "Copper compatible barium titanate thin films for embedded passives." *J. Electroceram* **14** 95-102 (2005).

[2]B. Laughlin, J. Ihlefeld and J.-P. Maria, "TEM and electrical analysis of sputtered barium strontium titanate (BST) thin films on flexible copper substrates." *Materials Research Society Proceedings* **784** 301-06 (2004).

[3]M. D. Losego, L. H. Jimison, J. F. Ihlefeld and J.-P. Maria, "Ferroelectric response from lead zirconate titanate thin films prepared directly on low-resistivity copper substrates." *Appl. Phys. Lett* **86** 172906 (2005).

[4]R. J. Ong, J. T. Dawley and P. G. Clem, "Chemical solution deposition of biaxially oriented (Ba,Sr)TiO_3 thin films on < 100 > Ni." *J. Mater. Res* **18** 2310-17 (2003).

[5]J. T. Dawley and P. G. Clem, "Dielectric properties of random and < 100 > oriented $SrTiO_3$ and (Ba,Sr)TiO_3 thin films fabricated on < 100 > nickel tapes." *Appl. Phys. Lett* **81** 3028-30 (2002).

[6]M. P. Siegal, P. G. Clem, J. T. Dawley, J. Richardson, D. L. Overmyer and T. G. Holesinger, "Optimizing SrTiO3 films on textured Ni substrates using chemical solution deposition." *J Mater Res* **20** 910-21 (2005).

[7]B. Laughlin, J. Ihlefeld and J.-P. Maria, "Preparation of sputtered (Ba_xSr_{1-x})TiO_3 thin films direcly on copper." *J. Am. Ceram. Soc* **88** 2652-54 (2005).

[8]J. P. Maria, K. Cheek, S. Streiffer, S. H. Kim, G. Dunn and A. Kingon, "Lead zirconate titanate thin films on base-metal foils: An approach for embedded high-permittivity passive components." *J. Am. Ceram. Soc* **84** 2436-38 (2001).

[9]P. G. Mercado and a. P. Jardine, "Thin-Film multilayers of ferroelastic TiNi-ferroelectric PZT - fabrication and characterization." *J Intel Mat Syst Str* **6** 62-70 (1995).
[10]S. Seifert, D. Sporn, T. Hauke, G. Muller and H. Beige, "Dielectric and electromechanical properties of sol-gel prepared PZT thin films on metallic substrates." *J Eur Ceram Soc* **24** 2553-66 (2004).
[11]Q. Zou, H. E. Ruda, B. G. Yacobi, K. Saegusa and M. Farrell, "Dielectric properties of lead zirconate titanate thin films deposited on metal foils." *Appl. Phys. Lett* **77** 1038-40 (2000).
[12]J. Ihlefeld, J.-P. Maria and W. Borland, "Dielectric and microstructural properties of barium titanate zirconate thin films on copper substrates." *J. Mater. Res* **20** (2005).
[13]S. H. Kim, D. J. Kim, J. Hong, S. K. Streiffer and A. I. Kingon, "Imprint and fatigue properties of chemical solution derived $Pb_{1-x}La_x(Zr_yTi_{1-y})_{1-x/4}O_3$ thin films." *J. Mater. Res* **14** 1371-77 (1999).
[14]R. A. Assink and R. W. Schwartz, " ^{1}H and ^{13}C NMR investigations of $Pb(Zr,Ti)O_3$ thin-film precursor solutions." *Chem. Mater* **5** 511-17 (1993).
[15]J. S. Hwang, W. S. Kim, H. H. Park and T. S. Kim, "The effect of intermediate anneal on the ferroelectric properties of direct-patternable PZT films." *Sensor Actuat a-Phys* **117** 137-42 (2005).
[16]T. Kim, A. I. Kingon, J. P. Maria and R. T. Croswell, "Ca-doped lead zirconate titanate thin film capacitors on base metal nickel on copper foil." *J. Mater. Res* **19** 2841-48 (2004).
[17]R. W. Schwartz, T. Schneller and R. Waser, "Chemical solution deposition of electronic oxide films." *Cr. Chim* **7** 433-61 (2004).
[18]K. Kato, "Low-temperature synthesis of $SrBi_2Ta_2O_9$ ferroelectric thin films through the complex alkoxide method: Effects of functional group, hydrolysis and water vapor treatment." *Jpn. J. Appl. Phys. 1* **37** 5178-84 (1998).
[19]K. Kitaoka, K. Takahara, H. Kozuka and T. Yoko, "Sol-gel processing of transparent PLZT($(Pb,La)(Zr,Ti)O_3$) fibers." *J. Sol-Gel Sci. Techn* **16** 183-93 (1999).
[20]H. Suzuki, M. B. Othman, K. Murakami, S. Kaneko and T. Hayashi, "Low-temperature processing of ferroelectric $Pb(Zr_{0.53}Ti_{0.47})O_3$ thin film from molecular-designed alkoxide precursor solution." *Jpn. J. Appl. Phys. 1* **35** 4896-99 (1996).
[21]R. W. Schwartz, R. A. Assink, D. Dimos, M. B. Sinclair, T. J. Boyle and C. D. Buchheit, "Effects of acetylacetone additions on PZT thin film processing." *Materials Research Society Symposium Proceedings* **361** 377-87 (1995).

Electroceramic Materials and Applications
Edited by R. W. Schwartz

SOL-GEL ROUTES TO NANOSTRUCTURED PATTERNED FERROELECTRIC THIN FILMS WITH NOVEL ELECTRONIC AND OPTICAL FUNCTIONS

Makoto Kuwabara*,**,[1], Yong Jun Wu[1,2], Juan Li[1], and T. Koga[1]

[1]Department of Applied Science for Electronics and Materials, Kyushu University,
6-1 Kasuga-kouen, Kasuga 816-8580, Japan
[2]Department of Materials Science and Engineering, Zhejiang University,
Zheda road 38, Hangzhou 310027, China

ABSTRACT

We have successfully fabricated barium titanate ($BaTiO_3$: BTO) patterned thin films of < 1 μm in thickness on Pt/Ti/Si substrates by combining an electrophoretic deposition (EPD) method using a suspension of BTO nanoparticles and electron beam lithography, which was used to form resist molds on the substrates. A suspension of BTO nanoparticles (size=10-15 nm) for EPD was prepared by dispersing BTO nanocrystals, which were synthesized by a sol-gel process using a high concentration alkoxides solution (1.1 mol/L), in 2-methoxyethanol added with acetylacetone. BTO ceramic thin films obtained by firing the EPD green films at various temperatures between 600° and 800°C consisted of particles with diameters within 30 nm, and had a dielectric constant of about 100 and a dielectric loss < 1% and a very high dielectric strength of more than 500 kV/cm. The obtained nanostructured BTO ceramic thin films and sub-micrometer-sized ceramic pillars exhibited distinct piezoelectric butterfly curves and also clear ferroelectric hysteresis loops; this may prove that even BTO crystals of 30 nm in diameter can exhibit induced ferroelectric characteristics under high electric field.

INTRODUCTION

Deposition of nano-structured ceramic thin films on conducting substrates has attracted increasing attention due to novel electronic and optical functions that are provided exclusively by such materials, and moreover, from points of view of both materials science and processing technology. As one of the most important issues in electroceramics, preparation of barium titanate ($BaTiO_3$ or BTO) thin films has been investigated extensively because of its promising electrical and optical properties. $BaTiO_3$ thin films are of significant interest for a variety of technological applications, such as multilayer ceramic capacitors (MLCC), sensors, actuators, thermistors, dynamic random access memory (DRAM) and electroluminescent elements.[1-3] A variety of methods, such as sputtering, metal organic chemical vapor deposition (MOCVD), pulsed laser deposition (PLD), chemical solution deposition (CSD) have been investigated to prepare BTO thin films.[4-7] Compared with these methods, electrophoretic deposition (EPD) offers advantages of process simplicity, low cost, easily deposition of multilayer films of controlled thickness on complex shaped or patterned substrates.[8,9] It has been studied extensively for fabricating advanced coating, nanocomposites, laminated structures, and functional graded materials. Electrophoretic deposition of BTO thick films from micrometer or sub-micrometer-sized particles suspensions have been investigated by several groups[10-12] and EPD of BTO thin films has been reported in our previous paper.[13]

Meanwhile, the continuing trend of miniaturization in the microelectronics industry requires integration of micropatterned BTO thin films into electroceramic packages. Various methods, including top-down and bottom-up approaches have been used to prepare micropatterned electroceramics. Within the top-down approach, focused ion beam patterning, electron beam direct writing, dry-etching, wet-etching, nanoimprint lithography and other lithography techniques have been widely investigated. The bottom-up approach can be classified into two main routes, viz., physical and chemical.[14,15] However, because a relatively high sintering temperature is required to process $BaTiO_3$ ceramics by conventional methods, the development of new patterning techniques applicable to BTO is desired.

In this study, a templating method using the sol-gel process, EPD and electron beam lithography was proposed to fabricate micropatterned $BaTiO_3$ thin films at a relatively low temperature.

EXPERIMENTAL PROCEDURE

Figure 1 shows the schematic representation of preparing BTO nano-particles and micropatterned BTO thin films. The BTO nano-particles were prepared by a high-concentration sol-gel method.[16,17] High purity barium diethoxide ($Ba(OC_2H_5)_2$, >99%) was dissolved in a mixed solvent of methanol (CH_3OH, >99%) and 2-methoxyethanol ($CH_3OC_2H_4OH$, EGMME, >99.8%) with a volume ratio of 3:2. After stirring the solution for 24 hours, titanium tetra-iso-propoxide ($Ti(OiC_3H_7)_4$, >99%) was added in it to prepare a BTO precursor solution and then stirred for another 24 hours. The concentration of the precursor solution was 1.1 mol/l. A mixed solution of distilled water and EGMME with a volume ratio of 1:1 was used for hydrolysis. The molar ratio of H_2O and Ba was 10:1. All above processes were performed in a dry N_2 atmosphere. The hydrolyzed solution, after stirred at -30 °C for 10 min, was sealed in an autoclave and aged in a 90°C oven for 1 hour to allow gel formation. After discarding the liquid generated in association with the gel contraction, the BTO gel was dispersed into EGMME using sonication for several hours. Then acetylacetone ($CH_3COCH_2COCH_3$, Acac, >99%) was added and again sonicated to prepare BTO suspension. The volume ratio of EGMME and Acac was fixed as 9:1 and the concentration of BTO nanoparticles in the suspension was controlled to be 0.075 mol/l.

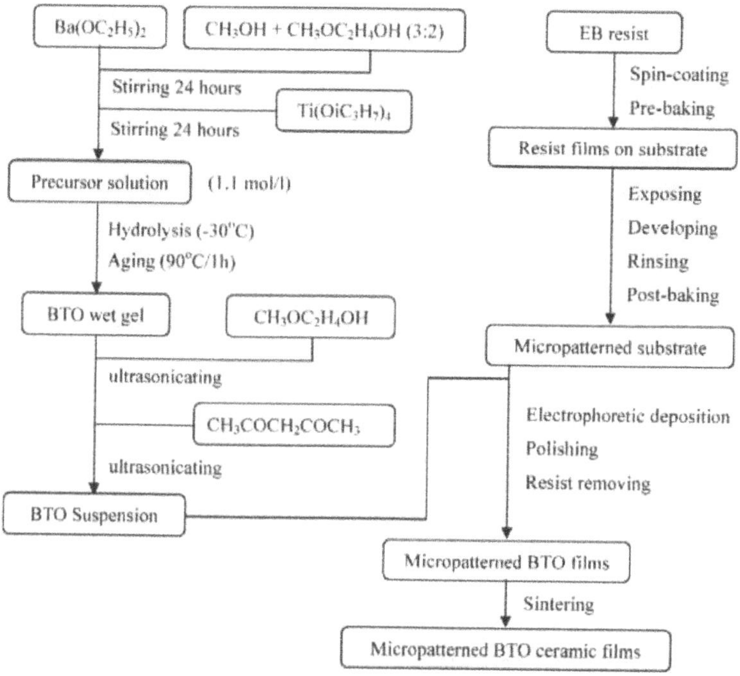

Fig. 1. Schematic representation of the method used for preparing micropatterned $BaTiO_3$ ceramic films.

For preparation of micropatterned BTO thin films, electron beam resist films were first formed on Pt/Ti/Si substrates by spin-coating at 3000 rpm for 20s, followed by pre-baking on a hot-plate at 180°C for 3 minutes. The spin-coating/pre-baking cycle was repeated several times to obtain resist films with a thickness of about 800 nm. An EB lithography system (ELS-5700, Elionex, Japan) was then used to create latent micropatterns in the resist films. After developed and baked, resist molds with different micropatterns on Pt/Ti/Si substrates were fabricated. These micropatterned substrates were used as cathodes and the same size plates of stainless steel were used as anodes in the EPD process. A schematic diagram of the EPD system employed is shown in Fig. 2. The distance between the cathode and the anode was 2 cm. A DC voltage of 5 V was applied for EPD. As-deposited samples were

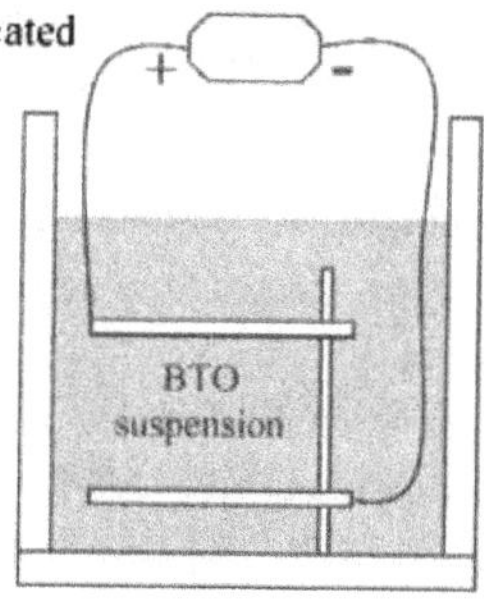

Fig. 2. Schematic diagram of EPD system.

dried in a 50°C oven in air. After removing $BaTiO_3$ nanoparticles layers superfluously deposited on the resist molds by mechanical polishing, the resist molds were dissolved in a remover to obtain nanoparticles patterns left on the substrates. The nanoparticles patterns were then calcined at 600°C for 30min to yield micropatterned $BaTiO_3$ ceramic thin films.

The microstructures of the sol-gel-derived BTO particles were evaluated by high resolution transmission electron microscopy (HR-TEM, EM-002BF, Topcon, H-9000NAR, Hitachi, Japan) and X-ray diffraction (XRD, M18XHF, Mac Science, Japan). The transmittance property of a BTO suspension in the wavelength of 300 to 900 nm was measured with a UV/VIS spectrophotometer (U-550, Jasco, Japan). The surface morphology and thickness of both an EB resist mold and a deposited thin film were evaluated by field emission scanning electron microscopy (FE-SEM, S5000, Hitachi, Japan); FE-SEM images were taken on the surface and cross section of each sample. Scanning probe microscopy (SPM, Nano-R system, Pacific Nanotechnology, USA) was also used to characterize the surface of a micropatterned $BaTiO_3$ thin film. The strain hysteresis loop of a single BTO pillar was measured using SPM (contact mode) equipped with a ferroelectric properties measurement system (FCE system, Toyo Technical, Japan). All the measurements were carried out at room temperature.

RESULTS AND DISCUSSION

Figure 3 shows the transmittance of a BTO suspension prepared in this study (slightly yellowish) with a high value in the visible region and an absorption edge near 400 nm (about 3.1 eV), indicating that the BTO particles were well dispersed and no significant size effect on the absorption edge was observed for the BTO nanoparticles. The BTO suspension was very stable and no agglomeration or sedimentation occurred for more than two months. An appropriate addition of Acac was found to be effective to obtain a suspension of well-dispersed BTO nanoparticles. The reason is explained by an improved electrostatic stabilization of BTO particles caused by adsorption of protons, generated in the keto-enol equilibrium reaction of Acac, on the particles' surfaces.

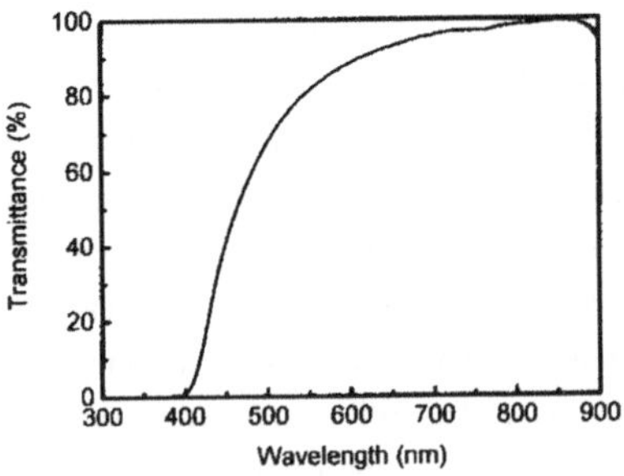

Fig. 3. Transmittance spectrum of a BTO suspension between 300 and 900 nm.

A representative TEM micrograph of BTO nanoparticles prepared in the present study is shown in Fig. 4, clearly indicating that the BTO particles were well crystallized and the average particle size was about 10 nm. From a selected area electron diffraction (ED) pattern of the BTO nanoparticles, shown in the inset of Fig. 4, it has been found that the BTO particles crystallized in the perovskite cubic phase. This was also confirmed from an analysis of the XRD peak broadening. By using the Scherrer equation, the crystallites size was calculated to be 8.6 nm. This result is consistent with that of TEM observations. For EPD experiments, a BTO suspension with a BTO particles concentration of 0.075 mol/l was used, and the BTO particles in the suspension were found to have an average particle size of ~10 nm and a very narrow particle size distribution, measured with a laser particle size analyzer. This suggests that the primary BTO nanoparticles formed no significant agglomeration in the suspension. The small size of BTO nanocrystallites in the well dispersed suspension makes it possible to deposit BTO into micropatterned resist mold formed on Pt/Ti/Si substrates.

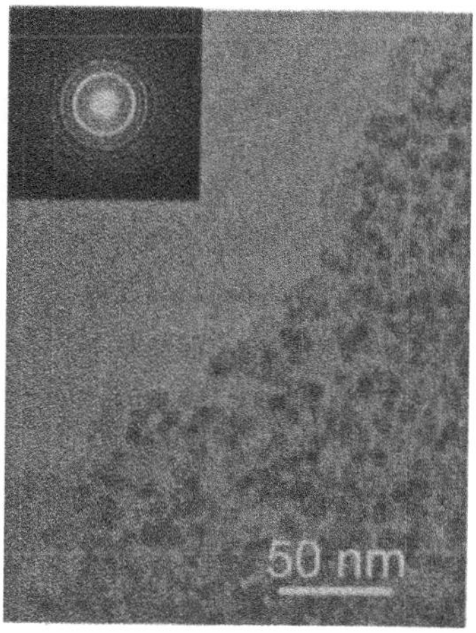

Fig. 4. TEM image of BTO nanoparticles and an ED pattern of the particles (in the inset).

SEM micrographs of BTO films deposited by EPD at different conditions are shown in Fig. 5. The BTO films deposited from the nanoparticles suspension had much flatter surfaces than those from the sub-micron particle suspension.[11,12] A uniform microstructure and a very smooth surface were observed for the BTO films. The thickness of the deposited film was able to be easily controlled by adjusting the EPD conditions. BTO thin films with a desired thickness from ~100 nm to about 1 μm have been fabricated in one step by adjusting the EPD voltage and duration time. Dielectric property measurements of BTO ceramic thin films prepared at temperatures between 600 and 800 °C demonstrated that the films had a dielectric constant of about 100 and a dielectric loss <1%. The films were also found to have a very high dielectric strength >500 kV/cm. The cause that the BTO ceramic thin films exhibited such a relatively low dielectric constant is being investigated.

Figure 6 a) and b) show SEM micrographs of a resist mold with a hexagonal array of air holes in it, which was formed on a Pt/Ti/Si substrate. The micropatterned resist mold was prepared by electron beam lithography, which is one of the most advantageous techniques for preparing high quality micropatterns with high resolution. The thickness of the resist film (the depth of air holes) was measured to be about 800 nm, and the edge of the air holes was confirmed to be very sharp. Figure 6 c) and d) show different views of a hexagonal array of BTO nanoparticles pillars as prepared by EPD. The BTO nanocrystallites are seen to have been compactly deposited and the

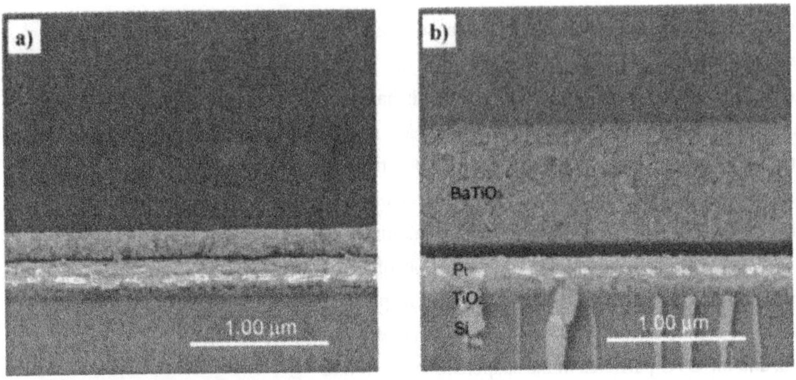

Fig. 5. FE-SEM micrographs of BTO films prepared under different EPD conditions: a) 5v/1min; b) 15V/20min.

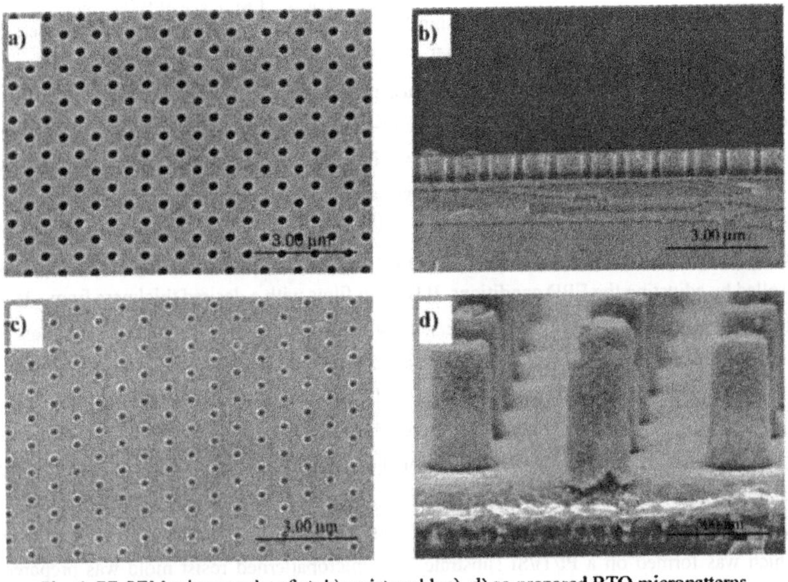

Fig. 6. FE-SEM micrographs of a), b) resist mold; c), d) as-prepared BTO micropatterns.

BTO pillars almost had the same diameter as that of air hole in the resist mold. The height of the pillars can be controlled in the polishing process.

Other micropatterns of microlines or microgrids also have been successfully prepared. Figure 7

Fig. 7. FE-SEM micrographs of a), c) resist mold; b), d) BTO micropatterns.

a) shows an SEM micrograph of a resist mold with microlines on a Pt/Ti/Si substrate. The resist microlines had a width of 400 nm with a spacing of 400 nm in between, and in Fig. 7 b) is shown an SEM micrograph of BTO microlines produced from the mold. In Fig. 7 c) and d) are shown an SEM micrograph of a resist mold with an array of square pillars with a width of 2.4 μm and a spacing of 600 nm in between and an SEM micrograph of a BTO microgrid produced from the resist mold, respectively.

Figure 8 shows a 3D image of an as-deposited $BaTiO_3$ nanoparticles microgrid, evaluated with SPM over an area of 10 μm×10 μm, indicating that the BTO microgrid had a height of about 200 nm and a robust structure with a sharp edge.

Fig. 8. SPM 3D image of the as-deposited $BaTiO_3$ micropatterns.

Sintering of BTO

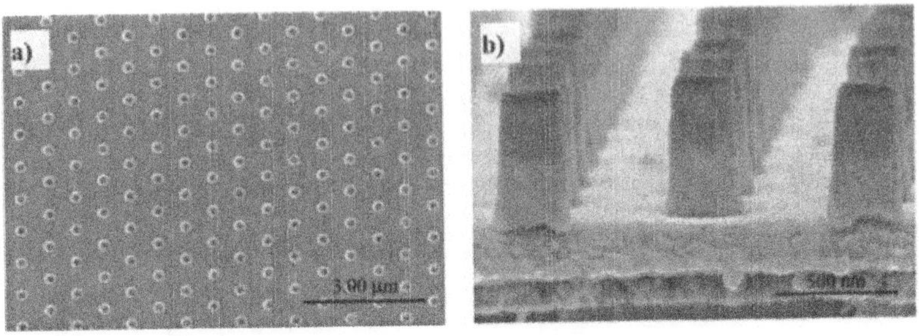

Fig. 9. FE-SEM micrographs of sintered BTO micropatterns.

nanoparticles pillars as deposited by EPD, whose SEM micrographs are shown in Fig. 6 c) and d), at 600°C for 30 min yielded robust BTO ceramic pillars without any appreciable disordering on the substrate, as shown in Fig. 9. To examine the ferroelectric properties of the BTO ceramic pillars, measurements of strain-DC characteristics were carried out for some of the ceramic pillars. Figure 10 shows a quadratic strain hysteresis loop observed for a single BTO ceramic pillar, demonstrating that the BTO pillars sintered at a low temperature of 600°C exhibited induced ferroelectric characteristics under high electric field.

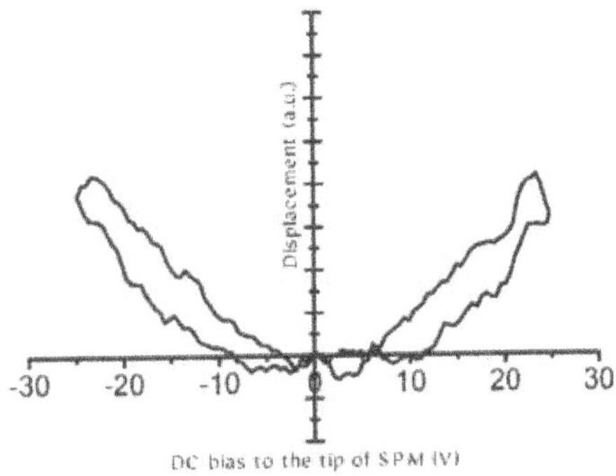

Fig. 10. Strain hysteresis loop of a sintered $BaTiO_3$ pillar.

CONCLUSION

Well crystallized BTO nanoparticles in a pseudo-cubic perovskite phase with an average particle size of ~10 nm and a narrow size distribution were synthesized by a high concentration sol-gel process. By using a mixed solvent of EGMME and Acac as a dispersion medium, transparent and stable BTO suspensions were successfully obtained. Crack-free BTO thin films of controlled thickness from ~100 nm to about 1 μm were electrophoretically deposited on Pt/Ti/Si substrates. A new method of sol-gel process and nanoparticles EPD combined with an EB lithography technique were investigated to prepare BTO ceramic micropatterns at a relatively low temperature. Confirmation of good feature qualities and good adhesion between the micropattern and the substrate suggested that this method is promising as one for micropatterning electroceramic thin films on conductive substrates. Nanoparticles EPD-derived BTO ceramic pillars, sintered at a low temperature of 600°C, exhibited induced ferroelectric characteristics under high electric field.

FOOTNOTES

* Corresponding author. Email: kuwabara@asem.kyushu-u.ac.jp.

** This work is partially supported by a Grant-in-Aid (No. 16206065) from the Japan Society for the Promotion of Science (JSPS). Y. J. Wu is grateful to the JSPS for a Fellowship.

REFERENCES

[1]Y. Yamashita, H. Yamamot, Y. Sakabe, "Dielectric Properties of $BaTiO_3$ Thin Films Derived from Clear Emulsion of Well-Dispersed Nanosized $BaTiO_3$ Particles," *Jpn. J. Appl. Phys.*, **43,** 6521-24 (2004).

[2]K. Tanaka, K. Susuki, D. Fu, K. Nishizawa, T. Miki, and K. Kato, "Grain Size Effect on Dielectric and Piezoelectric Properties of Alkoxy-Derived $BaTiO_3$-Based Thin Films," *Jpn. J. Appl. Phys.*, **43,** 6525-29 (2004).

[3]M. E. Marssi, F. L. Marrec, I. A. Lukyanchuk, M. G. Karkut, "Ferroelectric transition in an epitaxial barium titanate thin film: Raman spectroscopy and x-ray diffraction study," *J. Appl. Phys.*, **94**, 3307-12 (2003)

[4]H.Basantaka Sharma, H.N.K. Sarma, and A. Maningh, "Ferroelectric and Dielectric Properties of Sol-Gel Processed Barium Titanate Ceramics and Thin Films," *J. Mater. Sci.*, **34,** 1385-90 (1999).

[5]C.L. Jia, M. Siegert, and K. Urban, "The Structure of the Interface Between $BaTiO_3$ Thin Films and MgO Substrates," *Acta. Mater.*, **49,** 2783-89 (2001).

[6]Y. Sakabe, Y. Takeshima, and K. Tanaka, "Multilayer Ceramic Capacitors with Thin $(Ba,Sr)TiO_3$ Layers by MOCVD," *J. Electroceramics,* **3,** 115-21 (1999).

[7]H. Kozuka, and A. Higuchi, "Single-layer Submicro-thick $BaTiO_3$ Coating from Poly(vinylpyrrolidone)-coating Sols: Gel-to-ceramic Film Conversion, Densification, and Dielectric Properties," *J. Mater. Res.*, **16,** 3116-23 (2001).

[8]P. Sarkar, and P. S. Nicholson, "Electrophoretic Deposition (EPD): Mechanisms, Kinetics, and Application to Ceramics," *J. Am. Ceram. Soc.*, **79,** 1987-2002 (1996).

[9]I. Zhitomirsky, and A. Petric, "Electrolytic and Electrophoretic Deposition of CeO_2 Films," *Mater. Lett.*, **40,** 263-68 (1999)

[10]S. Okamura, T. Tsukamoto, and N. Koura, "Fabrication of Ferroelectric $BaTiO_3$ Films by

Electrocphoretic Deposition," *Jpn. J. Appl. Phys.*, **32,** 4182-85 (1993).

[11]R. Louh, and Y. Hsu, "Fabrication of Barium Titanate ferroelectric Layers by Electrophoretic Deposition Technique," *Mater. Chem. and Phys.*, **79,** 226-69 (2003).

[12]M. Nagai, K. Yamashita, T. Umegaki, and Y. Takuma, "Electrophoretic Deposition of Ferroelectric Barium Titanate Thick Films and Their Dielectric Properties," *J. Am. Ceram, Soc.*, **76,** 253-55 (1993).

[13]J. Li, Y.J. Wu, H. Tanaka, T. Yamamoto and M. Kuwabara, "Preparation of Mono-dispersed Suspension of Barium Titanate Nanoparticles and Electrophoretic Deposition of Thin films", *J. Am. Ceram. Soc.*, **87,** 1578-81 (2004).

[14]M. Alexe, C. Harnagea, and D. Hesse, "Non-conventional Micro- and Nanopatterning Techniques for Electroceramics," *J. Electrceramics*, **12,** 69-88 (2004).

[15]J. Baborowski, "Microfabrication of Piezoelectric MEMS," *J. Electrceramics*, **12,** 33-51 (2004).

[16]H. Shimooka, and M. Kuwabara, "Preparation of Dense $BaTiO_3$ Ceramics from Sol-Gel-Derived Monolithic Gels," *J. Am. Ceram. Soc.*, **78,** 2849-52 (1995).

[17]H. Shimooka, and M. Kuwabara, "Crystallinity and Stoichiometry of Nano-structured Sol-Gel-Derived $BaTiO_3$ Monolithic Gels," *J. Am. Ceram. Soc.*, **79,** 2983-85 (1996).

Electroceramic Materials and Applications
Edited by R. W. Schwartz

PATTERNING OF CERAMICS -SITE-SELECTIVE DEPOSITION OF CERAMIC THIN FILMS USING SELF-ASSEMBLED MONOLAYERS

Yoshitake Masuda
Graduate School of Engineering, Nagoya University,
Furo-cho, Chikusa-ku,
Nagoya 464-8603, Japan

ABSTRACT

Site-selective deposition (SSD) and nano/micropatterning of ceramic thin films were realized in the solutions using self-assembled monolayers(SAMs). The processes were developed based on scientific knowledge obtained from investigations of interactions and chemical reactions between functional groups of SAMs and ions, clusters and homogeneously nucleated particles in solutions. Several conceptual processes for SSD using SAMs as templates were proposed, and nano/micropatterns of ceramic thin films such as TiO_2, ZnO, Fe_3O_4, $SrTiO_3$, Ta_2O_5, SnO_2, ZrO_2, hydroxyapatite, etc. were successfully fabricated. Molecular recognition of SAMs was effectively used to achieve high site-selectivity.

Our proposals and investigations will contribute to the development of SSD for the fabrication of nano/micropatterns of thin films. The studies showed the good performance of environment-friendly chemistry such as Green & Sustainable Chemistry for the fabrication of nano/micro ceramic devices.

INTRODUCTION

In the present production methods, many patterns of thin films are produced by photolithography and etching, in which films are deposited on entire areas of substrates and partially removed using various processes. The same methods are used in the production of many ceramic devices. Substances having the desired chemical compositions and crystal structures are prepared in the first step, and devices having the desired shapes are formed from the substances in the second step. Furthermore, many ceramic devices are produced using high temperature sintering.

By focusing our attention on nature, we can find many examples of environmentally friendly processes. Numerous creatures and plants fabricate nano/microstructures from inorganic and organic materials within their bodies under their natural living conditions.(*1*) We can learn many things from nature for the production of materials and devices. Bio-inspired processing and science will hopefully provide a solution to serious global problems.

Many techniques for the synthesis of ceramic thin films from aqueous solutions at low temperatures (25–100°C) have been reported.(*2, 3*) Originally used for sulfide and selenide thin films, such techniques have also been applied to oxides since the 1970s. Films of single oxides (e.g., transition metal oxides, In_2O_3, SiO_2, SnO_2) and multicomponent films (doped ZnO, Cd_2SnO_4, $ZrTiO_4$, ZrO_2-Y_2O_3, Li-Co-O spinel, ferrites, and perovskites) have been produced. Compared to vapor-deposition techniques, liquid-deposition methods offer lower capital equipment costs, lower processing temperatures, and greater flexibility in the choice of substrates with respect to topography and thermal stability. Compared to sol-gel techniques, liquid-deposition methods offer lower processing temperatures, less shrinkage, and (being based on aqueous precursors) lower costs in addition to the potential for reduced environmental impact.

The main techniques for non-electrochemical synthesis of polycrystalline ceramic films from aqueous solutions at low temperatures, and some of the materials produced using them, are as follows:(*1*)[1)] 1. Chemical bath deposition (CBD)—CoO, Co_3O_4, NiO, AgO, Ag_2O, ZnO, CdO, In_2O_3, SnO_2, Cd_2SnO_4; CdS, ZnS, Snx S, PbS, MnS, CoS, NiS, Cux S, Ag_2S, As_2S_3, Sb_2S_3, Bi_2S_3, MoS_2, and corresponding selenides. 2. Successive ion layer adsorption and reaction (SILAR)—MnO_2, FeOOH/Fe_2O_3, NiO, Cu_2O, CuO, ZnO and ZnO with Ni, Cu, or Cd doping, Tl_2O_3, SnO_2, $LaNbO_x$, CeO_{2+x} and Y, La and Eu hydroperoxide; ZnS, CdS, PbS, CoS, CuS, Ag_2S, Sb_2S_3, In_2S_3 and Bi_2Se_3. 3. Electroless deposition (ED) with catalyst (Ag^+, Sn^{2+}, or Pd^{2+})—MnO_2, La_{1-x} MnO_3, ZnO, In_2O_3, Tl_2O_3, α-PbO_2. 4. Liquid phase deposition (LPD)—SiO_2, TiO_2, ZrO_2, V_2O_5, β-FeOOH/α-Fe_2O_3, $NiFe_2O_4$, $LnMO_3$ (Ln = La, Nd; M = Cr, Mn, Fe, Co), SnO_2. Liquid-phase deposition is an aqueous technique for deposition of oxide films that has been widely used for SiO_2(*4, 5*) but is being increasingly studied as a route for films of oxides such as TiO_2(*6-13*) and others.(*2*)

We proposed site-selective deposition (SSD) of ceramic thin films and try to fabricate nano/micropatterns of thin films(*14-21*). Self-assembled monolayer (SAM), which can modify the surface of solids with various functional groups, was used as the template to enable molecular recognition for SSD. Solution systems were developed and applied to SSD of ceramic thin films on patterned SAMs by the proposed novel SSD processes.

Here, we report nano/micro-patterning of anatase TiO_2 and ZnO nanocrystals using self-assembled monolayers. The concept of the method for TiO_2 nanopatterning is the use of the difference in adhesive strength of the depositions to the substrate. This difference is caused by the difference in chemical bonds between depositions and substrates. Nanopatterning of anatase TiO2 was successfully realized in an aqueous solution with this method. We further fabricated micropatterns of UV- and visible-light-emitting ZnO crystals in an aqueous solution without the use of the annealing process or catalyst such as Pd. ZnO crystals homogeneously nucleated in the solution were deposited and grown on hydrophobic regions of patterned SAMs selectively by molecular recognition of SAMs. Site-selective nucleation and growth of ZnO crystals progressed simultaneously without the use of catalyst. Furthermore, the morphology of ZnO crystals were controlled by solution conditions with this method.

SELF-ASSEMBLED MONOLAYER

SAM preparation

An Si wafer (p-type Si [100]) was sonicated in water, ethanol or acetone for 10 min, respectively, and exposed for 15 min to VUV light (vacuum ultraviolet light) (184.9 nm and 253.7 nm) (low-pressure mercury lamp 200 W, PL21-200, 15 mW/cm^2 for 254 nm, SEN Lights Co.) to clean the surface. The OTS(octadecyltrichlorosilane, $C_{18}H_{37}SiCl_3$)-SAM was prepared by immersing the Si substrate in an anhydrous toluene solution containing 1 vol% OTS for 5 min under an N_2 atmosphere (Fig. 1). The substrate with the SAM was baked at 120°C for 5 min to remove residual solvent and promote chemisorption of the SAM.

SAMs of DTS (dodecyltrichlorosilane, $C_{12}H_{25}SiCl_3$), HTS (hexyltrichlorosilane, $C_6H_{13}SiCl_3$), PTS(propyltrichlorosilane, $F_7OC_6H_6SiCl_3$), MTS(methyltrichlorosilane, CH_3SiCl_3), PTCS (phenyltrichlorosilane, $C_6H_5SiCl_3$) or APTS (aminopropyltriethoxysilane, $NH_2C_3H_6Si(OC_2H_5)_3$) were employed) were also prepared by immersion of Si substrates into the solutions containing SAM reagents.

The SAMs on the silicon substrates were exposed for 15 min to VUV light through a photomask to be used as a template for SSD. VUV-irradiated regions became hydrophilic due to silanol group formation, while the non-irradiated part remained unchanged. Formation of the SAM and its modification to silanol groups by VUV irradiation was verified using the water drop contact angle (θw). The initially deposited OTS-SAM, DTS-SAM, HTS-SAM, PTS-SAM, MTS-SAM, PTCS-SAM or APTS-SAM showed a static water contact angle of 105°, 104°, 101°, 82°, 81°, 81° or 63°, but the VUV-irradiated surface of SAM was wetted completely (contact angle < 5°)

Additionally, we used an old SAM for Fig. 1 (b) to check the influence of pinholes and defects in the SAM for TiO_2 deposition. An SAM was kept in air for 1 month after preparation and was patterned by UV irradiation. The patterned SAM, which probably had many pinholes and other defects, was then immersed in the solution to deposit TiO_2 thin films.

NANOPATTERINING OF TITANIUM DIOXIDE

In this study we developed a novel method to realize site-selective deposition of thin films using a site-selective elimination method. The concept of this method is the use of the difference in adhesive strength of the depositions to the substrate. This difference is caused by the difference in chemical bonds between depositions and substrates. In this method, the patterned SAM having OTS regions and silanol regions was used as a template. The substrate was immersed in the solution containing Ti precursor and subjected to ultrasonication for several hours. Ultrasonication eliminated many deposited particles from OTS regions but not as many from silanol regions because depositions adhered strongly on silanol regions compared to OTS regions due to the difference in the number of chemical bonds between depositions and substrates. Consequently, anatase TiO_2 was deposited on silanol regions selectively to form thin films and a pattern of anatase TiO_2 thin films having high feature edge acuity compared to our previous works(*9, 10*) was successfully fabricated.

Deposition of TiO_2 thin films

Ammonium hexafluorotitanate ($[NH_4]_2TiF_6$) (purity 96%, 1.031 g) and boric acid (H_3BO_3) (purity 99.5%, 0.932 g) were dissolved separately in deionized water (50°C, 50 ml) (Fig. 1). An appropriate amount of HCl was added to the boric acid solution to control pH, and ammonium hexafluorotitanate solution was added.(*9, 10*) Solutions (100 ml) with 0, 0.1 or 0.6 ml of HCl showed pH 3.8, 2.8 or 1.5, respectively. Supersaturation of solution can be changed by pH value as discussed in ref. 55. TiO_2 thin films can be formed fast by the deposition of homogeneously nucleated particles at high pH condition such as pH 3.88, and uniform films can be obtained slowly by heterogeneous nucleation at low pH condition. SAMs were immersed in the solution (100 ml) containing 0.05 M $(NH_4)_2TiF_6$ and 0.15 M (H_3BO_3) at pH 1.5, 2.8 or 3.8 and kept at 50°C for 4 h. to deposit anatase TiO_2. Ultrasonication was done during immersion period. Deposition of TiO_2 proceeded by the following mechanism(*6-8*):

$$TiF_6^{2-} + 2H_2O \rightleftarrows TiO_2 + 4H^+ + 6F^- \quad \cdots(a)$$

$$BO_3^{3-} + 4F^- + 6H^+ \longrightarrow BF_4^- + 3H_2O \quad \cdots(b)$$

Equation (a) is described in detail by the following two equations:

$$TiF_6^{2-} \xrightarrow{nOH^-} TiF_{6-n}(OH)_n^{2-} + nF^- \xrightarrow{(6-n)OH^-} Ti(OH)_6^{2-} + 6F^- \quad \cdots(c)$$

$$Ti(OH)_6^{2-} \longrightarrow TiO_2 + 2H_2O + 2OH^- \quad \cdots(d)$$

Fluorinated titanium complex ions gradually change into titanium hydroxide complex ions in an aqueous solution as shown in Eq. (c). Increase of F^- concentration displaces the Eq. (a) and (c) to the left, however, produced F^- can be scavenged by H_3BO_3 (BO_3^{3-}) as shown in Eq. (b) to displace the Eq. (a) and (c) to the right. Anatase TiO_2 was formed from titanium hydroxide complex ions ($Ti(OH)_6^{2-}$) in Eq. (d), and thus the supersaturation degree and the deposition rate of TiO_2 depend on the concentration of titanium hydroxide complex ions. The high concentration of H^+ displaces the equilibrium to the left in Eq. (a), and the low concentration of OH^-, which is replaced with F^- ions, suppresses ligand exchange in Eq. (c) and decreases the concentration of titanium hydroxide complex ions at low pH such as pH 1.5. The solution actually remained clear at pH 1.5, showing its low degree of supersaturation. On the other hand, the solution at high pH such as pH 2.8 or 3.8 became turbid because of homogeneously-nucleated anatase TiO_2 particles caused by a high degree of supersaturation. Anatase TiO_2 thin film was formed by heterogeneous nucleation in the solution at pH 1.5, while the film was formed by heterogeneous nucleation and deposition of homogeneously nucleated particles at pH 2.8 or 3.8.

Site-selective elimination and nano/micro-scaled pattern of thin films

After having been immersed in the solution with ultrasonic treatment, the substrates were rinsed with distilled water and observed by an optical microscope (BX51WI with CCD camera), an optical microscope (BX51WI Microscope, Olympus Optical Co., Ltd.) with a digital camera (DP50, 5.8 megapixels, Olympus Optical Co., Ltd.) and a computer for capturing data, a scanning electron microscope (SEM; S-3000N, Hitachi, Ltd.), and a scanning probe microscope (SPI 3800N, Seiko Instruments Inc.) that was operated in AFM (atomic force microscopy) tap mode to observe the topography of the surface. AFM scans were operated at room temperature under ambient air. Thin films were observed on the silanol group regions to form nano/micro-scaled patterns at pH 3.88, 2.8 or 1.5. The films deposited for 4 h from the solution without the addition of HCl are shown in Figs. 1–4. Thin films were observed as being dark in an optical micrograph. Separated parallel lines 200–400 nm in width at 100–200 nm intervals were successfully fabricated with this method (Fig. 1 (a-b)). The length of the separated parallel lines reached more than 100 μm (Fig. 1 (a-b)). A cross section of the lines was shown as a semicircle, and the thickness of the center of the lines was estimated to be about 100 nm by AFM observation (Fig. 2). Feature edge acuity of the pattern was higher than that of the pattern fabricated by our lift-off process or by the site-selective immersion method.(*17*) Site-selective deposition was realized at any pH conditions such as pH 3.88, 2.8 or 1.5. Patterns which have higher feature edge acuity can be obtained at low pH conditions because films were formed slowly without the deposition of homogeneously nucleated particles.

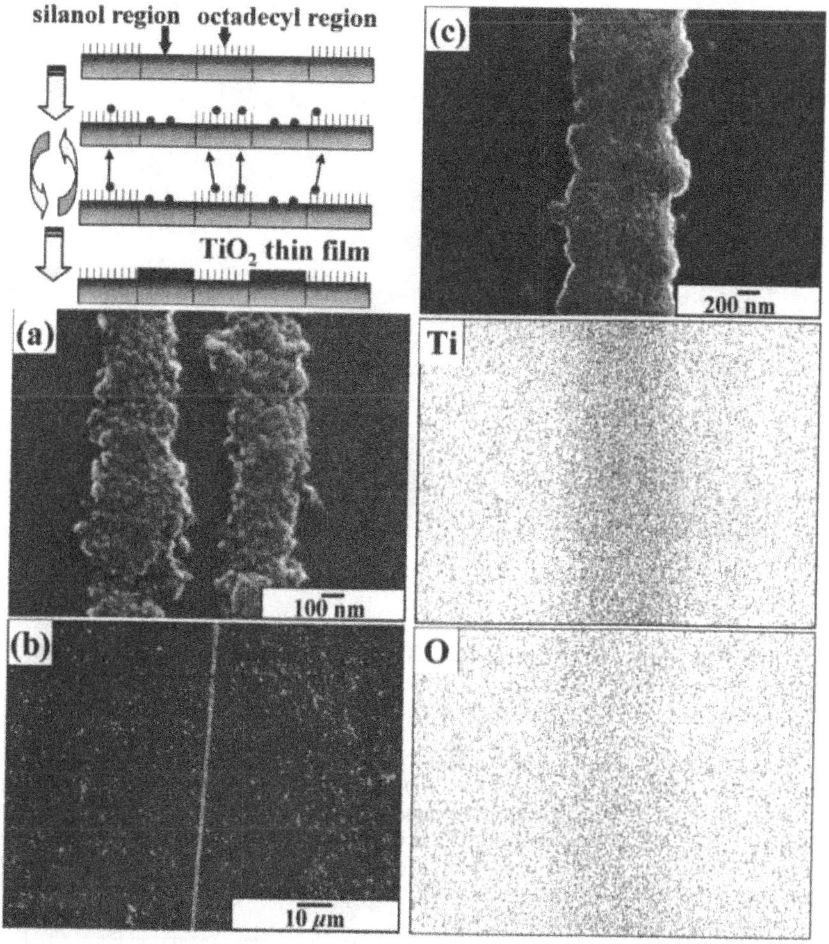

Figure 1. Conceptual process for nanopatterning of anatase TiO_2. (a-c) SEM micrographs and EDX images of (c) for Ti and O. (The films were deposited for 4 h from the solution without the addition of HCl.)

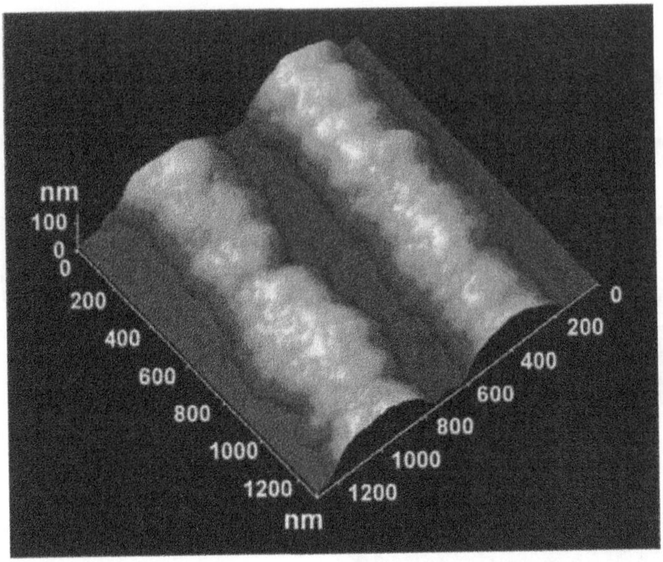

Figure 2. AFM image of micropattern of anatase TiO_2 thin films on OTS / OH-patternd SAM fabricated by LPD using ultrasonication.

For the adhesion of TiO_2 films to silanol groups, the pH of the deposition solution is critical. Pizem *et al.*(*3*) reported that adherent TiO_2 films formed from solutions similar to those used here at pH=3.9 (*i.e.*, the pH used in the films of Figures 2-4), but that the films were less adherent at pH=2.9. They related this difference in adherence to an increased electrostatic attraction of TiO_2 to the oxidized surface of silicon at the higher pH. In our experiment, there also were less adherence of TiO_2 at pH 2.8 and 1.5 than that at pH 3.5 due to low supersaturaiton shown in Eq. (a) and low electrostatic attraction.(*3*)

However, some depositions were observed on octadecyl group regions in Fig. 1 (b). One probable cause is that pinholes and other defects in the films provide at least some degree of access of water to underlying unreacted OH groups in the OTS films. Once exposed to the solution, these sites can act as nucleation points for TiO_2 growth. Because the depositions are performed at elevated temperatures, it is likely that pinholes and defects will continually open and close on the OTS film surfaces due to thermal motions of alkyl chains in the films. The TiO_2 precursors formed in these defects would act as points for eventual growth of TiO_2 over the entire SAM-covered region. This would provide a weakly bound TiO_2 film on the OTS film regions due to the limited number of connections to the underlying silanol sites in the film regions. In fact, Sagiv and others have shown that macroscopic defects induced in alkylsiloxane films can readily be accessed by solution species.(*22-24*) More recently, Dressick and coworkers demonstrated that solvent accessibility to underlying substrates in aromatic siloxane films is also important(*25-27*) and may be an even greater factor in controlling the properties of those films, which may account for our previous selectivity observation using phenylsiloxane films, as shown in the lift-off process(*10*).

In the lift-off process,(*10*) thin films were formed on the entire area of patterned SAM that has silanol group regions and phenyl (or octadecyl) group regions. After being dried, the substrate was sonicated in water to lift off thin films on phenyl (or octadecyl) group regions selectively. Thin films on phenyl (or octadecyl) group regions were peeled off along the cracks that formed during the drying process. Thin films on phenyl (or octadecyl) group regions without cracks were not peeled off because depositions strongly connected to each other to form solid timber (monolith). The lift-off along cracks decreased the feature edge acuity of the pattern in this method. Thin films were formed on silanol group regions selectively and site-selective deposition was realized with our newly developed method. This resulted in high feature edge acuity of the patterns compared to our previous works.(*10*)

Additionally, the micropattern of thin films was also fabricated by the site-selective immersion method(*17*). A solution containing a Ti precursor contacted the hydrophilic regions during the experiment and briefly came into contact with the hydrophobic regions. The solution on the hydrophilic surface was replaced with a fresh solution by continuous movement of bubbles. Thus TiO_2 was deposited and a thin film was grown on the hydrophilic regions selectively. This technique can be applied for the formation of many kinds of films from any solution and to fabricate micropatterns for many kinds of thin film because the technique creates the difference in contact time of the solution between hydrophilic regions and hydrophobic regions. However, it is difficult to form a solution layer on nano-scaled hydrophilic regions selectively and replace it with a fresh solution by continuous movement of bubbles while avoiding contact of the solution on hydrophobic regions. This prevents fabrication of nano-scaled pattern with this method. On the other hand, site-selective deposition was realized in the solution with our newly developed method using the difference of adhesive strength of depositions to substrates. Heterogeneously nucleated deposition and homogeneously nucleated particles and/or clusters can be removed from octadecyl group regions even if these regions are designed in nano-scale order in which depositions are smaller. This allowed us to realize high feature edge acuity of the patterns compared to site-selective immersion.(*17*)

Characterizations of thin films

The distribution of elements on the surface of the substrates was evaluated by energy dispersive X-ray analysis (EDX; EDAX Falcon, EDAX Co. Ltd.), which is built into SEM. Titanium was detected from thin films selectively and oxygen was detected mainly from silanol group regions by EDX (Fig. 1). Other elements, except for silicon from the substrate, were not observed from the thin film and substrate by EDX. Oxygen was detected from not only the deposited thin film but also from the natural oxide layer (amorphous SiO_2 layer) formed on all surface areas of a silicon substrate. These observations showed predominant deposition of titanium oxide on silanol group regions.

The deposited thin films were also investigated using an X-ray diffractometer (XRD; RAD-C, Rigaku) with CuKα radiation (40 kV, 30 mA) and Ni filter plus a graphite monochromator. Thin films deposited at pH 3.8 for 4 h showed an XRD pattern of anatase-type TiO_2 having orientation similar to that of films deposited in the solution at pH 1.5 or 2.8(*10, 17*)(Fig. 3). The diffraction from parallel to c-plane such as (004) was observed as being strong compared to that of the randomly oriented powder diffraction pattern (Fig. 3). Pizem *et al.*(*3*) postulated that the commonly observed [00l] orientation of anatase films could be due to the slight polarity of the planes parallel to the [001] axis, unlike other low-index planes of this structure such as {100}, (110), and (210). The orientation and crystal growth mechanism are further discussed in a separate article.(*9*)

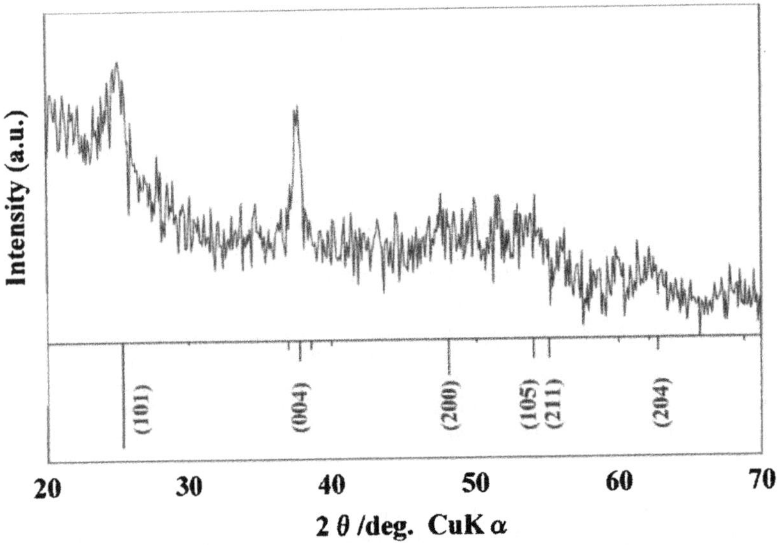

Figure 3. XRD pattern of deposited thin films on silanol group regions from an aqueous solution.

Thin films were further evaluated by X-ray photoelectron spectroscopy (XPS; ESCA-3200, Shimazu Corporation, 1×10^{-5} Pa). The X-ray source (MgKα, 1253.6 eV) was operated at 8 kV and 30 mA. The spectral peaks corresponding to Ti 2p (458.7 eV) were observed from thin films deposited on the silanol region (Fig. 4). This binding energy is higher than that of Ti metal (454.0 eV), TiC (454.6 eV), TiO (455.0 eV), TiN (455.7 eV) and Ti_2O_3 (456.7 eV), and similar to that of TiO_2 (458.4 - 458.7 eV).(*28-30*) This suggests that the titanium atoms in thin films are positively charged relative to that of titanium metal by formation of direct bonds with oxygen. On the other hand, this spectrum was not observed from octadecyl group regions. The O 1s spectrum was observed from the silanol regions and divided into O 1s (530.2 eV) and O 1s (532.3 eV). O 1s (532.3 eV) can be assigned to the silicon oxide layer on the surface of the silicon wafer (532.0 eV(*29*)), whereas the binding energy of O 1s (530.2 eV) is similar to that of TiO_2 (529.9 eV(*30*), 530.1 eV(*29*)) as observed by Shin *et al.(31)*. This shows that oxygen is negatively charged compared with neutral oxygen molecules (531.0 eV), possibly through the formation of chemical bonds with Ti. The ratio of titanium to oxygen was estimated from the Ti $2p_{3/2}$ (458.7 eV) spectrum and O 1s (530.2 eV) spectrum to be Ti:O = 1:2.0.

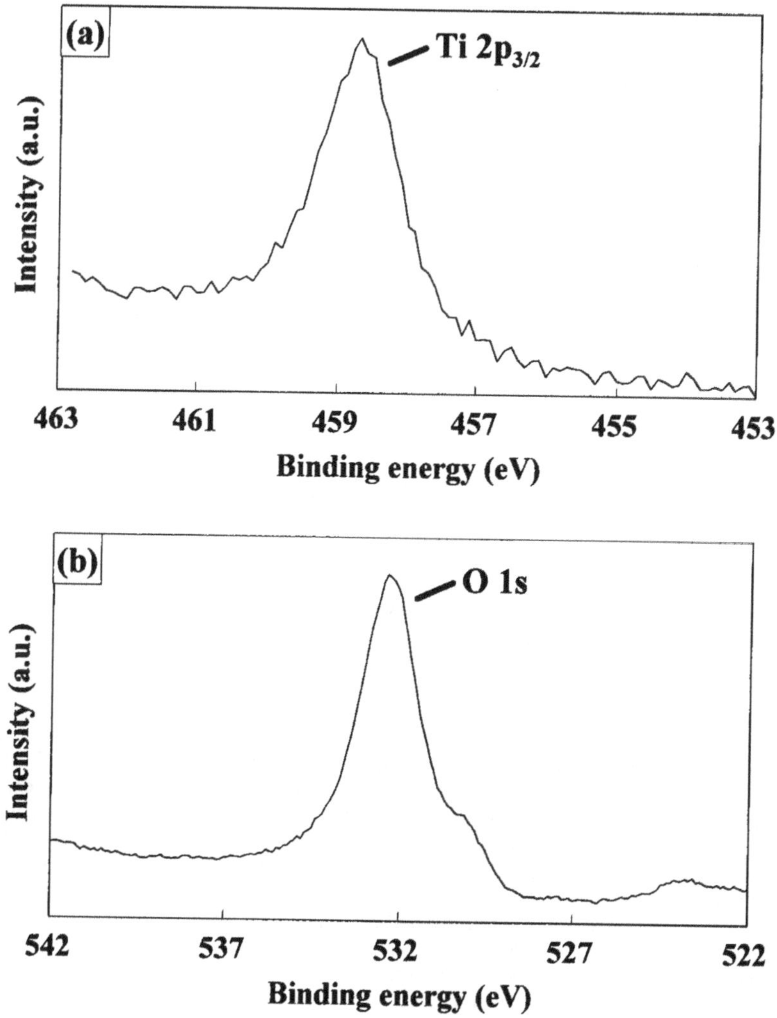

Figure 4 XPS spectra of (a) Ti 2p and (b) O 1s for deposited thin films from an aqueous solution.

MORPHOLOGY CONTROL AND MICROPATTERING OF ZINC OXIDE

In this study, we propose a novel process to fabricate micropatterns of UV- and visible-light-emitting ZnO crystals in an aqueous solution without the use of the annealing process or catalyst such as Pd. Patterned SAMs having hydrophilic silanol regions and hydrophobic SAM regions were effectively used as a template for site-selective deposition of ZnO crystals. ZnO crystals homogeneously nucleated in the solution were deposited and grown on hydrophobic regions of patterned SAMs selectively by molecular recognition of SAMs. Site-selective nucleation and growth of ZnO crystals progressed simultaneously without the use of catalyst. Furthermore, the crystallinity and morphology of ZnO crystals were found to affect the photoluminescence properties and micropatterns of UV- and visible-light-emitting ZnO crystals were successfully fabricated.

Preparation and pattering of ZnO crystals

Zinc acetate ($Zn(CH_3COO)_2$, Kishida Chemical Co., Ltd.) was dissolved into water to be 15 mM at 50°C, and ammonia (28% solution, Kishida) was then added to be 30, 60 or 90 mM ($[NH_3] / [Zn] = 2.0$, 4.0 or 6.0) with stirring as complexing agent. These solutions showed pH = 7.04, 7.50 or 8.93, respectively. Zinc ions reacted with ammonium ions (NH_4^+) formed from ammonia to form tetra amine zinc(II)(*32*) $[Zn(NH_3)_4]^{2+}$. ZnO was crystallized from the reaction between $[Zn(NH_3)_4]^{2+}$ and OH^-. The solution became clouded shortly after adding ammonia due to homogeneous nucleation of ZnO crystals. Morphology of ZnO crystals was controlled by the ratio of ammonia to zinc acetate, i.e., super-saturation degree for crystallization.

Patterned OTS-SAMs were immersed downward into the solution containing zinc acetate (15 mM) and ammonia (30 mM) as complexing agent ($[NH_3] / [Zn] = 2.0$) at 50°C for 3 h (Fig. 1). After having been immersed in the solution, SAMs were rinsed with distilled water and dried in air. Patterns of ZnO crystals were evaluated with a scanning electron microscope (SEM; S-3000N, Hitachi, Ltd.), an X-ray diffractometer (XRD; RINT-2100, Rigaku, in-plane x-ray diffraction profiles) with CuKα radiation (40 kV, 30 mA) and Ni filter plus a graphite monochromator, an optical- and photoluminescence-microscope (BX51WI Microscope, excitation wavelength for photoluminescence image 330–385 nm, Olympus Optical Co., Ltd.) with a digital camera (DP50, 5.8 megapixels, Olympus Optical Co., Ltd.) and a computer for capturing data, and a fluorescence spectrometer (F-4500, excitation wavelength 350 nm, Xe lamp, Hitachi, Ltd.).

Site-selective deposition of ZnO crystals and their morphology control

ZnO crystals having long hexagonal cylinder shape (inset of Fig. 5-1b) were homogeneously nucleated to make the solution turbid shortly after adding ammonia. Crystals showed sharp hexagonal facets of about 100 nm in diameter and larger than 500 nm in length. The morphology indicated high crystallinity of ZnO nanoparticles. The nanoparticles were deposited and further grown on hydrophobic octadecyl group regions of a patterned SAM selectively (Fig. 5-1). Consequently, a micropattern of light-emitting ZnO crystals was successfully fabricated in an aqueous solution without Pd catalyst. ZnO crystals were also deposited on hydrophobic regions of patterned SAMs such as DTS-SAM, HTS-SAM, PTS-SAM, MTS-SAM, PTCS-SAM or APTS-SAM. This showed that the method is highly versatile and offers good potential for the fabrication of devices.

ZnO crystals were deposited on hydrophobic SAM regions such as OTS-, APTS- or other SAMs rather than hydrophilic silanol regions. Zeta potential of ZnO crystals deposited on a silicon substrate was measured to be 10 mV at pH 8.1 and ZnO crystals should thus have positive zeta potential not less than 10 mV in the solution at pH 7.04. SAM of OTS, silanol and APTS showed zeta potential of –3 mV, –38.2 mV or +22.0 mV, respectively. ZnO having positive zeta potential should be deposited on silanol regions having negative zeta potential rather than other SAMs, if the site-selective deposition was caused only by electrostatic interactions. The site-selective deposition of ZnO crystals would be caused by not only electrostatic interactions as shown by the relation of zeta potentials. ZnO crystals having long hexagonal cylinder shape were deposited on a hydrophilic silicon substrate to evaluate the surface of crystals. The substrate covered with many deposited ZnO crystals exhibited high water contact angle (WCA 140°). The deposited ZnO crystals were found from the experiment to have hydrophobic surfaces. Surfaces of naked ZnO crystals would be hydrophilic because of surface hydroxyl groups and they would become hydrophobic by being covered with organic molecules having hydrophobic functional groups. CH_3COO^- ions coming from $Zn(CH_3COO)_2$ might be adsorbed to ZnO crystal surfaces by the interaction between Zn and $-COO^-$ to cover the surface with hydrophobic $-CH_3$ groups and some of $Zn(CH_3COO)_2$ would exist in the surface layer of ZnO crystals. Additionally, deposited ZnO crystals having long hexagonal cylinder shape became hydrophilic (< 10°) and their zeta potential shifted positively by UV irradiation in air. ZnO crystals deposited on a silicon substrate showed zeta potential of 10 mV at pH 8.1, 0 mV at pH 8.8 and –15 mV at pH 9.2, while they shifted to 20 mV at pH 8.1, 10 mV at pH 8.8 and 7 mV at pH 9.2 by UV irradiation. The decomposition of CH_3COO^- ions and the breakage of the bond between CH_3COO and Zn would be caused by light excitation, ozone and active oxygen by UV irradiation in air. This finding suggests that organic molecules, such as CH_3COO^- ions, which show negative zeta potential and can be removed by UV irradiation, would be absorbed onto the surfaces of ZnO crystals. Furthermore, ZnO crystals were confirmed to deposit on a hydrophobic polyethylene terephthalate surface rather than on a hydrophilic polyethylene terephthalate surface modified by UV irradiation in the same solution. Additionally, organic molecule was reported to adsorb to growing ZnO crystals, in which poly (ethylene oxide)-blockpoly (methylacrylic acid) (PEO-b-PMAA) was adsorbed preferentially to {0001} face of ZnO to retard crystal growth perpendicular to this face(*33*). Consequently, site-selective deposition was achieved by the effective molecular recognition caused by combination of the forces composed mainly of hydrophobic interactions between functional groups of SAMs and ZnO crystal surfaces.

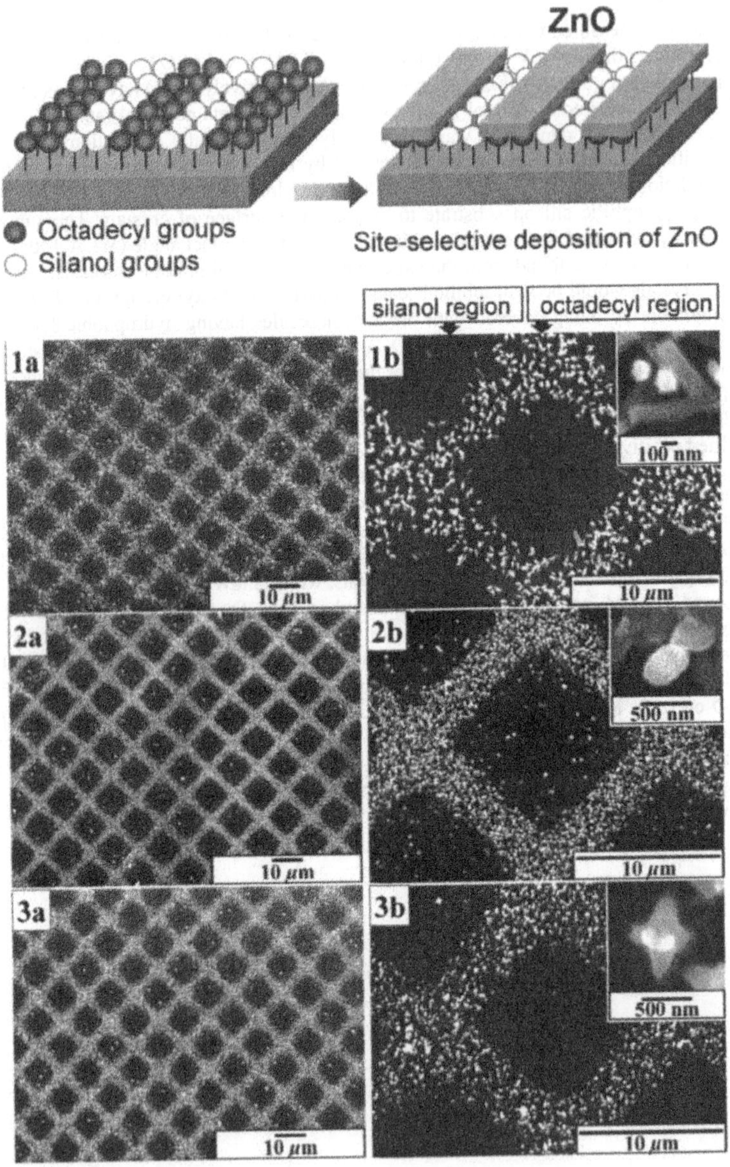

Figure 5. Conceptual process for self-assembly patterning of light-emitting crystalline ZnO nanoparticles in an aqueous solution.

SEM micrographs of (1a) patterned crystalline ZnO (long hexagonal cylinder, $[NH_3] / [Zn] = 2.0$), (1b) magnified area of (1a), (2a) patterned crystalline ZnO (ellipse, $[NH_3] / [Zn] = 4.0$), (2b) magnified area of (2a), (3a) patterned crystalline ZnO (multi-needles, $[NH_3] / [Zn] = 6.0$) and (3b) magnified area of (3a).

Patterned SAMs were also immersed into the solution containing zinc acetate (15 mM) and ammonia (60 mM or 90 mM) as complexing agent ($[NH_3] / [Zn] = 4.0$ or 6.0) for 3 h. ZnO crystals having ellipse or multi-needle shape (two large needles and four small needles) (insets of Fig. 5-2b, 3b) were homogeneously nucleated to make the solution turbid shortly after adding ammonia. Nucleation and deposition of ZnO crystals were accelerated by addition of ammonia. Each ZnO crystal was about 500 nm in size. The crystals were deposited and further grown on hydrophobic regions of patterned SAMs selectively (Fig. 5). Micropatterns of light-emitting ZnO crystals having ellipse or multi-needle shape were fabricated on patterned SAMs such as OTS-SAM, DTS-SAM, HTS-SAM, PTS-SAM, MTS-SAM, PTCS-SAM or APTS-SAM in aqueous solutions.

XRD spectra of ZnO crystals having ellipse or multi-needle shape showed dominant peaks corresponding to ZnO (0002) planes revealing that ZnO crystals were deposited with a high degree of orientation of their c-axes perpendicular to the substrate (Fig. 6). Enhanced (0002) and $(10\bar{1}0)$ peaks from ZnO crystals having long hexagonal cylinder shape showed that crystals were deposited to make (0002) or $(10\bar{1}0)$ planes parallel to the substrate. Crystals having high crystallinity and high purity with no additional phase were shown to be prepared in an aqueous solution with precise control of their morphologies without the use of Pd catalyst. The aqueous solution system showed high ability for fabricating nano/micro devices composed of crystalline materials. ZnO crystals are well known to grow along the c-axis. The c-axes of ZnO crystals deduced from their shapes and the deposited directions revealed from XRD spectra are shown in Fig. 6. The orientations evaluated from XRD patterns were consistent with SEM observations and were shown to be controlled precisely by the solution conditions.

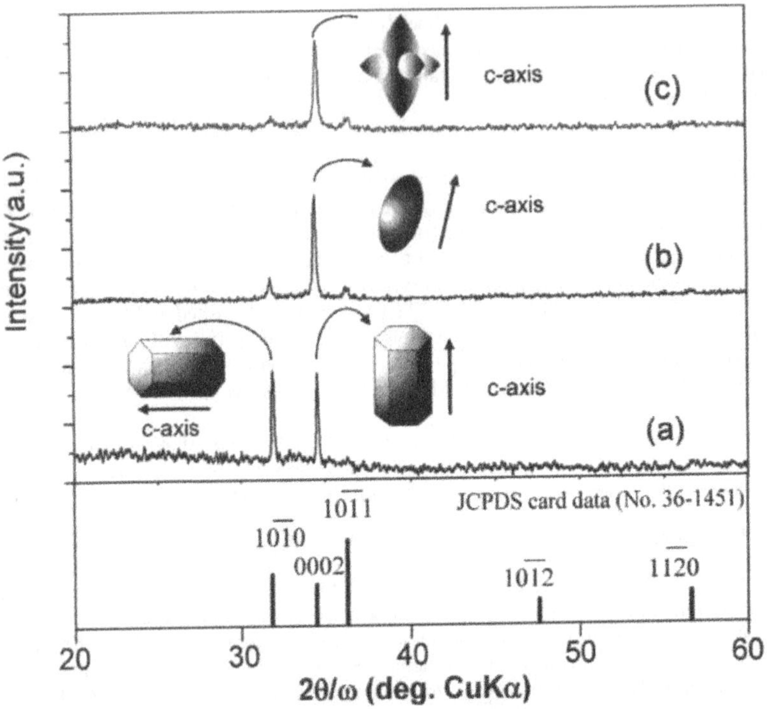

Figure 6. The in-plane x-ray diffraction profiles of ZnO changing with the morphology: (1) long hexagonal cylinder, (2) ellipse, (3) multi-needles.

Photoluminescence of patterned ZnO crystals

Photoluminescence properties of ZnO crystal patterns were further evaluated. Micropatterns of ZnO crystals were observed by an optical microscope (Fig. 7) and strong visible-luminescence from ZnO crystals excited by 330–385 nm light was observed by a photoluminescence microscope. ZnO crystals showed strong UV luminescence (around 390 nm) attributed to band-edge luminescence and visible-light luminescence caused from oxygen vacancy (450–600 nm)(*34, 35*). All of the crystals showed photoluminescence (Fig. 7) due to high purity and high crystallinity with optimal oxygen vacancy, and this caused the bright visible-photoluminescence image. Luminescence properties can be controlled by changing the crystalline morphologies. ZnO crystals deposited from an aqueous solution were shown to have high visible-light-emitting properties.

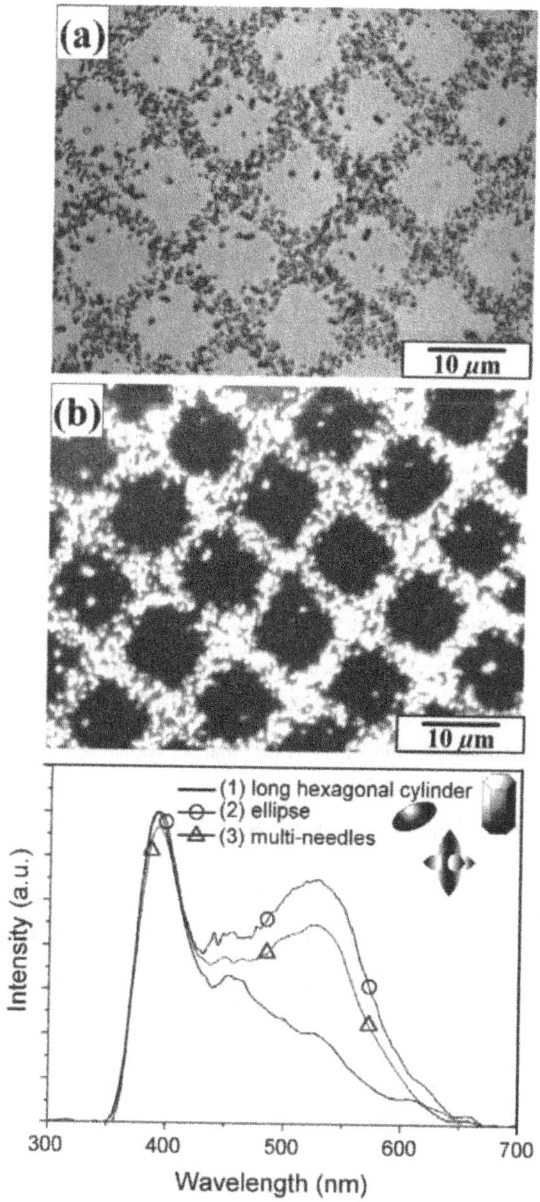
(a)
10 μm
(b)
10 μm
(1) long hexagonal cylinder
(2) ellipse
(3) multi-needles
Intensity (a.u.)
300
400
500
600
700
Wavelength (nm)

Figure 7. (a) Optical microscope image and (b) photoluminescence image of patterned ZnO particles (ellipse) under white light or UV light (330–385 nm). Photoluminescence properties for ZnO crystals; (1) long hexagonal cylinder, (2) ellipse, (3) multi-needles.

CONCLUSIONS

We realized nano/micro-patterning of anatase TiO_2 thin films and ZnO nanocrystals based on scientific knowledge obtained from investigations of interactions and chemical reactions between functional groups of SAMs and ions, clusters and homogeneously nucleated particles. Our proposals and investigations will contribute to the development of "Solution Chemistry for Inorganic Materials". This study also showed the good performance of environment-friendly chemistry such as Green & Sustainable Chemistry for the fabrication of nano/microdevices.

REFERENCES

1. S. Mann, Ed., *Biomimetic Materials Chemistry* (VCH Publishers, Weinheim, 1996), pp.
2. T. P. Niesen, M. R. DeGuire, *J. Electroceram.* **6**, 169 (2001).
3. H. Pizem, C. N. Sukenik, U. Sampathkumaran, A. K. McIlwain, M. R. De Guire, *Chem. Mater.* **14**, 2476 (2002).
4. P. H. Chang, C. T. Huang, J. S. Shie, *J. Electrochem. Soc.* **144**, 1144 (1997).
5. H. Nagayama, H. Honda, H. Kawahara, *J. Electrochem. Soc.* **135**, 2013 (1988).
6. S. Deki, Y. Aoi, O. Hiroi, A. Kajinami, *Chem. Letters* **6**, 433 (1996).
7. S. Deki *et al.*, *J. Mater. Chem.* **6**, 1879 (1996).
8. K. Shimizu, H. Imai, H. Hirashima, K. Tsukuma, *Thin Solid Films* **351**, 220 (1999).
9. Y. Masuda, T. Sugiyama, W. S. Seo, K. Koumoto, *Chem. Mater.* **15**, 2469 (2003).
10. K. Koumoto, S. Seo, T. Sugiyama, W. S. Seo, *Chem. Mater.* **11**, 2305 (1999).
11. H. Imai, Y. Takei, K. Shimizu, M. Matsuda, H. Hirashima, *J. Mater. Chem.* **9**, 2971 (1999).
12. H. Imai, M. Matsuta, K. Shimizu, H. Hirashima, N. Negishi, *J. Mater. Chem.* **10**, 2005 (2000).
13. S. Yamabi, H. Imai, *Chem. Lett.*, 220 (2001).
14. Y. Masuda, T. Sugiyama, H. Lin, W. S. Seo, K. Koumoto, *Thin Solid Films* **382**, 153 (2001).
15. Y. Masuda, W. S. Seo, K. Koumoto, *Langmuir* **17**, 4876 (2001).
16. Y. Masuda, Y. Jinbo, T. Yonezawa, K. Koumoto, *Chem. Mater.* **14**, 1236 (2002).
17. Y. Masuda, T. Sugiyama, K. Koumoto, *J. Mater. Chem.* **12**, 2643 (2002).
18. Y. Masuda, S. Ieda, K. Koumoto, *Langmuir* **19**, 4415 (2003).
19. Y. Masuda, S. Wakamatsu, K. Koumoto, *J. Euro. Ceram. Soc.* **24**, 301 (2004).
20. N. Shirahata, Y. Masuda, T. Yonezawa, K. Koumoto, *Langmuir* **18**, 10379 (2002).
21. T. Nakanishi, Y. Masuda, K. Koumoto, *Chem. Mater.* **16**, 3484 (2004).
22. J. Gun, J. Sagiv, *J. Colloid Interface Science* **112**, 457 (1986).
23. M. E. McGovern, K. M. R. Kallury, M. Thompson, *Langmuir 1994* **10**, 3607 (1994).
24. J. Duchet *et al.*, *Langmuir* **13**, 2271 (1997).
25. W. J. Dressick, M.-S. Chen, S. L. Brandow, *J. Amer. Chem. Soc.* **122**, 982 (2000).
26. W. J. Dressick *et al.*, *Appl. Phys. Lett.* **78**, 676 (2001).
27. W. J. Dressick, P. F. Nealey, S. L. Brandow, *Proc. SPIE 2001* **4343**, 294 (2001).

28. R. J. Collins, H. Shin, M. R. De Guire, A. H. Heuer, C. N. Sukenik, *Appl. Phys. Lett.* **69**, 860 (1996).
29. D. Huang, Z. D. Xiao, J. H. Gu, N. P. Huang, C. W. Yuan, *Thin Solid Films* **305**, 110 (1997).
30. F. Zhang *et al.*, *Thin Solid Films* **310**, 29 (1997).
31. H. Shin, R. J. Collins, M. R. DeGuire, A. H. Heuer, C. N. Sukenik, *J. Mater. Res.* **10**, 699 (1995).
32. R. L. Call, N. K. Jaber, K. Seshan, J. R. Whyte, *Solar Energy Mater.* **2**, 373 (1980).
33. M. Oner, J. Norwig, W. H. Meyer, G. Wegner, *Chem. Mater.* **10**, **460** (1998).
34. X. L. Wu, G. G. Siu, C. L. Fu, H. C. Ong, *Appl. Phys. Lett.* **78**, 2285 (2001).
35. J. S. Kang, H. S. Kang, S. S. Pang, E. S. Shim, S. Y. Lee, *Thin Solid Films* **443**, 5 (2003).

Electroceramic Materials and Applications
Edited by R. W. Schwartz

ELECTRICAL PROPERTIES OF $Pt/(Y,Yb)MnO_3/HfO_2/Si$ STRUCTURE CONSTRUCTED THROUGH CHEMICAL SOLUTION PROCESS

Kazuyuki Suzuki, Yiping Guo, Kaori Nishizawa, Takeshi Miki, Kazumi Kato
National Institute of Advanced Industrial Science and Technology
2266-98 Anagahora, Shimoshidami, Moriyama-ku,
Nagoya 463-8560, Japan

ABSTRACT

The MFIS structures were constructed using $Y_{0.5}Yb_{0.5}MnO_3$ as ferroelectrics and HfO_2 as insulator. The $Y_{0.5}Yb_{0.5}MnO_3$ ferroelectric film and HfO_2 insulating layer were prepared by chemical solution deposition. The effect of the preparation condition of HfO_2 layer on the crystallographic appearance and electrical properties was investigated. HfO_2 insulating layers crystallized on Si(100) substrates consisted of uniform grains and had smooth surfaces. The $Y_{0.5}Yb_{0.5}MnO_3$ films prepared on the insulator layers had preferred orientation along c-axis. The $Y_{0.5}Yb_{0.5}MnO_3$ films consisted of uniform grains and had smooth surface. The leakage current density of MFIS structures was 10^{-7}~10^{-8}A/cm^2 at 5V. The clockwise C-V hysteresis induced by ferroelectric polarization switching was observed in the MFIS structures. The memory window of MFIS using the improved HfO_2 layer was about 2V and the retention time was over 10^4s.

INTRODUCTION

$YMnO_3$-based material having the hexagonal structure exhibits ferroelectric properties and a single polarized direction along the c-axis. The $YMnO_3$ has a low dielectric constant of 20 and a spontaneous polarization of 5.0 μC/cm^2 in the form of a single crystal.[1] In addition to its low dielectric constant, the chemical characteristic of being volatile-element-free is attractive for application to the FET (field effect transistor) -type ferroelectric memory. Ferroelectric random access memories (FeRAM) using polarization switching of the ferroelectrics is one of the nonvolatile memories. Particularly, FET (field effect transistor) -type ferroelectric memory has received great attention as next-generation memory because it has an advantage of high speed, low power consumption, reducing the memory cell size and nondestructive readout.[2-3] This type of memory uses the remanent polarization of the ferroelectrics to control the surface conductivity of silicon. One promising structure is the metal/ferroelectrics/insulator/semiconductor (MFIS) structures which have been studied using several ferroelectric and insulator. The MFIS structure has been proposed because it is difficult to obtain ferroelectric/Si structure having good interface. In this structure, the ferroelectric film must have small dielectric constant in order to apply the sufficient voltage on the ferroelectric film for polarization switching.

Conventionally, the $YMnO_3$ films were synthesized by sputtering, pulse laser deposition (PLD), molecular beam epitaxy (MBE) and metalorganic chemical vapor deposition (MOCVD) [4-6]. Also, some groups reported the synthesis of $YMnO_3$ films via chemical solutions [7-8]. In our previous work, the solid solution of $YMnO_3$ and $YbMnO_3$ has been studied for the purposes that increase the stability of the hexagonal structure and lower the processing temperature [9-11]. The synthesis of $(Y,Yb)MnO_3$ thin films using alkoxy-derived solutions and the effects of crystallization condition on the crystallographic properties have been investigated in detail. On the basis of fundamental studies, we were able to find that crystallization in Ar was inevitable for the improvement of electrical properties. Additionally, the ferroelectric properties of the (Y,Yb)MnO_3 thin films with metal-ferroelectrics-metal (MFM) and MFIS structures were reported [12-13].

In this paper, the MFIS structures were constructed using $Y_{0.5}Yb_{0.5}MnO_3$ as ferroelectrics and HfO_2 as insulator though the chemical solution process. And the effects of the preparation condition of HfO_2 layer on the crystallographic appearances and electrical properties of the MFIS structures are investigated.

EXPERIMENTAL PROCEDURE

HfO_2 and insulating layers were prepared by spin coating alkoxy-derived solutions. The alkoxy-derived solutions were prepared using hafnium iso-propoxide and ethylene glycol monomethyl ether (EGMME). The concentration was adjusted to 0.1 or 0.05 mol/L. The *p*-Si(100) substrate was chemically cleaned using HF acid and then soaked in EGMME and dried prior to coating. The films were deposited in two steps of 1000 rpm for 3 s and 3000 rpm for 30 s. The as-deposited thin film was dried at 150°C and calcined at 350°C for 10 min in air. Then, the film was heated by rapid thermal annealing at 750°C for 10 min in O_2.

For the preparation of precursor solutions for (Y,Yb)MnO_3 films, yttrium iso-propoxide, ytterbium iso-propoxide and manganese iso-propoxide were selected as starting chemicals. EGMME was selected as a solvent. Since alkoxides are extremely sensitive to moisture, the entire procedure was conducted in dry nitrogen. Yttrium iso-propoxide was dissolved in EGMME. Manganese iso-propoxide was added to the solution with atomic ratios of (Y+Yb):Mn=1:1. The concentrations of the solutions were adjusted to 0.2 mol/L for (Y,Yb)MnO_3 film. The solutions were then reacted at a reflux temperature of 124°C for 2 h.

(Y,Yb)MnO_3 films were prepared on the insulating layers using the (Y,Yb)MnO_3 precursor solutions. The films were deposited, dried and calcined in the same manner as the insulating films. Then, the film was heated by rapid thermal annealing at 750°C for 10 min in Ar.

The crystalline phase of the films was identified by using X-ray diffraction (XRD) measurements. The surfaces of the films were observed by atomic force microscopy (AFM). Pt top electrodes with 150μm diameter were deposited on the surface of the insulating layers on Si and also the ferroelectric films through a metal mask by electron beam evaporation method for measurements of electrical properties. The leakage current densities of the films were measured using an electrometer (Keithley 6517). The capacitance-voltage (C-V) characteristics were measured at 1 MHz using an impedance analyzer (Agilent 4294A).

RESULTS AND DISCUUSION

Preparation and electrical properties of the Pt/HfO_2/Si structure

The HfO_2 insulating layers were prepared using 0.1 and 0.05 mol/L solutions in order to investigate the effect of the concentration of the solutions. The thickness of the HfO_2 insulating layers were about 20nm. Figure 1 show XRD profiles of HfO_2 insulating layers crystallized on Si(100) substrates at 750°C. HfO_2 insulating layers had monoclinic structure and showed (-1 1 1) preferred orientation. Figure 2 shows AFM images of the surfaces of HfO_2 insulating layers. These insulating layers consisted of uniform grains and had smooth surfaces. The RMS of the HfO_2 insulating layers prepared using 0.1 and 0.05 mol/L solutions were 2.6nm and 2.9nm, respectively.

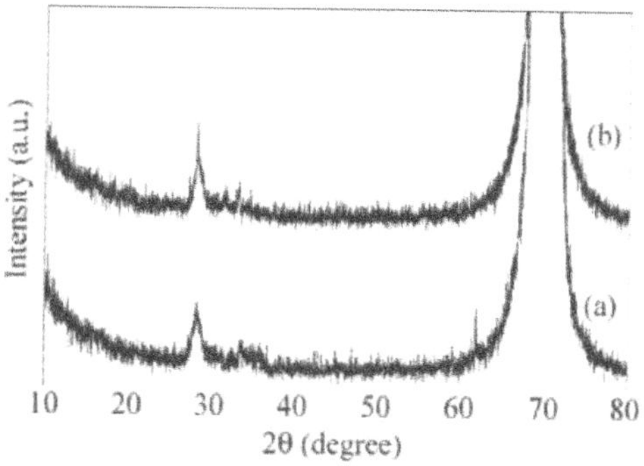

Figure 1 XRD profiles of HfO$_2$ insulating layers crystallized on Si(100) substrates at 750°C via (a)0.1 and (b)0.05 mol/L precursor solutions.

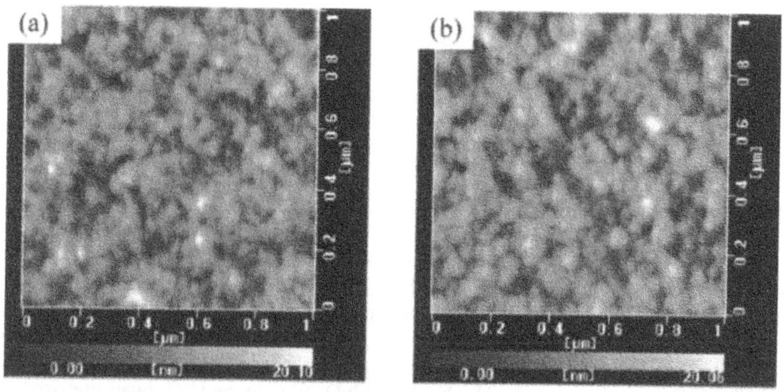

Figure 2 AFM images of HfO$_2$ insulating layers crystallized on Si(100) substrates at 750°C via (a)0.1 and (b)0.05 mol/L precursor solutions.

The electrical properties of HfO$_2$ insulating layers were evaluated in the metal-insulator-semiconductor (MIS) structure. The leakage current densities of the Pt/HfO$_2$/Si structures were about 10^{-7}A/cm^2. Figure 3 shows C-V characteristics of the Pt/HfO$_2$/Si structures measured at 1MHz with a sweep rate of 0.2 V/s from -5 V to +5 V and vice versa. The hysteresis was observed for the Pt/HfO$_2$/Si structure prepared using 0.1mol/L solution and it was considered to be caused by charge injection from the interface. The dielectric constant of the HfO$_2$ insulating layers calculated from the capacitance of the accumulation region were 12~13. This value was small compared with the dielectric constant of the bulk HfO$_2$. And the thicker HfO$_2$ layers showed larger dielectric constant. Therefore, it seems that the thin SiO_2 layer formed during the annealing process.

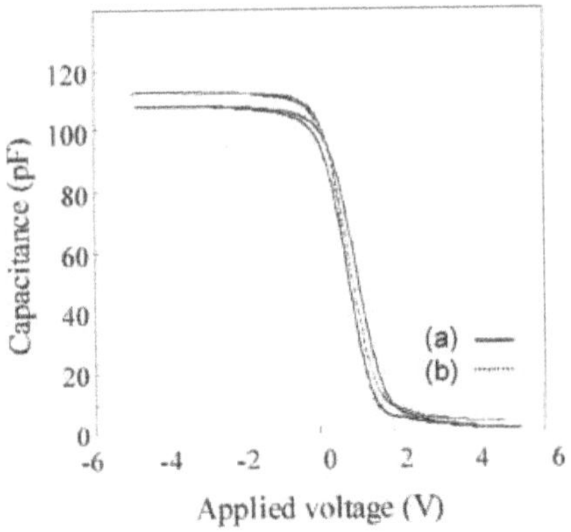

Figure 3 C-V characteristics of HfO_2 layers in Pt/HfO_2/Si structures prepared using (a)0.1 and (b)0.05 mol/L precursor solutions.

Preparation and electrical properties of the Pt/(Y,Yb)MnO_3/HfO_2/Si structure

The $Y_{0.5}Yb_{0.5}MnO_3$ films with thickness of 200nm were prepared on the HfO_2 insulating layers. Figure 4 shows the XRD profiles of the $Y_{0.5}Yb_{0.5}MnO_3$ films prepared on the HfO_2/Si. Hexagonal $Y_{0.5}Yb_{0.5}MnO_3$ was crystallized on the both insulating layers and no secondary phase was observed. It was found that the $Y_{0.5}Yb_{0.5}MnO_3$ films showed preferred orientation along c-axis. This preferred orientation was advantage to the application because the polarization axis of the (Y,Yb)MnO_3 was along c-axis. The $Y_{0.5}Yb_{0.5}MnO_3$ film on the HfO_2 layer prepared using 0.05mol/L solution has higher degree of c-axis orientation. Figure 5 shows AFM images of the surfaces of the $Y_{0.5}Yb_{0.5}MnO_3$ films prepared on the HfO_2/Si . These films consisted of uniform grains and had smooth surfaces. The grain sizes were below 100nm in the both films. The RMS of the $Y_{0.5}Yb_{0.5}MnO_3$ films on the HfO_2 layers prepared using 0.1 and 0.05mol/L solutions were 2.5nm and 2.7nm, respectively. The $Y_{0.5}Yb_{0.5}MnO_3$ film on the HfO_2 layer prepared using 0.05mol/L solution consisted of larger grains. It was considered that the surface uniformity of HfO_2 affected the crystallization of $Y_{0.5}Yb_{0.5}MnO_3$ film, and thereby the crystallinity and orientation of the $Y_{0.5}Yb_{0.5}MnO_3$ film changed.

The electrical properties such as leakage current density and capacitance-voltage characteristics were evaluated in the MFIS structure. The leakage current densities of the Pt/$Y_{0.5}Yb_{0.5}MnO_3$/HfO_2/Si structures were in range of 10^{-8}~10^{-7}A/cm^2 at 5V of applied voltage. Figure 6 shows the C-V characteristics of the Pt/$Y_{0.5}Yb_{0.5}MnO_3$/HfO_2/Si structures measured at 1 MHz with a sweep rate of 0.2 V/s from -10 V to +10 V and vice versa. The clockwise C-V hysteresis loops induced by ferroelectric polarization switching were observed in the both structure. The $Y_{0.5}Yb_{0.5}MnO_3$ on the HfO_2 layer using prepared using 0.05mol/L solution showed larger memory window. The electrical property depended on the crystallinity and orientation of the $Y_{0.5}Yb_{0.5}MnO_3$ ferroelectric film, and the uniformity of the interface between the $Y_{0.5}Yb_{0.5}MnO_3$ ferroelectric film and HfO_2 insulating layer. It is important to control the surface structure of the HfO_2 layer in order to obtain the advantage for crystallization of the $Y_{0.5}Yb_{0.5}MnO_3$ ferroelectric film and uniform interface.

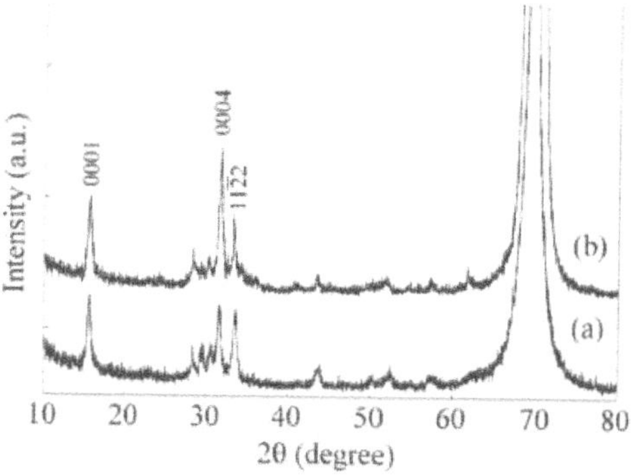

Figure 4 XRD profiles of $Y_{0.5}Yb_{0.5}MnO_3$ films on HfO_2 layers prepared using (a)0.1 and (b)0.05 mol/L precursor solutions.

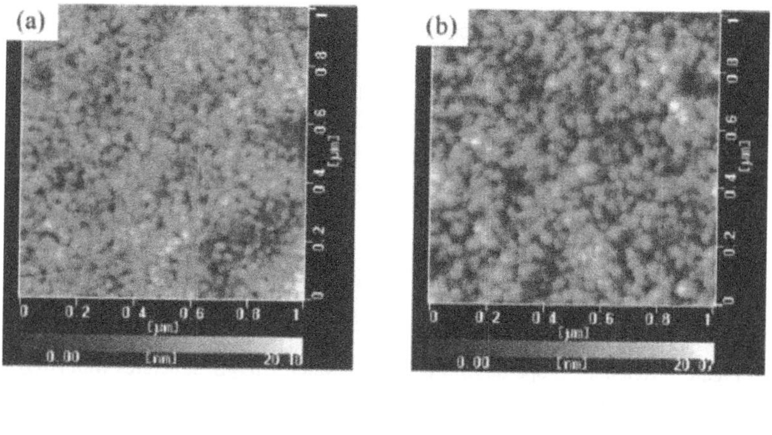

Figure 5 AFM images of $Y_{0.5}Yb_{0.5}MnO_3$ films on HfO_2 layers prepared using (a)0.1 and (b)0.05 mol/L precursor solutions.

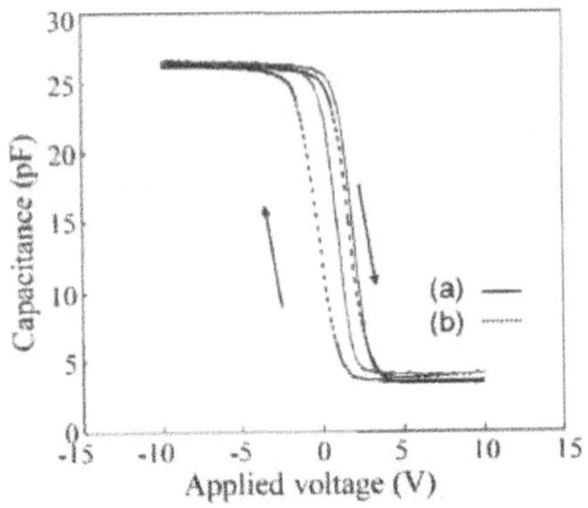

Figure 6 C-V characteristics of Pt/ $Y_{0.5}Yb_{0.5}MnO_3/HfO_2$/Si structures with HfO_2 layers prepared using (a)0.1 and (b)0.05 mol/L precursor solutions.

Figure 7 shows retention properties at the write voltage of +10V or -10V for the Pt/$Y_{0.5}Yb_{0.5}MnO_3/HfO_2$/Si structures. In the case of the Pt/$Y_{0.5}Yb_{0.5}MnO_3/HfO_2$/Si structures with HfO_2 layer prepared using 0.1mol/L, the capacitance changed with time and the margin of the capacitance after 10^4 s decreased below half of the initial value. In contrast, the Pt/$Y_{0.5}Yb_{0.5}MnO_3/HfO_2$/Si structures using HfO_2 layer prepared using 0.05mol/L the capacitances were almost constant over 10^4s.

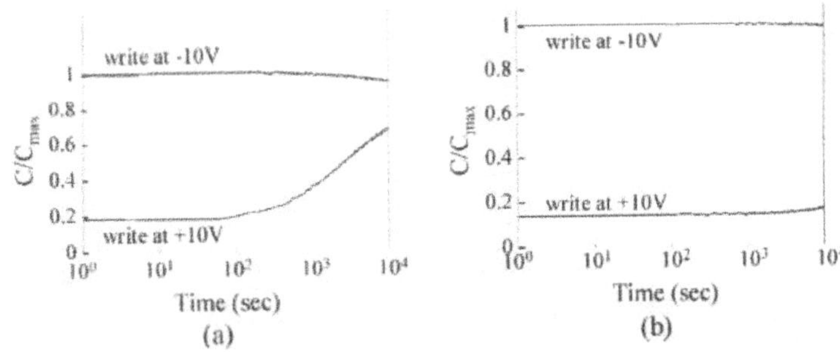

Figure 7 Retention properties of Pt/ $Y_{0.5}Yb_{0.5}MnO_3/HfO_2$/Si structures with HfO_2 layers prepared using (a)0.1 and (b)0.05 mol/L precursor solutions.

CONCLUSION

$Y_{0.5}Yb_{0.5}MnO_3$ (200nm)/HfO_2 (20nm)/Si structures were constructed by chemical solution deposition. HfO_2 insulating layers crystallized on Si(100) substrates consisted of uniform grains and had smooth surfaces. It seems that the thin SiO_2 layer formed between HfO_2 and Si substrate during the annealing process. The $Y_{0.5}Yb_{0.5}MnO_3$ film on the HfO_2 insulating layer prepared using the dilute solution had higher preferred orientation along c-axis. The $Y_{0.5}Yb_{0.5}MnO_3$ film consisted of uniform grains and had smooth surface. The electrical properties such as capacitance-voltage characteristics and retention of the Pt/(Y,Yb)MnO_3/HfO_2/Si (MFIS) structure were improved by the dilution of the solution for the HfO_2 layer. The leakage current density of MFIS structures was 10^{-7}~10^{-8}A/cm^2 at 5V. The clockwise C-V hysteresis induced by ferroelectric polarization switching was observed in the MFIS structures. The memory window of MFIS using improved HfO_2 layer was about 2V and the retention time was over 10^4s.

REFERENCES

[1]G.A. Smolenskii and V.A. Bokov, "Coexistence of magnetic and electric ordering in crystals", *J. Appl. Phys.*, **35**, 915-18 (1964).

[2]T. Hirai, Y. Fujisaki, K. Nagashima, H. Koike and Y. Tarui. "Preparation of $SrBi_2Ta_2O_9$ Film at Low Temperatures and Fabrication of a Metal/Ferroelectric/Insulator/Semiconductor Field Effect Transistor Using Al/$SrBi_2Ta_2O_9$/CeO_2/Si(100) Structures", *Jan. J. Appl. Phys.*, **36**, 5908-11 (1997).

[3]E. Tokumitsu, G. Fujii and H. Ishiwara, "Nonvolatile ferroelectric-gate field-effect transistors using $SrBi_2Ta_2O_9$/Pt/$SrTa_2O_6$/SiON/Si structures", *Appl. Phys. Lett.*, **75**, 575-77 (1999).

[4]H.N. Lee, Y.T. Kim and S.H. Choh, "Comparison of memory effect between $YMnO_3$ and $SrBi_2Ta_2O_3$ ferroelectric thin films deposited on Si substrates", *Appl. Phys. Lett.*, **76**. 1066-1068 (2000).

[5]N. Fujimura, T. Ishida, T. Yoshimura and T. Ito, "Epitaxially grown $YMnO_3$ film: New candidate for nonvolatile memory devices", *Appl. Phys. Lett.*, **69**, 1011-1013 (1996).

[6]S. Imada, T. Kuraoka, E. Tokumitsu and H. Ishiwara, " Ferroelectricity of $YMnO_3$ thin films on Pt(111)/Al_2O_3(0001) and Pt(111)/Y_2O_3(111)/Si(111) structures grown by molecular beam epitaxy", *Jpn. J. Appl. Phys.* **40**, 666-671 (2001).

[7]N. Fujimura, H. Tanaka, H. Kitahata, K. Tadanaga, T. Yoshimura, T. Ito and T.Minami, "$YMnO_3$ thin films prepared from solutions for non volatile memory devices", *Jan. J. Appl. Phys.*, **36**, L1601-L1603 (1997).

[8]W.C. Yi, J.S. Choe, C.R. Moon, S.I. Kwun and J.G. Yoon, "Ferroelectric characterization of highly (0001)-oriented $YMnO_3$ thin films grown by chemical solution deposition". *Appl. Phys. Lett.*, **73**. 903-905 (1998).

[9]K. Suzuki, K. Nishizawa, T. Miki and K. Kato, " Synthesis of ferroelectric $YMnO_3$ thin film by chemical solution deposition" *Key. Eng. Mater.*, **214-215**, 151-156 (2002).

[10]K. Suzuki, K. Nishizawa, T. Miki and K. Kato, " Effects of annealing conditions on crystallization of hexagonal manganite films", *Ferroelectrics.*, **270**, 99-104 (2002).

[11]K. Suzuki, D. Fu. K. Nishizawa, T. Miki and K. Kato, "Effects of substrates on alkoxy-derived (Y,Yb)MnO_3 thin films", *Integ. Ferroelectr.*, **47**, 91-100 (2002).

[12]K. Suzuki, D. Fu, K. Nishizawa, T. Miki and K. Kato, "Preparation of (Y,Yb)MnO_3/Y_2O_3/Si (MFIS) structure by chemical solution deposition method", *Jpn. J. Appl.*

Phys., **42**, 6007-6010 (2003).

[13]K. Suzuki, D. Fu, K. Nishizawa, T. Miki and K. Kato, "Ferroelectric properties of (Y,Yb)MnO_3 thin films prepared using alkoxide solutions", *Key. Eng. Mater.*, **248**, 77-80 (2003).

Electroceramic Materials and Applications
Edited by R. W. Schwartz

CONTROLLING OF SURFACE AND HUMIDITY DETECTING PROPERTIES OF POTASSIUM TANTALATE FILMS FABRICATED BY A HYDROTHERMAL ELECTROCHEMICAL METHOD

Takashi HASHIZUME, Atsushi SAIKI and Kiyoshi TERAYAMA
Department of Material Systems Engineering and Life Science, Faculty of Engineering, Toyama University
Gofuku 3190, Toyama, 930-8555, Japan

Takeshi TORISHIMA
DIJET, Co., LTD
Kamihigashi 2-1-18, Hiranoku, Osaka, 547-0002, Japan

ABSTRACT

Potassium tantalate film (KT film) having a good humidity detecting property was synthesized at relatively lower temperature such as 100-110°C and a pressure of about 0.1MPa. KT film has been grown on Ta electrode in aqueous solution of KOH by hydrothermal electrochemical method for a short time. A current source was used to maintain the current density of $1mA/cm^2$. KT films were fabricated in the concentration range of 2-5mol/L KOH. The film formed on Ta substrate had smooth surface and blue color fabricated in case of the 2mol/L KOH solution used. A porous KT film was formed at the range of 3-4mol/L KOH solution. This film was grown in the shape of islands. At more concentration of 5.0M KOH solution, the film was barely formed. The XRD results showed a specific crystallographic face of the tantalum was reacted preferentially. The thickness on the formed films was evaluated about 5.0μm.

In the porous material, water molecule is adsorbed into the capillary pores. As dielectric constant of the water is very high, a capacitance of the porous material increases with humidity. The amount of water absorbed depends upon the pore size. This KT film showed an increase in electrostatic capacity when it was brought into contact with humidity. The capacitance increase versus the relative humidity of the sensor was linear and showed good sensitivity in the lower humidity. The film can be applied as a humidity sensor for gas analysis system.

INTRODUCTION

Tantalum double oxide films are interested for applications in high temperature resistance materials, oxygen sensors and capacitor dielectric materials. Tantalum double oxide films can be obtained by several methods such as sputter[1,2], vapor deposition[3], sol-gel[4,5] and anodization. In recent years, a hydrothermal electrical synthesis method is in the creation technology of the thin film used for electronic material such as a capacitor. This synthesis method belongs to called soft solution processing [6,7], concerning the fabrication of pure, shaped, sized and oriented ceramic materials in a solution.

Materials used as humidity sensors are ceramic, polymer and composite. Compared with other materials, the humidity sensors based on porous ceramics have received much attention due to their chemical and physical stability. The purpose of this paper is to investigate the humidity sensor properties of fabricated potassium tantalate oxide by a hydrothermal electrochemical method.

EXPERIMENTAL PROCEDURE

Preparation of thin films

Tantalum metal plates with purity of 99.5% and potassium hydroxide with 85.5% were used as starting materials. Tantalum plate was cut into specimens with size of 10.0 mm x 25.0 mm x 0.4 mm, the polished by diamond and alumina suspensions successively to a mirror finish, and etched for 30s at anodic current density of 1.5 mA/cm^2 in solution of 10 vol.% H_2SO_4, 20 vol.% HF, and 70 vol.% ethylene glycol [8]. The plate was then ultrasonicated for 10 minutes in acetone and water solution individually. A Teflon case containing the electrolyte was used for hydrothermal-electrochemical method. This case was equipped with two-electrodes and a thermocouple. Ta and Pt plates are prepared as the working electrode and counter electrode, respectively. The concentration of Potassium hydroxide electrolyte solution was changed from 2 to 5 mol/L. The current source was used to maintain the current density of 1 mA/cm^2. The autoclave was sealed and maintained at about 373 K for 3600 s without staking or stirring, and cooled with cold water. After experiment, the Ta plate was washed with deionized water and dried in air.

The product on a Ta plate was characterized by X-ray diffraction (XRD) (MSJ-D006, Mac science, Yokohama, Japan). XRD patterns were recorded from 20° to 80° at a step scan of 0.02° with Cu-Kα radiation. Surface morphology were observed by optical microscopy, scanning electron microscopy (SEM) (S-3500, Hitachi, Tokyo, Japan), energy dispersive X-ray spectrometer (EDS) (ISIS-300, Oxford Instruments, England) and atomic force microscopy (AFM) (Nanoscope, Digital instruments, USA) .

Fabrication of Humidity Sensor and Measurements

For the fabrication of humidity sensor, tantalum wire of 99.5% purity was cut into specimens with size of 1.0mm in diameter and 15.0mm long. Potassium tantalum oxide films on tantalum wire were fabricated with same conditions. The humidity sensor image is shown in Fig1. Vacuum evaporation was used for the fabrication of the Au thin film of 0.14 μm in thickness. The one end of this wire was polished up to remove the Au film for lower electrode and bonded lead wire. The relative humidity (RH) in the test cell was controlled by mixing wet air with dry air in a given ratio. The flow rate of the gas was 100 ml/min. The sensitivity is defined as the ratio of the capacitance in the mixing wet air to the dry air.

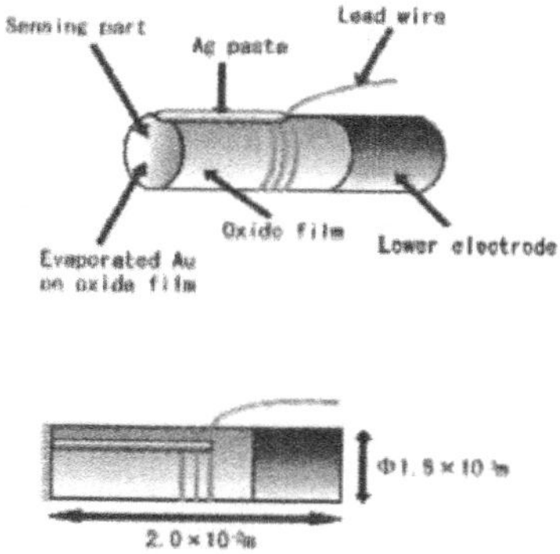

Fig1. Humidity sensor image.

RESULTS AND DISCUSSION

Film characterization

Fig.2 shows the SEM images of film on Ta substrates. Potassium tantalate (KT) films were fabricated using hydrothermal electrochemical method in the potassium hydroxide solution of 2-4 mol/L. For the solution of 5 mol/L, a tantalum plate is dissolved. The XRD patterns of the film on tantalum substrates were shown in Fig.3. Potassium tantalate oxide or tantalum oxide were not detected in the case of 2 mol/L or less of concentration and 5 mol/L or more. The phase in the potassium tantalate oxide film measured by XRD was only pyrochlore type.

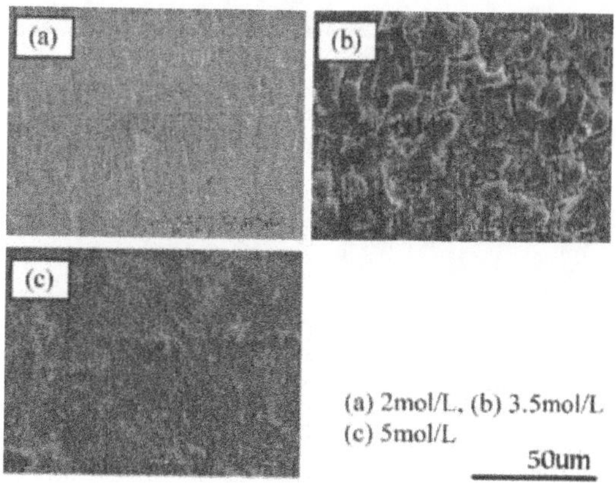

Fig.2 SEM micrograph of the surface of KT films fabricated with (a) 2 mol/L, (b) 3.5 mol/L, (c) 5mol/L KOH solutions (1.0mA/cm^2, 373K, 3600s)

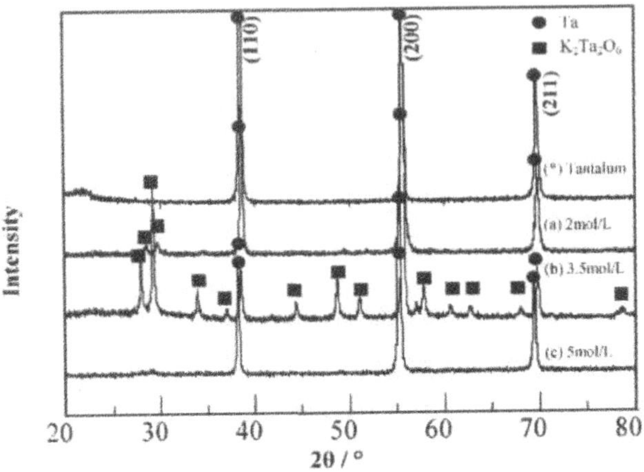

Fig.3 XRD patterns of hydrothermal synthesized in (a) 2 mol/L KOH, (b) 3.5 mol/L KOH and (c) 5 mol/L KOH.

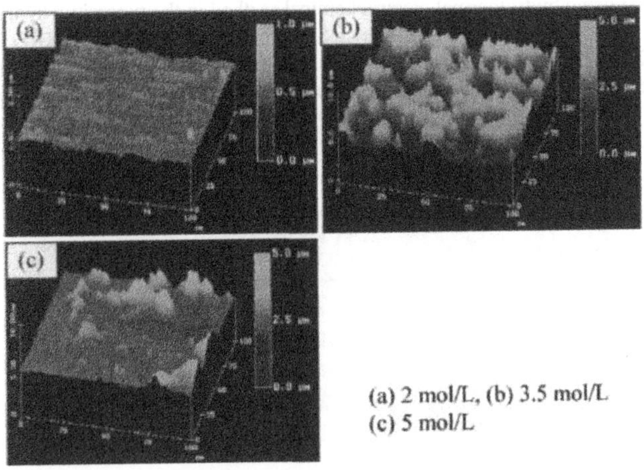

Fig.4 AFM image of the films

The film was uneven in the case of 3.5 mol/L by AFM image. The maximum height difference of the film was 5 micrometers. The crystal grown in the shape of an island was observed. Fig.4(a) shows that potassium tantalate films are hardly formed at a KOH concentration of 2 mol/L[8]. In the case of 5 mol/L, growth of a crystal was checked partly. If the speed of dissolving tantalum is higher than a fabricating film speed, the potassium tantalate films cannot be grown. Fig.5 shows the result of the X-ray diffractions of the samples which were fabricated with different reaction time. As the fabrication was proceeded, the peak of (110) and (211) of a tantalum substrate became smaller preferentially.

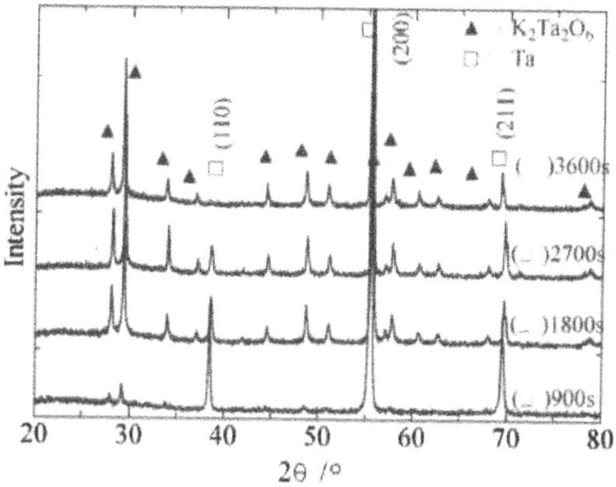

Fig.5 XRD patterns of the film fabricated for different reaction time.

Humidity sensing behaviors

Fig.6 shows the change in capacitance of the film fabricated at increasing and degreasing humidity levels. Humidity sensor using potassium tantalate film fabricated by 3.5 mol/L KOH appeared the change in capacitance only. In porous materials, water molecule gets adsorbed into the capillary pores. The capacitance of films increased almost linearly with increasing relative humidity. Fig.7 shows the response time of this sensor in the case of adsorption and recovery time in the case of desorption. The time taken by this sensor to achieve 90% value of total change is defined as the response time. The measurement response and recovery times were less than 15s and 150s, respectively. This sensor showed a rapid increase in electrostatic capacitance when brought into contact with moisture and a reversion to the original value when in dry air. In the results, the films from the hydrothermal electrochemical method have sensitivity to humidity. The porous nature of the films prepared by the hydrothermal methods has substantial effect on the sensing properties.

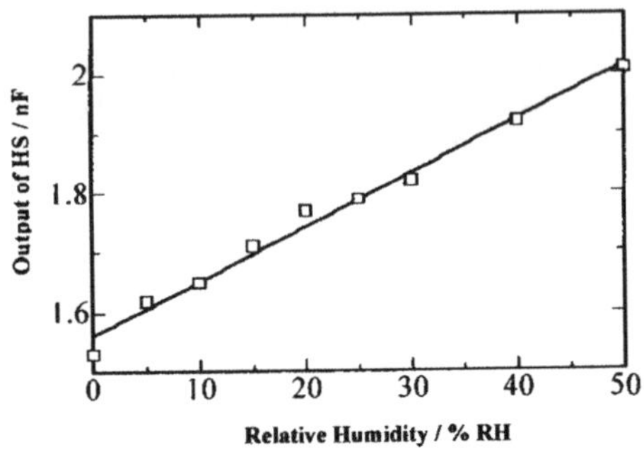

Fig.6 Relation between capacitance of humidity sensor and Relative Humidity.

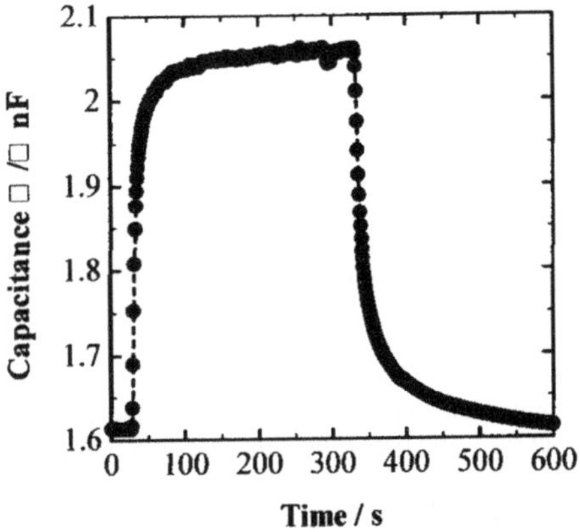

Fig.7 Response of humidity sensor in adsorption and desorption (0-50%RH,)

CONCLUSION

Potassium tantalate films fabricated by hydrothermal electrochemical method are sensitive to humidity. This methods is relatively much simple and easy as compared to other methods such as sol-gel, sputtering and CVD, to fabricated tantalum double oxide thin films at low temperature. When the reaction temperatures were about 110 °C and the KOH concentrations were 3.5 mol/L, only rough films on the substrate were obtained. The change in capacitance with humidity, of the potassium tantalate thin film, shows that these films can be used as a humidity sensor through the entire range of humidity. Humidity detection is considered to be essentially due to water condensation in the pores. This technique provides an inexpensive and environmentally friendly route to direct synthesizing for porous films for humidity sensor.

REFERENCES

[1]Ngaruiya, J.M., Venkataraj, S.a , Drese, R.a , Kappertz, O.a , Leervad Pedersen, T.P.a , Wuttig, M., " Preparation and characterization of tantalum oxide films produced by reactive DC magnetron sputtering", *Phys. Status Solid A*,**198**, 99-110(2003)

[2]Peng-Heng Chang, Hung-Yu Liu, "Structures of tantalum pentoxide thin films formed by reactive sputtering of Ta metal", *Thin Solid Films*, **258**, 56-63(1995)

[3]Textor, M.c , Ruiz, L.a c c , Hofer, R.c , Rossi, A.b c c , Feldman, K.c , Hähner, G.c , Spencer, N.D., " Structural chemistry of self-assembled monolayers of octadecylphosphoric acid on tantalum oxide surfaces", *Langmuir*, **16**, 3257-3271(2004)

[4]Weon-Pil tai, Jun-Gyu Kim, Jae-Hee Oh, Young-Sung Kim, "Preparation and humidity sensing behaviors of nanostructured potassium tantalate:titania films", *Sens. Actuators B*, **105**,199-203(2005).

[5]Weon-Pil tai, Jun-Gyu, Jae-Hee Oh, Chongmu Lee, Dong-wha Park, Wha-seung Ahn, "Preparation and humidity-sensing properties of nanostructured potassium tantalate thin films", J. Mater. Sci.: Mater. Electron., **15**, 25-28(2004).

[6]Zhibin Wu, M. Yoshimura, "Investigations on procedures of the fabrication of barium titanate ceramic films under hydrothermal-electrochemical conditions", *Solid state Ionics.*, **122**,161-172(1999).

[7]Zhibin Wu, M. Yoshimura, "Formation of pyrochlore potassium tantalate thin films by soft solution processing", *The Solid Films.* **375,** 46-50 (2000).

[8]Z-B.Wu, T.Tsukada, M.Yoshimura, "Direct synthesis of potassium tantalate thin films by hydrothermal-electrochemical method", *J.Mater.Sci.*, **35**, 2833-2839(2000).

Electroceramic Materials and Applications
Edited by R. W. Schwartz

PREPARATION OF PREFERENTIALLY ORIENTED $BaTiO_3$ THIN FILM BY HYDROTHERMAL SOFT CHEMICAL PROCESS

Koji Kajiyoshi and Kazumichi Yanagisawa
Research Laboratory of Hydrothermal Chemistry, Faculty of Science, Kochi University
2-5-1, Akebono-cho, Kochi 780-8520, Japan

Qi Feng
Department of Advanced Materials Science, Faculty of Engineering, Kagawa University
2217-20, Hayashi-cho, Takamatsu 761-0396, Japan

Richard E. Riman and George A. Rossetti
Department of Ceramic and Materials Engineering, Rutgers University
607 Taylor Road, Piscataway, NJ 08855-0909, USA

ABSTRACT

Epitaxial crystal growth technique is usually utilized to obtain preferentially oriented thin films of ceramics. In this technique, single crystal substrates are indispensable and lattice matching between the substrate crystal and the thin-film crystal is crucial, leading to high costs for the preparation of epitaxial thin films. In contrast, if polycrystalline or amorphous materials can be used as substrates, oriented thin films can be prepared on them at a low cost with a high degree of flexibility in their size and shape. In the present study, a hydrothermal soft chemical processing has been established as a method to prepare $BaTiO_3$ thin films with preferred orientation on a variety of substrates. A layered H^+-form titanate $H_{1.07}Ti_{1.73}O_4 \cdot nH_2O$ (HTO) was prepared from a layered titanate of $K_{0.8}Ti_{1.73}Li_{0.27}O_4 \cdot nH_2O$ (KTLO) with a lepidocrocite-like layered structure and was used as a precursor. The HTO precursor powder was exfoliated into the nanosheets of elementary layer of its structure by reacting with a 0.1 M *n*-propylamne solution. A preferentially oriented HTO thin film was then prepared on the metal substrate of Ti, Pt, or Au by spin-coating with the HTO nanosheet colloidal solution, followed by heating at 200°–400°C. $BaTiO_3$ thin films with high [110] preferred orientation (Lotgering factor ≈ 1.0) were successfully prepared by hydrothermal treatment of the HTO precursor thin film in $Ba(OH)_2$ solutions at 150°–250°C. The direction of preferred orientation of the titanate thin film was dominated by the structure of the HTO precursor thin film and was not dependent on the properties of the substrate.

INTRODUCTION

Preferentially oriented thin films of metal oxides are high performance materials for electronic devices. For the preparations of the preferentially oriented thin films, usually epitaxial crystal growth techniques are utilized.[1-3] In these cases, a single crystal substrate and lattice matching between the substrate crystal and the film crystal are necessary, meaning the high cost for the preparation of the epitaxial films. If a polycrystalline or amorphous substrate can be used for the oriented thin films, a low cost thin film process is possible. Some studies have indicated that the oriented thin films can be obtained on polycrystalline substrates or single crystal substrates without lattice matching by controlling growth direction of film crystal.[4–6] However, the control of crystal growth direction on a polycrystalline substrate is not easy in normal cases.

For the preparations of layered metal oxide thin films with preferred orientation, exfoliation and layer-by-layer assembly techniques have developed.[7,8] Titania nanostructured films have been obtained by heat-treatment of a layered titanate film prepared by using the layer-by-layer assembly technique.[9] Recently, we have studied the preparation of $BaTiO_3$ and anatase from a titanate with a layered structure, and found that plate-like $BaTiO_3$ and anatase particles could be obtained by hydrothermal treatment of the plate-like titanate particles in $Ba(OH)_2$ solution and distilled water, respectively.[10] The structural transformations from the layered titanate to $BaTiO_3$ and anatase were topotactic reactions. Further, we have successively prepared $BaTiO_3$ and TiO_2 thin films with preferred orientation from the layered titanate precursor thin film formed on the stainless steel substrate by using an exfoliation-restacking technique and topotactic structural transformation reactions (hydrothermal soft chemical process).[11] The dielectric constants and the dielectric losses of the $BaTiO_3$ thin film were 570 and 11% at 10^2 Hz, and 440 and 18% at 10^4 Hz of frequency, respectively. In the present study, we describe our recent developments of preparation and characterization of the $BaTiO_3$ thin films on polycrystalline metal substrates by the hydrothermal soft chemical process.

EXPERIMENTAL PROCEDURE

Experimental procedure to prepare $BaTiO_3$ thin film with preferred orientation by using the hydrothermal soft chemical method is outlined in Fig. 1. A layered titanate powder of $K_{0.8}Ti_{1.73}Li_{0.27}O_4 \cdot nH_2O$ (KTLO) with a lepidocrocite-like layered structure was synthesized by a flux method, as described previously.[10] The orthorhombic structure of the KTLO is composed of corrugated host layers of edge-shared octahedra of (Ti, Li)O_6 and interlayers of K^+, as shown in Fig. 2. A layered H^+-form titanate powder of $H_{1.07}Ti_{1.73}O_4 \cdot nH_2O$ (HTO) with the lepidocrocite-like layered structure was prepared from KTLO powder by an ion-exchange in 1.0 M HNO_3 and was used as a precursor.

The HTO precursor powder was exfoliated into the nanosheets of elementary layer of its structure by reacting with a 0.1 M *n*-propylamne solution.[12] The titanium content in the resulted nanosheet colloidal solution was adjusted to 0.1 M. The HTO thin film was prepared on mirror-polished metal substrate of Ti, Pt, or Au by spin-coating with the HTO nanosheet solution, followed by heating at 200°–400°C to prevent the dissolution of the HTO thin film by re-exfoliation reaction in the subsequent hydrothermal treatment.

Thin films of $BaTiO_3$ were prepared by hydrothermal treatment of the HTO thin films in $Ba(OH)_2$ solutions contained in a titanium lined autoclave shown in Fig. 3.

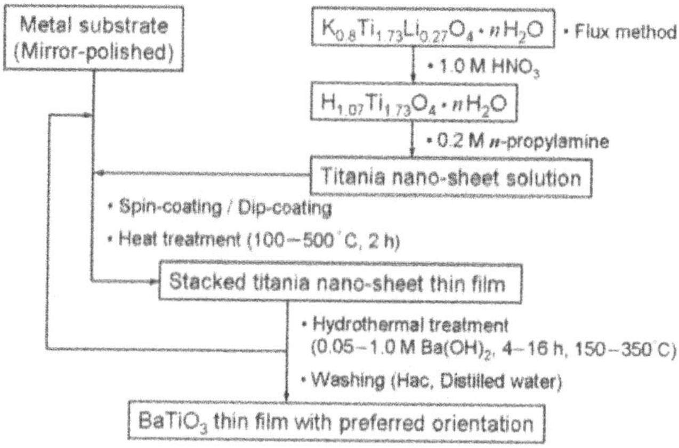

Figure 1. Experimental procedure to prepare $BaTiO_3$ thin film with preferred orientation by hydrothermal soft chemical method.

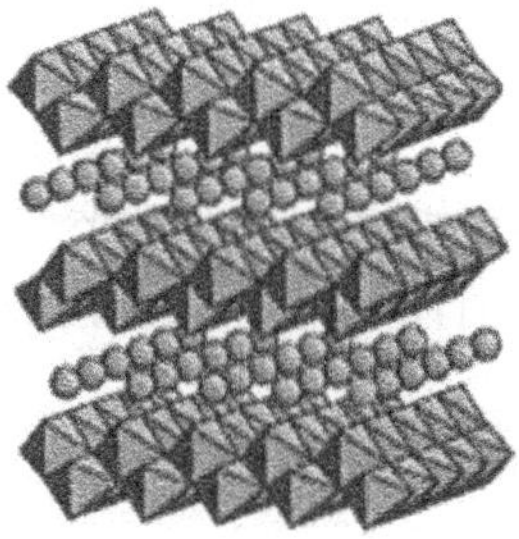

Figure 2. Lepidocrocite (γ-FeOOH) -type layered structure of $K_{0.8}Ti_{1.73}Li_{0.27}O_4$ (KTLO) and $H_{1.07}Ti_{1.73}O_4 \cdot nH_2O$ (HTO) (after ref. 7).

RESULTS AND DISCUSSION

Figure 4 shows XRD patterns (CuKα) of KTLO and HTO powders. The KTLO and HTO powders have a layered structure with a basal spacing of 0.7544 and 0.9225 nm, respectively. The resulting HTO precursor thin films on polycrystalline Ti metal substrates were confirmed to have the layered structure with a basal spacing similar to that of the HTO precursor powder.

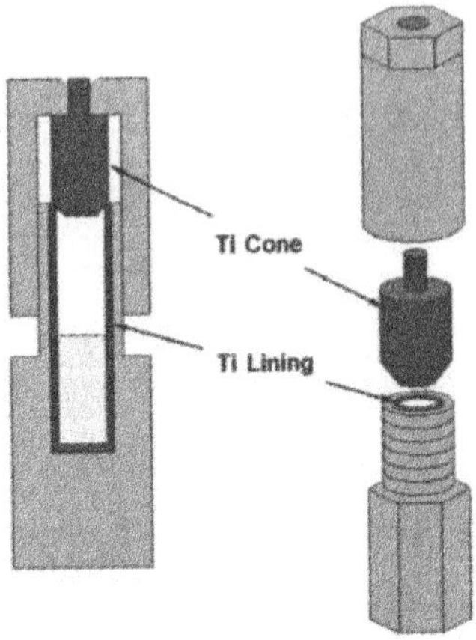

Figure 3. Titanium lined autoclave used for hydrothermal synthesis.

Figure 5 shows XRD patterns (CuKα) of $BaTiO_3$ thin films formed at various hydrothermal temperatures. The X-ray diffraction patterns showed very strong 110 peak only at hydrothermal temperatures up to 250°C, suggesting that $BaTiO_3$ thin film was obtained with a preferred orientation along [110] direction. Above 300°C, very weak peaks for other diffractions were also observed. Lotgering's factor[13] for the [110] orientation was calculated to be 1.0 for $BaTiO_3$ thin films formed at hydrothermal temperatures below 250°C and 0.85 for those formed above 300°C.

Pole figures of (110) and (111) planes for the $BaTiO_3$ thin film formed hydrothermally at 350°C are depicted in Figs. 6 and 7. The (110) pole was highly concentrated at the stereographic pole and the (111) pole was distributed coaxially around the stereographic pole. This indicates that microcrystals of $BaTiO_3$ are uniaxially oriented with their [110] direction being normal to the substrate surface.

Figures 8 and 9 show surface morphologies of the $BaTiO_3$ thin films formed hydrothermally at 250° and 350°C, respectively. The $BaTiO_3$ thin film formed at 250°C had a smooth surface, whereas the thin film formed at 350°C consisted of rectangular microcrystals faceted by $BaTiO_3$ {100} faces. It is likely that dissolution-precipitation reaction proceeds more extensively at 350°C, leading to a turbulence of preferred orientation as seen in Fig. 5.

The $BaTiO_3$ thin films could be prepared also on polycrystalline Pt and Au metal substrates similarly with preferred orientation along [110] direction, indicating the orientation direction is not dependent on the properties of the substrate, but on the structure of HTO precursor. These results suggest that the $BaTiO_3$ lattices are formed through a topotactic structural transformation reaction, where Ba^{2+} ions migrate into the stacked HTO nanosheet crystals through their interlayer pathways and react topotactically with the TiO_6 octahedral layers.[10]

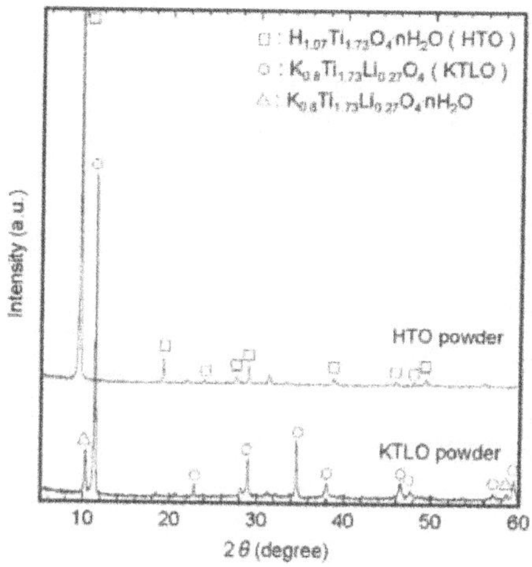

Figure 4. XRD patterns (CuKα) of KTLO and HTO powders.

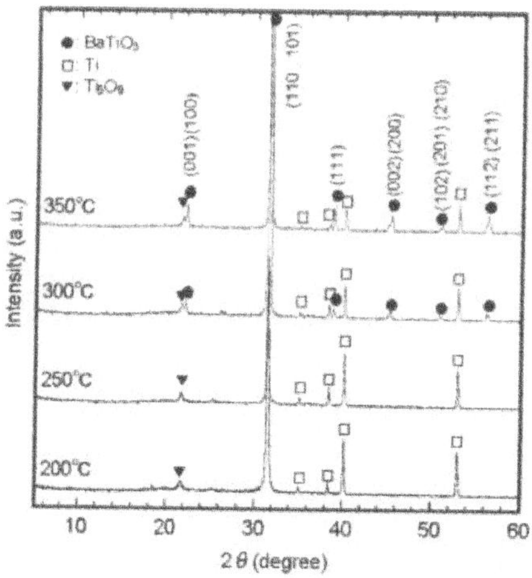

Figure 5. XRD patterns (CuKα) of $BaTiO_3$ thin films formed at various hydrothermal temperatures.

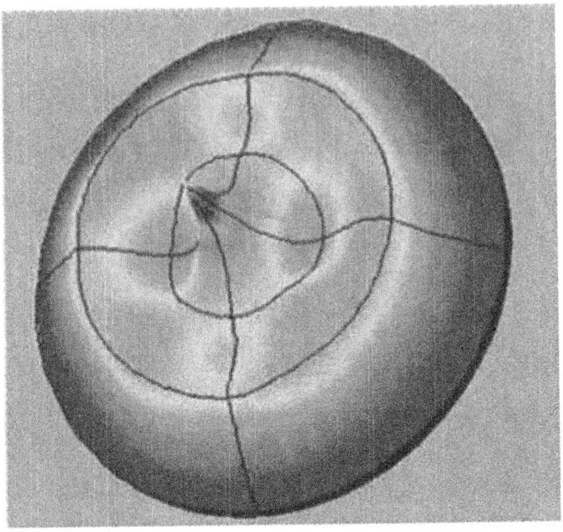

Figure 6. Pole figure of (110) plane for $BaTiO_3$ thin film formed hydrothermally at 350°C.

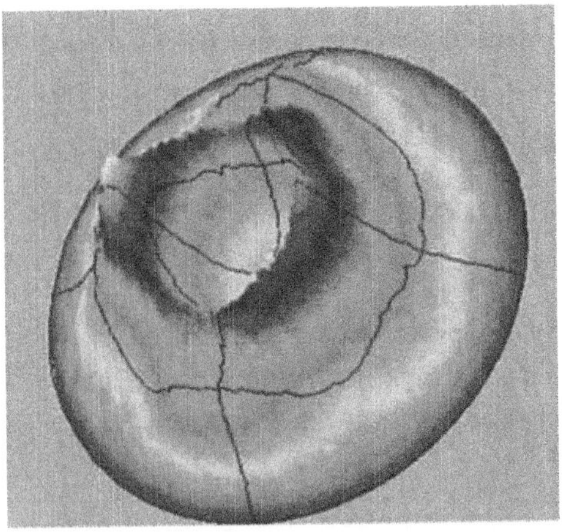

Figure 7. Pole figure of (111) plane for $BaTiO_3$ thin film formed hydrothermally at 350°C.

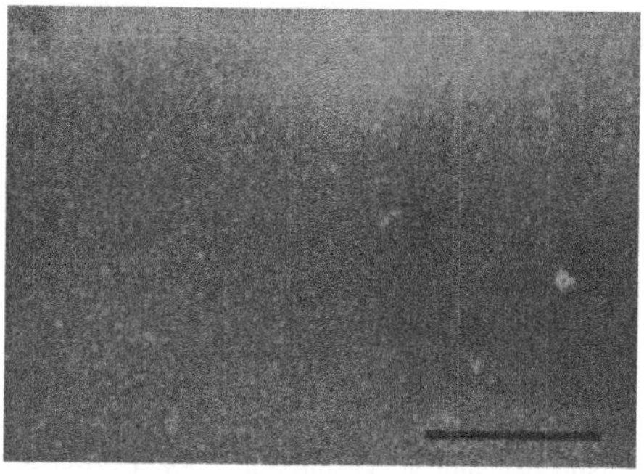

Figure 8. SEM image of surface of $BaTiO_3$ thin film formed hydrothermally at 250°C (bar = 10 μm).

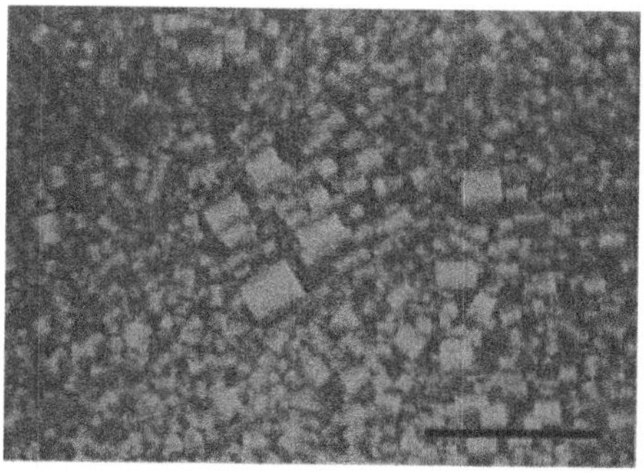

Figure 9. SEM image of surface of $BaTiO_3$ thin film formed hydrothermally at 350°C (bar = 10 μm).

CONCLUSIONS

$BaTiO_3$ thin films have been prepared on polycrystalline metal substrates such as Ti, Pt, and Au with highly preferred orientation (Lotgering's factor ≈ 1.0). The orientation is uniaxial along [110] direction. The direction is dominated by the structure of the $H_{1.07}Ti_{1.73}O_4 \cdot nH_2O$ (HTO) precursor thin film, and not dependent on the properties of the substrate. Dissolution-recrystallization becomes dominant under hydrothermal conditions above ca. 300°C, leading to a turbulence of preferred orientation. $BaTiO_3$ lattices are considered to be formed through a topotactic structural transformation under hydrothermal conditions.

ACKNOWLEDGEMENTS

We thank Ms. Emi Mentaki and Mr. Naoto Futenma of Kochi University for their experimental assistance.

REFERENCES

[1]K. Iijima, Y. Tomita, R. Takayama, and I. Ueda, *J. Appl. Phys.*, **60**, 361 (1986).

[2]K. Kanno, S. Hayashi, R. Takayama, and T. Hirano, *Appl. Phys. Lett.*, **68**, 328 (1996).

[3]H.-M. Christen, L. A. Boatner, J. D. Budai, M. F. Chishoim, L. A. Gea, P. J. Marrero, and D. P. Norton, *Appl. Phys. Lett.*, **68**, 1488 (1996).

[4]C. H. Lee and S. J. Park, *J. Mater. Sci.: Mater. Electron.*, **1**, 219 (1990).

[5]B. Yang, X. J. Zhang, S. T. Zhang, X. Y. Chen, Z. C. Wu, Y. F. Chen, Y. Y. Zhu, Z. G. Liu, and N. B. Ming, *Appl. Phys. Lett.*, **79**, 4559 (2001).

[6]K. Kajiyoshi, N. Ishizawa, and M. Yoshimura, *J. Am. Ceram. Soc.*, **74**, 369 (1991).

[7]T. Sasaki, F. Kooli, M. Iida, Y. Michiue, S. Takenouchi, Y. Yajima, F. Izumi, B. C. Chakoumakos, and M. Watanabe, *Chem. Mater.*, **10**, 4123 (1998).

[8]R. E. Schaak and T. E. Mallouk, *Chem. Mater.*, **12**, 3427 (2000).

[9]T. Sasaki, Y. Ebina, K. Fukuda, T. Tanaka, M. Harada, and M. Watanabe, *Chem. Mater.*, **14**, 3524 (2002).

[10]Q. Feng, M. Hirasawa, and K. Yanagisawa, *Chem. Mater.*, **13**, 290 (2001).

[11]Q. Feng, K. Kajiyoshi, and K. Yanagisawa, *Chem. Lett.*, **32**, 48 (2003).

[12]M. Iida, T. Sasaki, and M. Watanabe, *Chem. Mater.*, **10**, 3780 (1998).

[13]F. K. Lotgering, *J. Inorg. Nucl. Chem.*, **9**, 113 (1959).

Electroceramic Materials and Applications
Edited by R. W. Schwartz

SYNTHESIS AND CHARACTERIZATION OF ORGANIC–INORGANIC HYBRID LAYER PHOTOCATALYSTS

Shuji Ichimura
Department of Environmental Technology and Urban Planning, Nagoya Institute of Technology
Department of Research and Development, NICHIBAN CO., LTD.
3-1-10 Nihongi-Shinmachi
Anjo, Japan, 446-0075

Hiroshi Ebisu
Department of Electrical and Computer Engineering, Nagoya Institute of Technology
Gokiso-cho, Showa-ku
Nagoya, Japan, 466-8555

Kazumi Kato
Department of Environmental Technology and Urban Planning, Nagoya Institute of Technology
Advanced Manufacturing Research Institute, National Institute of Advanced Industrial Science and Technology
2266-98 Anagahora, Shimoshidami, Moriyama-ku
Nagoya, Japan, 463-8560

ABSTRACT

Organic-inorganic hybrid layer photocatalysts were synthesized by irradiating hybrid material consisted of urethane acrylic oligomer and titanium lactate with ultraviolet light. A hybrid reaction of photo-polymerization of urethane acrylic oligomer and the sol-gel reaction of titanium lactate proceeded at relatively low temperature as 100 °C by the irradiation. Therefore, the substrate like plastic films, which do not have high temperature resistance, can be used. The functional layer with a thickness of 5 μm on polyethylene terephthalate film showed a high rubbing resistance and a high transparency of 80-90% in the wavelength range of 400 to 800 nm. Surface hardness was identified as more than H. The photo-enhanced reactivity of the functional layer was analyzed by ESR using spin trapping experiment. 5,5-dimethyl-1-pyrroline N-oxide (DMPO) was used as a spin-trap reagent. Photo-generated radicals were detected as DMPO-OH when irradiated by metal halide light. Photocatalytic property was confirmed by measuring bleaching of methylene blue solution.

INTRODUCTION

Titanium dioxide (TiO_2) shows photocatalytic activity upon absorbing light.[1)] The electron in the 2p orbital of O, which exists at a valence band, absorbs light with a wavelength of 380-400 nm or less (equivalent to 3-3.2 eV of the band gap), is excited to a 3d orbital of Ti, and produces e^-(electron) and h^+(hole) inside TiO_2. These photoproduced e^- and h^+ react with oxygen and water to generate $O_2^{-}\cdot$ and $OH\cdot$ on the surface, respectively, and then induce other reactions.[2-4)]

$$TiO_2 + h\nu \rightarrow e^- + h^+ \quad (1)$$
$$e^- + O_2 \rightarrow O_2^{-}\cdot \quad (2)$$

$$h^+ + H_2O \rightarrow OH\cdot + H^+ \quad (3)$$
$$O_2^{-}\cdot + H^+ + e^- \rightarrow HO_2^- \quad (4)$$
$$2O_2^{-}\cdot + H^+ \rightarrow HO_2^- + O_2 \quad (5)$$
$$OH\cdot + OH\cdot \rightarrow H_2O_2 \quad (6)$$

Since the photocatalytic properties of TiO_2 were established as being due to the Honda-Fujishima effect[5)] early in the 1970s, many studies have been carried out. Various organic compounds were found to decompose by reaction with the photogenerated radicals at the surface of TiO_2. Therefore, this photocatalyst is used in many environmental applications such as pollution control, deodorization and antibacterial uses.[5-7)]

Usually, TiO_2 fine powders were used as photocatalysts. For practical use, they often have some difficulties such as separation from the reaction system. TiO_2 films or coatings are much more convenient. However, high temperature must be applied for preparation of crystalline TiO_2 films and coatings.

Transparent organic titanium compound is attractive instead of the usual TiO_2 photocatalysts. And the urethane acrylic oligomer that is used as floor wax and becomes the transparent hard layer by irradiating with ultraviolet light is also attractive for a matrix material.[8)] In this paper, a hybrid of the urethane acrylic oligomer and the organic titanium compound was synthesized. The characteristics, microstructure and photocatalytic activities were discussed.

EXPERIMENT

Preparation of organic-inorganic hybrid layer

Organic-inorganic hybrid layer was prepared by a following process. Urethane acrylic oligomer (GEN Maintenance Technology, WINUP ST3-A) and organic titanium compound (Matsumoto Chemical Industry, Orgatix) were mixed with solvent in a specific weight ratio as shown in Table I. The mixture was coated on polyethylene terephthalate film and was dried at 100 °C for 30 s and was irradiated with a high pressure mercury lamp (irradiation intensity of 320-390nm: 0.25W/cm^2, center of the peak: λ = 365 nm) for 90 s. In addition, the organic layer without titanium was also produced on polyethylene terephthalate film by irradiating urethane acrylic oligomer with a high pressure mercury lamp as a reference.

Characterization of surface hardness and transparency of organic-inorganic hybrid layer

Surface hardness of organic-inorganic hybrid layer was evaluated in accordance with Japanese Industrial Standards (JIS K5600). Transparency of organic-inorganic hybrid layer at a wavelength of 400-800 nm was measured using a spectrophotometer (Jasco V-570).

Characterization of microstructure of organic-inorganic hybrid layer

The microstructure of organic-inorganic hybrid layer was analyzed by scanning electron microscopy (SEM, Shimadzu SSX-550 and Hitachi S-4300, acceleration voltage: 15kV and 10 kV respectively). The distribution of Ti was analyzed by energy-dispersive X-ray spectroscopy (EDS, Horiba EXMA-3770).

Evaluation of photocatalytic activity of organic-inorganic hybrid layer

Table I. Chemical components and characterization of organic-inorganic hybrid layer.

		Sapmle A	Reference
Chemical components of organic-inorganic hybrid solutions	Urethane acrylic oligomer [WINUP ST3-A]	38	100
	Titanium lactate [Orgatix TC-310] (Ti/element proportion)	62 (5.1/27.3)	
	Toluene		100
	Isopropyl alcohol	114	
	Methyl ethyl ketone	266	
	The weight ratio of element proportion/whole solution	13.6 %	50.0 %
Characterization	The ratio of Ti/element proportion	7.8 %	0.0 %
	Surface hardness	H	2H
	Transparency [400-800 nm]	81-90 %	82-90 %
	Photocatalytic activity : Difference of absorbance at a wavelength of 660 nm by irradiation for 60 min [100-$Abs_{60\ min}/Abs_{0\ min} \times 100$]	21.8 %	1.3 %

ESR measurements

Radicals generated at the surface of photoaccelerated organic-inorganic hybrid layer was analyzed by ESR (JEOL JES-FE1XG X-band Spectrometer) using $MgO:Mn^{2+}$ as a reference. Because the radicals decay rapidly, 5,5-dimetyl-1-pyrroline-N-oxide (DMPO) was used as a spin-trap reagent.[4,9-13] A distilled water (30 μL) and DMPO (6.54 mg) were mixed. Then the solution and a specimen of organic-inorganic hybrid layer coated on polyethylene terephthalate film (1mm × 35mm) were placed in a thin flat quartz cell for liquids (inner size 0.4 × 4.2 × 50 mm^3), which was set in an ESR apparatus for the measurements. A metal halide lamp (Ushio, US510) was used for photoexcitation of the specimen of organic-inorganic hybrid layer coated on polyethylene terephthalate film. The characteristics of the lamp are shown in Fig. 1 (center of the peak: λ = 420 nm). TiO_2 is photo-activated by irradiation of the UV light with wavelength of 380-400 nm and less. Therefore, both of the black light and the metal halide lamp are suitable for the photo-activation of TiO_2.

Methylene blue bleaching

The photocatalytic activity of organic-inorganic hybrid layer was evaluated by the methylene blue bleaching method.[4,14] An aqueous solution of methylene blue (1.5 mL, concentration: 10 ppm) were placed into a flat quartz cell (optical path length: 5 mm) and the absorbance at a wavelength of 660 nm was measured using a spectrophotometer (Jasco V-570). Then a specimen of organic-inorganic hybrid layer coated on polyethylene terephthalate film (9mm × 30mm) was placed into a flat quartz cell. The cell was then irradiated with black light (27 W), the characteristics of which are shown in Fig. 1 (center of the peak: λ = 370 nm) for 60 min. After irradiation, organic-inorganic hybrid layer coated on polyethylene terephthalate film was removed from the cell, and the absorbance at a wavelength of 660 nm was measured again.

RESULTS AND DISCUSSION

Characteristics and microstructure of organic-inorganic hybrid layer

Sample A is the organic-inorganic hybrid layer synthesized by irradiating hybrid material consisted of urethane acrylic oligomer and titanium lactate with a high pressure mercury lamp. Reference is the organic layer without titanium synthesized by irradiating urethane acrylic oligomer with a high pressure mercury lamp. Table I shows surface hardness and transparency of Sample A and Reference. Sample A and Reference showed a high rubbing resistance and a high transparency of 80-90%. Surface hardness of Sample A was identified as H. Surface hardness of Reference was identified as 2H.

Figure 2 shows SEM images of Sample A (organic-inorganic hybrid layer) and Reference (organic layer). The surface of Sample A was relatively rough compared with Reference. The layer was approximately uniform, and coherence of titanium compound was not observed.

Figure 3 shows titanium mapping images of Sample A and Reference analyzed by SEM-EDS. In the titanium mapping image of Sample A, the precipitation of titanium was not observed, and titanium dispersed uniformly. The quantity of titanium contained in the system was about 8 weight %, it is almost identical with the chemical component of the mixture for coating. On the other hand, in the titanium mapping image of Reference, no titanium was observed.

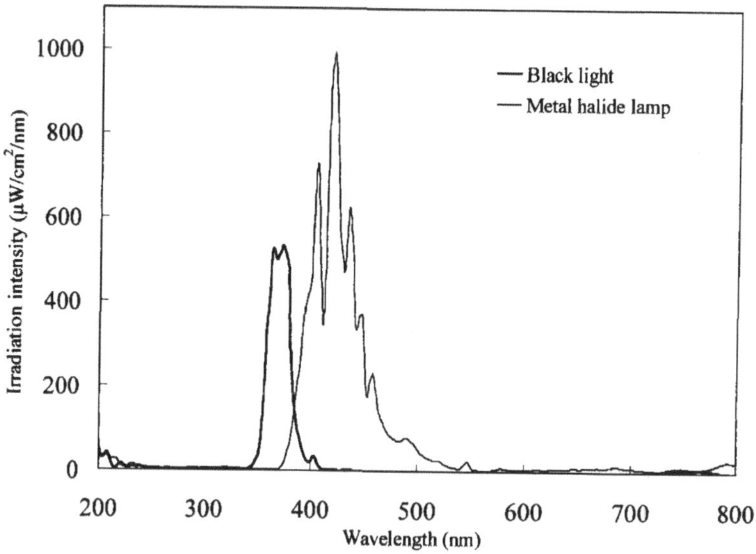

Fig. 1. Intensity of photoirradiation for lamps used in evaluation of methylene blue bleaching and ESR analysis.

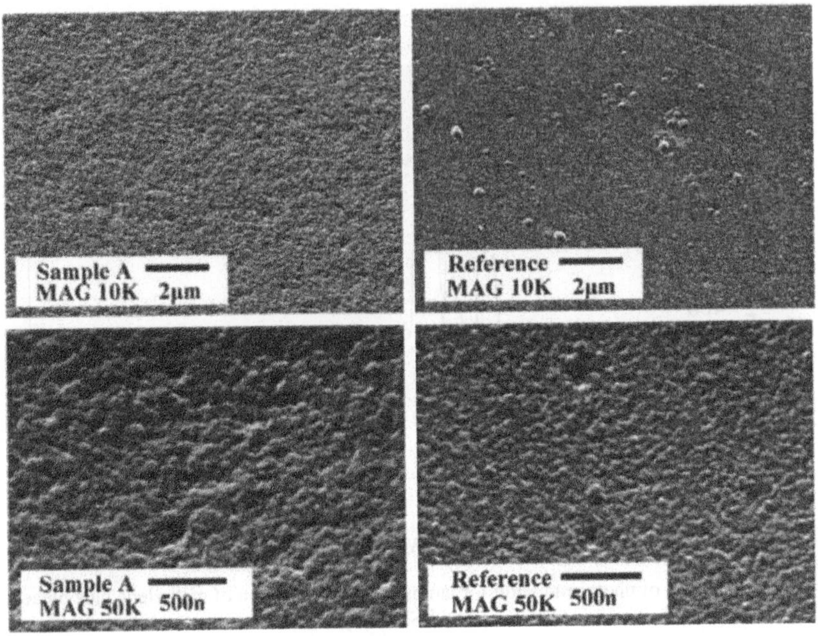

Fig. 2. SEM images of Sample A (Organic-inorganic hybrid layer) and Reference (Organic layer).

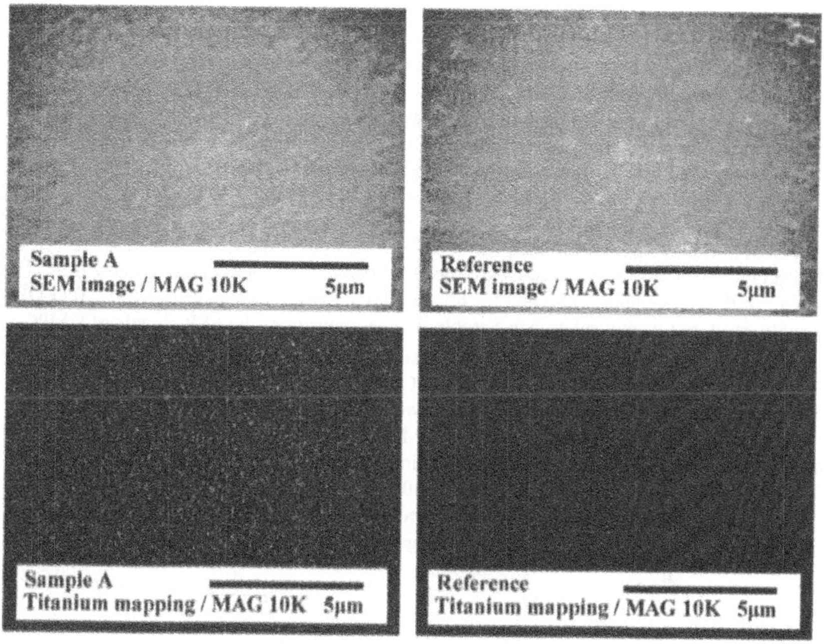

Fig. 3. SEM and titanium mapping images of Sample A (Organic-inorganic hybrid layer) and Reference (Organic layer).

Photocatalytic activity of organic-inorganic hybrid layer by ESR measurements

Radicals generated at the surface of Sample A and Reference were analyzed by ESR in order to compare the photocatalytic activity of the organic-inorganic hybrid layer and the organic layer. We focused our attention on OH· radicals on the surface of Sample A generated by the reaction of a hole with water, described by eq. (3). DMPO was used as a spin-trapping reagent to detect the OH· radicals by observing the stable ESR signals of DMPO-OH generated by chemical reaction (7). ESR spectra of Sample A in distilled water with DMPO (2 M) added as a spin-trapping reagent with photoirradiation times of 0, 60 and 500 s, were shown in Fig. 4. In the spectra of Sample A photoirradiated for 60 and 500 s, four signals (g=2.005(7), a^N=a^H=1.49 mT) due to DMPO-OH were observed clearly near H=334.4, 335.9, 337.4, and 338.9 mT.[4,9-13] In the spectra of Reference photoirradiated for 60 and 500 s, these four signals due to DMPO-OH were also observed, but were weaker than Sample A. The intensity of the ESR signals of DMPO-OH is evaluated by comparison of $h{\cdot}\Delta H_{pp}^2$, where h and ΔH_{pp} are the peak-to-peak amplitude and width of the signal respectively, with that of MgO:Mn^{2+} simultaneously observed in Fig. 4 as a reference.[15] The net amount of DMPO-OH with respect to photoirradiation time was calculated from the difference in the intensity ratios of DMPO-OH to MgO:Mn^{2+} before and after irradiation, as shown in eq. (8). Figure 5 shows changes in [DMPO-OH] for Sample A and Reference with DMPO (2 M) under photoirradiation for 0-500 s. Photogenerated [DMPO-OH] of Sample A (organic-inorganic hybrid layer) increased with irradiation time. On the other hand, Photogenerated [DMPO-OH] of Reference (organic layer) was little.

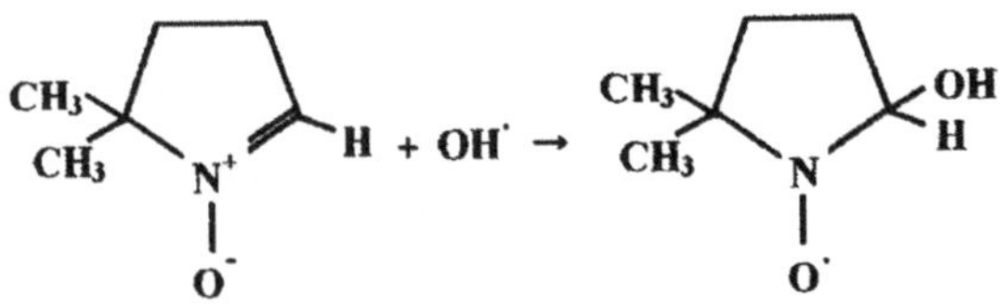

DMPO DMPO-OH (7)

$$[\mathrm{DMPO\text{-}OH}]_{IT=0\text{-}500} = [I_{\mathrm{DMPO\text{-}OH}}/I_{\mathrm{MgO:Mn^{2+}}}]_{IT=0\text{-}500} - [I_{\mathrm{DMPO\text{-}OH}}/I_{\mathrm{MgO:Mn^{2+}}}]_{IT=0} \quad (8)$$

(where $I = h{\cdot}\Delta H_{pp}^2$, IT: photoirradiation time)

Photocatalytic activity of organic-inorganic hybrid layer: methylene blue bleaching

Table I also shows photocatalytic activities of Sample A and Reference for methylene blue bleaching.[4,14] The difference of absorbance ($100\text{-}Abs_{60\,min}/Abs_{0\,min} \times 100$) at a wavelength of 660 nm for methylene blue solutions of Sample A with photoirradiation for 0-60 min were compared with that of Reference. The values of Sample A and Reference were 21.8% and 1.3%, respectively. While the value of the commercial product which declared the photocatalytic activity was 18.2%. Therefore, it was confirmed that Sample A had higher photocatalytic properties.

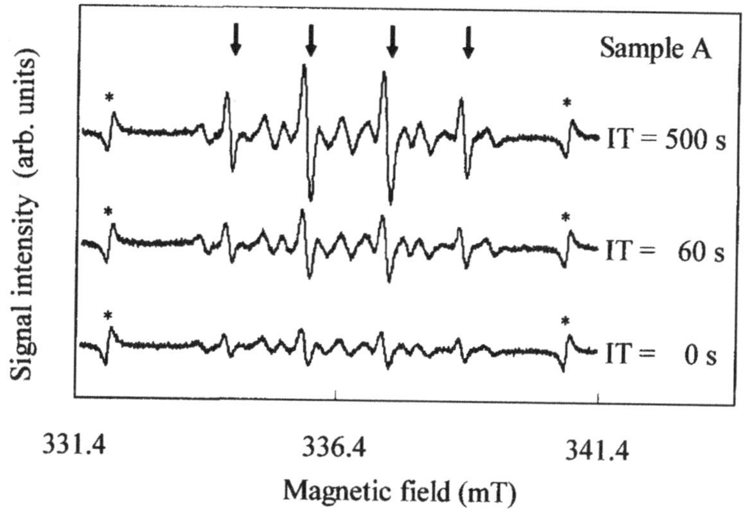

Fig. 4. ESR spectra of Sample A (Organic-inorganic hybrid layer) with DMPO (2.0 M). ↓: signals of DMPO-OH. IT is photoirradiation time.

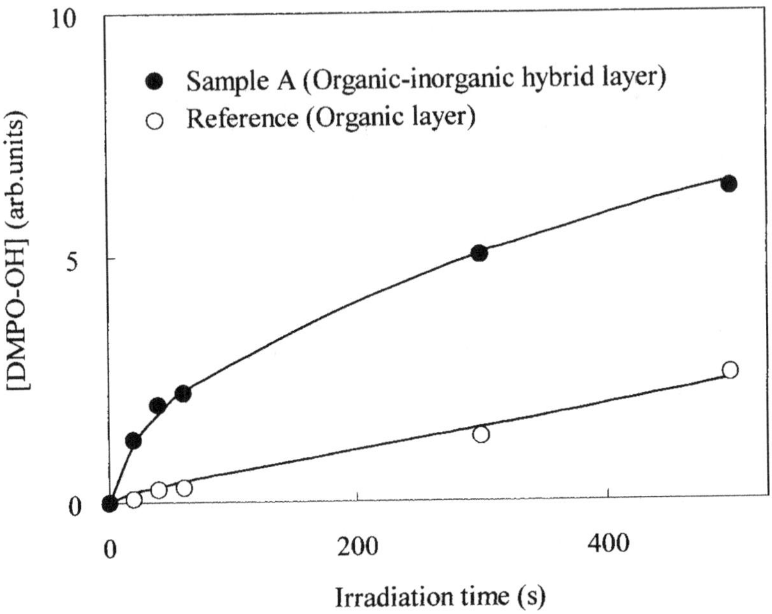

Fig. 5. Changes in [DMPO-OH] for Sample A (Organic-inorganic hybrid layer) and Reference (Organic layer) with DMPO (2.0 M) during photoirradiation for 0-500s.

CONCLUSIONS

Organic-inorganic hybrid layer photocatalysts were synthesized by irradiating hybrid material consisted of urethane acrylic oligomer and titanium lactate with a high pressure mercury lamp. A hybrid reaction of photo-polymerization of urethane acrylic oligomer and the sol-gel reaction of titanium lactate proceeded at relatively low temperature as 100 °C by the irradiation. Therefore, the substrate like plastic films, which do not have high temperature resistance, can be used. The functional layer with a thickness of 5 μm on polyethylene terephthalate film showed a high rubbing resistance and a high transparency of 80-90 % in the wavelength range of 400 to 800 nm. Surface hardness was identified as H. The surface was approximately uniform, and coherence of titanium compound was not observed. The quantity of titanium contained in the hybrid layer was about 8 weight %. The photo-enhanced reactivity of the functional layer was analyzed by ESR using spin trapping experiment. Photo-generated radicals were detected as DMPO-OH when irradiated by metal halide light. Photocatalytic property was confirmed by measuring bleaching of methylene blue solution.

REFERENCES

[1]A. Fujishima and K. Honda, "Electrochemical Photolysis of Water at a Semiconductor Electrode," Nature, **238**, 37 (1972).

[2]M. R. Hoffmann, S. T. Martin, W. Choi and D. W. Bahnemann, "Environmental Applications of Semiconductor Photocatalysis", Chem. Rev. **95**, 69 (1995).

[3]T. Hirakawa, H. Kominami, B. Ohtani and Y. Nosaka, "Mechanism of Photocatalytic Production of Active Oxygens on Highly Crystalline $Ti0_2$ Particles by Means of Chemiluminescent Probing and ESR Spectroscopy," J. Phys. Chem. B, **105**, 6993 (2001).

[4]S. Ichimura, H. Ebisu, T. Nonami and K. Kato, "Photocatalytic Activity of Titanium Dioxide Coated with Apatite," Jpn. J. Appl. Phys. **44**, No.7A, 5164 (2005).

[5]A. Fujishima and K. Honda, "Electrochemical Evidence for the Mechanism of the Primary Stage of Photosynthesis," Bull. Chem. Soc. Jpn., **44**, 1148 (1971).

[6]T. Hisanaga, K. Harada and K. Tanaka, "Photocatalytic degradation of organochlorine compounds in suspended TiO_2," J. Photochem. Photobiol. A, **54,** 113 (1990).

[7]K. Kato, A. Tsuzuki, H. Taoda, Y. Torii, T. kato and Y. Butsugan , "Crystal structures of TiO_2 thin coatings prepared from alkoxide solution via the dip-coating technique affecting the photocatalytic decomposition of aqueous acetic acid," J. Mater. Sci., **29**, 5911 (1994).

[8]US patent US6372340B1.

[9]Y. Nosaka, "Behavior of Chemical Species in Photocatalytic Reaction at the Surface of Titanium Oxides," Photochemistry, **34**, 14 (2003) [in Japanese].

[10]Y. Nosaka, S. Komori, K. Yawata, T. Hirakawa and A. Y. Nosaka, "Photocatalytic ·OH radical formation in TiO_2 aqueous suspension studied by several detection methods, "Phys. Chem. Chem. Phys., **5**, 4731 (2003).

[11]M. A. Grela, M. E. J. Coronel and A. J. Colussi, "Quantitative Spin-Trapping Studies of Weakly Illuminated Titanium Dioxide Sols. Implications for the Mechanism of Photocatalysis," J. Phys. Chem., **100**, 16940 (1996).

[12]S. Horikoshi, H. Hidaka and N. Serpone, "Hydroxyl radicals in microwave photocatalysis. Enhanced formation of OH radicals probed by ESR techniques in microwave-assisted photocatalysis in aqueous TiO_2 dispersions," Chem. Phys. Lett., **376**, 475 (2003).

[13]H. Hidaka, J. Zhao, E. Plzzetti and N. Serpone, "Photodegradation of Surfactans. **8**. Comparison of Photocatalytic Processes between Anionic Sodium Dodecylbenzenesulfonate and Cationic Benzyldodecyldimethylammonium Chloride on the TiO2 Surface," J. Phys. Chem., **96**, 2226 (1992).

[14]K. Takami, T. Nakasone, K. Hashimoto and A. Fujishima, " Proposal of a method to evaluate photocatalytic activity about prevention of pollution," Eng. Mater., **46**, No. 5, 102 (1998) [in Japanese].

[15]C. Poole, "Electron Spin Resonance," John Wiley & Sons, New York, 2nd ed., Chap. 12. (1983).

Electroceramic Materials and Applications
Edited by R. W. Schwartz

EFFECTS OF pH CONTROL ON PREPARATION OF Z-TYPE HEXAGONAL FERRITE $Ba_3Co_2Fe_{24}O_{41}$ BY POLYMERIZED COMPLEX METHOD

Makoto Nakanishi, Asako Akiyama, Tatsuo Fujii, Jun Takada
Division of Chemstry and Biochemistry, Graduate School of Natural Science and Technology, Okayama University
Tsushima-naka 3-1-1
Okayama 700-8530, Japan

Takeyuki Kikuchi
Department of Materials Science and Chemistry, Graduate School of Engineering, University of Hyogo
Syosya 2167
Himeji 671-2201, Japan

Yoshihiro Kusano
Department of Applied Art, College of the Arts, Kurashiki University of Science and the Art
Nishinoura 2640, Tsurajima-cho
Kurashiki 712-8505, Japan

ABSTRACT

The formation process of Z-type hexagonal ferrite, $Ba_3Co_2Fe_{24}O_{41}$, prepared by polymerized complex method and the effects of pH control of citrate complex solution were investigated. By heating a precursor, a pyrolyzed complex polymer, spinel phase, M-type ferrite and Y-type ferrite were formed sequentially with increasing the heating temperature. By heating up to 1200 °C, monophasic Z-type ferrite was not synthesized in usual preparation procedures. So pH of citrate complex solution was changed and the effects on formation process were examined. As ammonia solution was added to change the complex solution from acidic to basic, the phases in the pyrolysate changed form barium carbonate to barium nitrate and the crystallinity of the constituent phases decreased. By controlling pH to 11, monophasic Z-type ferrite was obtained by heating at 1200 °C. On the other hand, pH control of the complex solution toward basic condition suppressed the formation of Y-type ferrite. It was revealed that the optimum conditions for preparing the complex solution depended on phases even for same Ba-Co-Fe-O system.

INTRODUCTION

Hexagonal barium ferrite has been developed and used for permanent magnetic materials. Recently ferroxplana-type hexagonal ferrites have also been investigated intensively for high frequency electronic devices[1]. It is hard to prepare ferroxplane-type hexagonal ferrites by solution processes and they are usually prepared by heating at high temperature. So synthesis processes of ferroxplane-type hexagonal ferrites have been investigated for low temperature synthesis, control of particle sizes and so on. Polymerized complex method is one of the sol-gel methods and widely used for synthesis of high-T_c superconductors with complicated compositions or instability[2,3]. The advantages of this method are low cost and easy handling, compared with usual sol-gel method using metal alkoxides as starting materials. We investigated the formation process of Y-type ferrite, $Ba_2Co_2Fe_{12}O_{22}$ or Co_2Y, prepared by polymerized complex method and revealed the effects of starting materials on the morphologies of the particles. Z-type ferrite, $Ba_3Co_2Fe_{24}O_{41}$ or Co_2Z, also belongs to ferroxplana-type hexagonal ferrite and exhibits higher permeability and lower resonant frequency than Y-type ferrite. We also tried to prepare Co_2Z by polymerized complex method but could not be synthesized in the same procedures for Co_2Y. Recently it was reported that the pH of solution influenced to the formation processes through the complex formation between metal cation and citric acid in the case of M-type ferrite[4]. So we investigated the effects of pH of the citrate complex solution on formation processes of Z-type ferrite. It was revealed that the optimum preparation conditions depended on phases even for the same Ba-Co-Fe-O system.

EXPERIMENTAL PROCEDURES

Sample preparation by polymerized complex method was performed as follows. A stoichiometric amount of barium nitrate, iron nitrate and cobalt nitrate for the composition of $Ba_3Co_2Fe_{24}O_{41}$ was dissolved in distilled water. Citrate acid was added to the solution and pH of the solution was adjusted by adding ammonia solution to it. The metal-citrate complexes are stable in aqueous solution, so no precipitates were observed by adjusting the pH of the solution up to 11.Ethylene glycol is usually added to the citrate solution in polymerized complex method, but for some systems it was not included in preparation processes[5,6]. In this experiment some samples were prepared without ethylene glycol. After ethylene glycol was added, the solution was condensed at 150 °C for 12 hours, and then heated at 180 °C for ester polymerization of citrate complexes and ethylene glycol. The polymerized sample was pyrolized at 360 °C for 12 hours in air. The obtained precursor powder was heated at 1200 °C for 6 hours in air. For a comparison Y-type ferrite samples, $Ba_2Co_2Fe_{12}O_{22}$, were also prepared by the same procedures, and the effects of pH control were examined. The identification of phases in the samples were performed by means of powder X-ray diffraction measurements (Rigaku RINT-2500) using a graphite-monochromized Cu-Kα radiation.

RESULTS AND DISCUSSION

Fig.1 shows the XRD patterns of the precursors and the heated samples prepared from the metal complex solution with the different pH. The pH of the solution was about 1 before adjusting by ammonia solution. From the width of the diffraction peaks, it was revealed that the crystallinity of the constituent phases in the precursor decreased with increasing pH value of the solution. Though barium nitrate was used as a starting reagent, barium carbonate was also formed in the precursor. It is expected that an atmosphere during the pyrolysis included the carbon dioxide generated by the combustion of the polymer matrix. The barium compounds mainly included in the precursor changed from barium carbonate to barium nitrate as the pH became higher. The heated samples prepared from these precursors consisted of several phases of hexagonal ferrites. Z-type ferrite was included as a main phase in all samples, but M-type ferrite coexisted in the samples in the lower pH region and U-type ferrite was observed in the higher pH region. These results are summarized in Table I.

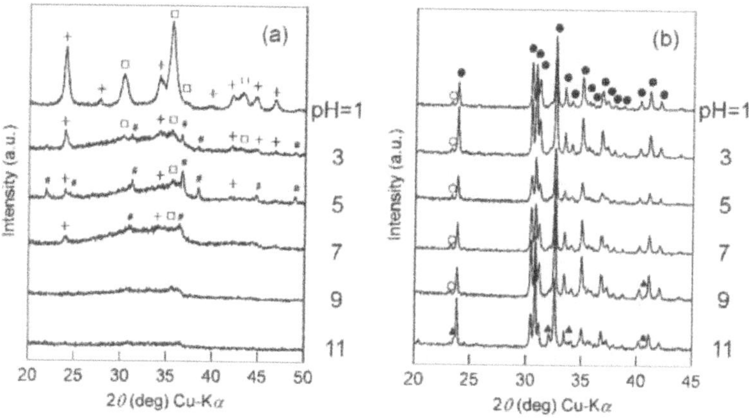

Fig.1 XRD patterns of the precursors (a) and the heated samples (b) prepared from the solution with the different pH. The symbols in the figure indicate the diffraction peaks from the constituent phases as follows; • Z-type, ○ M-type, ▲ U-type, □ spinel type, + barium carbonate and # barium nitrate.

Table I Constituent phases in the samples prepared with ethylene glycol

pH	Precursors	Heated samples
1	Spinel, $BaCO_3$	Z-type, M-type
3	Spinel, $BaCO_3$, $Ba(NO_3)_2$	Z-type, M-type
5	Spinel, $BaCO_3$, $Ba(NO_3)_2$	Z-type, M-type
7	Spinel, $BaCO_3$, $Ba(NO_3)_2$	Z-type, M-type
9	-	Z-type, M-type, U-type
11	-	Z-type, U-type

To prepare monophasic Co_2Z the preparation procedures was reexamined and the effects of pH of the solution without ethylene glycol (EG) on the formation of Co_2Z were investigated. The preparation procedures were same as the previous experiments except for the addition of no EG. Fig.2 shows the XRD patterns of the precursors and the heated samples prepared from the solution including no EG with the different pH. In this series the crystallinity of the constituent phases in the precursor did not change regularly with pH value of the solution. The hematite was detected in the precursor derived from higher pH solutions, which is different from the precursors prepared from the solution with EG.

The main phase in all samples was Z-type ferrite and M-type ferrite coexisted with Co_2Z in the case of the solutions with lower pH, same as the case without EG. On the other hand Y and U-type ferrite were observed with Co_2Z by heating the highly crystallized precursors. Finally the monophasic Co_2Z was prepared from the low crystallized precursor derived from the most basic solution without EG. These results are summarized in Table II.

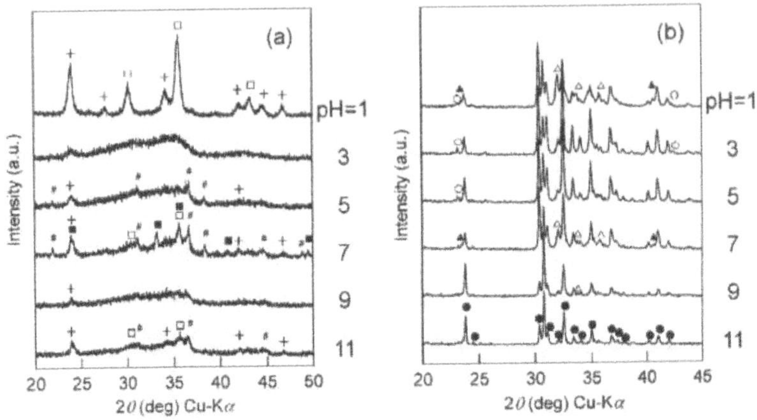

Fig.2 XRD patterns of the precursors (a) and the heated samples (b) prepared from the solution without ethylene glycol. The symbols in the figure indicate the constituent phases as follows; • Z-type, ○ M-type, ▲ U-type, Δ Y-type, □ spinel type, ■ hematite, + barium carbonate and # barium nitrate.

Table II Constituent phases in the samples prepared without ethylene glycol

pH	Precursors	Heated samples
1	Spinel, $BaCO_3$	Z-type, M-type, U-type, Y-type
3	$BaCO_3$	Z-type, M-type
5	$BaCO_3$, $Ba(NO_3)_2$	Z-type, M-type
7	Spinel, Hematite, $BaCO_3$, $Ba(NO_3)_2$	Z-type, U-type, Y-type
9	$BaCO_3$	Z-type, Y-type
11	Spinel, $BaCO_3$, $Ba(NO_3)_2$	Z-type

The same experiments were performed for preparing the Y-type ferrite, but the results were completely different from that for Z-type ferrite. Figure 3 shows the XRD patterns of the heated samples of Y-type ferrite prepared from the solution of pH = 1 with EG and pH = 11 without EG. For comparison the XRD patterns of Z-type ferrite prepared with the same conditions are also shown in the same figure. Only in the case of the lowest pH, namely without adjustment of pH, and with EG, monophasic Y-type was obtained. The high crystallinity of the precursor seemed to promote the formation of Y-type ferrite and suppress the coexistence of other types of ferrite. From these considerations it was revealed that the optimum preparation conditions depended on phases even for the same Ba-Co-Fe-O system.

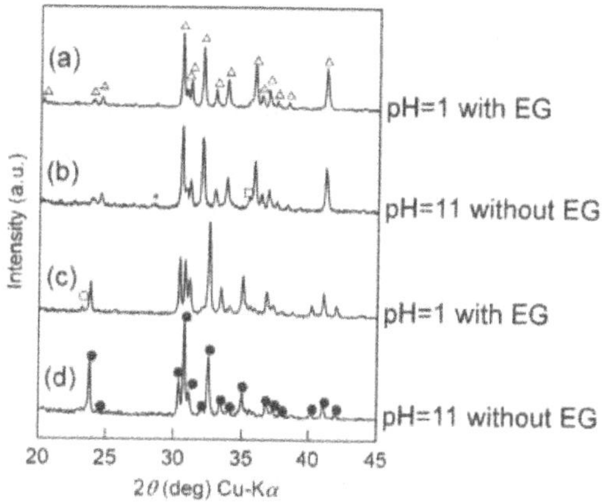

Fig.3 XRD patterns of the heated samples of Y-type ferrite ((a) and (b)) and Z-type ferrite ((c) and (d)) prepared with different conditions. The symbols in the figure indicate the constituent phases as follows; • Z-type, ○ M-type, Δ Y-type, □ spinel type and * $BaFe_2O_4$.

CONCLUSIONS

We investigated the effects of pH of the citrate complex solution on formation processes of Z-type ferrite. As ammonia solution was added to change the complex solution from acidic to basic, the phases in the pyrolysate changed form barium carbonate to barium nitrate and crystallinity of the constituent phases decreased. The monophasic Z-type ferrite was obtained without EG by controlling pH to 11. Compared with the case of Y-type ferrite, it was revealed that the optimum preparation conditions depended on phases even for the same Ba-Co-Fe-O system.

ACKNOWLEDGMENTS

The authors are grateful to Dr. Takeda of Osaka University for sample preparations and fruitful discussion.

REFERENCES

[1]T. Nakamura, K. Hatakeyama, "Complex permeability of polycrystalline hexagonal ferrites", *IEEE Trans. on Magnetics*, **36**, 3415-7 (2000).

[2]M. Kakihana, M. Yoshimura, H. Mazaki, H. Yasuoka, L. Börjesson, "Polymerized complex synthesis and intergranular coupling of bismuth lead strontium calcium copper oxide superconductors characterized by complex magnetic susceptibility", *J. Appl. Phys.*, **71**, 3904-10 (1992).

[3]P. Berastegui, M. Kakihana, H. Mazaki, H. Yasuoka, S. Eriksson, L. G. Johansson, L. Börjesson, M. Kall, "Polymerized complex synthesis of a pure 93 K yttrium barium copper oxide ($Y_2Ba_4Cu_7O_{15-\delta}$) superconductor without the need of high oxygen pressure and additive catalysts", *J. Appl. Phys.*, **73**, 2424-8 (1993).

[4]H. F. Yu, K. C. Huang, "Effects of pH and citric acid contents on characteristics of ester-derived $BaFe_{12}O_{19}$ powder", *J. Magn. Magn. Mater.*, **260**, 455-61 (2003).

[5]M. Kakihana, S. Kato, V. Petrykin, J. Bäckström, L. Börjesson, M. Osada "A simple and reproducible way to synthesize $PrBa_2Cu_4O_8$ under 1 atm of oxygen by amorphous citrate method", *Physica C*, **321**, 74-80 (1999).

[6]Y. Sato, V. Petrykin, M. Kakihana, M. Osada, A. Yamaguchi, S. Tochihara, Y. Tanaka, H. Yasuoka, H. Mazaki, "Synthesis of $(Ca_xLa_{1.00-x})(Ba_{1.75-x}La_{0.25+x})Cu_3O_z$ tetragonal superconductor by amorphous metal complex method", *Solid State Ionics*, **151**, 299-304 (2002).

Electroceramic Materials and Applications
Edited by R. W. Schwartz

EFFECT OF PYROLYSIS TEMPERATURES ON ZNO FILMS ON SODA-LIME-SILICA GLASS SUBSTRATES PREPARED BY CHEMICAL SOLUTION DEPOSITION

Kyu-Seog Hwang, Jun-Hyung An
Department of Applied Optics and Institute of Photoelectronic Technology, Nambu Univeristy
864-1 Wolgye-dong Gwangsan-gu
Gwangju, 506-824, Republic of Korea

Hyun-Wook Ryu, Young-Sun Jeon, Kyung-Ok Jeon
Department of Materials Science and Engineering, Chonnam National University
300 Yongbong-dong Buk-ku
Gwangju, 500-757, Republic of Korea

ABSTRACT

The influence of pyrolysis temperature on the properties of chemical solution derived zinc oxide films has been investigated. As-deposited films were pyrolysis at 300°C for 30 min and at 500°C for 10 min. Final heat treatment was done at 600°C for 30 min in air. The as-grown films were highly c-axis oriented. Distinct grain structure was present in the film pyrolyzed at 500°C, while the surface of the film pyrolyzed at 300°C was smooth and no observed texture. The surface of ZnO pyrolyzed at 300°C was covered with needle-like grain growth. With increasing pyrolysis temperature at 500°C, a three-dimensional grain out growth was appeared.

INTRODUCTION

In the field of optoelectronics, wide band gap semiconductor such as GaN has attracted enormous attention because of their potential for use in light-emitting diodes and short wavelength diode lasers. Recently, ZnO has also emerged as a promising material. The letter has a large direct band gap of 3.37 eV, a low power threshold for optical pumping at room temperature, and emits ultraviolet (UV) radiation as the result of a large exciton binding energy of 60 meV at room temperature. [1,2]

A variety of methods have been reported for the preparation of ZnO thin film. For example, films have been deposited by rf-sputtering, chemical vapor deposition, laser molecular - beam epitaxy and chemical solution deposition (CSD). [3-6] Among them, CSD has received a fair bit of attention because of its simplicity and consequent economics, as it does not require a high vacuum apparatus. However, in CSD process, pores and cracks due to vaporization of organics in precursor were easily recognizable in the film. Since the defects of final annealed film were largely affected by the elimination modes of the organics in the precursor for the chemical-solution-based process, we studied the relation between the pyrolysis condition and the film's character for films prepared by using CSD. [7,8] In this work, to investigate the effect of pyrolysis temperature on the properties of ZnO film, we prepared ZnO films from zinc naphthenate.

EXPERIMENTAL

The starting solution was prepared by mixing of commercially available zinc naphthenate (Nihon Kagaku Sangyo Co., Ltd., Japan) and toluene. The concentration and viscosity of the starting solution were adjusted by toluene for depositing smooth films by spinning (4wt% Zn / 100mL starting solution). The solution was spin-coated onto the cleaned soda-lime-silica slide glass (SLSG, 2.5cm × 2.5cm × 0.2cm^t) substrates for 10 sec at 1500 rpm in air. In the atmosphere of air, decomposed and evolved organic components should be oxidized, so that, random nucleation and heterogeneous crystal growth, void or crack formation due to the excessive and local heating of the film are supposed to take place in CSD. [10] Thus, following two prefiring conditions; (1) 300°C for 30 min, and (2) 500°C for 10 min were adopted to investigate the effect of burning of organic component. Thermogravimetric analysis (TGA, TGA2950, TA Instruments, U. S. A.) of the starting solution was also performed for each of the prefiring conditions. The spin coating and pyrolysis were repeated five times to adjust the thickness of the precursor films. The films were subsequently heat treated at a final annealing temperature of 600°C for 30 min in air by directly inserting the samples into a preheated tube-type furnace.

The thickness of annealed ZnO films was 0.60 ~ 0.65μm, confirmed by observation of the cross section of the fractured films with a field emission – scanning electron microscope (FE-SEM, S-4700, Hitachi, Japan). The crystallinity of the films was investigated by using a high-resolution X-ray diffraction (HRXRD, X'pert-PRO, Philips, Netherlands). The surface morphology of the films was evaluated from the FE-SEM. The surface roughnesses of the films were studied by using a scanning probe microscope (SPM, XE200, PSIA, Korea). All the SPM measurements were performed in air using the tapping mode. The transmittance in the visible wavelength range was observed by using a UV-visible-NIR spectrophotometer (Cary 500 Scan, Varian, Australia). The transmittance was automatically calibrated against that of a bare glass substrate as a reference sample, and the absorption coefficient was obtained from the transmittance curve.

RESULTS AND DISCUSSION

Fig. 1 shows TGA curve of the starting solution (heating rate: 3°C / min). Weight loss corresponding to vaporization of the solvent and pyrolysis of the zinc naphthenate began at below 100°C and completed at above 460°C. Therefore, pyrolysis of the starting zinc naphthenate solution is concluded to complete at about 500°C. As seen in Fig. 1, the temperature of 300°C corresponds to the intermediate stage of the pyrolysis. The weight decrease at 300°C was found and weight loss about 55.3wt% was obtained. This value is smaller than that 96.9wt% by gradual heating to 500°C. So larger amount of hydrocarbon is supposed to still remain in the precursor film after pyrolysis at 300°C.

The precursor films, pyrolyzed through the above two annealing paths, were found amorphous by XRD and smooth surfaces without cracks or voids by FE-SEM observation and significant difference was not recognized between these films. There were differences, however, about the content of residual carbon or carbon hydroxides in the precursor films, as shown in Fig. 1.

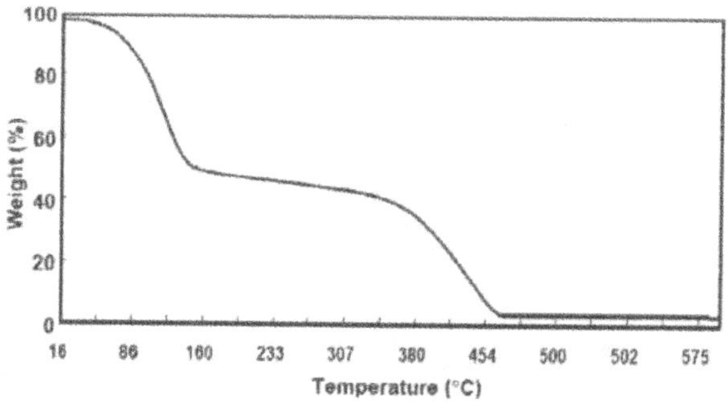

Figure 1

Fig. 2 shows XRD spectra of the films pyrolyzed at 300°C and 500°C, followed by final annealing at 600°C. The *c*-axis oriented ZnO thin films were obtained for all the prefiring temperature. With increasing the prefiring temperature from 300°C to 500°C, strong (002) peak was seen at $2\theta = 34° \sim 35°$. It should be noted that the peak intensity of ZnO films was significantly affected by pyrolysis temperature although the final annealing temperature was the same. The lower peak intensity of resultant film pyrolyzed at 300°C may be attributed to the presence of residual organic components. The above result of TGA indicated that the ZnO film pyrolysis at 300°C is assumed to contain some residual carbon or carbon hydroxides. In this case, crystallization of film and decomposition of organic components concurrently proceeded during final heat treatment. [9] Crystal growth may be suppressed by residual organic components during the final heat treatment, resulting in lower peak intensity of ZnO film pyrolyzed at 300°C.

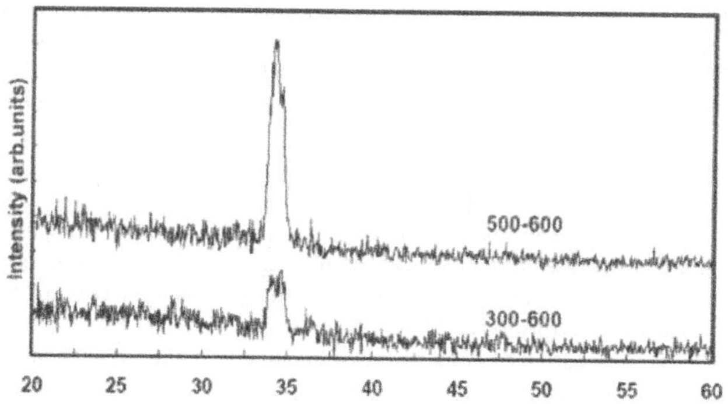

Figure 2

FE-SEM photographs of the free surfaces of the films are shown in Fig. 3. Films after pyrolyzing at 300°C and 500°C exhibited flat and smooth surface morphologies without cracks or distinct grain boundaries [see Fig. 3 (a) and (b)]. After annealing at 600°C, distinct grain structure was present in the film pyrolyzed at 500°C, while the surface of the film pyrolyzed at 300°C was smooth and no observed texture [compare Fig. 3(c) and (d)].

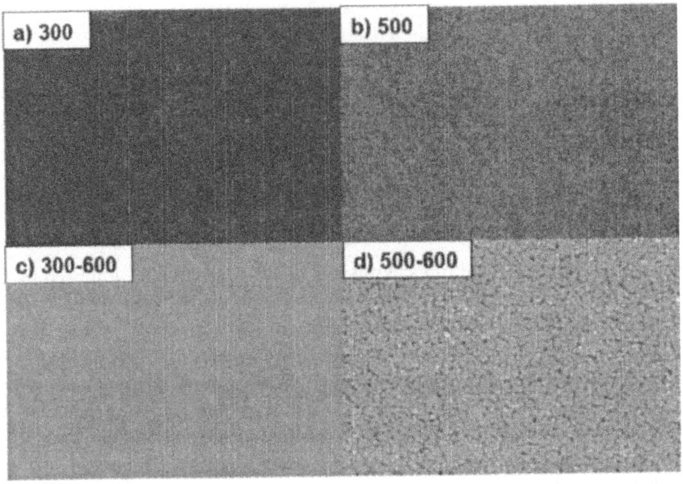

Figure 3

In order to elucidate surface roughness and morphology of the films according to pyrolysis temperature, SPM analysis was performed. SPM images ($10 \times 10\mu m^2$) for the ZnO films pyrolyzed at 300°C and at 500°C are shown in Fig. 4 (a) and (b), respectively. Pyrolyzed films showed smooth surfaces without 3-dimentional grain growth. From the surface morphological results by FE-SEM and SPM, we can conclude that there is no distinct difference between the surface structures of the films pyrolyzed at different ways. However, after annealing at 600°C, the surface of ZnO pyrolyzed at 300°C was wholly covered with needle-like grain growth. With increasing pyrolysis temperature at 500°C, a three-dimensional grain out growth was appeared on the surface of the films annealed at 600°C.

Furthermore, we studied how the surface roughness characteristics of ZnO thin films were varied with pyrolysis temperature by using root mean square (RMS) roughness obtained from SPM measurements. Fig. 5 shows the SPM top-view images and surface roughness profiles of the annealed ZnO films pyrolyzed at 300°C (a) and at 500°C (b). The surface roughness of the annealed ZnO pyrolyzed at 500°C was increased due to larger grain growth and the three-dimensional grain outgrowth. Thus, we assume that heterogeneous crystal growth may be suppressed by pyrolysis at 300°C, resulting in lower RMS roughness than that of film pyrolyzed at 500°C.

Figure 4

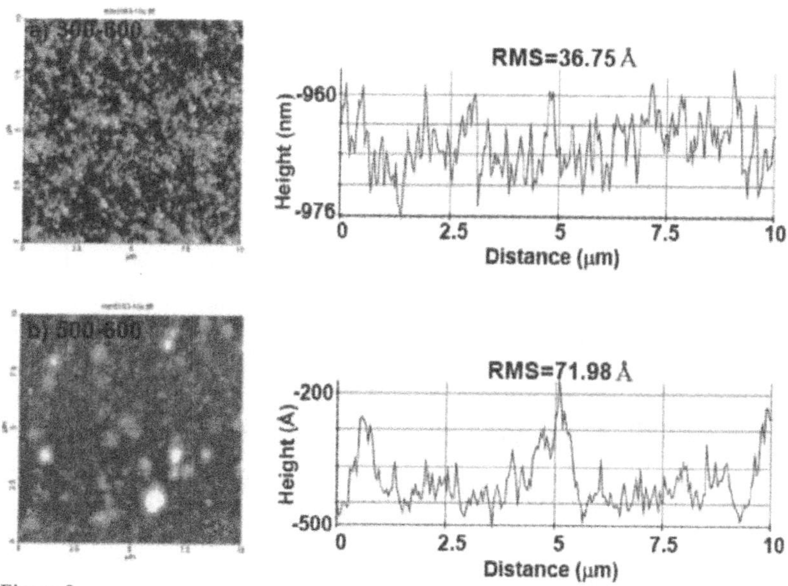

Figure 5

Fig. 6 shows transmission spectra of the annealed ZnO thin films pyrolyzed at different temperatures. All the films exhibit a high transmittance (>90%) in the visible region, and show a sharp fundamental absorption edge at about 0.3 ~ 0.4μm, which is very close to the intrinsic band-gap of ZnO. Further, the film pyrolyzed at 300°C showed a shift towards the UV range, as shown in Fig. 7.

Generally, the absorption of light occurs when the electrons in a valance band are activated up to a conduction band. The band gap (E_g) between them can be estimated by the following equation:

$$\alpha = (h\nu - E_g)^{1/2}$$

where α is the absorption coefficient, h is Plank's constant, and ν is the frequency of the incident photon. Fig. 8 shows a plot of α^2 *vs* $h\nu$. A linear relation between α^2 and $h\nu$ is obtained, and the extrapolation of the line ($\alpha^2 = 0$) gives E_g values as shown in Fig. 8. [6] The optical band gaps, E_g, of the present films pyrolyzed at 300°C and 500°C are 3.29 eV and 3.28 eV, respectively. The estimated value of the band gap for the film pyrolyzed at 300°C is larger than that of the film pyrolyzed at 500°C. Generally, a film consisting of fine crystallites shows a 'blue shift'. [10] As shown in Figs. 7 and 8, a comparison of optical energy gaps and transmittance spectra between the films treated at different pyrolysis temperatures shows an obvious red shift in the pyrolyzed ZnO at high temperature. The red shift in optical energy gaps is considered to be due to a larger crystal size of the film pyrolyzed at 500°C.

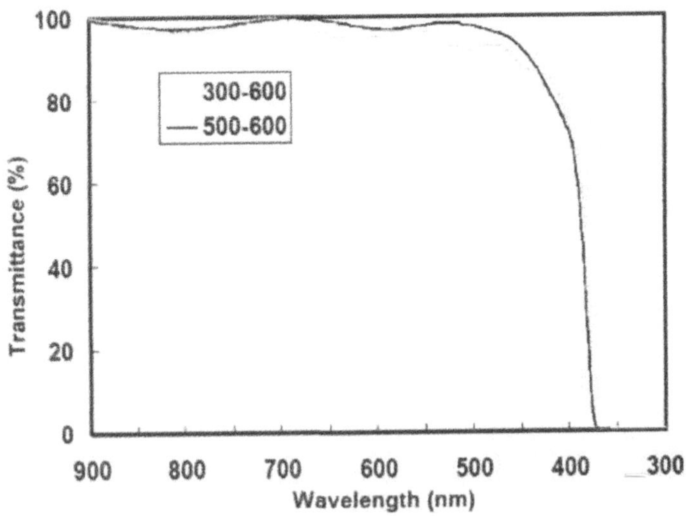

Figure 6

CONCLUSIONS

Highly *c*-axis oriented ZnO films were prepared by CSD process with a zinc naphthenate precursor. With increasing the prefiring temperature from 300°C to 500°C, strong XRD peak was seen. The lower peak intensity of resultant film pyrolyzed at 300°C may be attributed to the presence of residual organic components. The surface roughness of the ZnO pyrolyzed at 500°C was increased due to larger grain growth and the three-dimensional grain outgrowth. All the films exhibit a high transmittance (>90%) in the visible region, and show a sharp fundamental absorption edge at about 0.3 ~ 0.4μm.

REFERENCES

[1]A. B. M. Almamun Ashrafi, B. Zhang, N. T. Binh, K. Wakatsuki, and Y. Segawa, "High-Quality ZnO Layers Grown on 6H-SiC Substrates by Metalorganic Chemical Vapor Deposition," Jpn. J. Appl. Phys., **43**, 1114-17 (2004)

[2]B. H. Kim, J. H. An, Y. S. Jeon, J. T. Jeong, B. A. Kang, and K. S. Hwang, "Highly c-axis Oriented ZnO Film on Silica Glass Using Zinc Naphthenate Precursor," J. Mater. Sci., **40**, 237-39 (2005).

[3]M. Chen, Z. L. Pei, X. Wang, C. Sun, and L. S. Wen, "Dependence of Structural, Electrical, and Optical Properties of ZnO:Al Films on Substrate Temperature," J. Mater. Res., **16**, 2118-23 (2001).

[4]J. Nishino, and Y. Nosaka, "Preparation of ZnO by a Nearby Vaporizing Chemical Vapor Deposition Method," J. Mater. Res., **18**, 2029-32 (2003)

[5]Y. Chen, S. Hong, H. Ko, M. Nakajima, T. Yao, and Y. Segawa, "Plasma-Assisted Molecular-Beam Epitaxy of ZnO Epilayers on Atomically Flat $MgAl_2O_4$(111) Substrates," Appl. Phys. Lett., **76**, 245-47 (2000).

[6]D. Shimono, S. Tanaka, T. Torikai, T. Watari, and M. Murano, "Preparation of Transparent and Conductive ZnO Films Using a Chemical Solution Deposition Process," J. Ceram. Proc. Res. **2**, 184-88 (2001).

[7]K. Hwang, T. Manabe, I. Yamaguchi, S. Mizuta, and T. Kumagai, "Preparation of Epitaxial $Pb(Zr,Ti)O_3$ Thin Films on MgO(100) Substrates by Dipping-Pyrolysis Process," J. Ceram. Soc. Jpn., **105**, 952-56 (1997).

[8]B. Kim, J. An, K. Hwang, B. Kang, K. Nishino, and T. Tsuchiya, "AFM Analysis of Chemical-Solution-Derived Epitaxial PZT Films Prepared by Using Oxidizing or Non-Oxidizing Pyrolysis," J. Kor. Phys. Soc., **44**, 346-49 (2004).

[9]K. Hwang, Doctor Thesis, Tokyo University of Science (2003).

[10]S. A. Syudenikin, N. Golego, and M. Cocivera, "Optical and Electrical Properties of Undoped ZnO Films Grown by Spray Pyrolysis of Zinc Nitrate Solution," J. Appl. Phys., **83**, 2104-111 (1998).

Electroceramic Materials and Applications
Edited by R. W. Schwartz

EFFECT OF Ba SUBSTITUTION ON STRUCTURE AND ELECTRICAL PROPERTIES OF $SrRuO_3$

Hidetoshi Miyazaki, Yoko Suyama

Department of Material Science, Faculty of Science and Engineering, Shimane University, 1060, Nishikawatsu-cho, Matsue, Shimane, 690-8504, Japan.

miya@riko.shimane-u.ac.jp

Hisao Suzuki

Department of Materials Science, Faculty of Engineering, Shizuoka University, 3-5-1, Johoku, Hamamatsu, Shizuoka 432-8561, Japan

Keywords: Perovskite, Chemical solution deposition, SrRuO3, thin film

ABSTRACT

$Sr_{1-x}Ba_xRuO_3$ (x=0 to 0.5, SBRO) ceramics were synthesized by traditional solid state reaction. Increasing Ba contents in SBRO increased continuously the lattice constant of SBRO ceramics at x=0.1 to 0.5. SBRO ceramics showed hexagonal single phase Ba contents of 0.2 to 0.5 in SBRO. SBRO films were prepared by chemical solution deposition using Ba metal, Sr metal and $RuCl_3{\cdot}2H_2O$ as starting materials. The resultant SBRO film were good crystallinity and dense, and the resistivity of the SBRO film were 2.235 to $4.696\times10^{-3}\Omega$cm.

1. INTRODUCTION

Ferroelectric thin films of PZT have been widely investigated for their applications of a ferroelectric memory or piezoelectric devices. A Pt film is usually used as electrodes in the fabrication of integrated ferroelectric capacitors for memory applications. However, there have been several problems such as high leakage current and especially short polarization fatigue life span with the PZT/Pt capacitor.[1-3]

To improve the fatigue property of a ferroelectric devise using PZT and Pt electrode, conductive oxides such as $LaNiO_3$,[4,5] $YBa_2Cu_3O_{7-x}$,[6,7] $SrRuO_3$ (SRO)[8-10] have been investigated instead of the Pt electrode. These oxides are a perovskite structure and a low electrical resistivity which is about 1×10^{-3} to $10^{-4}\Omega$cm. SRO is one of the most promising candidate because of their similar lattice constants (the lattice constants for a $Pb_{0.52}Zr_{0.48}TiO_3$: a=4.036Å, b=4.146Å; and the lattice constants for a SRO with a pseudo cubic symmetry: a=3.930Å, for the distorted perovskite structure: a=5.573Å, b=7.856Åand c=5.538Å (JCPDS: 430472) and Resistivity is about $2.8\times10^{-4}\Omega$cm. The difference of the lattice constant of PZT and SRO was about 0.1Å. Therefore, SRO is mentioned as a candidate of electrode for the PZT films.

The lattice misfit between PZT and SRO was about 2.5%. To substitute a larger ion than Sr to Sr site in SRO, it is assumed that a lattice constant of SRO become large and close to the PZT lattice constant. In this study, Ba solid solution SRO ceramics were synthesized by solid state reaction, and the lattice constant of $Sr_{1-x}B_xRuO_3$ (SBRO) ceramics was investigated. In previous, we reported that the SRO film was successfully synthesized by chemical solution deposition method (CSD) using Sr metal and $RuCl_3{\cdot}2H_2O$ as starting materials.[10] The resultant SRO film was good crystallinity, dense, no cracks with the electrical resistivity of about $10^{-3}\Omega$cm. In this paper, SBRO films with various Ba contents were also fabricated by CSD method and an electrical resistivity of SBRO film were investigated.

2. EXPERIMENTAL

2.1 Bulk preparation

Commercial $SrCO_3$, $BaCO_3$ and RuO_2 powders were used as starting materials. These were weighed the composition from x = 0 to 5 in SBRO, and these powders were mixed. The mixture was calcined at 1000°C for 2hr and sintered at 1200°C for 20hr in air.

2.2 Film preparation

Ba metal, Sr metal and $RuCl_3{\cdot}2H_2O$ were used as starting materials and 2-methoxyethanol was used as a solvent. A stable SRO or BRO precursor solution of 0.1M were prepared using Ba metal, Sr metal, $RuCl_3{\cdot}2H_2O$ as Ba, Sr and Ru sources, respectively. The SBRO thin films were spin-coated using SBRO mixture solution on Si (100) substrate at 3000rpm and 30sec. The as deposited thin film was dried at 150°C for 10min and pre-annealed at 420°C for 10min to remove residual organics. Finally, the film was annealed at 700°C by RTA method. These processes were repeated 10 times, and then, the precursor film was annealed at 700°C for 2hr in an electric furnace.

2.3 Characterization

The crystalline phases of the SBRO ceramics and films were characterized by XRD with Rigaku RINT-2200. Microstructure of the SBRO films was observed by SEM using Jeol JSM-5600. Resistivity of the films was measured by a four-point probe measurement using HP 34401A.

3. RESULTS AND DISCUSSION

3.1 SBRO Bulk

Fig.1 shows the XRD patterns for the SBRO ceramics. SRO showed the orthorhombic structure with the JCPDS card 25-0912. The x=0.1 in SBRO was cubic phase with the JCPDS card 80-1529, and included a few hexagonal phase. The mixing ratio x in SBRO was more than 0.2, all the SBRO ceramics showed hexagonal single phase (6H) with the JCPDS card 19-0137. Longo et al. observed the structural change of SBRO with varying Ba contents in SBRO ceramics.[11] They reported that increasing Ba contents in SBRO caused structural change from orthorhombic phase to hexagonal phase, and these results agreed with our investigation.

Fig.2 indicates the lattice constant of the perovskite block for the resultant SBRO ceramics. The lattice constant of the orthorhombic SRO was calculated approximated as a pseudo cubic structure, and that of the hexagonal SBRO was estimated with the perovskite block part.[12] The lattice constant of 0.1SBRO was smallest in the SBRO ceramic system. We assumed for this result that the 0.1SBRO structure was cubic because of Ba substitution for the SRO host structure.

However, more investigations are needed to clarify this phenomenon. In the system of the x =0.1 to 0.5 in SBRO, increasing the x in SBRO increased the lattice constant of the SBRO ceramics. This suggested that the lattice constant of SBRO could be changed continuously by the quantity of the Ba contents in SBRO ceramics.

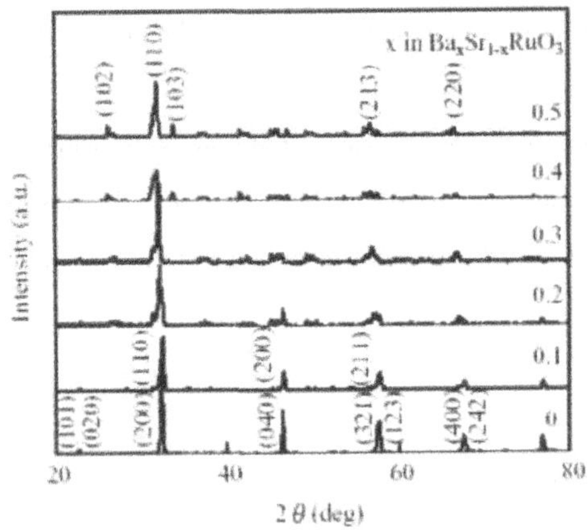

Fig. 1. XRD patterns for the SBRO ceramics Ba

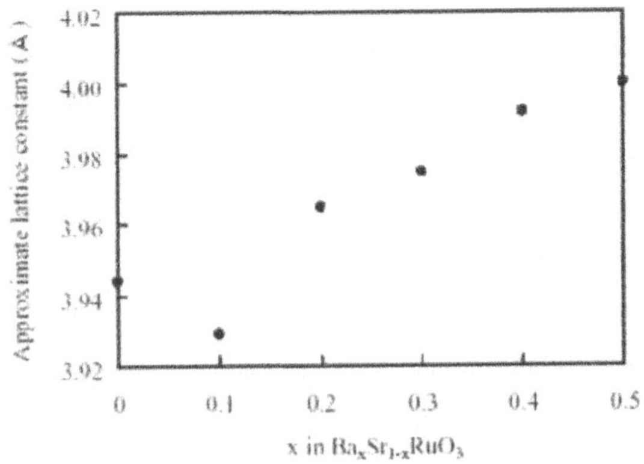

Fig. 2. Effect of Ba contents on the lattice constant of SBRO ceramics.

3.2 SBRO Film

Fig.3 depicts the XRD pattern for the SRO film. The SRO film indicated (200) preferred orientation, and assigned an orthorhombic structure of SRO. The peak in all the resultant SBRO films were observed near $2\theta=32$ for the perovskite SRO (JCPDS 25-0912) or SBRO (cubic: JCPDS 80-1529 or hexagonal: JCPDS 19-0137), therefore, we assumed that the obtained SBRO films could be characterized same as the bulk SBRO ceramics.

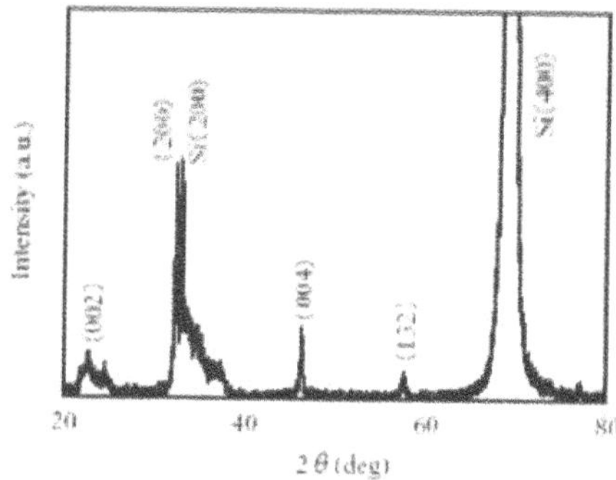

Fig. 3. XRD pattern for the SRO film on Si (100) substrate.

Fig.4 shows the surface and cross sectional micrographs of the resultant SRO film. The resultant SRO film was very dense with no crack from the cross sectional microstructure. In addition, the resultant SRO film had no pore or no crack on the film surface from the observation of surface morphology. All the SBRO films indicated the similar microstructure and surface morphology.

Fig.5 illustrates the electrical resistivity for the SBRO films. The resistivity of the SRO film was $2.866\times10^{-3}\Omega$cm. Generally, a resistivity of SRO was about $2\times10^{-4}\Omega$cm, and the resistivity of SRO in the present work was 1 order as large as the reported value. To fabricate the SRO film in this study, we used $RuCl_3$ as a starting material, and chloride ion maybe remain in the SRO film. It was assumed from this reason that the resistivity of the resultant SRO film was inferior to the resistivity of the common SRO ceramics or films.

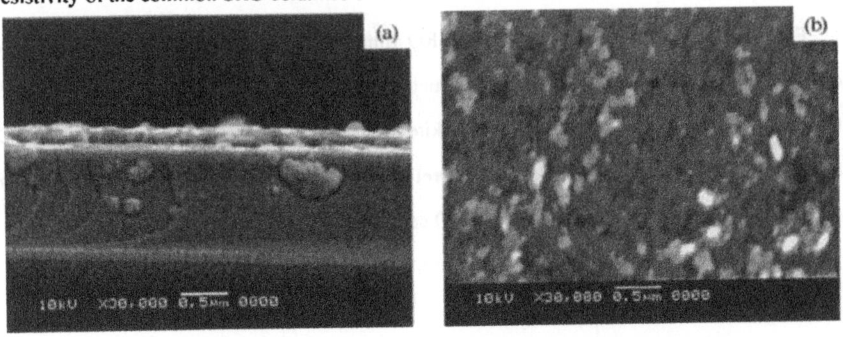

Fig. 4. SEM images of (a) cross section and (b) surface of the SRO film.

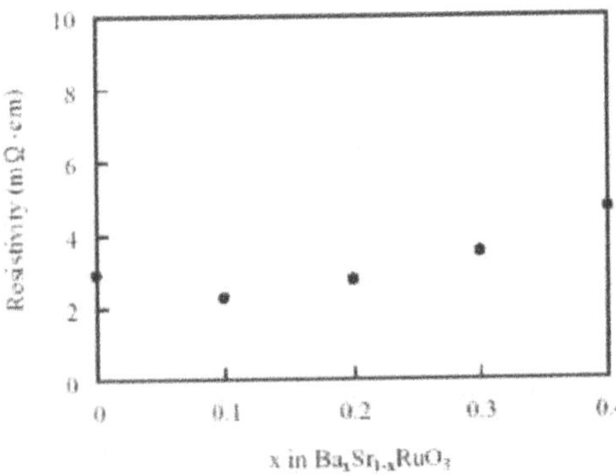

Fig. 5. Resistivity of the SBRO films Ba contents from 0 to 0.4 in $Sr_{1-x}Ba_xRuO_3$.

The ratio Ba / (Sr + Ba) was 0.1, the resistivity of the resultant SBRO film was $2.235\times10^{-3}\Omega cm$, and this value was lowest in the all SBRO films. Referring to the result of figure 2, we assumed that this result was attributed to the good symmetry (cubic) of the 0.1SBRO structure. Increasing Ba mixing ratio in SBRO increased the resistivity of the SBRO films. The resistivity of the 0.4 SBRO film was $4.696\times10^{-3}\Omega cm$.

N. Fukushima et al. prepared the SBRO films by rf magnetron sputtering.[13] They reported that increasing Ba ratio in the SBRO system increased resistivity of the resistivity of the SBRO films, these result agreed with our results.

It is concluded that the conductive SBRO film is successfully deposited with a CSD, which can be used as an electrode for the ferroelectric materials.

4. CONCLUSION

SBRO ceramics were fabricated by solid state reaction. Increasing Ba mixing ratio in SBRO ceramics increased the lattice constants of SBRO, and the SBRO structure changed from orthorhombic to hexagonal. SBRO films were prepared by CSD method using Ba metal, Sr metal and $RuCl_3{\cdot}nH_2O$ as starting materials. The resultant SBRO film was good crystallinity with no crack and no pore. Changing Ba mixing ratio in SBRO films from 0 to 0.4, the resistivity of the films changed from $2.235\times10^{-3}\Omega$cm to $4.696\times10^{-3}\Omega$cm.

REFERENCES

[1]S.Y. Chen and C.L. Sun, "Ferroelectric characteristics of oriented $Pb(Zr_{1-x}Ti_x)O_3$ films" *J. Appl. Phys.*, **90** 2970–4 (2001).

[2]T. Mihara, H. Watanabe and C.A.P. Araujo, "Characteristic Change Due to Polarization Fatigue of Sol-Gel Ferroelectric $Pb(Zr_{0.4}Ti_{0.6})O_3$ Thin-Film Capacitors" *Jpn. J. Appl. Phys.*, **33** 5281–6 (1994).

[3]H.M. Duiker, P.D. Beale, J.F. Scott, C.A.P. Araujo, B.M. Melnick, J.D. Cuchiaro and L.D. McMillian, "Fatigue and switching in ferroelectric memories: Theory and experiment" *J. Appl. Phys.*, **68** 5783–91 (1990).

[4]B.G. Chae, Y.S. Yang, S.H. Lee, M.S. Jang, S.J. Lee, S.H. Kim, W.S. Beak and S.C. Kwon, "Comparative analysis for the crystalline and ferroelectric properties of $Pb(Zr,Ti)O_3$ thin films deposited on metallic $LaNiO_3$ and Pt electrodes" *Thin Solid Films*, **410** 107–13 (2002).

[5]H. Miyazaki, T. Goto, Y. Miwa, T. Ohno, H. Suzuki, T. Ota and M. Takahashi, "Preparation and Evaluation of $LaNiO_3$ thin film Electrode with chemical solution deposition" *J. Eur. Ceram. Soc.*,

24 1005-8 (2004).

[6]R. Ramesh, W.K. Chan, B. Wilkens, H. Gilchrist, T. Sands, J.M. Tarascon, V.G. Keramidas, D.K. Fork, J. Lee and A. Safari, "Fatigue and retention in ferroelectric Y-Ba-Cu-O / Pb-Zr-Ti-O / Y-Ba-Cu-O heterostructures" *Appl. Phys. Lett.*, **61** 1537–9 (1992).

[7]J. Lee, L. Johnson, A. Safari, R. Ramesh, T. Sands, H. Gilchrist and V.G. Keramidas, "Effects of crystalline quality and electrode material on fatigue in $Pb(Zr,Ti)O_3$ thin film capacitors" *Appl. Phys. Lett.*, **63** 27–9 (1993).

[8]Y.K. Wang, T.Y. Tseng and P. Lin, "Enhanced ferroelectric properties of $Pb(Zr_{0.53}Ti_{0.47})O_3$ thin films on $SrRuO_3/Ru/SiO_2/Si$ substrates" *Appl. Phys. Lett.*, **80** 3790–2 (2002).

[9]T. Morimoto, O. Hidaka, K. Yamakawa, O. Arisumi, H. Kanaya, T. Iwamoto, Y. Kumura, I. Kunishima and S. Tanaka, "Ferroelectric Properties of $Pb(Zi, Ti)O_3$ Capacitor with Thin $SrRuO_3$ Films within Both Electrodes" *Jpn. J. Appl. Phys.*, **39** 2110–2113 (2000).

[10]H. Suzuki, Y. Miwa, H. Miyazaki, T. Ota and M. Takahashi, "Chemical solution deposition of conductive $SrRuO_3$ thin film on Si substrate" *Ceram. Int.*, **30** 1357-1360 (2004).

[11]J.M. Longo and J.A. Kafalas, "Pressure-induced structural changes in the system $Ba_{1-x}Sr_xRuO_3$" *Mater. Res. Bull.*, **3** 687-92 (1968).

[12]D-K. Choi, B.-S. Kim, S.-Y. Son, S.-H. Oh and K.-W. Park, "Evaluation of tailored electrode $(Ba,Sr)RuO_3$ for $(Ba,Sr)TiO_3$" *J. Appl. Phys.*, **86** 3347-51 (1999).

[13]N. Fukushima, K. Sano, T. Schimizu, K. Abe and S. Komatsu, "Lattice deformation and magnetic properties in epitaxial thin films of $Sr_{1-x}Ba_xRuO_3$" *Appl. Phys. Lett.*, **73** 1200-2 (1998).

Electroceramic Materials and Applications
Edited by R. W. Schwartz

ALKOXY-DERIVED PHOTOCHROMIC ZrO_2 PRECURSOR

Kaori Nishizawa, Takeshi Miki, Kazuyuki Suzuki, Kazumi Kato
National Institute of Advanced Industrial Science and Technology (AIST)
2266-98 Anagahora, Shimoshidami, Moriyama-ku, Nagoya, 463-8560 Japan

ABSTRACT

Photochromic ZrO_2 precursor solution was prepared using zirconium alkoxide, azobenzene derivatives and ethyleneglycol monomethylether. The precursor solution was irradiated with ultraviolet light (UV) at room temperature. After that, UV-irradiated precursor solution was irradiated with visible light (Vis) at room temperature. UV-Vis spectra were measured before irradiation, after UV irradiation and Vis irradiation. Changes of UV-Vis spectra indicated that the ZrO_2 precursor solution prepared using 4-(phenylazo)benzoic acid showed photochromism. The phenomena have synchronized with reversible photoisomerization of 4-(phenylazo)benzoic acid in the precursor solution. FT-IR spectra indicated that the zirconium alkoxides and 4-(phenylazo)benzoic acid interacted chemically in the solution and the chemical bonding state changed reversibly accompanied with photoisomerization of 4-(phenylazo)benzoic acid in the precursor solution.

INTRODUCTION

In the fields of polymer science, a lot of photoresponsive polymer gels have been prepared by grafting photochromic molecules to the main chain or branches of polymer gels. Photochromism means the phenomenon that the structure of compounds changes reversibly by UV irradiation or Vis irradiation. Thus, the color of many photochromic molecules changes reversibly by photoirradiation. In addition, the physical and chemical properties of prepared photoresponsive polymers change reversibly by photoirradiation. Therefore, many photoresponsive polymer gels have been used as optics, artificial materials and so on. Azobenzene is one of the most famous photochromic molecules. Azobenzene derivatives with azo units in the molecules, such as 4-(phenylazo)benzoic acid, are also known to show the reversible cis-trans photoisomerization by photoirradiation and the physical and chemical properties change[1].

By the way, ZrO_2 thin films have high potential for many applications, such as a buffer layer for integration of ferroelectric thin films in nonvolatile ferroelectric memories, the gate dielectric for complementary metal oxide semiconductor and oxygen sensors[2-4]. It is important for the various applications of ZrO_2 thin films to control the crystallinity, orientation and surface morphology. The precise design of chemical structures of precursor is effective for control of crystallinity of the ZrO_2 thin films. In our previous study, we paid attention to the effects of β-diketone addition to ZrO_2 precursor solution and UV-assisted sol-gel process[5]. As the results, control the structure of ceramic precursor by addition of organic compounds is useful for the

UV-assisted sol-gel process in order to control of crystallinity and surface morphology of ceramics. In this paper, we report on the preparation of novel photochromic ZrO_2 precursor applied the technique for polymer materials described above. The final aim is the control of the crystallinity of ZrO_2 thin films by use of the novel photochromic ZrO_2 precursor.

EXPERIMENTAL

Some 0.1M ZrO_2 precursor solutions were prepared using a zirconium tetra-*n*-butoxide ($Zr(O\text{-}n\text{-}Bu)_4$), azobenzene derivatives shown in Table 1 and ethyleneglycol monomethylether as a solvent without hydrolysis process. Fourier-transform infrared spectroscopy measurements were performed to analyze the specific chemical bonds in the concentrated ZrO_2 precursors by KBr method (FT-IR; FT/IR 610, JASCO, Co. Ltd., Tokyo, Japan).

Next, the precursor solutions were diluted with the solvent. Then the precursor solution was irradiated with ultraviolet light (UV, main wavelength: 365 nm) through a glass filter using a 250 W ultrahigh pressure mercury lamp (UHPML, multilight UIV-270, Ushio Co. Ltd., Tokyo, Japan, irradiance: 70 mW/cm^2) at room temperature for 15 min. After that, the precursor solution was irradiated with visible light (Vis, main wavelength: 436 nm) through a glass filter using a same lamp at room temperature for 15 min. The relative irradiance intensity in each wavelength was 100% at 365 nm and 50% at 436 nm. The optical absorption spectra of the solutions before UV irradiation, after UV irradiation, and after Vis irradiation were measured using an ultraviolet-visible light spectrometer (UV-Vis; U-4100, HITACHI, Co. Ltd., Tokyo, Japan).

Table 1. Components of ZrO_2 precursor solutions

	Kinds of azobenzene derivatives	$Zr(O\text{-}n\text{-}Bu)_4$: azobenzene derivatives
A	Non	----
B	Azobenzene 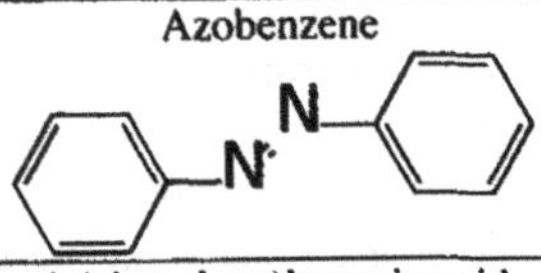	20:1
C	4-(phenylazo)benzoic acid 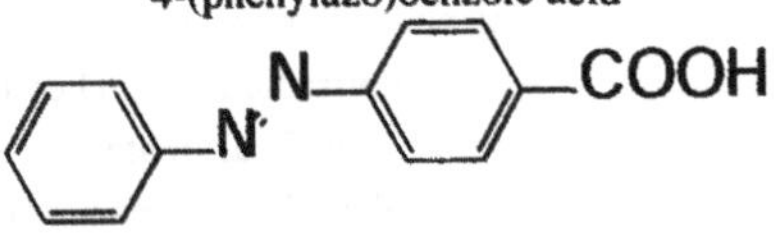	20:1

RESULTS AND DISCUSSION

Fig.1 showed the UV-Vis spectra of ZrO_2 precursor solution prepared using zirconium alkoxide without organic additives. The thin line shows the spectrum before UV irradiation. The thick line shows the spectrum after UV irradiation for 15 min. The dotted line shows the spectrum after Vis irradiation for 15 min. The absorption peak at 209 nm was assigned to Zr-O charge transfer (CT) transition in zirconium alkoxide. All spectra were consisted with one another completely. In this way, it was found that the absorption band assigned to Zr-O CT transition in zirconium alkoxide did not change by photoirradiation.

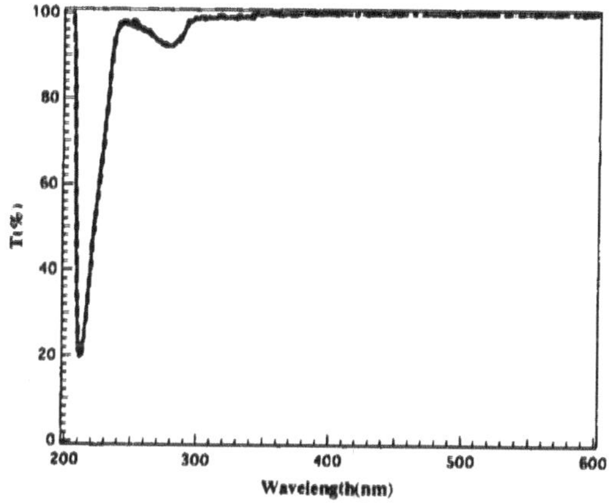

Fig.1. UV-Vis spectra of ZrO_2 precursor solution prepared using zirconium alkoxide without organic additives. The thin line (——) shows the spectrum before UV irradiation. The thick line (▬) shows the spectrum after UV irradiation for 15 min. The dotted line (.....) shows the spectrum after Vis irradiation for 15 min.

Fig.2 showed the UV-Vis spectra of azobenzene in ethyleneglycol monomethylether. The thin line shows the spectrum before UV irradiation. The large absorption band assigned to π-π^* transition of azo unit appeared at 320 nm. After UV irradiation for 15 min, the spectrum of thin line changed to that of the thick line. The strength of absorption band at 320 nm got lower. On the contrary, the strength of absorption band at 440 nm assigned to n-π^* transition of azo unit got higher. Then, the spectrum of thick line changed to that of the dotted line after Vis irradiation for 15 min. The spectrum of the dotted line got almost the same as that of the thin line. However, these were not consisted with each other completely.

Azobenzene in organic solvent changes from trans forms to cis forms by UV irradiation and changes from cis forms to trans forms by Vis irradiation as shown Fig.3. In organic solvent, a part of cis forms transform to other phenanthrene-like compounds. Thus, all cis forms do not return back to trans forms when cis forms were irradiated by Vis light. Anyway, the UV-Vis spectrum of azobenzene in organic solvent dramatically and reversibly changes in the range of 250-500 nm accompanied with the structure change of azobenzene by photoirradiation.

The UV-Vis spectra of ZrO_2 precursor solution including azobenzene were showed in Fig.4(a). The spectra also showed the reversible change by photoirradiation. In particular, the absorption band at around 320 and 440 nm originated from azobenzene changed reversibly.

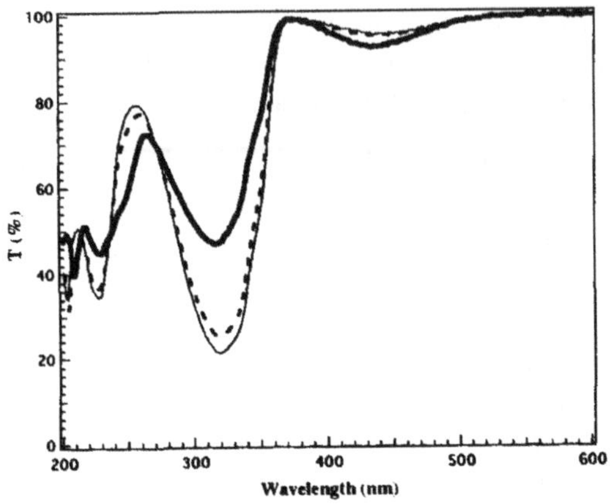

Fig.2. UV-Vis spectra of azobenzene in ethyleneglycol monomethylether. The thin line (—) shows the spectrum before UV irradiation. The thick line (—) shows the spectrum after UV irradiation for 15 min. The dotted line (····) shows the spectrum after Vis irradiation for 15 min.

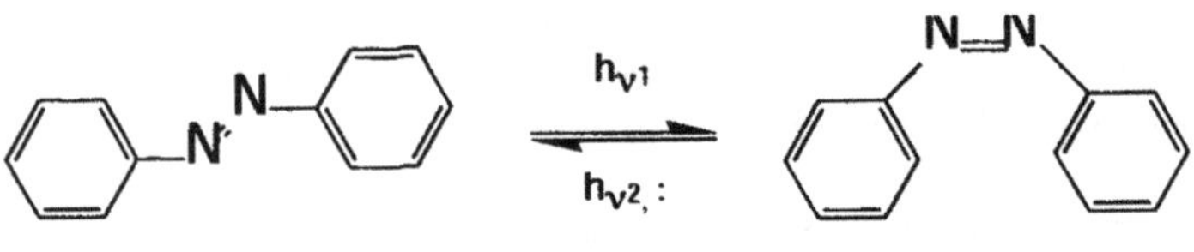

Fig. 3. Schematic illustration of the structure change of azobenzene by photoirradiation

Additionally, the spectrum of the dotted line was consisted with that of thin line completely. This is considered to be because zirconium alkoxide, which coexisted with azobenzene, inhibited side reactions of cis forms. However, the absorption band at 209 nm assigned to Zr-O CT transition in zirconium alkoxide did not change by photoirradiation as shown Fig.4(b). In this way, azobenzene did not interact with zirconium alkoxide.

Fig.5 showed the UV-Vis spectra of 4-(phenylazo)benzoic acid in ethyleneglycol monomethylether. The thin line shows the spectrum before UV irradiation. The large absorption band assigned to π-π* transition of azo unit appeared at 330 nm. After UV irradiation for 15 min, the spectrum of thin line changed to that of the thick line. The strength of absorption band at 330 nm got lower. On the contrary, the strength of absorption band at around 450 nm assigned to n-π* transition of azo unit got higher. Then, the spectrum of thick line changed to that of the dotted line after Vis irradiation for 15 min. The spectrum of the dotted line was consisted with that of the thin line.

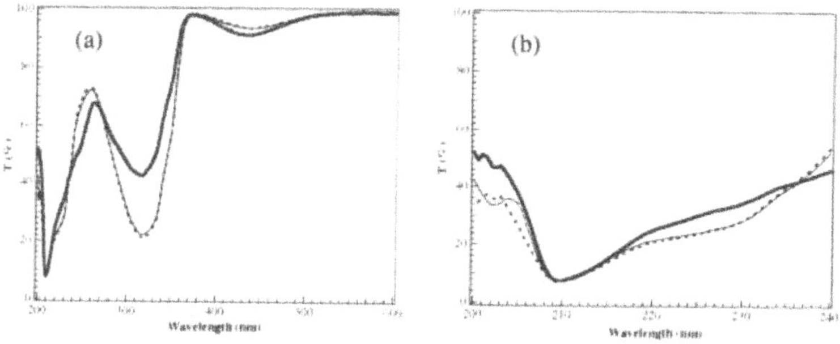

Fig.4. UV-Vis spectra of ZrO_2 precursor solution including azobenzene. (a) wide range spectra, (b) expanded spectra of Fig.4(a). The thin line (—) shows the spectrum before UV irradiation. The thick line (━) shows the spectrum after UV irradiation for 15 min. The dotted line (....) shows the spectrum after Vis irradiation for 15 min.

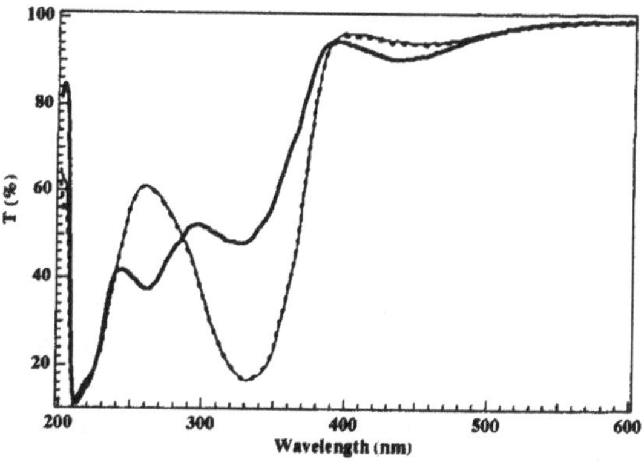

Fig.5. UV-Vis spectra of 4-(phenylazo)benzoic acid in ethyleneglycol monomethylether. The thin line (—) shows the spectrum before UV irradiation. The thick line (━) shows the spectrum after UV irradiation for 15 min. The dotted line (.....) shows the spectrum after Vis irradiation for 15 min.

Fig.6(a) shows the UV-Vis spectra of ZrO_2 precursor solution including 4-(phenylazo)benzoic acid. The absorption bands originated from 4-(phenylazo)benzoic acid also showed the reversible change by photoirradiation. Additionally, the spectrum of the dotted line was consisted with that of the thin line completely. Furthermore, the absorption band at around 330 nm in the spectrum of thick line got smaller than that in the spectrum of thick line in Fig. 4(a). And the absorption band at around 450 nm in the spectrum of thick line got larger than that in the spectrum of thick line in Fig. 4(a). This is considered to be because cis forms were stabilized well by addition of 4-(phenylazo)benzoic acid.

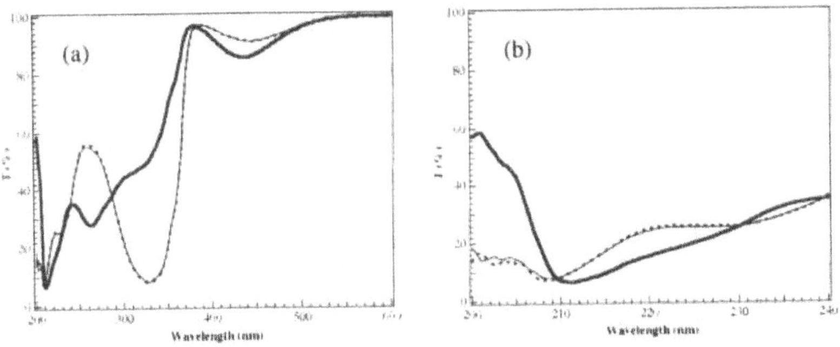

Fig.6. UV-Vis spectra of ZrO_2 precursor solution including 4-(phenylazo)benzoic acid. (a) wide range spectra, (b) expanded spectra of Fig.6(a). The thin line (—) shows the spectrum before UV irradiation. The thick line (▬) shows the spectra after UV irradiation for 15 min. The dotted line (·····) shows the spectra after Vis irradiation for 15 min.

The expanded spectra of Fig.6(a) was shown in Fig.6(b). The absorption band at 209 nm assigned to Zr-O CT transition in zirconium alkoxide changed accompanied with the spectral change of 4-(phenylazo)benzoic acid by photoirradiation as shown Fig.6(a). That is, the absorption band at around 209 nm slightly shifted to the longer wavelength region by UV irradiation and the shifted band returned back to the same position in the spectrum of thin line by visible light irradiation. In this way, the absorption band originated from Zr-O in ZrO_2 precursor solution including 4-(phenylazo)benzoic acid showed the photochromism. It suggested that 4-(phenylazo)benzoic acid interacted with zirconium alkoxide, and therefore, the absorption band originated from Zr-O CT transition showed photochromic-like behavior accompanied with the reversible spectral change of 4-(phenylazo)benzoic acid by photoirradiation.

To confirm the interaction between zirconium alkoxide and 4-(phenylazo)benzoic acid, the FT-IR spectra of ZrO_2 precursor without organic additives, 4-(phenylazo)benzoic acid and ZrO_2 precursor solution with 4-(phenylazo)benzoic acid were measured and shown in Fig.7. These spectra have differed one another. Thus, zirconium alkoxide is expected to interact with 4-(phenylazo)benzoic acid chemically. The main three peaks at 1680, 1420 and 1290 cm^{-1} in Fig.7(b), that originate from carboxyl group of 4-(phenylazo)benzoic acid, were disappeared in Fig.7(c). Furthermore, the new peaks appeared at 1550 and 1420 cm^{-1} in Fig.7(c), which originate from carboxylate group. It was confirmed that 4-(phenylazo)benzoic acid reacted with zirconium alkoxide and a new calboxylate compound formed in the solution. Based on these results, the new calboxylate compound was suggested to show photochromism in organic solvent.

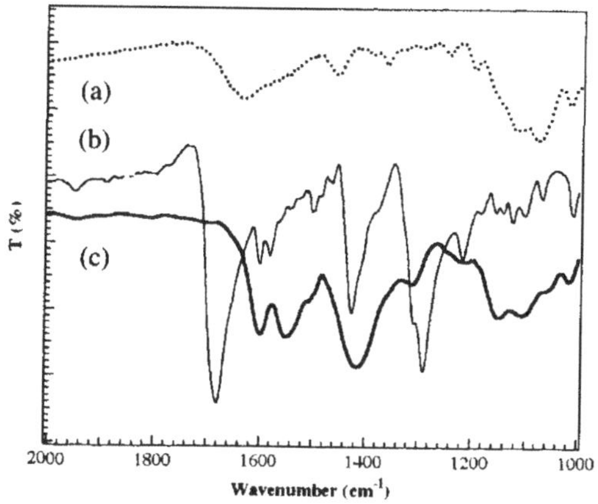

Fig.7. FT-IR spectra of (a) ZrO_2 precursor without organic additives, (b) 4-(phenylazo)benzoic acid, (c) ZrO_2 precursor including 4-(phenylazo)benzoic acid.

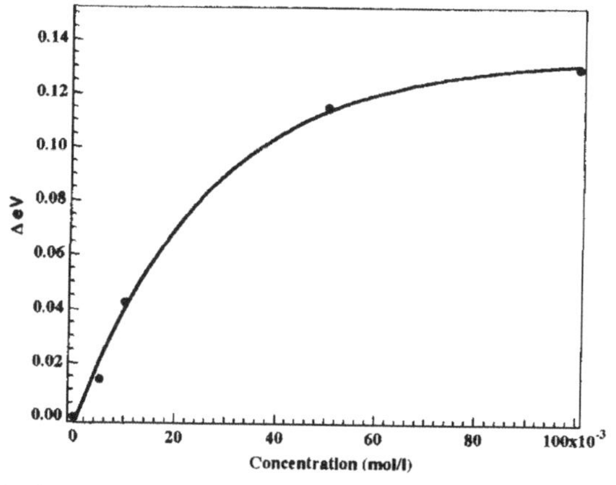

Fig.8. Relationship between concentrations of 4-(phenylazo)benzoic acid added in the ZrO_2 precursor solution and degrees of peak shift, which was originated from CT transition of Zr-O, between before UV irradiation and after UV irradiation in the UV-Vis spectra.

Lastly, the relationship between concentrations of 4-(phenylazo)benzoic acid added in the ZrO_2 precursor solution and differences of peak position, which was originated from CT transition of Zr-O, between before UV irradiation and after UV irradiation in the UV-Vis spectra was investigated. The relation was shown in Fig.8. The differences of peak position between before UV irradiation and after UV irradiation increased with increasing the concentrations of 4-(phenylazo)benzoic acid. That is, the structure of ZrO_2 precursor was expected to change greatly by addition of large amounts of 4-(phenylazo)benzoic acid.

CONCLUSION

The ZrO_2 precursor solution prepared using azobenzene did not show the photochromism. However, the ZrO_2 precursor solution prepared using 4-(phenylazo)benzoic acid, which has a carboxyl group in a terminal of azobenzene molecule, showed the photochromism. That is, the absorption band originated from Zr-O CT transition in UV-Vis spectra of the precursor changed reversible by UV irradiation and Vis irradiation. The FT-IR spectra indicated that the carboxyl group in a terminal of 4-(phenylazo)benzoic acid molecule interacted with zirconium alkoxide chemically.

REFERENCES

[1] A.H.Cook, "The preparation of some cis-Azo-compouds", *J.Chem.Soc.*, 876(1938).

[2] H. S. Choi, E. H. Kim, I. H. Choi, Y. T. Kim, J. H. Choi and J. Y. Lee, "The effect of ZrO_2 buffer layer on electrical properties in $Pt/SrBi_2Ta_2O_9/ZrO_2/Si$ ferroelectric gate oxide structure", *Thin Solid Films*, **388**, 226(2001).

[3] C. M. Perkins, B. B. Triplett, P. C. McIntyre, K. C. Saraswat, S. Haukka and M. Tuominen, "Electrical and materials properties of ZrO_2 gate dielectrics grown by atomic layer chemical vapor deposition", *Appl. Phys. Lett.*, **78 [16]**, 2357(2001).

[4] G. L. Tan, X. J. Wu, "Electronic conductivity of a ZrO_2 thin film as an oxygen sensor", *Thin Solid Films*, **330**, 59(1998).

[5] K.Nishizawa, T. Miki, K. Suzuki, D. Fu, K. Kato, "Effects of β-diketone addition on crystallinity of photo-assisted alkoxy-derived zirconia thin films", *Key Engineering Materials*, **269**, 125-128(2004).

Electroceramic Materials and Applications
Edited by R. W. Schwartz

Fundamental Properties of Conducting and Magnetic Electroceramics

Electroceramic Materials and Applications
Edited by R. W. Schwartz

TRANSPARENT CONDUCTING PROPERTIES IN LAYERED OXYCHALCOGENIDES

Kazushige UEDA
Kyushu Institute of Technology
1-1 Sensui, Tobata,
Kitakyushu, 804-8550, JAPAN

Hidenori HIRAMATSU, Masahiro HIRANO, Toshio KAMIYA and Hideo HOSONO
Tokyo Institute of Technology / ERATO-SORST JST
4259 Nagatsuta, Midori,
Yokohama, 226-8503, JAPAN

ABSTRACT

Transparent and conducting properties are found in not only oxides but also several layered oxychalcogenides. For example, a layered oxysulfide, LaCuOS, composed of $(La_2O_2)^{2+}$ and $(Cu_2S_2)^{2-}$ layers is known as a transparent *p*-type conducting material. The energy gap of LaCuOS is larger than 3 eV even though the energy gap of Cu_2S, which is structurally similar to the $(Cu_2S_2)^{2-}$ layers, is less than 2 eV. LaCuOSe and YCuOSe are materials that have the same crystal structure as that of LaCuOS. To investigate the influence of the $(Ln_2O_2)^{2+}$ (*Ln*=La and Y) layers on their energy gaps, the basic optical and electrical properties of YCuOSe were examined. The energy gap of YCuOSe was found to be 2.58 eV, which is smaller than the energy gap of LaCuOSe (2.82 eV). This smaller energy gap of YCuOSe was attributed to smaller ionic radii of Y ions, which lead to shorter Cu - Se distances resulting in the larger interaction between Cu and Se ions.

INTRODUCTION

Most transparent conducing oxides (TCOs) such as ZnO and In_2O_3 are *n*-type conductors and their utilization is usually restricted to transparent electrodes due to their mono-polar conduction. To open new applications of TCOs, several *p*-type TCOs were developed in delafossite materials such as $CuAlO_2$ and $CuGaO_2$.[1,2] Since the electrical conductivity of these *p*-type TCOs is not sufficient and its control by doping is rather difficult, the materials development was extended from oxides to oxysulfides. As a result, LaCuOS was found as a new type of transparent *p*-type conducting materials.[3]

The crystal structure of LaCuOS is a layered structure composed of $(La_2O_2)^{2+}$ and $(Cu_2S_2)^{2-}$ layers.[4] LaCuOS is transparent in the visible region and its electrical conductivity can be widely controlled by Sr or Mg substitution at La^{3+} sites, and degenerate conduction is achieved without losing transparent properties.[3,5] In addition to these wide-gap and conducting properties, this material shows excitonic absorption and emission even at room temperature.[6,7] Since La and S sites can be substituted by lanthanide ions and chalcogen ions, respectively, the optical and electrical properties of analogous materials *Ln*CuO*Ch* (*Ln*=lanthanides, *Ch*=chalcogen) were examined for the band gap engineering.[8-10]

In the series of *Ln*CuOS (*Ln*=La, Pr and Nd), the energy gap decreases as the atomic number of the lanthanide ions increases. Similar energy gap narrowing occurs in the series of LaCuO*Ch* (*Ch*=S, Se and Te). The changes of the energy gap in LaCuO*Ch* are basically understood from the energy levels of the *Ch* ions because the *Ch p* orbitals form the top of the valence band with Cu 3*d* orbitals and influence the energy gap directly. On the other hand, the energy band derived from the lanthanide ions does not affect the energy gap straightaway and the origin of the band gap variation in *Ln*CuOS is not the same with that in LaCuO*Ch*. In this study, basic optical properties of YCuOSe were examined as well as its electrical transport properties and compared with those of LaCuOSe to understand the variation of the energy gap in these materials.

EXPERIMENTS

Sample synthesis

Cu_2Se precursor, Y_2Se_3 precursor and Y_2O_3 were used as the starting materials for YCuOSe. The precursors of Cu_2Se and Y_2Se_3 were prepared using Cu and Se or Y and Se in the following chemical reaction. The reaction was conducted at 300 °C for 12 h in the evacuated silica glass tubes.

$$2\,Cu + Se \rightarrow Cu_2Se$$

$$2\,Y + 3\,Se \rightarrow Y_2Se_3$$

The starting materials of Cu_2Se, Y_2Se_3 and Y_2O_3 were mixed and disked by cold isostatic press. The disks were heated at 700 °C, 750 °C or 900 °C for 12 h in the evacuated silica glass tubes. After the following reaction, YCuOSe samples were obtained.

$$3\,Cu_2Se + Y_2Se_3 + 2\,Y_2O_3 \rightarrow 6\,YCuOSe$$

After grinding the disks, the purity of the powdered samples was examined by the power x-ray diffraction (XRD) measurements. After pelletizing the powder and sintering the pellets, electrical properties were analyzed using the sintered pellets.

Evaluation of optical and electrical properties

Electrical conductivity, Seebeck coefficients, diffuse reflection and photoluminescence were measured to examine basic optical and electrical properties. Energy gaps were estimated approximately from absorption edges observed in the diffuse reflection. The types of conduction (*p*/*n*) were roughly evaluated from the sign of the Seebeck coefficients and the type of band gap (direct/indirect) was done by the observation of the band edge emission in the photoluminescence measurements. The 4ω of a Nd:YAG laser was used as an excitation source in the photoluminescence measurements.

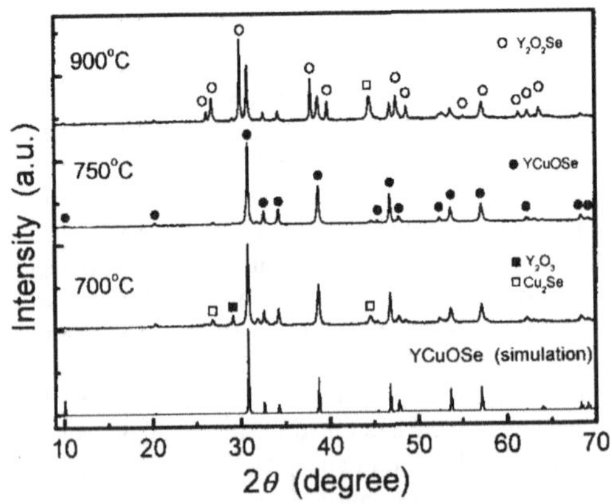

Figure 1 XRD patterns of samples heated at 700 °C, 750 °C or 900 °C.

RESULTS

Sample preparation and XRD measurements

The XRD patterns of the samples heated at 700 °C, 750 °C or 900 °C are shown in Figure 1. The starting materials of Cu_2Se and Y_2O_3 were observed in the sample heated at 700 °C, indicating 700 °C is not high enough to obtain an YCuOSe single phase. On the other hand, the XRD pattern for the sample heated at 900 °C shows that the sample was almost decomposed into Y_2O_2Se and Cu_2Se. As shown in the XRD pattern for the sample heated at 750 °C, the single phase of YCuOSe was obtained in the limited temperature range around 750 °C. The measurements of optical and electrical properties were performed against the sample prepared at 750 °C.

Electrical conductivity

The temperature dependence of the electrical conductivity in YCuOSe is shown in Figure 2 along with those for LaCuOSe and LaCuOS. YCuOSe shows semiconducting properties near room temperature and degenerate semiconducting behavior at low temperatures. The electrical conductivity of YCuOSe higher than those of LaCuOSe or LaCuOS is probably due to the high density of defects such as Cu deficiency. Table I lists the electrical conductivities and Seebeck coefficients of these materials at room temperature. The Seebeck coefficients are all positive values, indicating that all samples are *p*-type conductors.

Table I Electrical conductivities and Seebeck coefficients of YCuOSe, LaCuOSe and LaCuOS at room temperature.

	σ (Scm^{-1})	α (mVK^{-1})
LaCuOS	6.7×10^{-6}	+381
LaCuOSe	8.5×10^{-3}	+405
YCuOSe	1.4×10^{-1}	+13

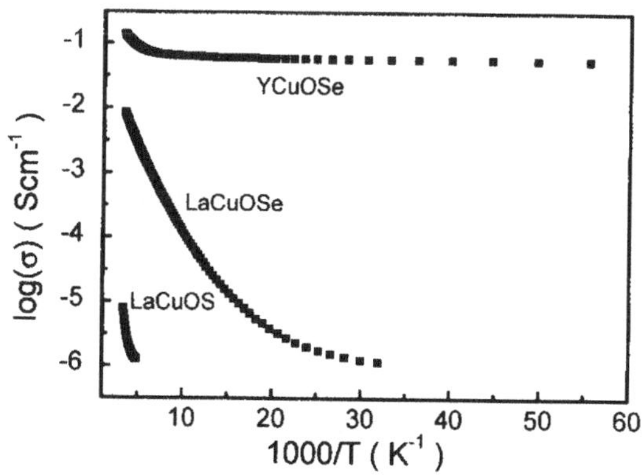

Figure 2 Electrical conductivity of YCuOSe as a function of reciprocal temperatures, along with those of LaCuOSe and LaCuOS.

Diffuse reflection

The diffuse reflection spectra of YCuOSe, LaCuOSe and LaCuOS are shown in Figure 3. The reflection abruptly drops at 395 nm for LaCuOS, at 440 nm for LaCuOSe and at 480 nm for YCuOSe. These reflection drops are considered to originate from the fundamental absorption. Therefore, substitution of Y for La causes a red shift of the absorption edge. After the conversion of the diffuse reflection spectra into absorption spectra, the energy gap of YCuOSe was estimated to be 2.58 eV.

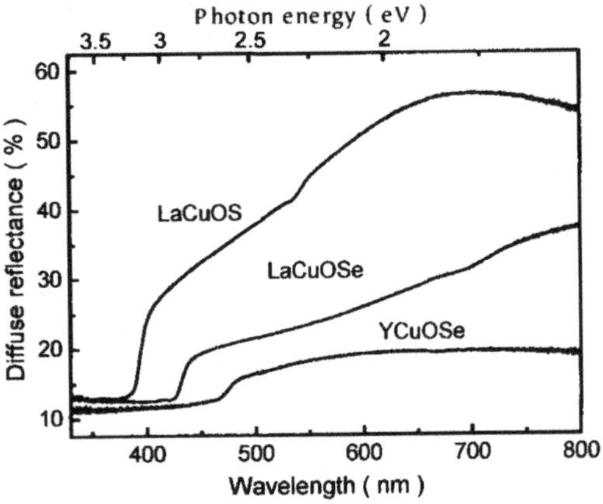

Figure 3 Diffuse reflection spectra of YCuOSe, LaCuOSe and LaCuOS.

Photoluminescence

The photoluminescence spectra of YCuOSe, LaCuOSe and LaCuOS measured at

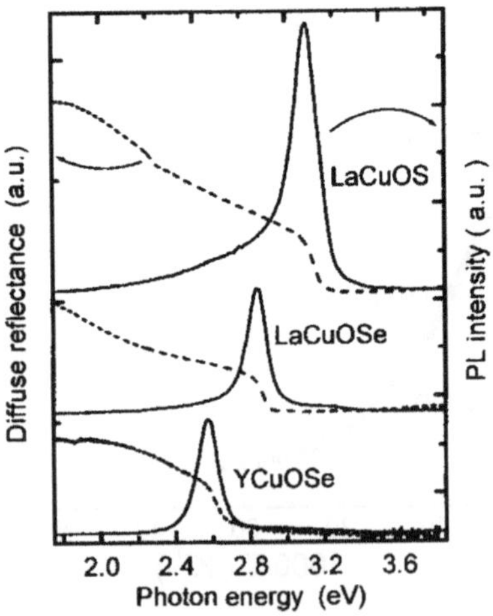

Figure 4 Photoluminescence spectra of YCuOSe, LaCuOSe and LaCuOS along with their diffuse reflection spectra shown in dashed lines.

room temperature are shown in Figure 4 along with their diffuse reflection spectra. As observed in LaCuOSe and LaCuOS, the sharp band edge emission was observed in YCuOSe. The observation of the band edge emission demonstrates that the energy gap of YCuOSe is a direct type with allowed transition similar to the La oxychalcogenides.

DISCUSSION

Energy gap of *Ln*CuO*Ch* (*Ln* = La, Pr, Nd, Y and *Ch* = S, Se)

To understand the variation of energy gaps in *Ln*CuO*Ch*, the energy gap of *Ln*CuO*Ch* and Ln_2O_3 are summarized in Table II.[11,12] In the comparison between lanthanides, the energy gaps of the simple oxides, Ln_2O_3, decrease as large as 1eV when the lanthanide ions vary from La to Nd. However, the energy gaps of *Ln*CuOS (*Ln* = La, Pr, Nd) very slightly decrease as La is substitute with Pr or Nd. In the comparison between La and Y, the energy gap of Y_2O_3 is larger than that of La_2O_3 by 0.5 eV. But the energy gap of YCuOSe is smaller than LaCuOSe. Therefore, the energy gaps of Ln_2O_3 seem to have no correlation with those of *Ln*CuO*Ch*. Namely, it is considered that the species of lanthanide and Y ions themselves do not influence the energy gaps of *Ln*CuO*Ch* directly.

Table II Energy gap of *Ln*CuO*Ch* (*Ln* = La, Pr, Nd, Y and *Ch* = S, Se) and Ln_2O_3

Ln / *Ch*	E_g (eV)	
	*Ln*CuO*Ch*	Ln_2O_3
La / S	3.14	5.5
Pr / S	3.03	4.6
Nd / S	2.98	4.4
La / Se	2.82	5.5
Y / Se	2.58	6.1

Figure 5 shows the variation of the lattice constants in *Ln*CuO*Ch* (*Ln* = La, Pr, Nd, Y and *Ch* = S, Se), which are all in tetragonal system.[13] In *Ln*CuOS (*Ln* = La, Pr, Nd), the

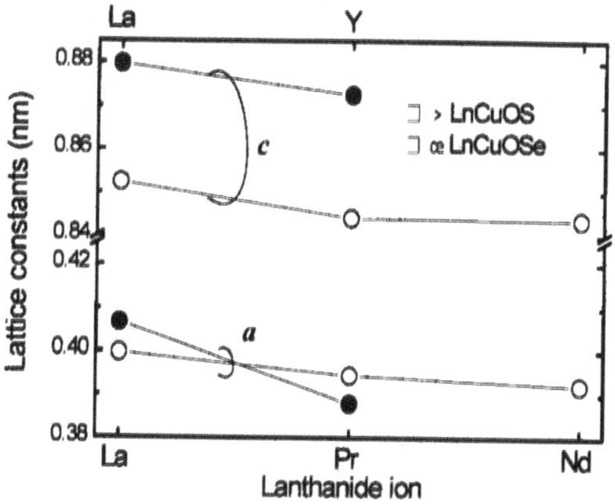

Figure 5 Variation of lattice constants in *Ln*CuO*Ch* (*Ln* = La, Pr, Nd, Y and *Ch* = S, Se)

lattice constants of both *a* and *c* decrease due to the lanthanide contraction when the lanthanide ions vary from La to Nd. In *Ln*CuOSe (*Ln* = La, Y), the lattice constants decrease when La ions are substituted by Y ions. These decreases of the lattice constants seem to correlate with the decreases of the energy gap in *Ln*CuO*Ch*. Therefore, it is suggested that the variation of the energy gaps in *Ln*CuO*Ch* correlates with the size of *Ln* ions rather than their species.

Electronic structure of *Ln*CuO*Ch*

In the previous study on the electronic structure of LaCuOSe,[14] the top of the valence band is composed of the Cu 3*d* - Se 4*p* hybridized band and the bottom of the conduction band is of the Cu 4*s* band. The band edge emission occurs by the transition or recombination of electrons and holes between these two bands. The La 5*d* band and O 2*p* band are located above the Cu 4*s* band and below the Cu 3*d* - Se 4*p* hybridized band, respectively.

Since the band edge emission was observed in YCuOSe, it is considered that the electronic structure of YCuOSe is basically the same as that of LaCuOSe. Taking account of the energy position and bandwidth of each energy band, the electronic structures of LaCuOSe and YCuOSe are schematically illustrated in Figure 6.

The energy separation between the Y 4*d* and O 2*p* bands will be larger than the separation between the La 5*d* and O 2*p* bands because the band gap of Y_2O_3 is larger than that of La_2O_3. On the other hand, the energy separation between the Cu 3*d* - Se 4*p* and Cu 4*s* bands, that is the band gap of these materials, will become smaller when Y ions substitute La ions. Since the components of these latter two bands are the same between LaCuOSe and YCuOSe, the central energy position of the two bands will not differ significantly. However, smaller lattice constants in YCuOSe result in decreases of Cu - Cu and Cu - Se distances, which will cause larger overlaps of their orbitals and lead to the larger bandwidth of the Cu 3*d* - Se 4*p* and Cu 4*s* bands. This bandwidth variation is considered to be the origin of the smaller band gap of YCuOSe.

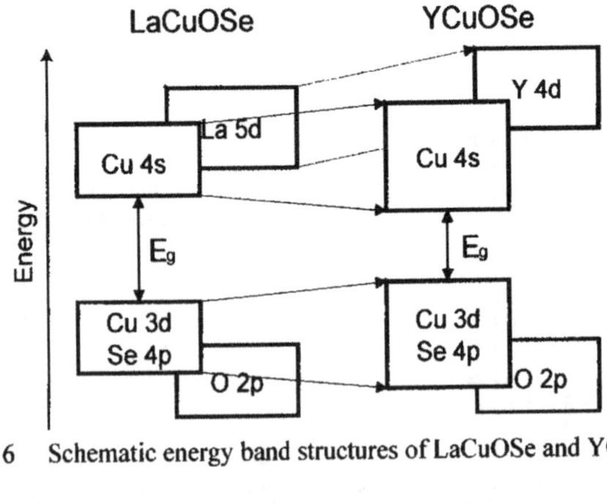

Figure 6 Schematic energy band structures of LaCuOSe and YCuOSe

CONCLUSION

The samples of YCuOSe were prepared by solid-state reaction at 750°C for 12 h in evacuated silica glass tubes. They showed *p*-type semiconducting properties and the energy gap of YCuOSe was 2.58 eV, which is smaller than that of LaCuOSe. The band edge emission in the photoluminescence spectrum demonstrated that YCuOSe has a direct-type energy gap and electronic structure similar to those of LaCuO*Ch* (*Ch*=S and Se). The smaller energy gap of YCuOSe originates from larger interaction between Cu and Se due to shorter Cu - Se distances or smaller ionic radii of Y ions.

REFERENCES

[1]H. Kawazoe, M. Yasukawa, H. Hyodo, M. Kurita, H. Yanagi and H. Hosono, "*P*-type electrical conduction in transparent thin films of $CuAlO_2$," *Nature*, **389**, 939-942 (1997).

[2]H. Kawazoe, H. Yanagi, K. Ueda, and H. Hosono, "Transparent *p*-Type Conducting Oxides: Design and Fabrication of *p-n* Heterojunctions," *MRS Bulletin*, **25**, 28-36 (2000).

[3]K. Ueda, S. Inoue, S. Hirose, H. Kawazoe and H. Hosono, "Transparent *p*-type semiconductor: LaCuOS layered oxysulfide," *Appl. Phys. Lett.*, **77**, 2701-2703 (2000).

[4]M. Palazzi, "Préparation et affinement de la structure de (LaO)CuS," *C. R. Acad. Sci. Ser. 2*, **292**, 789-791 (1981).

[5]H. Hiramatsu, K. Ueda, H. Ohta, M. Orita, M. Hirano, and H. Hosono, "Heteroepitaxial growth of a wide-gap *p*-type semiconductor, LaCuOS," *Appl. Phys. Lett.*, **81**, 598-600 (2002).

[6]K. Ueda, S. Inoue, N. Sarukura, M. Hirano and H. Hosono, "Room-temperature excitons in wide-gap layered-oxysulfide semiconductor: LaCuOS," *Appl. Phys. Lett.*, **78**, 2333-2335 (2001).

[7]K. Ueda, H. Hiramatsu, H. Ohta, M. Hirano, T. Kamiya, and H. Hosono, "Single-atomic-layered quantum wells built in wide-gap semiconductors *Ln*CuO*Ch* (*Ln*=lanthanide, *Ch*=chalcogen)," *Phys. Rev. B*, **69**, 155305.1-4 (2004).

[8]K. Ueda, and H. Hosono, "Bandgap engineering, band edge emission and *p*-type conductivity in wide-gap $LaCuOS_{1-x}Se_x$ oxychalcogenides," *J. Appl. Phys.*, **91**, 4768-4770 (2002).

[9]K. Ueda, K. Takafuji, H. Hiramatsu, H. Ohta, T. Kamiya, M. Hirano, and H. Hosono, "Electrical and Optical Properties and Electronic Structures of *Ln*CuOS (*Ln*=La~Nd)," *Chem. Mater.*, **15**, 3692-3695 (2003).

[10]K. Ueda, H. Hosono and N. Hamada, "Energy band structure of LaCuO*Ch* (*Ch*=S, Se and Te) calculated by the full-potential linearized augmented plane-wave method," *J. Phys.: Condens. Matter*, **16**, 5176-5186 (2004).

[11]A. V. Prokofiev, A. I. Shelykh, and B. T. Melekh, "Periodicity in the band gap variation of Ln_2X_3 (X=O, S, Se) in the lanthanide series," *J. Alloys Comp.*, **242**, 41-44 (1996)

[12]V. N. Abramov and A.I. Kuznetsov, "Fundamental absorption of Y_2O_3 and $YAlO_3$," *Sov. Phys. Solid State*, **20**, 399-402 (1978).

[13]K. Ueda, K. Takafuji, and H. Hosono, "Preparation and crystal structure analysis of CeCuOS," *J. Solid State Chem.*, **170**, 182-187 (2003).

[14]K. Ueda, H. Hosono and N. Hamada, "Valence-band structures of layered oxychalcogenides, LaCuO*Ch* (*Ch*=S, Se and Te), studied by ultraviolet photoemission spectroscopy and energy-band calculations," *J. Appl. Phys.*, **98**, 043506.1-7 (2005).

Electroceramic Materials and Applications
Edited by R. W. Schwartz

ZnO:Ga-BASED TRANSPARENT CONDUCTIVE FILMS: AN ATTRACTIVE POTENTIAL FOR USE IN FLAT PANEL DISPLAY

Tetsuya Yamamoto [1], Takahiro Yamada [1], Toshiyuki Sakemi [2], Sho Shirakata [3], Minoru Osada [4], Seiichi Kishimoto [1], Kiyoshi Awai [1], Hisao Makino [1], and Tooru Mitsunaga [5]

[1] Kochi University of Technology, Research Institute, Materials Design Center,
185 Miyanokuchi, Tosayamada-cho, Kamigun, Kochi 782-8501, Japan

[2] Research and Development center, Sumitomo Heavy Industries Ltd.,
5-2 Soubiraki-cho, Niihama, Ehime 792-8588, Japan

[3] Faculty of Engineering, Ehime University,
3 Bunkyo-cho, Matsuyama, Ehime 790-9577, Japan

[4] Advanced Materials Laboratory, National Institute for Materials Science,
Tsukuba, Ibaraki 305-0044, Japan

[5] Application Laboratory, Rigaku Corporation,
3-9-12 Matsubara-cho, Akishima-shi, Tokyo 196-8666, Japan

ABSTRACT

The advantageous features of polycrystalline gallium (Ga)-doped ZnO (GZO) films, being cost-effective and easy to fabricate on a production scale, make them very attractive potential for use in transparent conductive electrode in flat display panel and thin film solar cells. In this work, GZO films have been prepared by reactive plasma deposition with non-traveling and traveling substrate mechanism. We have succeeded in fabricating 180 nm GZO films with low resistivity of 2.7×10^{-4} Ωcm and high transmittance more than 90 % in the visible wave length region on a glass substrate at temperature of 200 □ with a size up to 1m × 1m. On the basis of the analysis of data obtained by theoretical prediction and experimental study concerning micro-structural properties for GZO films, such as which sites Ga donors occupy, we have shown how to obtain electrical and optical properties required for application controlling the main process parameters, oxygen partial pressure.

INTRODUCTION

Zinc oxide (ZnO) with a wurtzite structure is a versatile material with a wide band gap of 3.37 eV at room temperature. *n*-type ZnO thin films have in recent years been rediscovered as a subject of considerable research interest due to their very unique physical properties (high conductivity of 2×10^{-4} Ωcm, high visible transmittance (90 %) and high infrared (IR) reflectance and absorptance in the microwave region) and a wide range of possible electronic and optical application. Important uses of ZnO include catalysis, luminescent materials, fungicides, surface acoustic wave devices, varistors, light emitting diodes (LED), and transparent conductors. Special care is directed to the blue to UV wavelength LED because of the wide band gap of ZnO and a highly efficient ultraviolet photoluminescence[1-3], and transparent conducting electrodes of solar cells, such as *a*-Si and $CuInGaSe_2$-based solar cells[4] because ZnO films are stable in hydrogen plasmas [5], and flat display panel (FPD) [6]. Very recently, Tamura and et al., reported InGaN-based LED with Ga-doped ZnO (GZO) as ohmic *p*-contact [7].

The transparent conducting electrode is a necessary component of all types, such as liquid crystal display (LCD), of FPD. It consists usually of a transparent conducting film and a glass plate substrate. The commercially most important material for transparent conducting films nowadays is Sn-doped In_2O_3 (ITO). While high rates of growth in the display industry as mentioned above will generate remarkably increased demand for indium, indium is a critical raw material. GZO thin films are very attractive on account that they are abundant raw material, cost-effective and nontoxic, they have comparable electrical and optical properties to ITO and they have been successfully achieved on a commercialized production scale, such as 1m×1m [8].

In this paper, a comprehensive review of our results of the fabrication and properties of GZO thin films on an alkali-free glass substrate will be given. We introduce two types of reactive plasma deposition (RPD). In these methods, the plasma caused by a DC-arc discharge using a pressure-gradient plasma-gun (Uramoto-Tanaka type gun) [9] is utilized for both the sublimation of the ZnO source material and the activation of the evaporated species. The GZO films are analyzed using high resolution x-ray diffraction (HR-XRD, Cu $K\alpha_1$), scanning electron microscope (SEM), atomic force microscope (AFM), secondary ion-mass spectroscopy (SIMS), Raman measurements, Hall effect (van der Pauw method) and a double-beam spectrometer in order to optical transmittance and reflectance measurements. The lattice constants of GZO thin films were theoretically calculated by a first-principle electronic-band-structure calculation method.

EXPERIMENTAL

A series of polycrystalline GZO thin films were prepared, on alkali-free glass substrates at 200°C using sintered ZnO source precursors, by RPD. Schematic diagrams of RPD apparatus are shown in Figs. 1a and 1b. This method enables the deposition of ZnO films: (i) at high rate, (ii) with low plasma damage, and (iii) on a large-area substrate.

The outline and the operating principle of the apparatus (Fig. 1b) with the traveling substrate mechanism for the industrial production are similar to those of the prototype RPD apparatus (Fig. 1a) used for films with a size up to 0.2 m × 0.2 m reported previously [10-16].

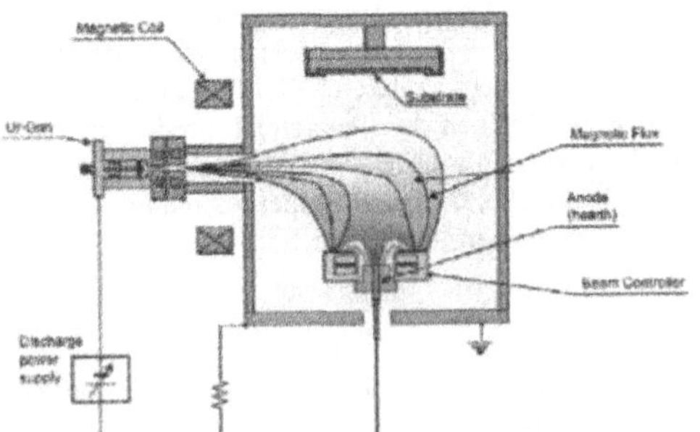

Figure 1a. Schematic diagram of reactive plasma deposition (RPD) system without the traveling substrate mechanism.

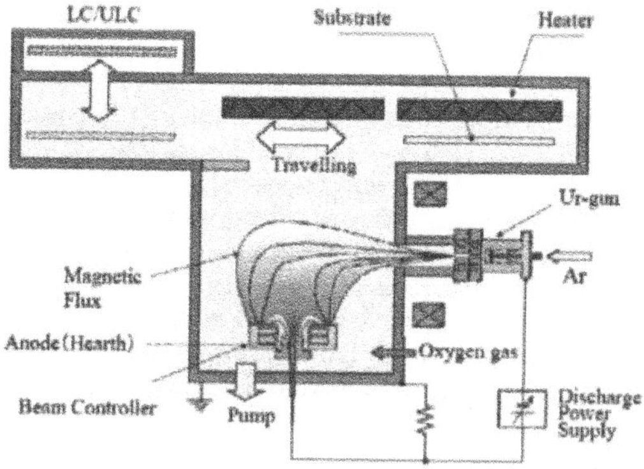

Figure 1b. Schematic diagram of reactive plasma deposition (RPD) system with the traveling substrate mechanism.

When we fabricated ZnO films with a size up to 1m × 1 m [17], we used the RPD designed for the deposition of the 1m-wide film using two plasma-guns with the same traveling substrate mechanism as the apparatus in Fig. 1 b. The distance between the two plasma guns was 700 mm. The non-uniformity of the film thickness, measured by a surface profile measurement system (DEKTAK3), along the 1m width of film is within ± 5 %.

The chamber was evacuated to 2.5×10^{-6} Torr. Ar gas is injected into the growth chamber through the Uramoto-gun, and an Ar plasma formed by the magnetic fields evaporates the Ga-doped ZnO tablet: ZnO:Ga_2O_3 tablets of various Ga_2O_3 content (0.003-10 wt%) were mounted on an anode hearth. The ringed magnets in the plasma beam controller form a cusp magnetic field and correct an incident angle of the plasma beam to be perpendicular to an anode. The distance between the source tablets and the substrate was 0.6 m. The oxygen gas was introduced into the deposition chamber in order to compensate the oxygen deficiencies. When the oxygen gas flow rate was changed from 0 to 30 sccm, the pressure in the growth chamber varied over the range of $1.2 \times 10^{-3} - 5.6 \times 10^{-4}$ Torr. The growth rate increased with increasing discharge current. In this work, the discharge current was 100 A. Under those conditions, the growth rate was found to be approximately 7-11 μm/h. The GZO films of various thickness were obtained by varying the deposition time. In this paper, the growth time for all GZO films prepared using prototype RPD (Fig. 1 a) has been fixed at 70 s; the averaged thickness of the GZO films is approximately 200 nm.

The structure, surface properties and thickness of the films were determined by means of SEM (Hitachi S4700 scanning electron microscope), using an acceleration voltage of 10 kV. The structural properties were analyzed with a high-resolution XRD (ATX-G system, Rigaku Co.). The out-of-plane XRD pattern (synchronous scan of 2θ and ω in the horizontal plane), in-plane XRD pattern (synchronous scan of $2\theta\chi$ and ϕ in the azimuth plane) and out-of-plane rocking curve (2θ fixed ω scan) were obtained. The Ga depth profiles in the GZO films were measured by the SIMS system (SIMS4100, ATOMIKA Instruments). The SIMS analysis was made using O_2^+ primary ions. The scan area was $87 \times 75\ \mu m^2$. Ga was quantified using the relative sensitivity factor (RSF), which was obtained by measuring the Ga-implanted ZnO reference sample. Raman measurements were performed in a backward micro-configuration, using the 514.5 nm line from an Ar^+ laser (~ 1m W) focused to ~ 1 μm-diameter spot on the sample surface. The scattered light was dispersed by a subtractive triple spectrometer (HORIBA-Jobin Yvon T64000) and collected with a liquid-nitrogen-cooled charge-coupled device (CCD) detector. The electrical resistivity ρ, carrier concentration n and Hall mobility μ of the GZO films at room temperature were determined by the van der Pauw method and by Hall-effect measurements (ACCENT, HL5500PC). For optical transmittance and reflectance measurements, a double-beam spectrometer (Hitachi, model U-4000) was used. The value of the transmittance used in this work is that measured with a reference glass substrate, and the effect of the substrate is subtracted.

COMPUTATIONAL

In order to calculate the lattice constant of both the *a*- and *c*- axes of the GZO films, we use the density-functional theory (DFT) [18] within the generalized gradient approximation (GGA) [19], as implemented with a plane-wave basis in the Vienna *ab initio* simulations package (VASP) [20]. Results were obtained using the projector-augmented plane wave (PAW) [21,22] potentials provided with the VASP. A plane-wave basis set with a kinetic-energy cutoff of 500 eV was used. Reciprocal space sampling was performed with a 3×3×5 *k*-point grid. The ions are relaxed toward equilibrium until the Hellmann-Feynman forces are less than 10^{-3} eV/Å.

Note that Ga concentration is $1.12 \times 10^{21}\ cm^{-3}$ if all the Ga species in the source ZnO tablets prepared by sintering ZnO powder containing Ga_2O_3 (3 wt %) are incorporated into the GZO thin films. In the calculations, we used supercells having 72 atoms; we replaced one of the 36 sites of Zn atoms with a Ga site in the model supercells. In such supercell, Ga concentration is found to be $1.17 \times 10^{21}\ cm^{-3}$ obtained using the theoretically optimized structural parameters, as discussed below.

RESULTS AND DISCUSSION

1. GZO films prepared by RPD with non-traveling substrate mechanism

In this work, we focus on ZnO thin films with Ga content of 3 wt%. In such sample, we find the resistivity minimized by optimizing O_2 gas flow rate from all samples with different Ga content from 0.003 wt% to 10 wt%.

Figure 2 SEM image of the cross-sectional view of GZO films on a glass substrate prepared at oxygen gas flow rate of 15 sccm.

1.1 *Structural characterization*

Figure 2 shows the SEM image of the cross-sectional view of GZO films prepared at oxygen (O_2) gas flow rate of 15 sccm. For the other GZO films, the SEM image does not change significantly. It shows a columnar structure which indicates an orientation parallel *c*-axis (002), as discussed below.

Figures 3 (a) and 3 (b) show the XRD patterns for the GZO thin films with different oxygen gas flow rates. They revealed that all the deposited GZO films were polycrystalline with the hexagonal structure. For all samples, on the basis of the analysis of the out-of-plane XRD patterns, the wurtzite GZO (002) peak (i.e., a 2θ value of $\sim$ 34 degree) is found to be much higher than all other peaks. This means that all films predominantly exhibit one peak showing a very high level of orientation along the *c*-axis. No Ga_2O_3 phase was detected from the X-ray patterns. This may be due to gallium replacing zinc substitutionally in the hexagonal lattice or gallium segregating to the non-crystalline region in grain boundary. The problem concerning Ga location in GZO films will be determined on the basis of the analysis of data obtained by Raman measurements as below. The crystalline size of the films which can be estimated using both Scherrer formula and Williamson-Hall analysis[23] was less than 40 nm.

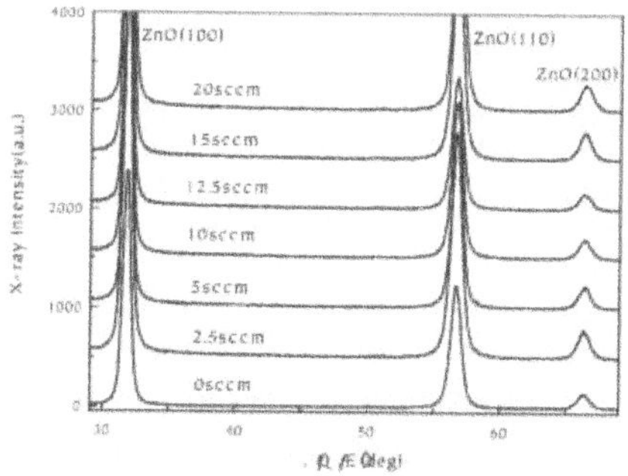

Figure 3a. In-plane XRD patterns for the GZO thin films with different O_2 gas flow rates.

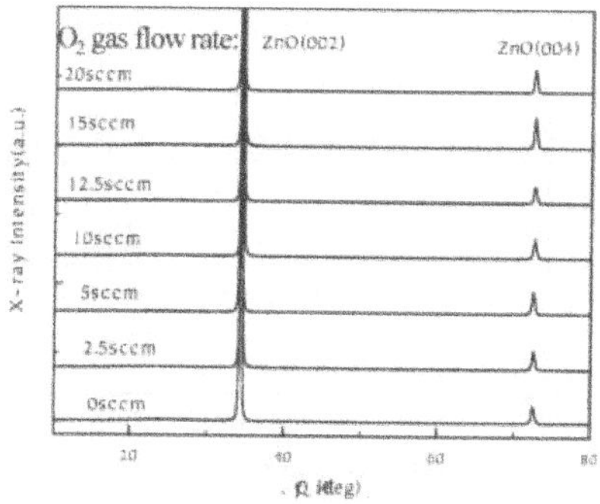

Figure 3b. Out-of-plane XRD patterns for the GZO thin films with different O_2 gas flow rates.

The lattice constant of the *a*- and *c*- axes of the GZO films calculated using the data of ZnO (006) and ZnO (300) peak obtained by both the out-of-plane and in-plane XRD measurements of GZO thin films grown at different O_2 gas flow rate are shown in Figs. 4a and 4 b. respectively. Note that the full-widths at half maximum (FWHMs) of ZnO (006) and ZnO (300) were 0.8° ~ 0.9° and 1.8° ~ 1.9°, respectively. Therefore, the error values, indicated in both Figs. 4a and 4b, of the lattice constant of the *a*- and *c*- axes are ~ 0.001 Å and ~ 0.002 Å, respectively. While Fig. 4b shows that the lattice constant of the *a*-axis for all samples remains almost the same value, the *c*-axis depends on O_2 gas flow rate. When O_2 gas flow rate varies from 0 to 10 sccm, the lattice constant of the *c*-axis remains almost constant. Note that the lattice constant of the *c*-axis of GZO films prepared under condition above is larger than that (5.207 Å) obtained by neutron diffraction for undoped ZnO single crystals [24]. With further increasing O_2 gas flow rate higher than 10 sccm, we find a decrease in the lattice constant of the *c*-axis.

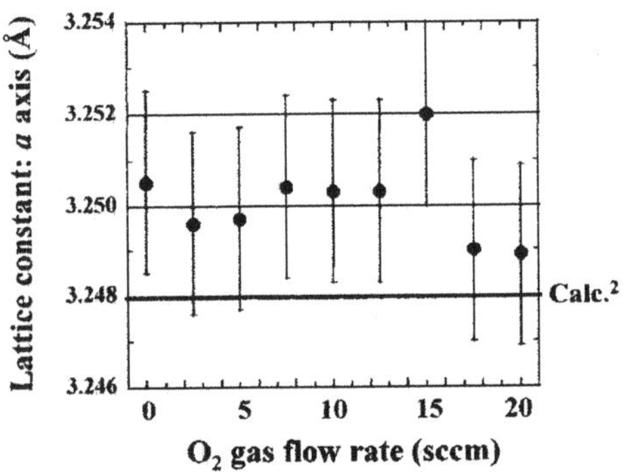

Figure 4a. Lattice constant of the *a*-axis of the GZO films as a function of O_2 gas flow rate.

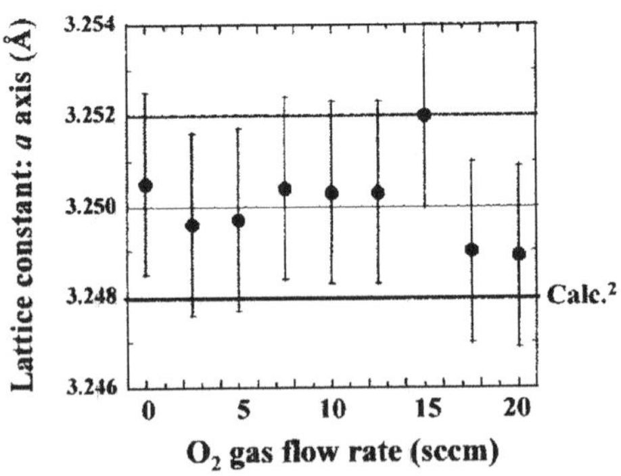

Figure 4b. Lattice constant of the *c*-axis of the GZO films as a function of O_2 gas flow rate.

Table 1. Lattice constants of undoped and GZO. Unit: Å.

axis	*undoped*		*GZO*	
	Ref (24)	calc.	calc.[1]	calc.[2]
a	3.250(= a_0)	3.282(= a)	3.280	3.248*
c	5.207(= c_0)	5.299(= c)	5.306	5.214**

*: 3.280 (a_0 / a), **: 5.306 (c_0 / c),

In order to investigate the effects of Ga doping on the lattice constant of ZnO films, we performed *ab initio* electronic-band-structure calculations. In Table 1, we summarize the lattice constant obtained by experiment [24] and GZO crystals obtained by theoretical calculations. GGA calculations usually overestimate cell volumes. For undoped ZnO crystals, as shown in Table 1, the lattice constant calculated by GGA approximation are found to be larger than those obtained by experiments [24]. Thus, when we calculate the lattice constant (calc.[2] in Table 1) of the GZO films to compare with experimental data, we multiply the lattice constants theoretically calculated that are referred to as calc.[1] in Table 1 by the ratios of the lattice constant (a_0 or c_0 in Table 1) obtained by experiments for undoped ZnO [24] to those (a or c in Table 1) calculated. The theoretical calculations reveal that the deviation between undoped ZnO and GZO (Ga content of 3 wt%) unit-cell dimensions is -0.06% in a and 0.13% in c. This is originated in that excess electrons with n-type doping using Ga species mainly distribute along the c-axis direction. The bold lines indicated by calc.[2] in Figs. 4a and 4b indicate the lattice constant of the a- and c-axes for the theoretical model with Ga at Zn sites (Ga_{Zn}), respectively.

Figure 4a shows that the lattice constant of the a-axis for the GZO films depends little on O_2 gas flow rate and is on good agreement with that (calc.[2] in Table 1) obtained by calculations. On the other hand, when O_2 gas flow rate varies from 0 to 10 sccm, the lattice constant of the c-axis is in good agreement with that obtained by calculations and remains almost constant. Both this and the above finding concerning the lattice constant of the a-axis mean that for the GZO films, the dominant defect is Ga_{Zn}. From calculations, we find that the formation of oxygen vacancies with a high concentration as well as Ga considerably reduces the lattice constant of both the a- and c-axes. The defect model that the dominant defect is oxygen vacancies cannot explain the behavior of the lattice constant of both the a- and c-axes discussed above. On the other hand, we find a decrease in the lattice constant of the c-axis with further increasing O_2 gas flow rate higher than 10 sccm. In such a case, we must consider two possible cases; (1) reduction in Ga_{Zn} concentration with segregation of nonconductive Ga oxides, such as Ga_2O_3, at the surface, grain boundary and/or intra-grains in the GZO films, (2) no change in Ga_{Zn} concentration with the formation of p-type defects associated with Zn vacancies (V_{Zn}) or O interstitials (O_i), which compensate n-type Ga_{Zn}. These two cases will give rise to a decrease in both carrier concentration and Hall mobility.

Then, we will discuss which of the two cases described above occurs in the GZO thin films prepared with further increasing O_2 gas flow rate. Ga profile and microstructures in GZO films were analyzed with SIMS and Raman measurements, respectively.

On the basis of the analysis of the SIMS data, we find almost the same Ga profiles for all the GZO films with different O_2 gas flow rates. Figures 5a and 5b show the depth profile of Ga concentration in the GZO films prepared at O_2 gas flow rates of 5 sccm and 15 sccm, respectively. As shown in Fig. 5, the Ga concentration in the two GZO thin films prepared with O_2 gas flow rate below/beyond 10 sccm was found to be almost constant from the GZO/substrate interface to the surface. We find very small difference in the averaged Ga concentration among the GZO films prepared at different O_2 gas flow rates.

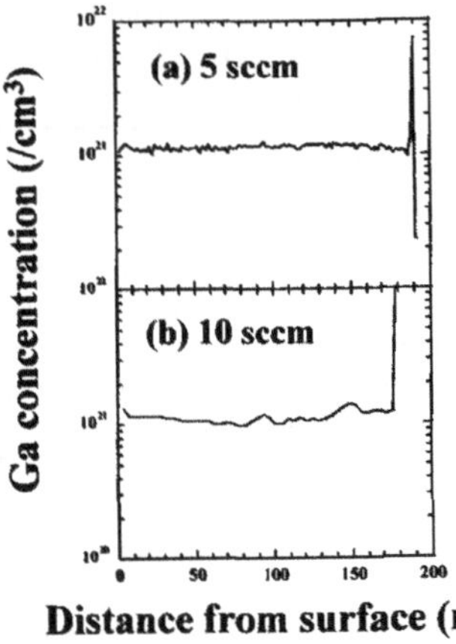

Figure 5. Ga depth profile in the GZO films prepared at O_2 gas flow rates of (a) 5 sccm and (b) 15 sccm.

ZnO with a wurtzite structure which belongs to the space group C_{6v} ($C6_3mc$) allows four sets of Raman-active phonons of symmetries A_1, E_1 and $2E_2$ [25], where A_1 and E_1 are polar and split into transverse optical (TO) and longitudinal optical (LO) modes with different frequencies. In this work, we concentrate on the upper E_2 phonon, located at 437 cm^{-1}. Generally, the nonpolar E_2 modes strongly appear in the unpolarized Raman spectra of bulk ZnO crystals under back-scattering geometry. Figure 6 shows unpolarized Raman spectra of GZO films with different O_2 gas flow rates [26]. It revealed vibrational modes at 575 and 630–660 cm^{-1} in addition to the host phonons of ZnO. These additional modes correspond to local vibrational modes associated with oxygen vacancy (V_O) [27] and Ga impurity (Ga_{Zn}) [28], respectively. We also found little changes in the Ga-impurity mode in various oxygen ambience. Note that, from the Raman mapping analysis on entire film surface, we confirmed no segregation of Ga_2O_3 for the GZO thin films prepared at higher O_2 gas flow rates than 10 sccm. This means that Ga concentration in GZO films depends little on O_2 gas flow rate. With increasing O_2 gas flow rate up to 10 sccm, the 575-cm^{-1} mode decreases in its intensity, indicating the reduced V_O concentration. Further increasing O_2 gas flow rate (>10 sccm), we find a substantial disorder apparent in host ZnO phonons and some additional modes. These results suggest that the oxygen-rich condition may cause the formation of compensating-defects, *n*-type killer, such as oxygen interstitials (O_i) and Zn vacancies (V_{Zn}).

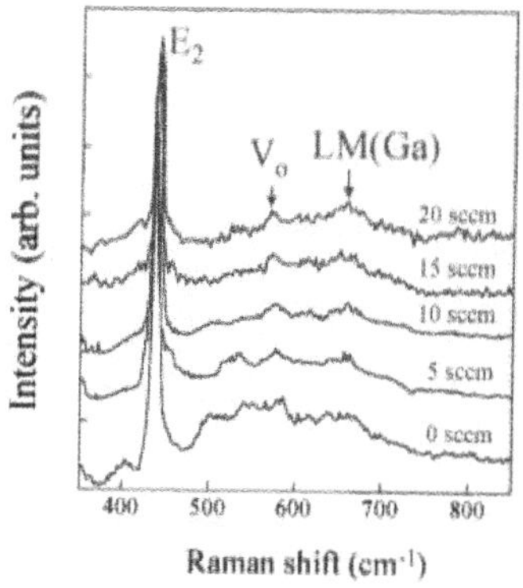

Figure 6. Unpolarized Raman spectra of GZO films prepared with different O_2 gas flow rates.

From those findings obtained by the XRD, SIMS, Raman measurements and theoretical calculations, we conclude that carrier concentration will exhibit an inflection point at O_2 gas flow rate of ~ 10 sccm: For GZO thin films prepared with increasing O_2 gas flow rate from 0 to 10 sccm, V_o concentration decreases with increasing O_2 gas flow rate, while the concentration of Ga_{Zn}, dominant defect, changes little; For GZO films deposited with further increasing O_2 gas flow rate, *n*-type killer will generate with increasing O_2 gas flow rate.

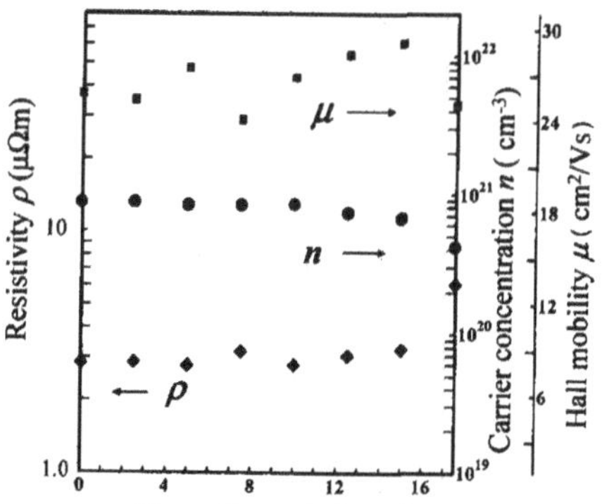

Figure 7. Resistivity, ρ, carrier concentration, n, and Hall mobility, μ, as a function of O_2 gas flow rate.

1.2 *Electrical properties*

All GZO thin films have *n*-type conductivity. The high conductivity of the films results mainly from a non-stoichiometry composition or intentionally doping. Carrier electrons in GZO films are supplied from donor sites associated with V_o and Ga_{Zn}.

We summarize the resistivity, carrier concentration and Hall mobility as a function of O_2 gas flow rate in Fig. 7. The lowest resistivity around 2.8×10^{-4} Ωcm with carrier concentration of 8.3×10^{20} cm^{-3} and Hall mobility of 27 cm^2/Vs was obtained at O_2 gas flow rate of 10 sccm. For comparison, Hirasawa and et al., reported that the lowest resistivity of GZO films (Ga content of about 4 at%) on a soda lime glass substrate at 350 □ prepared by DC-arc-discharge ionplating was 1.6×10^{-4} Ωcm [29]. Minami and Miyata reported that the lowest resistivity of GZO films with a film thickness of 319 nm on a OA-2 glass substrate at 350 □ deposited by the method mentioned above was 2.33×10^{-4} Ωcm [30].

From Figure 7, we confirmed that carrier concentration exhibits an inflection point at O_2 gas flow rate of 10 sccm, as discussed in previous section: while O_2 gas flow rate causes little influence on carrier concentration in GZO films prepared with O_2 gas flow rate from 0 to 10 sccm, carrier concentration substantially decreases with further increasing O_2 gas flow rate.

Note that oxygen-rich preparation condition gives rise to a reduction in both carrier concentration and Hall mobility for GZO films. As a result, we find a substantially increase in resistivity for the films. In such films, on the basis of the analysis of data obtained by Raman spectroscopy, the formation of *n*-type killer including O_i and/or V_{Zn} leads to the deterioration of the crystal quality.

1.3 *Optical properties*

Figures 8 show optical transmittance of GZO films. Note that the transmittance is almost more than 90 % for wavelength from 400 to 1000 nm for all films. The infrared reflectivity (wavelength longer than 1200 nm) is high for the corresponding films. For the wavelength longer than 1000 nm, the transmittance of GZO films decreases due to the plasma absorption especially for films with high carrier concentration prepared with O_2 gas flow rate from 0 to 10 sccm.

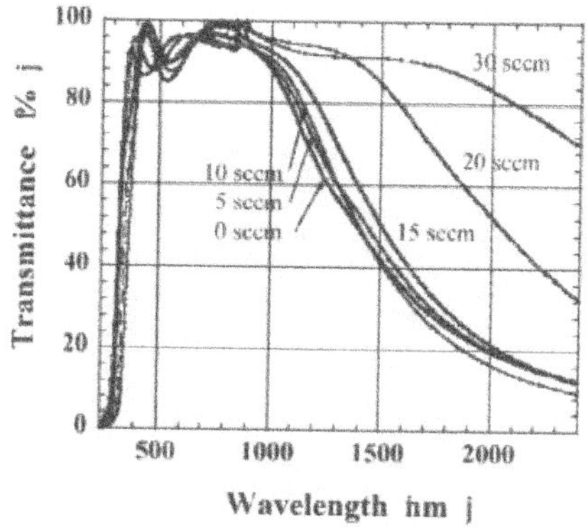

Figure 8. Optical transmittance for GZO films with different O_2 gas flow rate.

2. Large area GZO thin films

Very recently, we have succeeded in fabricating GZO films (Ga content of 3 wt%) on a large-area glass substrate with a size up to 1m × 1m prepared by RPD using two plasma guns with a traveling substrate technique.

2.1 *Electric properties*

Figure 9 shows a schematic diagram of RPD with two plasma guns. The distance between the two plasma guns was 700 mm. Figure 10 illustrates illustrates the SEM image (JSM-6700F, JEOL) of the cross-sectional view of GZO films prepared at oxygen (O_2) gas flow rate of 15 sccm. Dependences of the resisitivity, ρ, the carrier concentration, n, and the Hall mobility, μ, on O_2 gas flow rate is very similar to our previous reports [10-16]. The distribution of ρ, n and μ has been examined for GZO films grown at O_2 gas flow rate of 5 sccm, where the lowest resistivity of GZO films was obtained. The average of ρ, n and μ is $2.7 \times 10^{-4}\Omega$cm, 9×10^{20} cm^{-3} and 24 cm^2/Vs, respectively. The dispersion of ρ, n and μ along the 1m-width ofGZO film is within ± 5%, ± 8% and ± 7%, respectively. The distribution of n and μ clearly reflects the presence of two plasma guns locating at +/- 350 mm position from the center. The n value is somewhat high and μ is somewhat low at the gun positions. The plasma-gun-position related non-uniformity in n and μ cancels each other because ρ is inversely proportional to the product of n and μ. This yields the small non-uniformity in ρ (± 5 %) with the featureless in-plane distribution.

Figure 9. Schematic diagram of RPD with two plasma guns.

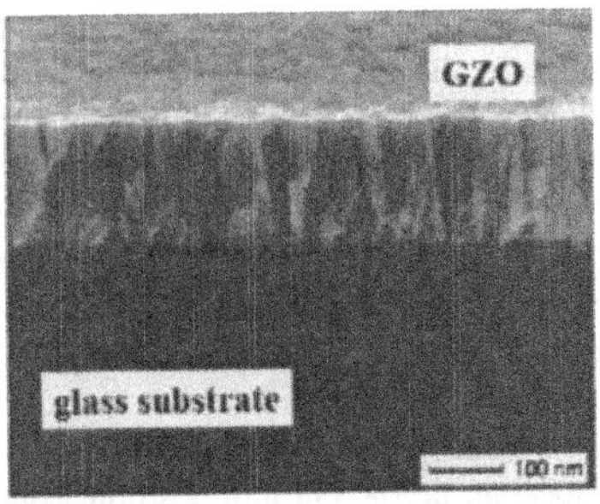

Figure 10. SEM image of the cross-sectional view of GZO films prepared at O_2 gas flow rate of 5 sccm by RPD with two plasma gun using traveling substrate mechanism (see Fog. 9).

2.2 Optical properties

Figures 11 shows optical transmittance (*T*) of GZO films, respectively. *T* is almost more than 90% for 400-1000 nm, and the value is comparable or slightly better than that for GZO films with small area (see Fig. 8)[14]. The blue-shift of the absorption edge with an increase in *n* is due to the Burstein-Moss effect. The decrease in *T* in the near-infrared region (wavelength longer than 1200 nm) is due to the plasma absorption. From Fig. 11, this is remarkable for GZO films with the high *n* value of GZO films with 7×10^{20} and 1×10^{21} cm^{-3}. From data obtained by reflectance spectra measurement, we confirmed that the infrared reflectivity (wavelength longer than 1200 nm) is high for the corresponding films.

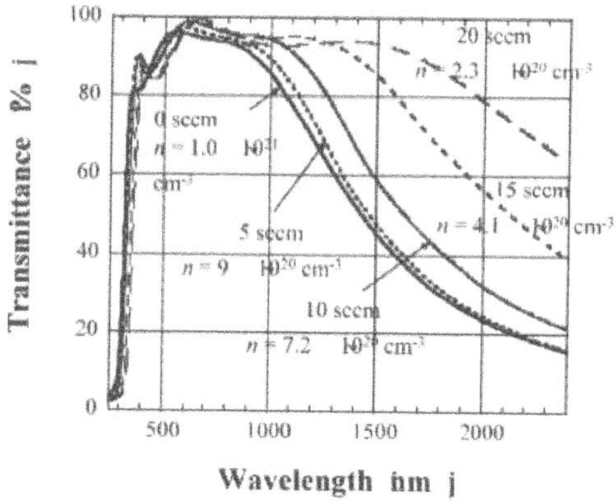

Figure 11. Optical transmittance for GZO films with different O_2 gas flow rate.

On the basis of the analysis of data concerning the in-plane transmittance distribution, we found that what determine the transmittance uniformity depends on wave region: Most important factor to be minimized the variation is the film thickness for short-wave length (< 500 nm) region and carrier concentration for the near-infrared wave-length region.

CONCLUSION

GZO-based transparent conducting thin films have been prepared by reactive plasma deposition. The substrate temperature and film thickness are almost 200 °C and 200 nm, respectively. We have succeeded in fabricating GZO films with low resistivity of 2.7×10^{-4} Ω cm and high transmittance more than 90 % in the visible range on a glass substrate with a size up to 1m × 1m by RPD using two plasma guns using a traveling substrate technique. The deviation of thickness, resistivity, carrier concentration and Hall mobility are limited within ± 5 %, ± 5 %, ± 8 % and ± 7 %, respectively.

In this work, the main process parameter for GZO films is oxygen partial pressure (O_2 gas flow rates). We established that the behavior of electrical and optical properties thorough variation of the process parameters depends little on the film size and substrate mechanism, traveling or non-traveling type. The advantageous features of cost-effective GZO films, being easy to fabricate on a production scale, make them very attractive potential for use in transparent conductive electrode in FDP and thin film solar cells.

ACKNOWLEDGEMENT

The authors acknowledges the financial support of the collaboration of Regional Entities for the Advancement of Technological Excellence of Japan Science and Technology Agency.

REFERENCES

[1]H. Cao, J. Y. Wu, H. C. Ong, J. Y. Dai, and R. P. H. Chang, "Second harmonic generation in laser ablated zinc oxide thin films", Appl. Phys. Lett., **73**, 572-74 (1998).
[2]S. Cho, J. Ma, Y. Kim, Y. Sun, G. K. L. Wong and J. B. Ketterson, "Photoluminescence and ultraviolet lasing of polycrystalline ZnO thin films prepared by the oxidation of the metallic Zn", Appl. Phys. Lett., **75**, 2761-63 (1999).
[3]M. H. Huang, S. Mao, H. Feick, H. Yan, Y. Wu, H. Kind, E. Weber, R. Russo, and P. Yang, "Room-Temperature Ultraviolet Nanowire Nanolasers", Science, **292**, 1897-99 (2001).
[4]A. E. Delahoy, M. Cheny, "Deposition Schemes for Low Cost Transparent Conductors for Photovoltaics", Mater. Res. Soc. Symp. Proc. **467**, 426-78 (1996).
[5]S. Major, Satyendra Kumar, M. Bhatnagar, and K. L. Chopra, "Effect of hydrogen plasma treatment on transparent conducting oxide", Appl. Phys. Lett., **49**, 394-96 (1986).
[6]T. Minami, "New n-type Transparent Conducting Oxides", MRS Bulletin, **25**, 38-43 (2000).
[7] K. Nakahara, K. Tamura, M. Sakai, D. Nakagawa, N. Ito, M. Sonobe, H. Takasu, H. Tampo, P. Fons, K. Matsubara, K. Iwata, A. Yamada and S. Niki, "Improved External Efficiency InGaN-Based Light-Emitting Diodes with Transparent Conductive Ga-Doped ZnO as p-Electrodes", Jpn. J. App. Phys. 43, L180-82 (2004).
[8]S. Shirakata, T. Sakemi, K. Awai and T. Yamamoto, " Electrical and optical properties of large area Ga-doped ZnO thin films prepared by reactive plasma deposition", European Materials Research Society 2005 Spring Meeting, Strasbourg, France, May 31 - June 3, 2005.
[9]M. Tanaka, U.S. Patent 5677012 (1997).
[10] T. Sakemi, S. Shirakata, K. Iwata, K. Matsubara, H. Tampo, P. Fons, S. Niki, K. Awai and T. Yamamoto,"High-Quality Transparent Conducting Oxide Films Deposited by a Novel Ion Plating Technique", MRS Proceedings Volume 763, Symposium B Compound Semiconductor Photovoltaics, Editors: Rommel Noufi, William N. Shafarman, David Cahen, Lars Stolt
[11]S. Shirakata, T. Sakemi, K. Awai, T. Yamamoto, "Optical and electrical properties of ZnO films prepared by URT-IP method", Thin Solid Films, **445**, 278-83 (2003).
[12]K. Iwata, T. Sakemi, A. Yamada, P. Fons, K. Awai, T. Yamamoto, K. Matsubara, H. Tampo, K.Sakurai, S. Ishizuka, S. Niki, "Growth and electrical properties of ZnO thin films deposited by novel ion plating method", Thin Solid Films, **445**, 274-77 (2003).
[13]K. Iwata, T. Sakemi, A. Yamada, P. Fons, K. Awai, T. Yamamoto, K. Matsubara, H. Tampo, K.Sakurai, S. Ishizuka, S. Niki, "Doping properties of ZnO thin films for photovoltaic devices grown by URT-IP (ion plating) method", Thin Solid Films, **451-452**, 219-23 (2004).
[14]S. Shirakata, T. Sakemi, K. Awai, T. Yamamoto, "Optical and electrical properties of URT-IP ZnO thin films for photovoltaic devices", Thin Solid Films, **451-452**, 212-18 (2004).
[15]T. Yamamoto, T. Sakemi, K. Awai, S. Shirakata, "Dependence of carrier concentrations on oxygen pressure for Ga-doped ZnO prepared by ion plating method", Thin Solid Films, **451-452**, 439-42 (2004).
[16]K. Iwata, T. Sakemi, A. Yamada, P. Fons, K. Awai, T. Yamamoto, S. Shirakata, K. Matsubara, H. Tampo, K. Sakurai, S. Ishizuka, S. Niki," Improvement of ZnO TCO film growth for photovoltaic devices by reactive plasma deposition (RPD)", Thin Solid Films, 480-481, 199-203 (2005).

[17] S. Shirakata, T. Sakemi, K. Awai, T. Yamamoto, 2005 Europe-MRS Spring Meeting, May 31-June 3, 2005, Strasbourg, France.
[18] P. Hohenberg, W. Kohn, Phys. Rev." Inhomogeneous Electron Gas", B **136**, 864-71 (1964).
[19] J. P. Perdew, K. Burke and Y. Wang, "Generalized gradient approximation for the exchange-correlation hole of a many-electron system", Phys. Rev. B **54**, 16533-39 (1996).
[20] G. Kresse and J. Hafner, "*Ab initio* molecular dynamics for liquid metals", Phys. Rev. B **47**, R558-61 (1993).
[21] P. E. Blöchl, "Projector augmented-wave method", Phys. Rev. B 50, 17953–79 (1994).
[22] G. Kresse and D. Joubert, "From ultrasoft pseudopotentials to the projector augmented-wave method", Phys. Rev. B **59**, 1758–75 (1999).
[23] G. K. Williamson and W. Hall, "X-ray line broadening from filed aluminium and wolfram", Acta Letallurgica **1**, 22-31 (1953).
[24] J. Albertsson, S. C. Abrahams, and A. Kvick, "Atomic displacement, anharmonic thermal vibration, expansivity and pyroelectric coefficient thermal dependences in ZnO", Acta Crystallographica **45**, 34-40 (1989).
[25] C.A. Arguello, D.L. Rousseau, S.P.S. Porto, "First-Order Raman Effect in Wurtzite-Type Crystals", Phys. Rev. **181**, 1351-63 (1969).
[26] M. Osada, T. Sakemi and T. Yamamoto, to be published in Thin Solid Films.
[27] J.N. Zeng, J.K. Low, Z.M. Ren, Thomas Liew, Y.F. Lu, "Effect of deposition conditions on optical and electrical properties of ZnO films prepared by pulsed laser deposition", Appl. Suf. Sci.**197-198** (2002) 362-67.
[28] C. Bundesmann, N. Ashkenov, M. Schubert, D. Spemann, T. Butz, E.M. Kaidashev, M. Lorenz, M. Grundmann, "Raman scattering in ZnO thin films doped with Fe, Sb, Al, Ga, and Li", Appl. Phys. Lett. **83**, 1974-76 (2003).
[29] H. Hirasawa, M. Yoshida, S. Nakamura, Y. Suzuki, S. Okada and K. Kondo, "ZnO:Ga conducting-films grown by DC-arc-discharge ionplating", Sol. Energy Mater. Sol. Cells **67**, 231-36 (2001).
[30] T. Minami and T. Miyata, "Transparent Conducting ZnO Thin Films Prepared by Vacuum Arc Plasma Evaporation", J. Vac. Soc. Jpn. **47**, 734-41 (2004).

Electroceramic Materials and Applications
Edited by R. W. Schwartz

COLORATION AND BLEACHING PHENOMENA OF AMORPHOUS WO_3 FILMS DUE TO THE ELECTROCHEMICAL INSERTION OF DIVALENT CATIONS

Y. Domori, T. Nanba, Y. Miura
Department of Environmental Chemistry and Materials,
Faculty of Environmental Science and Technology, Okayama University
3-1-1, Tsushima-Naka, Okayama-shi 700-8530, Japan

S. Sakida
Environmental Management and Safety Section,
Health and Environment Center, Okayama University
3-1-1, Tsushima-Naka, Okayama-shi 700-8530, Japan

ABSTRACT

It was confirmed that the divalent cations, Zn^{2+} and Cu^{2+} were electrochemically inserted into amorphous WO_3 films prepared by r.f.-magnetron sputtering method. Coloration of the films was also observed, and the films were, however, bleached gradually. The bleaching mechanism was discussed based on the ab-initio molecular orbital calculations, in which Li^+ or Zn^{2+} ions were introduced into the WO_3 frameworks. In the case of Zn insertion, electrons were trapped in a non-bonding Zn 4s orbital just below the conduction band. It was finally concluded that in the films with distorted WO_3 frameworks, Zn 4s orbital was mingled in the conduction band and electrons were free to be transferred between W^{6+} and W^{5+}. As the structural relaxation progressed, electrons were gradually trapped in Zn 4s orbital, and the films were finally bleached.

INTRODUCTION

Amorphous WO_3 thin films have been extensively studied because of their reversible electrochromic (EC) property, being applied to EC display and smart windows. Electrochromism of WO_3 is due to the formation of tungsten bronze according to the following reaction.[1]

$$WO_3 + xM^+ + xe^- \rightarrow M_xWO_3 \quad (M = H, Li, Na, \ldots etc.) \tag{1}$$

Where small-polaron[2] is formed by insertion of electron, and a part of W^{6+} changes into W^{5+}. A wide absorption peak ranging from visible to infrared region is caused by the transfer of an electron between W^{5+} and W^{6+}. In the previous researches, monovalent cations such as H^+ and Li^+ have been inserted to the films. Recently, it was found that the divalent cations, Zn^{2+} and Cu^{2+} were also inserted by means of the conventional electrochemical method.[3] After the coloration, gradual bleaching was observed in the case of the insertion of the divalent cations, whereas such the bleaching has not been observed in the monovalent cation insertion. According to the theoretical calculations, Nanba et al.[4] showed that the electron insertion brought about the distortion of WO_6 unit, which was consistent with the small-polaron absorption model.[2] In the case of insertion of divalent cations, two electrons are simultaneously inserted per one divalent cation. Furthermore, the divalent cations, Zn^{2+} and Cu^{2+} are quite different in the chemical bonding character with the monovalent cation, H^+ and Li^+. Then, in the present study, the coloration and bleaching phenomena due to the insertion of divalent cations were investigated under the various conditions, and the bleaching mechanism was discussed based on the ab-initio molecular orbital calculations.

EXPERIMENTAL

WO_3 films were prepared by an r.f. magnetron sputtering method on the indium tin oxide (ITO) coated glass substrates. According to Nanba et al.,[5] the hexagonal-WO_3 like structure is formed in the films prepared under a low O_2 partial pressure ($Ar/O_2 = 4/1$), and the tetragonal-WO_3 like framework is produced at a high O_2 pressure ($Ar/O_2 = 1/1$) (Fig.1). The electrochemical insertions of Li^+ or Zn^{2+} ions were performed using a conventional electrochemical cell; 1M $LiClO_4$/propylene carbonate solution (PC) or $ZnCl_2$/PC was used as the electrolyte. The amount of ions inserted into the films was estimated by the coulomb meter. X-ray photoelectron spectroscopic (XPS) measurements were carried out to obtain valence band spectra. Transmittance spectra were measured for the colored and bleached films.

(a) Tetragonal-WO_3

P4/nmm(No.129)
a_0=0.527nm
c_0=0.392nm

(2b)-O(1) = 0.27nm
(2b)-O(2) =0.26nm

(b) Hexagonal-WO_3
P63/mcm(No.193)
a_0=0.73nm
c_0=0.78nm

(2b)-O(1) = 0.33nm
(2b)-O(2) =0.37nm
(4d)-O(1) = 0.25nm
(4d)-O(2) =0.21nm

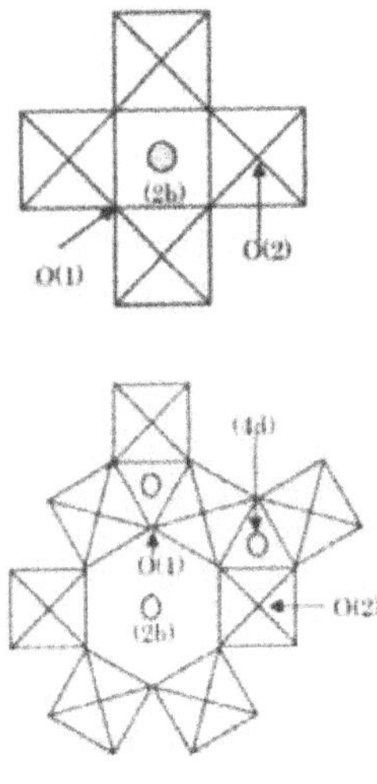

Fig. 1. Framework structures of (a) tetragonal- and (b) hexagonal-WO_3 crystals and the vacant sites for the cation insertion.[4]

The electronic states of WO_6 unit were examined by using the discrete variational $X\alpha$ (DV-$X\alpha$) molecular orbital (MO) method.[6] The atomic orbitals used in the MO calculations were W 1s-6s, O 1s-2p, Li 1s-2p, and Zn 1s-4p. Cluster models were constructed according to the previous study,[4] which were embedded in Madelung potentials generated by point charges outside the clusters to reduce the termination effects. In the previous study, only electrons were introduced to the vacant W 5d orbital, and in the present study, however, Li^+ and Zn^{2+} cations were also introduced to the vacant sites; 2b site in tetragonal-WO_3 and 4d site in hexagonal-WO_3 (Fig.1).

RESULTS AND DISCUSSION

Coloration and bleaching phenomena in WO_3 films

Films were amorphous even after the Zn insertion, which was confirmed by X-ray diffraction measurements. Fig. 2 shows the valence band XPS spectra.[7] Zn 3d peak at 10 eV increases in intensity with increasing the amount of Zn ions inserted to the films. Therefore, it is confirmed that Zn ions are successfully inserted. A small peak appears commonly at 0 eV in the Zn-inserted films, which is identified as the reduction of W ions; $W^{6+}(5d^0) \rightarrow W^{5+}(5d^1)$.

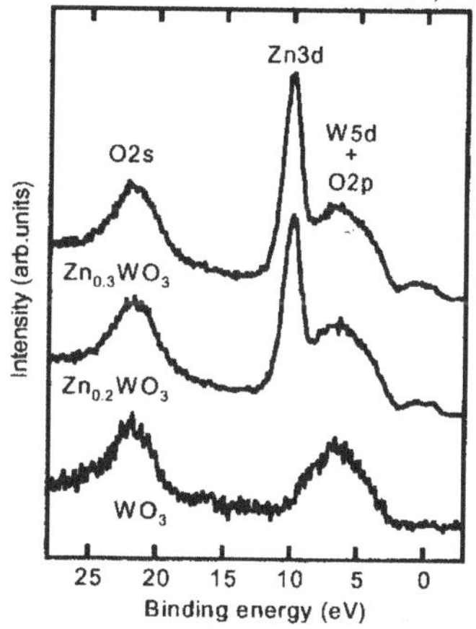

Fig. 2. Valence band spectra of the WO_3 films before and after Zn insertion.[7]

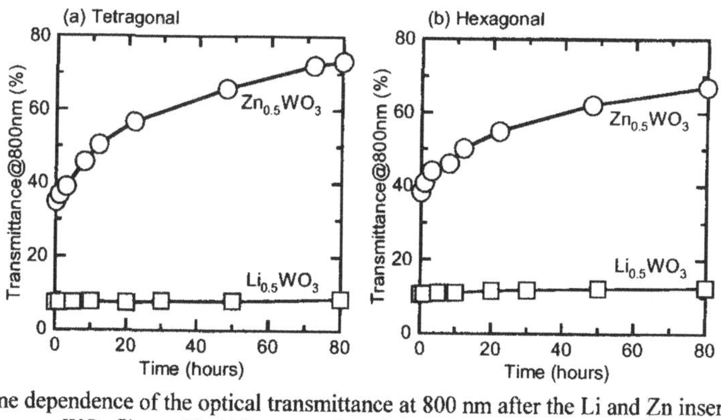

Fig. 3. Time dependence of the optical transmittance at 800 nm after the Li and Zn insertions into WO_3 films with (a) hexagonal- and (b) tetragonal-frameworks.[7]

WO_3 films were colored by the insertion of Zn ions as well as Li ions, in the case of Zn insertion, however, the films were gradually bleached. Fig. 3 shows the time dependence of the transmittance at 800 nm after the ionic insertions.[7] In the case of Li insertion, the transmittance is kept at almost constant. In the Zn insertion, however, the transmittance gradually increases and is restored to the initial state after 80 h.

Consideration on the bleaching mechanism

The ionic radii of Li^+ and Zn^{2+} are very close (Li^+ = 0.090 nm, Zn^{2+} = 0.088 in 6-coordination). The difference between Li and Zn ions is not only the charge but also the chemical bonding character. It is therefore supposed that the difference in chemical bonding character is the reason for the bleaching phenomena in the WO_3 films. Then, molecular orbital calculations were performed to examine the bonding character and electronic states in the WO_3 films. Fig. 4 shows the bond overlap population (BOP) of the W-O bonds in the central WO_6 units in the cluster models. BOP was estimated from the Mulliken population analysis of the results of MO calculations. It indicates the electronic population shared between two atoms, and is used as a measure of bond covalency and bond strength. In the MO calculation for the cluster model constructed from hexagonal-WO_3, when one Li or Zn ion was introduced to the cluster, the MO calculation did not reach to a convergence. The results shown in Fig, 4 were obtained by introducing two Li or Zn ions into the cluster.

In Fig. 4, "none" means the results without introducing the ions into $(WO_6)_7$ clusters, and "7e$^-$" indicates the results after introducing seven electrons into the clusters which simulates the reduction of all W ions in the clusters from 6+ to 5+ states. These results were given in Ref. 4. "O1-6" means oxygen in the central WO_6 units in the cluster models. After the insertions of ions and electrons, BOP decreases in most W-O bonds. As for Li and Zn insertions, larger change in BOP is commonly observed after Zn insertion.

Before the MO calculations, it was expected that Zn-O bonds showed much higher BOP than Li-O, because Zn has much higher electronegativity than Li (Zn = 1.65, Li = 0.98 in Pauling scale). Contrary to the expectation, Zn-O bonds in the WO_3 frameworks showed smaller BOP than Li-O bonds (Zn-O = 0.021, Li-O = 0.036 in tetragonal-WO_3, Zn-O = 0.130, Li-O = 0.226 in hexagonal-WO_3). This result is quite remarkable.

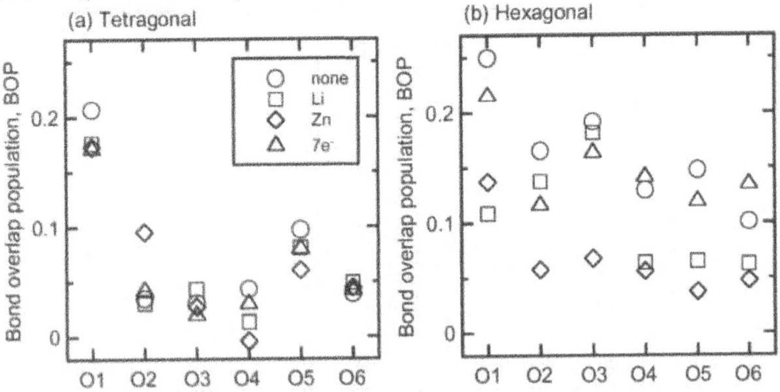

Fig. 4. Effect of cation and electron insertion on the bond overlap population of W-O bonds in the central WO_6 units in the cluster models constructed from (a) tetragonal- and (b) hexagonal-WO_3 frameworks.

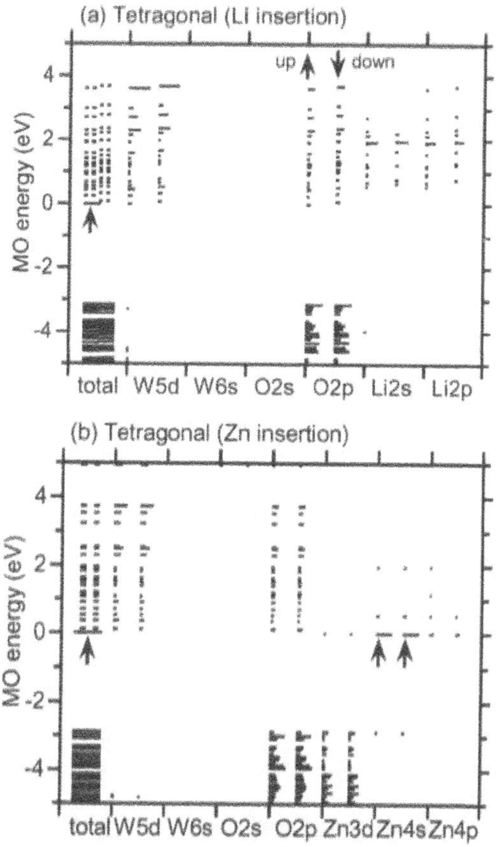

Fig. 5. Energy level diagrams for the cluster models constructed from the tetragonal-WO_3 framework associated with (a) Li and (b) Zn ions. On the extreme left, the MO levels occupied by electrons are drawn by continuous lines, and the vacant levels are by dotted lines. The contributions of atomic orbitals in each MO level are also shown.

Fig. 5 shows the energy level diagrams, in which the contributions of up spin and down spin are separately shown. As shown in Fig. 5a, Li insertion into the tetragonal-WO_3 framework introduces an electron into the bottom of the conduction band (0 eV), which is formed by W 5d and O 2p orbitals, and is in a conventional notation, anti-bonding t_{2g}^* level. Moreover, an energy splitting due to the spin polarization is also reproduced. It is noted that there is no contribution of Li in the t_{2g}^* level. As for the Zn insertion (Fig. 5b), the contribution of Zn is clearly observed in the MO level at the bottom of the conduction band. The level is located just below t_{2g}^* level. It is noted that there is no contribution of O 2p in this level, that is, this level is non-bonding Zn4s orbital. It is therefore indicated that due to the non-bonding Zn 4s orbital, BOP of Zn-O bonds in WO_3 frameworks is reduced. Moreover, it is identified as an "electron-trapped state." In such the state, electrons are not transferred to the surrounding W^{6+} ions through oxide ions, and the electron transfer between W^{6+} and W^{5+} must be restricted. The electronic states reproduced in Fig. 5b are the final states, that is, the bleached states. It may take a long time for Zn ions to settle into the stable sites because it must be accompanied by the relaxation of WO_3 frameworks and the migration of Zn ions within the WO_3 frameworks.

CONCLUSION

Coloration of WO_3 films was confirmed due to the electrochemical insertion of the divalent cations. But the films were gradually bleached only in the insertion of the divalent cations. The bleaching mechanism was theoretically interpreted. The molecular orbital calculations were performed by introducing Li and Zn ions into the WO_3 frameworks. In the case of Zn insertion, it was reproduced that electrons were trapped in a non-bonding Zn 4s orbital. The trapped level was located just below the conduction band. The following mechanism was finally proposed; if the WO_3 frameworks were distorted, electrons were transferred from Zn to W ions via neighboring O ions. As the structural relaxation progressed, electrons formerly transferred between W^{6+} and W^{5+} were gradually trapped in Zn 4s orbital, and the films were finally bleached.

REFERENCES

[1]B. W. Faughnan et al., *RCA. Rev.*, **36,** 177 (1975).

[2]O. F. Schirmer et al., *J. Electrochem. Soc.*, **124,** 749 (1977).

[3]M. Kita, "Electrochemical intercalation of Zn^{2+} ion into amorphous WO_3 thin films," Ms. Thesis, Okayama University, (2002).

[4]T. Nanba, M. Ishikawa, Y. Sakai, and Y. Miura, "Changes in atomic and electronic structure of amorphous WO_3 films due to electrochemical ion insertion," *Thin Solid Films*, **445**, 175-181 (2003).

[5]T. Nanba, T. Takahashi, S. Takano, J. Takada, A. Osaka, Y. Miura, T. Kudo, I. Yasui "X-ray diffraction study on the framework structure of amorphous tungsten trioxide films," *J. Ceram. Soc. Jpn.*, **103**, 222-9 (1995).

[6]H. Adachi, M. Tsukada, and C. Satoko, "Discrete Variational Xα Cluster Calculations. I. Application to Metal Clusters," *J. Phys. Soc. Japan*, **45,** 875-83 (1978).

[7]S. Umeda, "Effect on electrochromic property of electrochemical intercalation of divalent cations into amorphous WO_3 thin films," Ms. Thesis, Okayama University, (2004).

Electroceramic Materials and Applications
Edited by R. W. Schwartz

PREPARATION OF Z-TYPE HEXAGONAL FERRITE, Co_2Z, USING ALKALINE HALIDE METHOD

Akira Sasaki, Makoto Nakanishi, Tatsuo Fujii, and Jun Takada

Department of Material Chemistry, Graduate School of Natural Science and Technology, Okayama University
Tsushima-naka 3-1-1
Okayama 700-8530, Japan

ABSTRACT

In this study, low temperature preparation conditions for Co_2Z ($3BaO{\cdot}2CoO{\cdot}12Fe_2O_3$) particles were investigated by using alkaline halide method. The obtained particles were characterized by using X-ray diffraction and scanning electron microscopy techniques. Many factors such as the metal compositions, heating temperature, heating and cooling rates, and amount of flux reagents were changed for the preparation. The X-ray diffraction analysis indicated that (1) the metal compositions, (2) flux reagent, and (3) heating rate were important to obtain Co_2Z at low temperature using the alkaline halide method. Nearly monophasic Co_2Z was formed under the following conditions: metal compositional ratio of Ba : Co: Fe = 3.0 : 1.5 : 24 with 10wt% $BaCl_2$ flux and rapid heating to 1200 °C for 6 hours. Co_2Z powders mixed with an EPDM rubber were pressed to a sheet form. The electromagnetic properties of rubber sheets were examined to apply to use as an electromagnetic wave absorber.

INTRODUCTION

Recently, applied technologies of electromagnetic wave has been used in various field, for example wireless LAN, intelligent transport system (ITS), mobile-phone and so on, with development of electronics, information and communication technologies. The electromagnetic interferences (EMI) caused by mobile-phone and other electronic devices become one of serious problems. A solution for suppressing EMI is to develop electromagnetic wave absorbers working at microwave frequency. The Co_2Z ($Ba_3Co_2Fe_{24}O_{41}$) hexagonal ferrite is a candidate material for suppressing EMI in GHz range. This ferrite shows high permeability in the frequency range of 1-2 GHz [1], beyond the Snoek's limit for spinel-type ferrite [2]. A rubber ferrite is a composite type absorber widely used inside the electronic instruments. It is manufactured from a mixture of the ferrite powder and polymer by rolling into a sheet form.

A lot of methods have been proposed for preparing the hexagonal ferrite, such as solid state reaction [3], sol-gel method [4], molten salt synthesis [5], and coprecipitation method [6]. Among these methods, flux methods such as molten salt synthesis are most suitable for preparing planar particles. The planar shape of the particles is expected to promote a *c*-plane orientation in the rubber matrix to increase the permeability of the rubber ferrite.

In this study, alkaline halide method was used for preparing planar Co_2Z powders. In this method, alkaline halides are expected to play roles as flux and to promote crystal growth of hexagonal plate, as reported in the case of M-type ($BaFe_{12}O_{19}$) ferrite [7]. After the reaction, it is easy to remove alkaline halides from the synthesized ferrite powders by treating with an acid. The optimization of preparation conditions and characterization of electromagnetic properties of rubber ferrite sheet were investigated.

EXPERIMENTAL PROCEDURES

Sample preparations

The Co_2Z was prepared from reagent-grade $BaCO_3$, CoO and Fe_2O_3 and alkaline halide NaCl or $BaCl_2$ was added as a flux. The powders were mixed, pressed into cylindrical shape of $\phi 12$ mm and heated in alumina crucible. The heated samples were ground and treated with 1N HCl for 1 h under ultrasonic irradiation. The powders were filtered and dried in an oven for 1 h. The obtained powders were characterized by using X-ray diffraction (XRD, Rigaku RINT-2500) with Cu Kα radiation and field emission scanning electron microscopy (FESEM, Hitachi S-4300).

For the experimental method stated above, we discussed the effects of the following preparation conditions for Co_2Z; the heating temperature (1100-1300 °C), the kind of alkaline halides ($BaCl_2$, NaCl (m.p. 962, 801 °C)), the amount of alkaline halides (0-200wt%), the metal compositions (Ba:Co:Fe), the heating rate (constant (10 °C/min), rapidly) and the cooling rate (rapidly, slowly).

Characterization of electromagnetic properties

300, 800 and 1200 PHR (Per Hundred parts of Resin by weight) of the planar Co_2Z powders were mixed with an EPDM rubber and rolled to a sheet form. The rubber ferrite sheets were cut into the toroidal shape with outer and inner diameters of 7 mm and 3 mm, respectively. The reflection and transmission coefficients of cut sheets were measured by using network analyzer (Hewlett Packard HP8720D) in the frequency range from 100 MHz to 10 GHz. Relative complex permeability ($\mu_r = \mu_r' - j\mu_r''$) and permittivity ($\varepsilon_r = \varepsilon_r' - j\varepsilon_r''$) of the sheets were calculated from these coefficients by Nicolson-Ross-Weir procedure [8-9].

RESULTS AND DISCUSSIONS

Preparation conditions of planar Co_2Z

First the effects of heating temperature were examined in preliminary experiments. When the heating temperature was changed from 1100 °C to 1300 °C, sample was melted and adhered to alumina crucible above 1250 °C. So the heating temperature was fixed at 1200 °C in this study.

The X-ray diffraction patterns of the samples heated at 1200 °C for 18 h with the different flux reagents, NaCl or $BaCl_2$, are shown in Fig.1. The Z-type ferrite was observed as a main phase in both samples, but many unknown peaks appeared in the case of NaCl. It was revealed it was better to use $BaCl_2$ as a flux reagent to prepare nearly monophasic Co_2Z.

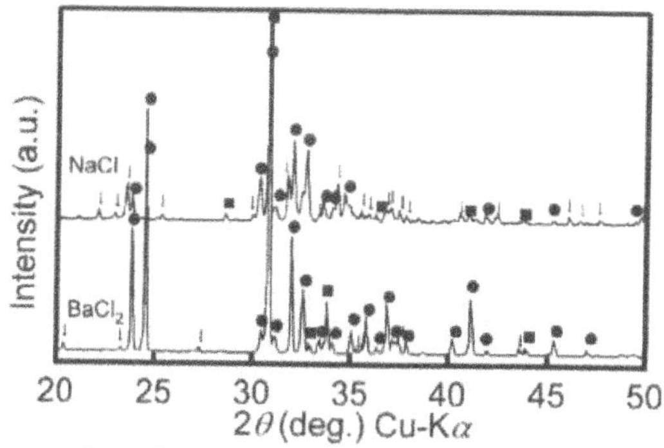

Fig.1 XRD patterns of samples heated at 1200 °C for 18 h with 200wt% of flux reagent.
●:$Co_2Z(Ba_3Co_2Fe_{24}O_{41})$, ▪:$Co_2Y(Ba_2Co_2Fe_{12}O_{22})$, ↓:Unknown

Then the effects of amount of flux reagent on formation of Z-type ferrite were examined. The XRD patterns of samples with different amount of $BaCl_2$ are shown in Fig.2. Sample with increasing the amount of $BaCl_2$ approached monophasic Co_2Z.

Even in the case of the sample with 200wt% of $BaCl_2$, it was found that Co_2Y coexisted with Co_2Z and the monophasic Co_2Z was not obtained. So, the metal compositions were changed around the ideal composition of Ba:Co:Fe=3:2:24, especially toward less Co content, according to the phase diagram of this system. From the results of the XRD measurements for the samples with various compositions, it was revealed that an optimum composition for this method was the metal composition ratio of Ba:Co:Fe=3.0:1.5:24. This composition was used to prepare the samples in discussions hereafter.

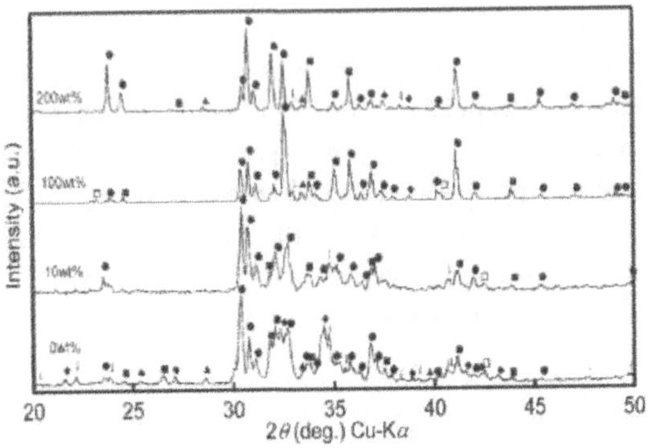

Fig.2 XRD patterns of samples heated at 1200 °C for 6 h with different amount of $BaCl_2$.
●:$Co_2Z(Ba_3Co_2Fe_{24}O_{41})$, •:$Co_2Y(Ba_2Co_2Fe_{12}O_{22})$,♦:$Co_2W(BaCo_2Fe_{16}O_{27})$,
□:$M(BaFe_{12}O_{19})$, ▲:$BaFe_2O_4$, ↓:Unknown

Next, the effects of the heating and cooling rates on the formation of Co_2Z were examined. Fig.3 shows the XRD patterns of the samples heated at different heating rates. Rapid heating means the process in which a sample is inserted into a furnace heated to objective temperature. When the sample was heated rapidly, the diffraction peaks from Co_2Y and unknown phases were decreased and almost monophasic Co_2Z was obtained. Rapid heating was effective to suppress the formation of the coexisting phases. However, the XRD patterns did not change by changing

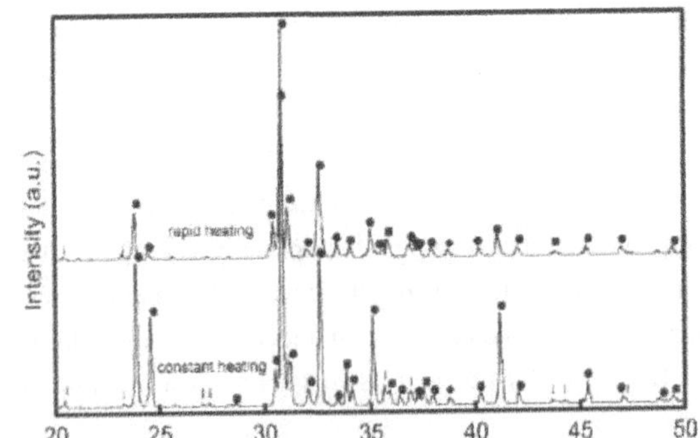

Fig.3 XRD patterns of samples heated at different heating rates.
●:$Co_2Z(Ba_3Co_2Fe_{24}O_{41})$, •:$Co_2Y(Ba_2Co_2Fe_{12}O_{22})$,♦:$Co_2W(BaCo_2Fe_{16}O_{27})$, ↓:Unknown

the cooling rates. So it was found that the cooling rates did not affect on the ratio of the constituent phases.

By considering the preparation conditions as mentioned above, almost monophasic Co_2Z could be obtained but the amount of $BaCl_2$ was very large, 200wt%. To reduce the amount of the flux reagent the samples were heated repeatedly at the same conditions. Fig.4 shows the XRD patterns of the samples with various amount of $BaCl_2$ heated twice at 1200 °C for 6 h. In the case of heating repeatedly, only 10wt% of $BaCl_2$ was enough to obtain monophasic Co_2Z. On the other hand, the samples with more than 50wt% of $BaCl_2$ included Co_2Y except for Co_2Z. It is revealed that the optimum amount of the flux reagent was affected by the repetition of heating. The optimum preparation conditions are summarized in Table 1.

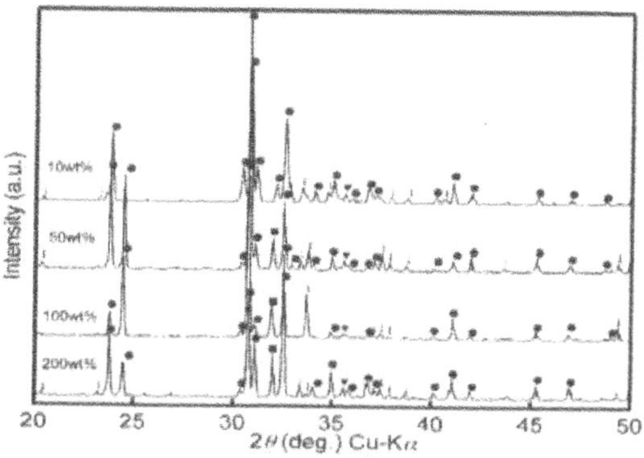

Fig.4 XRD patterns of samples with various amount of $BaCl_2$ heated twice at 1200 °C for 6 h.
●:$Co_2Z(Ba_3Co_2Fe_{24}O_{41})$, •:$Co_2Y(Ba_2Co_2Fe_{12}O_{22})$, ▼:$CoFe_2O_4$, ↓:Unknown

Table 1 Optimum preparation conditions for Co_2Z

metal compositions (Ba:Co:Fe)	flux		heating condition		
	reagent	amount	temperature	time	heating rate
3.0:1.5:24	$BaCl_2$	10wt%	1200 °C	6 h	rapid heating

Microstructures of Co_2Z powder

The particle size of the samples became larger with increasing the amount of $BaCl_2$, but those of the samples with more than 100wt% of $BaCl_2$, (c) and (d), did not change any more, as shown in Fig.5. It was reported that the planar Co_2Z particles with the size of 20 μm could be prepared by using Li_2SO_4-Na_2SO_4 flux [5]. The particles obtained in this study were as large as that in the literature.

Fig.5 SEM images of samples with different amount of $BaCl_2$.
These samples are same as shown in Fig.4 and the amount of $BaCl_2$ for sample (a), (b), (c) and (d) are 10, 50, 100 and 200wt%, respectively. The image (e) is enlargement of (d).

Electromagnetic properties of rubber ferrite sheets

Fig.6 shows the electromagnetic properties of the rubber ferrite sheets characterized by using network analyzer. The relative permeability of the rubber ferrite sheets was higher than that of bulk Co_2Z in high frequency range.

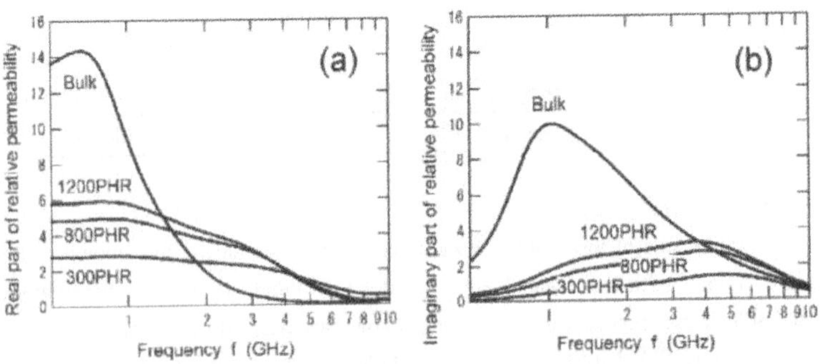

Fig.6 Real (a) and imaginary (b) part of relative permeability of rubber ferrite sheets.

Both a real and an imaginary parts of the permeability increased with increasing the volume fraction of the ferrite in the rubber sheets. Maximum of the imaginary part shifted to higher frequency with decreasing the volume fraction of the ferrite. This result suggests that a magnetic resonance of the samples with lower volume fraction occurred at higher frequency by the effect of a diamagnetic field [10]. From the planar Co_2Z powder, the rubber ferrite sheets with relatively high permeability were obtained.

CONCLUSIONS

The planar powders of almost monophasic Co_2Z were obtained through alkaline halide method by optimizing the preparation conditions. The amount of $BaCl_2$ was closely associated with grain growth of hexagonal planar particles. From the planar Co_2Z powder, the rubber ferrite sheets with relatively high permeability were obtained.

REFERENCES

[1] J. Smit, H. P. J. Wiji, *Ferrites*, Philips Technical Library, 177 (1959).

[2]J. L. Snoek, "Dispersion and Absorption in Magnetic Ferrites at Frequencies above one Megacycle" *Physica*, **14**, 207-217 (1948).

[3]X. Wang, L. Li, S. Su, Z. Yue, "Electromagnetic properties of low-temperature-sintered $Ba_3Co_{2-x}Zn_xFe_{24}O_{41}$ ferrites prepared by solid state reaction method" *J. Magn. Magn. Mater.*, **280**, 10-13 (2004).

[4]R. C. Pullar, A. K. Bhattacharya, "The synthesis and characterization of the hexagonal Z ferrite, $Sr_3Co_2Fe_{24}O_{41}$, from a sol-gel precursor" *Materials Research Bulletin*, **36**, 1531-1538 (2001).

[5]H. Hsiang, C. Chang, "Molten salt synthesis and magnetic properties of $3BaO{\cdot}2CoO{\cdot}12Fe_2O_3$ powder" *J. Magn. Magn. Mater.*, **278**, 218-222 (2004).

[6]S. Sugimoto, K. Haga, T. Kagotani, K. Inomata, "Microwave absorption properties of Ba M-type ferrite prepared by a modified coprecipitation method" *J. Magn. Magn. Mater.*, **290-291**, 1188-1191 (2005).

[7]Jpn. Patent, JP1128331.

[8]A. M. Nicolson and G. Ross, "Measurement of intrinsic properties of materials by time domain techniques" *IEEE Trans. Instrum. Meas.*, **19**, 377-382 (1970).

[9]W. B. Weir, "Automatic measurement of complex dielectric constant and permeability at microwave frequencies" *Proc. IEEE*, **62**, 33-36, (1974).

[10]T. Nakamura, T. Tsutaoka, K. Hatakeyama, "Frequency dispersion of permeability in ferrite composite materials" *J. Magn. Magn Mater.*, **138**, 319-328 (1994).

Electroceramic Materials and Applications
Edited by R. W. Schwartz

PREPARATION AND CHARACTERIZATION OF $FeTiO_3$-Fe_2O_3 SOLID SOLUTION SYSTEM.

Miki Yamashita, Makoto Nakanishi, Tatsuo Fujii, and Jun Takada,
Department of Material Chemistry, Graduate School of Natural Science and Technology, Okayama University, Tsushima-naka 3-1-1,
Okayama 700-8530, Japan

Hideki Yoshikawa,
National Institute for Material Science (NIMS), BL15 SPring-8, Mikazuki-cho, Sayou-gun, Hyogo 679-5198, Japan

Yuuji Saitoh, Shin-ichi Fujimori, and Keisuke Kobayashi
Synchrotron Radiation Research Center, Japan Atomic Energy Research Institute (JAERI), SPring-8, Mikazuki-cho, Sayou-gun,
Hyogo 679-5148, Japan

ABSTRACT

The ilmenite-hematite solid-solution, $xFeTiO_3$-$(1-x)\,\alpha$-Fe_2O_3 samples with x=0.7-1.0. were prepared by standard solid state reaction technique in an evaculated quartz tube. The sample pellets were heated at 1200°C for 24 hr, then furnance cooled to room temperature[1]. All samples consisted of a single phase of ilmenite structure. Lattice parameters of these samples estimated from the XRD patterns varied linearly with decreasing the Ti concentration on the basis of the Vegard's law. The saturating magnetization at 78 K was obtained at x=0.8 ($Fe_{1.2}Ti_{0.8}O_3$), and the sample with x=0.7 ($Fe_{1.3}Ti_{0.7}O_3$) was ferrimagnetic at room temperature. The carrier type measured by Seebeck coefficient was changed from p-type to n-type at about x=0.78 with decreasing the Ti concentration. Electronic states of Fe and Ti ions in $FeTiO_3$ seemingly had the characteristic features of Fe^{2+} and Ti^{4+} respectively. However the Ti $2p\rightarrow3p$ resonant valence band photoemission spectra of $FeTiO_3$ had a clear density assigned to the Ti 3 *d* derived states. The small density appeared just below the Fermi edge was intrinsic for the $FeTiO_3$ crystals because of identical x-ray absorption spectrum at Ti $L_{2,3}$ edges between constant initial state and total electron yield mode.

INTRODUCTION

In order to decrease the environmental risks of electronics related materials, new functional materials without toxic elements are strongly required to develop. Both iron and titanium minerals widely abound in earth and have no toxic effects. It is well-known that the solid solution series of ilmenite-hematite system, $Fe_{2-x}Ti_xO_3$, have interesting magnetic and electric properties[1-3]. Though both of the end members are antiferromagnetic insulator, the intermediate compounds between them are ferrimagnetic semiconductors[2-6]. Recent theoretical calculation predicted that the ilmenite-hematite system had possibility to lead a magnetic semiconductor with strongly spin-dependent transport properties and have high Curie temperature of about 1000 K[4] . Moreover, this system can produce a tailor-made semiconductor as p- or n-type by changing the Fe to Ti concentration ratio[3]. The interesting electric properties of the $Fe_{2-x}Ti_xO_3$ system were mainly caused from the formation of mixed valence states between

Fe^{2+} and Fe^{3+} ions. The electronic structure of Ti ions in $FeTiO_3$ was normally considered as Ti^{4+}. The Ti^{4+} ions partially substituted for Fe^{3+} in α-Fe_2O_3 produce the Fe^{2+} ions to keep the charge neutrality of the solid solutions. Moreover the ferrimagnetism appeared only when the solid solution system had the ilmenite structure with $R\bar{3}$ symmetry, in which the Fe-rich and Ti-rich cation layers were arranged alternately along the c-axis. In this study, we report the effects of the preparation conditions for the ilmenite-hematite solid solution system especially in the ilmenite-rich region, $Fe_{2-x}Ti_xO_3$ (x=0.7-1.0), on their stoichiometry and electric and magnetic properties. The electronic structures of $FeTiO_3$ were also discussed by using various spectroscopic techniques.

EXPERIMENTAL

The $Fe_{2-x}Ti_xO_3$ samples with x=0.7, 0.8, 0.9, 1.0 were prepared by standard solid state reaction technique. Commercially obtained ilmenite ($FeTiO_3$ 99.9%) and hematite (α-Fe_2O_3 99.0-100%) powders were used as starting materials. The mixtures were ground in a pebble mortar and then were pressed to form pellet shapes with a 10 mm diameter and 5 mm thickness under a pressure of 4.9 MPa at room temperature. The sample pellets were placed in quartz tubes which were evacuated to approximately 10^{-5} torr, and sealed[7]. The quartz tubes containing sample pellets were heated with a heating rate of 5 /min to 1200°C and held at 1200°C for 24 hr. The samples were furnance cooled to room temperature. The obtained sample pellets were characterized by powder X-ray diffraction technique (XRD:Rigaku RINT-2500) with CuKα radiation. The lattice parameters were estimated in comparison with diffraction lines of Si powders as an internal standard. The detailed crystal structures were refined by the Rietveld profile fitting using the program RIETAN 2000[8]. Oxygen nonstoichiometries of these samples were estimated by themo gravimetry analysis (TG:RIGAKU thermo plus TG 8120) up to 1000°C with a heating rate of 5°C /min and held for 10 min in O_2 gas flow (100 cc/min). The observed weight change was assumed due to the oxidation of Fe^{2+} ions to Fe^{3+} ions. To determine the carrier type of $Fe_{2-x}Ti_xO_3$, Seebeck measurements of prepared samples were carried out in the temperature range between room temperature and 673 K in air[4]. The magnetic hysteresis loops of these samples were measured at room temperature and 78 K by vibrating sample magnetometer (VSM:Toei VSM-5-15) [1,2]. Electronic states of Fe ions in $FeTiO_3$ were characterized by ^{57}Fe Mössbauer spectroscopy using a 925 MBq ^{57}Co source in Rh matrix. The velocity scale of obtained spectrum were calibrated by α-Fe foil. Ti ions in $FeTiO_3$ were analyzed by the valence-band resonant x-ray photoelectron spectroscopy (RPES), with an incidence photoenergy just at the Ti $L_{2,3}$ absorption edge[9,10]. The valence-band RPES was carried out on beamline 23SU of Japan Atomic Energy Reseach Institute beamline, for soft x-ray spectroscopy at SPring-8[11]. Before the RPES measurements, the sample was cleaned in situ in vacuum to reveal the clean surface. X-ray absorption spectrum of Ti $L_{2,3}$ edge for $FeTiO_3$ were measured by the total electron yield (TEY) mode[10] and constant initial state (CIS) mode[12].On- and off-resonance incidence photon energy for the RPES spectra were decided by the obtained XAS spectrum [12,13].

RESULTS AND DISCUSSION

The XRD patterns of prepared of $Fe_{2-x}Ti_xO_3$ samples are shown in Figure 1. All samples were a single phase of ilmenite structure with out other secondary phase. The lattice parameters estimated from the XRD patterns are plotted in Figure 2, as a function of the Ti concentration. The lattice parameters increased monatonically with decreasing the Ti concentration from $FeTiO_3$ to α-Fe_2O_3 on the basis of the Vegard's law.

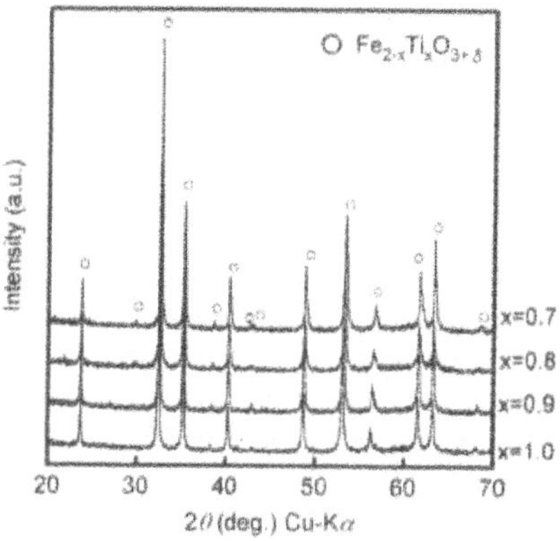

Figure 1 XRD patterns of $Fe_{2-x}Ti_xO_3$ with various Ti concentrations, x=0.7-1.0,.

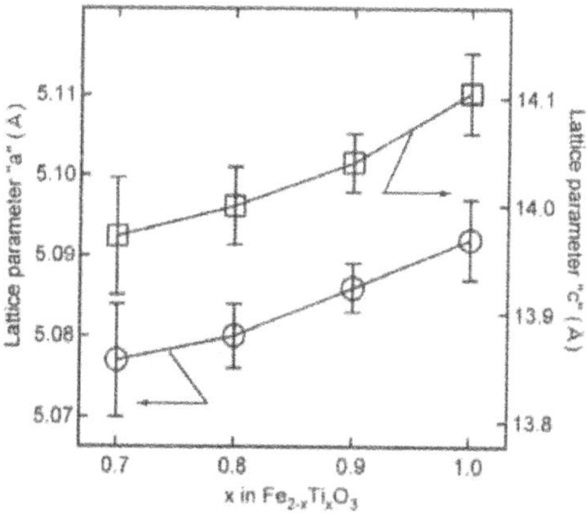

Figure 2 Lattice parameters of $Fe_{2-x}Ti_xO_3$ as function of the Ti concentration, x=0.7-1.0.

The oxygen concentration, *i.e.* nonstoichiometry of $Fe_{2-x}Ti_xO_{3+\delta}$, estimated from the amount of oxidized Fe^{2+} components to Fe^{3+} by the TG analysis, were tabulated in Table I. All samples were nearly stoichiometric, however the oxygen content of $Fe_{2-x}Ti_xO_{3+\delta}$ was gradually decreased with decreasing the Ti concentration. This trend was probably due to the insufficient oxygen partial pressure during the sample preparation. The vacuum pressure in the quartz tubes during the heat-treatment was almost the same for the all compositions, but the equilibrium oxygen pressure of $Fe_{2-x}Ti_xO_3$ increased with decreasing the Ti concentration[1]. Therefore the obtained $Fe_{2-x}Ti_xO_3$ with smaller Ti concentration were slightly reduced and had small oxygen concentration.

Table I. Estimated chemical formula and Fe/Ti order parameter for $Fe_{2-x}Ti_xO_3$ samples

Ti concentration:x	1.0	0.9	0.8	0.7
chemical formula	$FeTiO_{3.074}$	$Fe_{1.1}Ti_{0.9}O_{3.020}$	$Fe_{1.2}Ti_{0.8}O_{3.074}$	$Fe_{1.3}Ti_{0.7}O_{2.946}$
Fe/Ti order parameter (%)	96	91	100	100

Seebeck coefficients of various $Fe_{2-x}Ti_xO_3$ samples are plotted in Figure 3 as a function of the temperature. The Seebeck coefficients remained almost constant within the measured temperature ranges. But with decreasing the Ti concentration the sign of the Seebeck coefficient was systematically changed from positive values to negative ones. This means that the carrier type of the samples with x□0.8 was p-type, while that with x=0.7 was n-type. To interpolate the Seebeck coefficient values across zero, the carrier type change from p-type to n-type seemed to occur at about x=0.78. This value was nearly consistent with the one reported in the literature[4].

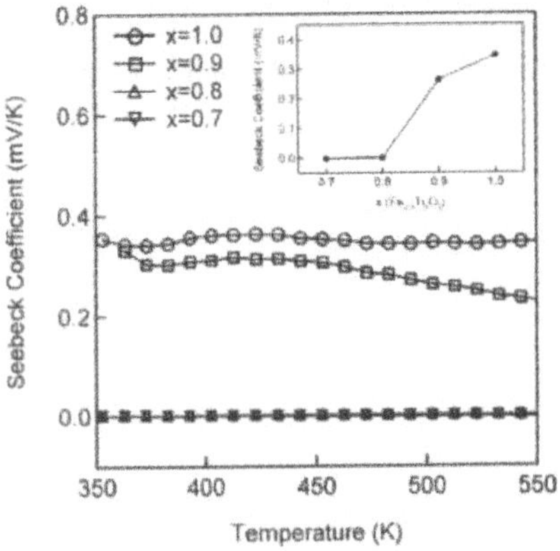

Figure 3. Seebeck coefficient versus temperature of $Fe_{2-x}Ti_xO_3$ (x=0.7-1.0). The inset shows the change of Seebeck coefficient at 400 K depending on the Ti concentration.

The magnetic hysteresis loops of the samples measured at room temperature and 78 K are shown in Figure 4. As is seen in the figure, the sample with x=1.0 had the paramagnetic loop for both temperatures. The samples with x=0.9 and 0.8 were ferrimagnetic at 78 K, while paramagnetic at room temperature. Finally, at the both temperatures, the sample with x=0.7 was ferrimagnetic. It was very consistent with the linearly increasing Curie temperature from 55 K for $FeTiO_3$ to 960 K for α-Fe_2O_3 with decreasing the Ti concentration. Moreover it was reported that the $Fe_{2-x}Ti_xO_3$ system had the maximum magnetization value at about x=0.8 at 4.2 K[2,3]. Saturation magnetization values measured at 78 K also had the maximum value at about x=0.8.

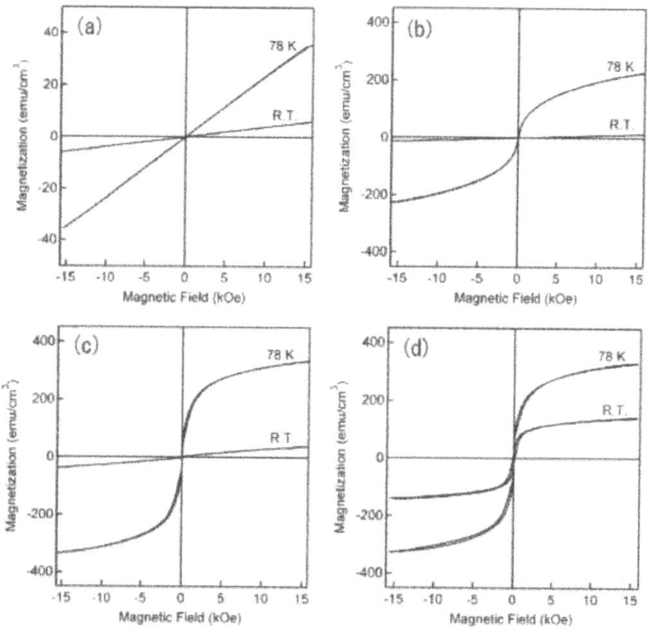

Figure 4 Magnetization curves of $Fe_{2-x}Ti_xO_3$ measured at room temperature and 78 K (a) x=1.0 ($FeTiO_3$) (b) x=0.9 ($Fe_{1.1}Ti_{0.9}O_3$) (c)x=0.8 ($Fe_{1.2}Ti_{0.8}O_3$) (d) x=0.7($Fe_{1.3}Ti_{0.7}O_3$)

The structural order parameters of the cation arrangement between Fe and Ti ions calculated by RIETAN 2000[8] are also shown in Table I. If all the Ti ions occupied only one of the two cation sites, which are arranged alternatively along the c-axis, the order parameter should be 100%., while if both Fe and Ti ions occupy the two cation sites, the order parameter should be 0%, resulting in a corundum structure with R$\bar{3}$c symmetry. The calculated order parameter suggested that the prepared $Fe_{2-x}Ti_xO_3$ samples had a nearly fully ordered structure with R$\bar{3}$ symmetry.

Figure 5 shows the room temperature Mössbauer spectrum of the $FeTiO_3$ sample. The Fe ions in $FeTiO_3$ had the electronic structures of Fe^{2+} states as known before.[14] The spectrum had clear one doublet peaks with an isomer shift of 1.08 mm/s. This value was characteristic of the octahedral Fe^{2+} ions with high spin states.

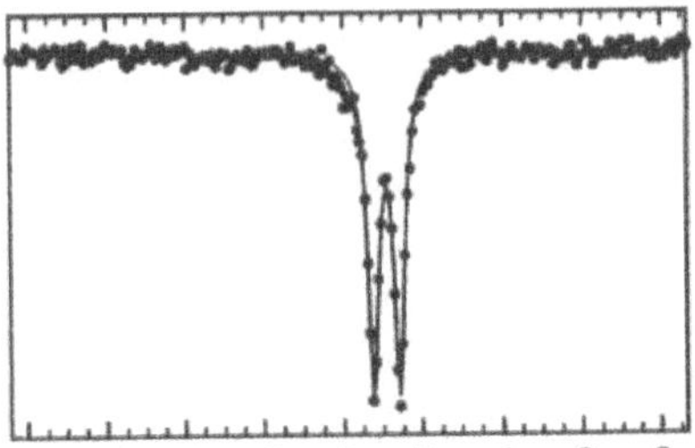

Figure 5 Room temperature Mössbauer spectrum of $FeTiO_3$

On the other hand, x-ray absorption spectrum (XAS) of Ti $L_{2,3}$ edges for $FeTiO_3$ is shown in Figure 6. The XAS spectrum was nearly consistent with that of other Ti^{4+} compounds such as TiO_2[12,13]. The Ti ions in $FeTiO_3$ could be assigned to Ti^{4+} states. However, in spite of no 3d electrons in Ti^{4+} states, the valence-band photoemission spectrum for $FeTiO_3$ had a clear density just below the Fermi edge. In order to survey the detailed electronic states of Ti ions in $FeTiO_3$, we measured Ti $2p \rightarrow 3d$ resonant photoemission spectra (RPES) of $FeTiO_3$ focusing on its valence-band region. If the valence-band structures contained the Ti $3d$ derived states, the structure should be intensified at the resonance conditions. Figure 7 shows the on- and off-resonance valence-band RPES spectra excited by the Ti $L_{2,3}$ edge. The small density marked A were considerably intensified at the resonance condition. This suggest that the Ti $3d$ derived states had a large contribution to the valence-band edge just below the Fermi edge. The Ti ions in $FeTiO_3$ could have some $3d$ electron populations.

The XAS spectrum measured by the constant initial state (CIS) mode using the photoelectrons near the Fermi edge (Peak A) was compared to that by the total electro yield (TEY) mode as shown in Figure 6. The detailed spectral futures obtained by CIS were essentially the same as those of TEY-XAS. This indicated that the valence-band intensity named as "A" was the intrinsic feature of $FeTiO_3$ crystals. The Ti ions in $FeTiO_3$ should have some Ti^{3+} character possessing the $3d$ valence electron.

CONCLUSIONS

Solid solution series of $Fe_{2-x}Ti_xO_3$ with x=0.7-1.0 were prepared by standard solid state reaction of $FeTiO_3$ and α-Fe_2O_3 powders. The palletized mixtures were sealed in quartz tubes in vacuum and heat-treated at 1200°C for 24 hr. After the furnace-cooling the phase formation was identified by x-ray diffraction technique. All samples formed to the single phase of the solid solution structure. The lattice parameters were linearly changed by following the Vegard's law. The solid solution seemed to contain small oxygen nonstoichiometry depending on the oxygen partial pressure during the reaction. The samples with well-ordered cation arrangements had maximum saturation magnetization at x=0.8 at 78 K. The carrier type of $Fe_{2-x}Ti_xO_3$ was changed from p- to n-type at x=0.78 with decreasing the Ti content. Fe ions in $FeTiO_3$ had typical Mössbauer patterns assigned to octahedral Fe^{2+} ions. While the Ti ions had unexpected valence band structures contributing to the clear density of states just below the Fermi edge, and were considerably intensified by the Ti $2p \rightarrow 3d$ resonance excitation conditions. Ti ions in $FeTiO_3$ seemed to have some Ti^{3+} characters with $3d$ electron.

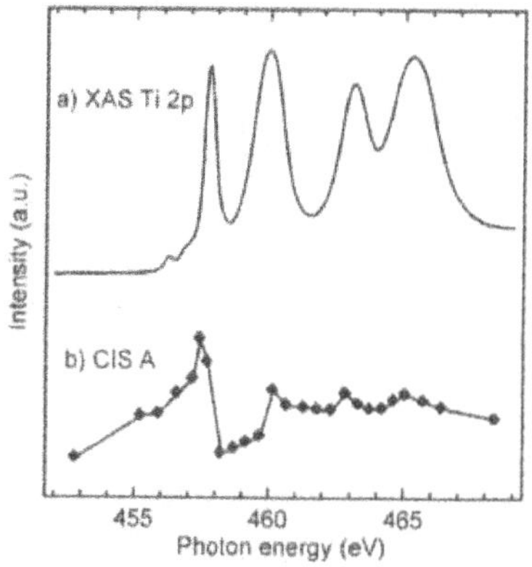

Figure 6 a)Ti $L_{2,3}$ edge XAS spectra of $FeTiO_3$ measured by total electron yield and b) constant initial state modes. The CIS spectrum was obtained by a peak A in Figure 7.

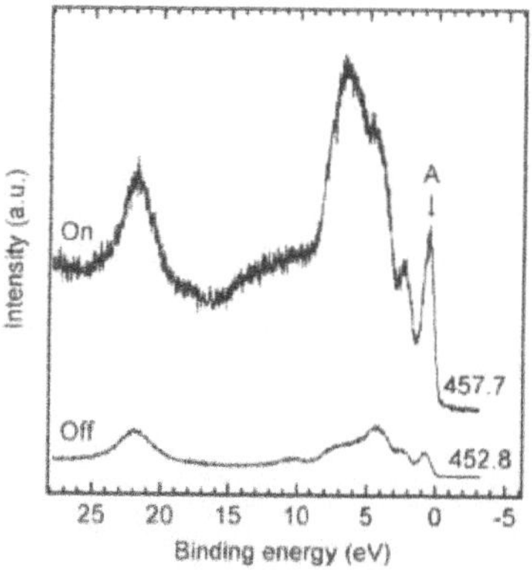

Figure 7 On and off valence band RPES spectra excited by the Ti $L_{2,3}$ edge of $FeTiO_3$. The incident photon energy is indicated in the figure.

ACKNOWLEDGEMENTS

We want to thank the staff of JAERI, Drs. T. Okane and K. Terai for technical support, and Prof. Ohtani for supporting to Seebeck coefficient measuments.

REFERENCES

[1]R. W. Taylor, "Phase Equilibria in the System FeO-Fe_2O_3-TiO_2 at 1300°C," *Am. Mineral.*, **49**, 1016-30 (1964)

[2]Y. Ishikawa and S. Sawada, "Magnetic Property and Crystal Chemistry of Ilmenite ($MeTiO_3$) and Hematite (Fe_2O_3) System II," *J.Phys. Soc. Japan*, **12**, 1298-1310 (1958).

[3]Y. Ishikawa, "Magnetic Properties of Ilmenite-Hematite System at Low Temperature, " *J.Phys. Soc. Japan*, **17**, 1835-44 (1962).

[4]Y. Ishikawa, "Electrical Properties of $FeTiO_3$-Fe_2O_3 Solid Solution Series, " *J.Phys. Soc. Japan*, **13**, 37-42 (1962).

[5]T. Fujii, M. Kayano, Y. Takada, M. Nakanishi, and J. Takada "Preparation and Characterization of Epitaxial $FeTiO_{3+\delta}$ films, " *J.Magn. Magn. Mat*, 272-276 (2004) 2010.

[6]T. Fujii, M. Kayano, Y. Takada, M. Nakanishi, and J. Takada "Ilmenite-hematite Solid Solution films for Novel Electronic Devices, " *Solid State Ionics*, **172**, 289-292 (2004)

[7]W. H. Butler, A. Bandyopadhyay, and R. Srinivasan, "Electronic and Magnetic Structure of a 1000 K Magnetic Semiconductor: α-hematite (Ti), " *J. Appl. Phys.* **93** 7882-84 (2003).

[8]F. Izumi and T. Ikeda, "A Rietveld-Analysis Program RIETAN-98 and its Applications to Zeolites "*Mater. Sci. Forum,* **321-324** 198-203 (2000)

[9]A. Agui, M. Mizumaki, Y. Saitoh, T. Matsushita, T. Nakatani, A. Fukuya, and E. Torikai, "Soft X-ray Absorption Spectra of Ilmenite Family, " *J. Synchrotron Rad.*, **8**, 907-909 (2001)

[10]Y. Ufuktepe, G. Akgül, J. and J. Lüningi, "X-ray Photoabsorption and Total Electron Yield of Fe thin films at the $L_{2,3}$ edge, " *J. Alloys Compd.* in press.

[11]Y. Saitoh, T. Nakatani, T. Matsushita, A. Agui, A. Yoshigoe, Y. Teraoka, and A. Yokoya, "First Resurts from the Actinide Science Beamline BL23SU at Spring-8, " *Nucl. Inst. Meth. A*, **474**, 253-258 (2001)

[12]M. Sánchez-Agudo. L. Soriano, J. F. Trigo, C. Quirós, G. G. Fuentes, C. Morant, E. Elizalde, and J. M. Sanz, "Factor Analysis Applied to the Study of Valence band Resonant Photoemission Spectra in Transition-metal Compounds "*Surf. Interface Anal.* **34** 224-227 (2002)

[13]K. C. Prince, V.R. Dhanak, P. Finetti, J. F. Walsh, R. Davis, C. A. Muryn, H. S. Dhariwal, G. Thornton, and G. van der Laan "2*p* Resonant Photoemission Study of TiO_2, " *Phys. Rev. B*, **55**, 9250-23 (1997)

[14]B. N. Warner , P. N. Shive, J. L. Allen, and C. Terry" A Study of the Hematite-Ilmenite Series by the Mössbauer Effect, " *J. Geomag. Geoelectr.*, **24**, 353-367 (1972)

Electroceramic Materials and Applications
Edited by R. W. Schwartz

OXYGEN RELAXATION OF $ZR_{1-X}Y_XO_{2-X/2}$ (X=0.04, 0.08, AND 0.15) CERAMICS OBSERVED BY INTERNAL FRICTION MEASUREMENT

Masakuni Ozawa, Tetu Kuwahara, Kenji Tuchita
Ceramics Research Laboratory, Nagoya Institute of Technology
Tajimi, Gifu, 507-0071, Japan

ABSTRACT

Dynamic modulus relaxation (internal friction) measurement is useful to determine the hopping rate (relaxation time) and activation energy of the oxygen vacancies around the local structures. In this work, the oxygen relaxation data of tetragonal and cubic Y_2O_3-doped ZrO_2 ceramics were provided by using the mechanical loss (internal friction) measurement at high and low frequencies. The internal-friction data of $Zr_{1-x}Y_xO_{2-x/2}$ (x=0.04, 0.08 and 0.15) concluded the existence of the same relaxation process, and consequently the possibility of similar local cation-anion (or vacancy) interaction in tetragonal and cubic zirconia. However, another relaxation appeared at higher temperature only in cubic zirconia (x=0.15).

INTRODUCTION

The zirconia-yttria solid solutions have been used as a high-temperature solid electrolyte in full cells and oxygen sensors. Since they show the superior ionic conductivity by doping oxides of lower-valent cations, oxygen vacancies are created as charge compensating defects, and should play the essential role for the conduction of oxide ions. The best ionic conductivity actually appears at around 8-10mol% Y_2O_3-doping to ZrO_2. The transport properties of them, i.e. ionic conduction or diffusion, are essentially determined by the defect structures as well as the kind and concentration of the dopant. At lower temperatures than about 1000□, the cations are found immobile. The charged oxygen vacancies become then trapped to form the complexes of cations with oxygen vacancies at low temperatures. The oxygen vacancy can locally jump between neighboring sites around the cations. This relaxation has been observed as mechanical and dielectric loss peaks [1] in several papers on the fluorite -structured ZrO_2 [2–5] and CeO_2 [6]. The long-range migration of the oxygen via vacancies, that is the basic process of ionic conduction and diffusion, requires a tunneling of the vacancy from the complex of a cation-vacancy pair as well as the hopping between sites. Although the local structure as its static state can be generally determined by EXAFS, X-ray and neutron diffractions, it is difficult to examine the dynamic response of the coordinated complex in the solid. Dynamic modulus relaxation (internal friction) measurement is useful to determine the hopping rate (relaxation time) and activation energy of the oxygen vacancies around possible local structures. In this work, the oxygen relaxation of typical three kind of (tetragonal and cubic) Y_2O_3-doped ZrO_2 ceramics are provided by using the mechanical loss (internal friction) measurement at low and high frequencies.

EXPERIMENTAL AND ANALYSIS

The starting powders of $Zr_{1-x}Y_xO_{2-x/2}$ (x=0-0.15) were obtained by a coprecipitation process using aqueous metal nitrates and ammonia solution. The aqueous $ZrOCl_2$ H_2O and YCl_3 H_2O solution of 1mol/l as total metal concentration (of Zr and Y) were prepared, then 5wt% (NH_3)

ammonia solution was added under stirring condition, by the final pH of 10. The precipitation was filtered, washed with distilled water, and calcined at 900□ for 3h in air. The pressed powder compact bodies were sintered at 1600□ for 3h in air. The specimens with $40\times4\times0.5mm^3$ were prepared through cutting and polishing procedures. Note that no additive was used in the sintering so that the grain boundary in the sintered body has no segregated composition as well as the secondary phase.

The mechanical loss measurements at 30-450□ were carried out with a torsion pendulum method in the frequency range of 0.3Hz-100Hz. The apparatus was controlled by PC-IEEE4880 system to detect the strain and load at different frequencies. The tanδ was evaluated by the direct measurement of the delay angle between stress and strain. At high frequency around 100kHz, the piezoelectric resonance method using a $LiNbO_3$ detector was applied to the measurement[9]

The anelastic relaxation (internal friction) profiles were analyzed by standard solid model to give the Debye-type relaxation peak,

$$\tan\delta = \Delta\omega\tau/(1+\omega^2\tau^2) \quad (1)$$

where δ is the loss angle for either the anelastic (Q-1 stands for t anδ) or dielectric causes, ω the angular frequency of the applied stress or electric field, and τ the relaxation time, Δ the relaxation strength. In the thermally-activated relaxation, the relaxation time τ obeys the Arrhenius equation,

$$\tau^{-1} = \tau_0^{-1}\Box\exp(-H/RT) \quad (2)$$

where H is the acivation enthalpy for relaxation, τ_0 is a pre-exponential factor, and RT is usual term. A plot of tanδ versus 1/T at a fixed frequency gives rise to a symmetric Debye peak with the maximum when ωτ=1.

RESULTS AND DISCUSSION

Relaxation of 2Y-TZP

Figure 1 shows the internal friction tanδof 2Y-TZP (2mol% Y_2O_3 doped tetragonal zirconia polycrystal; $Zr_{1-x}Y_xO_{2-x/2}$ (x=0.04)) versus temperature, measured at 32Hz and 85kHz. The internal friction peak (and anelastic relaxation) was observed at different temperatures when two measurement frequencies were applied to the same sintered sample (lot). The plot of tanδas a function of reciprocal temperature shows a symmetric peak. The experimental data on internal friction were analyzed by a parameter fitting procedure using a least square minimum method to obey the equation (1) and (2). Table 1 summarizes the results of peak temperature, relaxation strength, and activation enthalpy (H) by fitting experimental data at different frequencies. Peak shape showed good consistence to that of Debye-type (from equation (1) and (2)). The averaged H from half-width of each peak was broader than that where a peak indicated a single relaxation, which meaned the distribution of relaxation mode. Mastushita et al [7] have carried out the elastic measurement at 1kHz-5kHz about 2Y and 3Y-TZP by the flexural resonance method. They reported H=69±2 kJ/mol for 2Y-TZP and 81±5kJ/mol for 3Y-TZP. Weller and Schubert [3] have shown the Arrhenius plot using two values for 2.7 Hz torsion and 3.06 kHz flexural vibrations of 3Y-TZP samples, and calculated to be H=90±3 kJ/mol. They described that the internal friction peak observed at temperature was 2.8 times and 2.5 times broader than a single Debye peak. In this work, the Arrehnius plots in the wide range (6 order) of 0.3Hz to 120kHz gave H= 76±3 kJ/mol and τ_0=5±2x10^{-14}sec. If H=76kJ/mol is employed, the peak in Fig.1 is 2.2

times broader than a single Debye peak. This suggests the overlapping phenomena of relaxation time or energy for all relaxation peaks.

We measured the dielectric loss angle for various frequencies of 10kHz to 10MHz. The peak temperature of the dielectric loss peak was elevated when applied frequency was higher. An Arrehnius plot of ω vs. T^{-1} was represented by a straight line indicating thermally-activated relaxation to give H=79±4 kJ/mol and τ_0=6±3x10^{-13}sec. The activation enthalpy derived from an Arrehnius plot is the same for anelastic and dielectric relaxation within experimental error, and the H value is corresponding to that in oxide ionic conduction of ZrO_2. The results suggest that both anelastic and dielectic relaxation is due to the same origins of the movement and rearrangement of oxygen (or vacancy).

The Ce(4+)-doped zirconia shows no internal friction in general. As a reference experiment, we performed the heat-treatment of the 12.6mol% CeO_2-TZP sample at 1000□ in flowing H_2 for 8 hours. The reduction leads to the formation of oxygen vacancy, because the loss of weight, the slight expansion, the decease of tetragonality (c/a) and the increase of the lattice volume were observed. However, the internal friction of Ce-TZP after the reduction appeared as a small peak at 330□ at 100kHz, and also the Young's modulus showed non-linearity at the same temperature. This experiments of the reduced Ce-TZP directly indicates the significant contribution of the oxygen vacancy to the modulus relaxation of ZrO_2. In addition, our previous result of $Zr_{0.8-x}Ce_XY_{0.2}O_{1.9}$.suggest the local structure due to different caion radius induces the complex mode of relaxation of oxygen (vacancy) [10].

Relaxation of cubic zirconia

Figure 2 shows the mechanical loss spectra of three yttria-stabilized zirconia ceramics with different Y contents ($Zr_{0.96}Y_{0.04}O_{1.98}$, $Zr_{0.92}Y_{0.08}O_{1.96}$, $Zr_{0.85}Y_{0.15}O_{1.925}$) at a low (3.2Hz) frequency. A large feature is that when large amount of Y_2O_3 is doped ($Zr_{0.85}Y_{0.15}O_{1.925}$), the internal friction shows the two loss maxima at around 100□ (peak 1) and 250□ (peak 2). Weller has [5] reported this phenomenon at low frequency measurement and that H of peak 2 was 1.8 eV ($Zr_{0.82}Y_{0.18}O_{1.91}$). In our work (of $Zr_{0.85}Y_{0.15}O_{1.925}$), the activation enthalpies was determined as 120kJ/mol (1.24eV) for a peak 1, and 220kJ/mol (2.3eV) for peak 2. The internal friction behavior is largely different from that of 2Y-TZP. In 2Y-TZP, one kind of oxygen or vacancy relaxation (even with some distribution) was active, because of small content of a doping cation (oxygen vacancy). The characteristics of a peak 1 at low temperature have suggested that the same relaxation exists in cubic zirconia as in 2Y-TZP. However, more important fact is that the large internal friction peak (peak 2) appeared at higher content of vacancy in cubic zirconia with higher conductivity. The oxygen relaxation in cubic zirconia should be caused by two possible local interaction (or complexes) between cation and oxygen (or vacancy).

The local structure of cubic Y_2O_3-stabilized ZrO_2 has been investigated using an EXAFS technique by Catlow et al.[8]. They have suggested the average Zr^{4+} environment being more disordered than that of Y^{3+}, i.e.7 oxygen-coordinated Zr^{4+}. At elevated temperature, the increased oxygen mobility should result in an increasingly statistical distribution of oxygen vacancies. Also, they have concluded that the oxygen vacancy is located around Zr cation to form (Zr^{4+}-Vo") pair that decomposes at high temperature. Our relaxation data support the appearance of the local structure with an elastic dipole to give arise the elastic relaxation, and the movement of oxygen takes two kinds of relaxation mode in cubic zirconia.

Since the ionic conduction increases with the increase of oxygen vacancy concentration at x=0.04-0.15, the larger internal friction amplitude (peak 2) may correspond to the increase in mobility. Thus, it is suggested that the characteristic peak 2 of cubic zirconia is a measure of higher ionic conductivity, regarding with anelastic relaxation phenomena. In addition, since the static analyses of a local structure, such as EXAFS, detect no clear difference depending on Y content about local structure, this mechanical study provides one possibility to find the oxygen hopping for various zirconias with high oxide ion conduction.

CONCLUSION

The elastic measurement data of 2Y-TZP at 0.3Hz and 120kHz indicated an anelastic relaxation peak with the activation enthalpy H=76±3 kJ/mol and τ_0=5±2x10^{-14}sec. The relaxation is considered to be due to the same origins as that for the dielectric one. The comparative study on the internal friction for $Zr_{1-x}Y_xO_{2-x/2}$ (x=0.04 and 0.15) indicated the existence of the similar relaxation process, and consequently the possibility of similar local cation-anion (or vacancy) interaction. However, the other large relaxation appeared only in cubic zirconia with higher ionic conduction. The measurement of internal friction showed possibility to detect of a moving complex in zirconias.

REFERENCES

[1] G.E. Murch, A.S. Nowick (Eds.), "Diffusion in Crystalline Solids", Academic Press, New York, NY, 1984

[2] J.B. Wachtman, W.C. Corwin, "Internal friction in ZrO_2 containing CaO", *Phys. Rev.* **131** ,517(1963).

[3] M. Weller, H. Schubert, "Internal friction, dielectric loss, and ionic conductivity of tetragonal zirconia-3% yttria (Y-TZP)", *J. Am. Ceram. Soc.* **69**, 573-577(1986).

[4] A. Lakki, K. Herzog, M. Weller, H.Schubert, C.Reetz, O.Gorke, M.Kilo, and G.Borchardt, "Mechanical loss, creep, diffusion, and ionic conductivity of ZrO_2-8molY_2O_3 polycrystals"., *J. Eur. Ceram. Soc.*, **20**, 285-296(2000)

[5] M. Weller, "Mechanical loss measurement on yttria- and calcia-stanilized zirconia", *J. Alloys Compds.* **211-212**, 66-70(1994).

[6] M.P. Anderson, and A.S. Nowick,"Relaxation peaks produced by defect complexes in cerium dioxide doped with trivalent cations", *J. Phys. (France) Colloque* **C5-Pt.2** 823-828(1981).

[7] K.Matsusita,T.Okamoto, and M.Shimada,"Internal friction in partially stabilized zirconia", J.Physique **C10 suppl.46**,549-552 (1985)

[8]C.R.A Catlow,A.V.Chadwick,G.N.Greaves and L.M.Moroney, "EAXFS study of yttria-stabilized zirconia", *J.Amer.Ceram.Soc.* **69**, 272-77(1986)

[9] M.Ozawa, and M.Inagaki, "$LiNbO_3$ composite oscillator for internal friction and modulus measurement at elevated temperature" *Rev.Sci.Instrum.* **67**, 2419-20(1996)

[10] M.Ozawa, T.Ito, and E.Suda, "Mechanical loss of $Zr_{0.8-x}Ce_xY_{0.2}O_{1.9}$(x=0-0.4)", *Alloy Compds.* **374**,397-400 (2004)

Table 1 The data for the internal friction peaks of 2Y-TZP, analyzed by a Debye equation.

Applied frequency at peak (Hz)	Peak temperature (°C)	Half-width $x10^{-3}$ (K-1)	Relaxation amplitude
0.32	52	0.54	0.012
9	83	0.5	0.011
32	115	0.52	0.012
$85.5x10^3$	292 1	0.59	0.013
$112\ x10^3$	311 2	0.62	0.013
$120x10^3$	312 2	0.61	0.013

Table 2 The data for the internal friction peaks analyzed by Debye-type at 3.2Hz.

Content	Peak temperature (°C)	Half-width $x10^{-3}$ (K^{-1})	Relaxation strength $x10^{-3}$
$Zr_{0.92}Y_{0.08}O_{1.96}$			
peak1	81	0.6	0.014
peak2	239	1.35	0.005
$Zr_{0.85}Y_{0.15}O_{1.925}$			
peak1	101	0.86	0.019
peak2	253	0.44	0.024

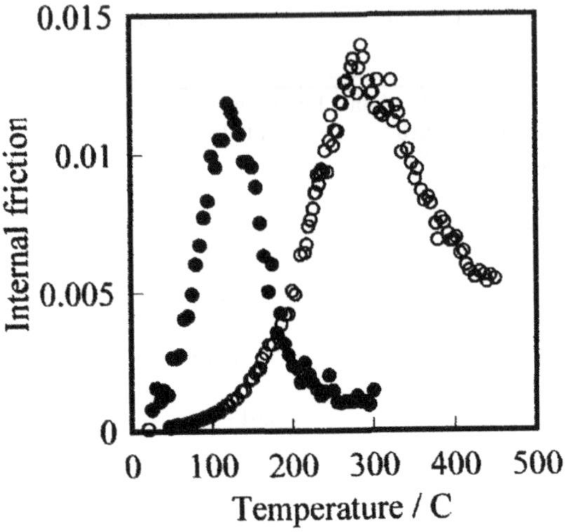

Fig.1 Internal friction versus temperature for 2Y-TZP measured at (•)32Hz and (○)85kHz

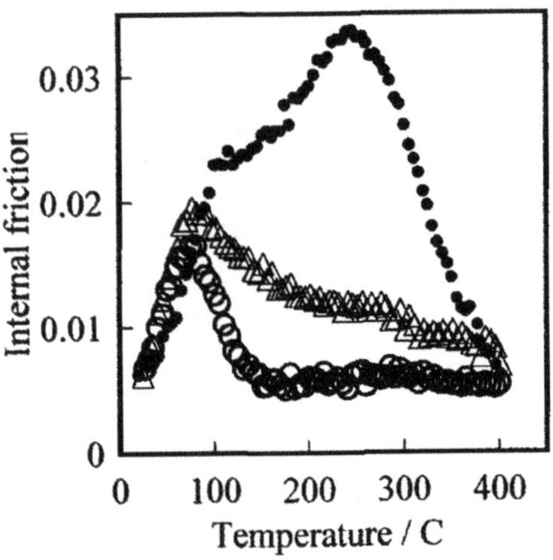

Fig.2 Three plots of internal friction versus temperature for (○) $Zr_{0.96}Y_{0.04}O_{1.98}$, (△) $Zr_{0.92}Y_{0.08}O_{1.96}$, and (•) $Zr_{0.84}Y_{0.16}O_{1.92}$ at 3.2Hz.

Electroceramic Materials and Applications
Edited by R. W. Schwartz

STUDY ON NONVOLATILE MEMORY BASED ON ELECTRIC-PULSE-INDUCED REVERSIBLE RESISTANCE CHANGE EFFECT

Qun Wang[a], Lidong Chen, Rui Dong, Dashan Shang, Xiaomin Li

The State Key Laboratory of High Performance Ceramics and Superfine Microstructure, Shanghai Institute of Ceramics, Chinese Academy of Sciences, Shanghai 200050, P.R.China
[a]email: wangqun@mail.sic.ac.cn

ABSTRACT

A persisting reproducible electric-pulse induced reversible resistance (EPIR) switching in Ag/$Pr_{1-x}Ca_xMnO_3$/Pt (x=0.3) sandwich structures was studied, in which the high-resistance state and low-resistance state can be assigned to logic 1 and 0 in resistance random access memory (RRAM) providing a new way for information storage. The pulse threshold voltage value, the relationship between the applied pulse voltage and the EPIR ratio between the two stable states were investigated. *I-V* measurement was employed to analyze the conductivity mechanism. For the fatigue and the retention behavior of EPIR materials, it is found that the retention properties can be modified by post-annealing of the materials and by the local and in-situ adjustment of the pulse applied mode. The EPIR devices can be optimized by initializing using voltage sweep or pulse sweep circles before application to possess both the appropriate original resistance and the stable resistance switching behavior.

INTRODUCTION

As a new style of information storage, electric-pulse induced reversible resistance (EPIR) change in Ag/$Ln_{1-x}Ca_xMnO_3$/Pt (Ln= Pr, La) arouses great interest in the field of information and physics [1-3]. In such kind of material and structural system, the high-resistance state (HRS) and the low-resistance state (LRS) used to store the information bit can be reversibly switched by applying electric pulses in nanoseconds with alternative polarities. Experiments show that the electric-pulse induced HRS and LRS differ in the conductivities by as much as ten times at room temperature without applied magnetic field. Thus, this resistance switch phenomenon may not only open a new research interest for fundamental physics and also provide a route of possible applications in high-density nonvolatile memory (NVM) calling resistance random access memory (RRAM). And the mechanism of the reversible resistance change and its emerging application has been paid much attention by the scientists and engineers of materials science [4-5].

Since its discovery, there is a dispute related to the bulk effect or the interfacial effect in the origin of reversible resistance change in Ag/$Ln_{1-x}Ca_xMnO_3$/Pt structural system. Although, models used to explain the mechanism were proposed, it is still not very clear about the intrinsic nature of the observed effects. Besides this, the reliability related to the fatigue and the retention behavior of the electric-pulse induced resistance states is another important issue and should be investigated [6-7]. The main hindrance to commercial applications of memory thin films due to their fatigue problems has been well recognized [8-9]. In the present paper, we report the EPIR effect and its fatigue, retention behavior in Ag/ $Pr_{1-x}Ca_xMnO_3$/Pt sandwiches. This research aims to present some recent experimental results and to improve the resistance switching performance for the future application in nonvolatile memory.

EXPERIMENTAL

Samples of $Pr_{1-x}Ca_xMnO_3$ (x=0.3) thin films were prepared on $Pt/Ti/SiO_2/Si$ substrate by pulse laser deposition (PLD) method [10]. X-ray diffraction has been performed on all prepared films demonstrating that the films are of single phase perovskite structures without other impurity phase. The film morphology and the thickness were examined using a field emission scanning electron microscopy (FE-SEM, JSM-6700F). The resistance measurements were carried out by the standard 2-wire I-V curve method using a 1 μA dc current as the measuring current. Pulse voltage was produced by an AVIR-3-B-P-QTKA pulse generator (Avtech Electrosystems Ltd.). Electric pulses (pulse width 100 ns, delay time 50 ns, burst time 49.98 μs) were applied across the film layers through the top and bottom Ag contact pads. The measuring current was applied in both positive and negative directions and the resistance was obtained from the average value in the two opposite directions. Along with the pulsing measurement, the relationship between current and voltage was also measured. All of the measurements were carried out at room temperature without applying magnetic field.

RESULTS AND DISCUSSION

1. *Electric pulse induced reversible resistance switching effect in Ag/ $Pr_{1-x}Ca_xMnO_3$/Pt sandwich structures*

Surface morphology of the film is imaged by field emission scanning electron microscopy (FE-SEM). Figure 1 shows a cross section of the sandwich structure. It is clearly shown that the obtained films are crack-free and homogenous in thickness. The thickness of the perovskite films ranges from 300 nm to 1000 nm depending on the deposition time.

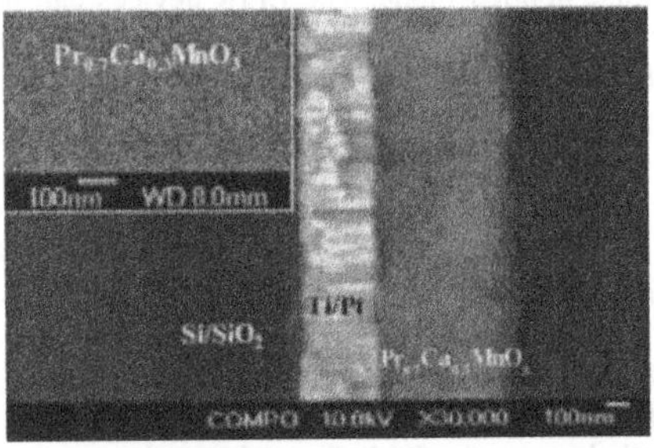

Fig. 1 Image of the cross section of the $Si/SiO_2/Ti/Pt/Pr_{1-x}Ca_xMnO_3$ structure

Fig.2 demonstrates the measurement setup, in which the electric pulse from the top Ag electrode to $Ln_{1-x}Ca_xMnO_3$ layer is defined as positive direction. In the same figure is the relationship between the pulse voltage and the reversible resistance change of the EPIR effect in Ag/ $Pr_{1-x}Ca_xMnO_3$/Pt sandwiches. It shows the relationship that the resistance versus applied pulse voltage for the Ag / $Pr_{1-x}Ca_xMnO_3$/Pt sandwiches after being applied a series of positive and negative electric pulses. The resistances of the Ag/ $Ln_{1-x}Ca_xMnO_3$/Pt samples kept almost unchanging unless the voltage of the applied electric pulse reached a critical value. When the applied voltage was higher than a certain critical value, the resistance showed a rapid increase with voltage. This critical voltage is defined as the threshold voltage. As shown in Fig 2, it is indicated that the ratio of HRS to LRS increases with the increasing of the pulse voltage before breakdown of the films.

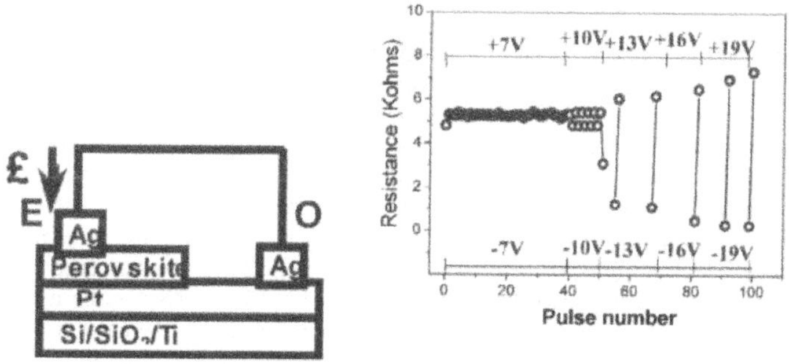

Fig.2 Schematic demonstration of the EPIR measurement and the threshold phenomenon

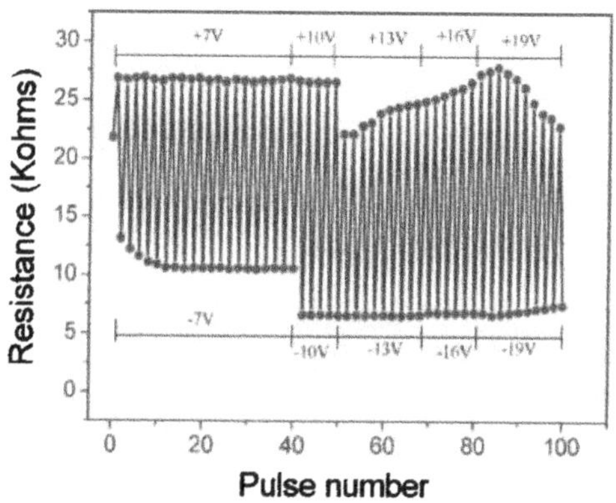

Fig. 3 EPIR phenomenon in the sandwich structure

When the applied pulse voltage was higher than the threshold voltage, the HRS was realized in sandwich structures using a single negative pulse and could be erased back to the LRS by a single positive pulse. The alternative resistance change phenomenon for Ag/ $Pr_{1-x}Ca_xMnO_3$/Pt sandwiches under positive and negative electric pulse excitation with the pulse voltage of 10V was shown in Fig.3. In Fig.3, the relationship between the resistance and the pulse number and pulse voltage was given. The resistance was measured after each electric pulse excitation. The resistance was decreased from the initial value of 27 kΩ to about 13 kΩ (low resistance state) after positive 7V voltage pulses excitation. After several alternative positive and negative pulses under the same voltage, the high resistance maintained its original value, while the low resistance was decreased first and then maintained at a fixed value of 10 kΩ. Such a high and low resistance switching circle repeated when the positive and negative pulses were applied alternately. It means that such a memory unit has excellent memory ability. When increasing the pulse voltage to 10V,13V,16V and 19V, the HRS is varying with pulse number and pulse voltage. The resistance drops from 27KΩ to 22KΩ, then increases with pulse number and increasing voltage. As shown in the figure, when the 13V pulse voltage was applied, initial three pulses resulted in increasing resistance, then the resistance decreased with pulse number. This phenomenon was not followed by the LRS after increasing the excited voltage, the LRS simply drops from 10 kΩ to 6 kΩ, and maintained at 6 kΩ, which does not vary with the pulse number and the pulse voltage. The ratio of HRS to LRS is increased from 2.6 to 3.6 and even higher. From this experiment, it is clearly shown that a stable ratio of HRS to LRS may be a critical issue, because in an actual bit operation process, different voltage values may be used in writing, erasing or reading process. Such a ratio depends on pulse number and pulse voltage, other factors affecting this should be investigated.

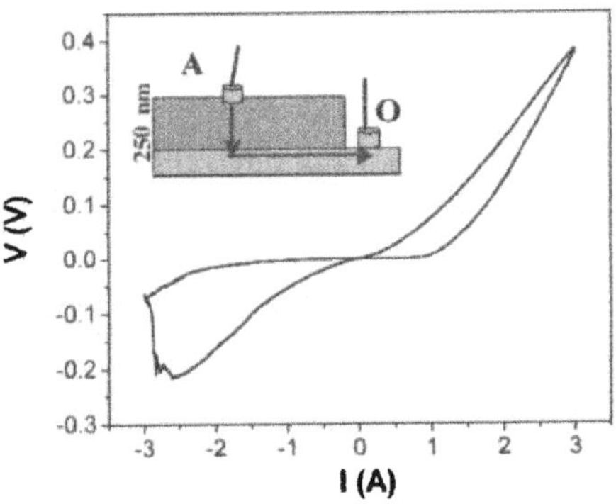

Fig.4. IV curves for Ag/ $Pr_{1-x}Ca_xMnO_3$/Pt sandwiches

The relationship between current and voltage (*I-V*) curve is presented in Fig.4. At the forward bias, the current does not vary with increasing voltage below 0.4V, which indicates an insulator state is obtained at low voltage. Over this point, the current increases nonlinearly with increasing current, approximately, an exponent relation is obtained. As known, this is very similar to the semiconductor junction effect. When arriving at the maximum 3V, the voltage was then withdrew, the current decreases when decreasing current, but with a hysteresis. When at the negative bias, it shows that there is an asymmetric trend between the positive and negative bias direction. This may be due to the different electrodes used as the top and bottom.

2. Improvement of the EPIR performance

The reliability of the reversible resistance is a very important issue, which has been a concern in the information field. If the stability of the reversible resistance with pulse voltage excitation can not be obtained, the resistance degeneration will seriously hinder the material and structural system from its application in information storage. This is a practical problem and should be solved. As known, the 'retention' problem of the reversible resistance switching was demonstrated, that is, for the high resistance state activated by electric-pulse, the resistance decreases slowly upon subsequent electric pulsing. Besides, the resistance change ratio decreases as the number of the high-low resistance switching circle increases, which is known as 'fatigue' behavior, as shown in Fig.5 and Fig.6.

Experiment shows that the fatigue was observed for the as-prepared samples. The high resistance drops after several cycles of pulse excitation as shows in Fig. 5. This fatigue can be improved by post annealing of the film. It is clearly demonstrated that the amplitude of the decrease in resistance for the as-prepared samples decreased more rapidly than the annealed ones, as inset in Fig. 5. After the post-annealing of the film in air, the sandwich resistance remained constant for a long period, i.e., the resistance of the annealed samples has a longer stable period than the as-prepared samples. It has previously been reported that post-annealing treatment can improve the physical properties of manganites and other kinds of materials [11–13].

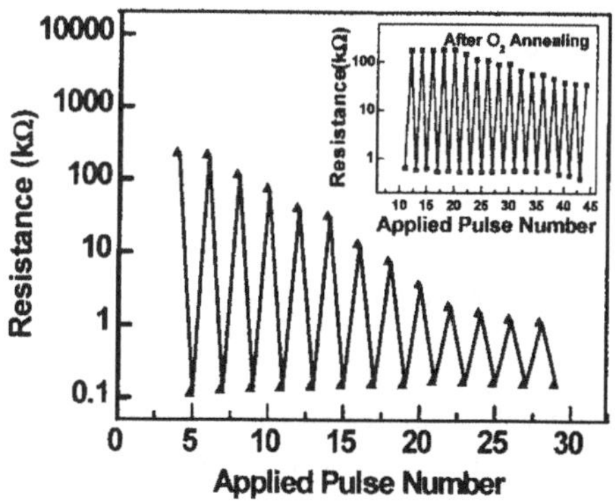

Fig. 5 Fatigue behavior and the improvement by post-annealing

It is found in our experiment that the EPIR devices can also be optimized by initializing pretreatment using voltage sweep or pulse sweep circles before operation. This is clearly shown in the inset in Fig.6, in which pretreatment results in the same improved reversible performance. This is very beneficial when local and in-situ modification is needed to change its respective state or other related properties.

As pointed out in the introduction, there exist divergent viewpoint about whether the resistance change effect is caused by the intrinsic property of CMR materials or occurs at the Ag–PCMO or Ag–LCMO interface [1-3]. In the present research, it is believed that both of these may have effects on the resistance change. Especially the junction effect may be a major factor responsible to the observed phenomenon, as from the evidence of shown in the above I-V curve. In our experiment, it is found that the low resistance states were easier to be achieved than the high resistance states in the electric pulse driven HRS-LRS switching. According to the viewpoint of Rozenberg, there existed three types of domains between the top and bottom electrodes [14]. The HRS and LRS can be assumed to be related to the two states with all domains filled or empty, respectively. Based above, we can demonstrate that the all domains-empty states were easier to realize compared to the all domains-filled states due to the HRS requires a energy to achieve it. Furthermore, the retention and fatigue behavior of the reported EPIR phenomena maybe originated from the traditional "domain walls pinning" or "hetero-charging" effects.

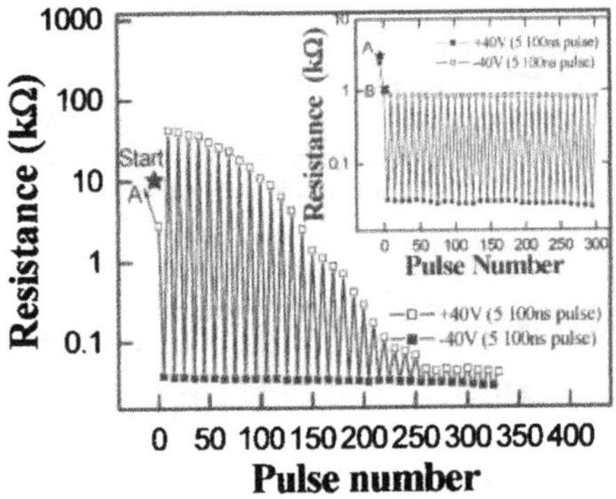

Fig. 6 Fatigue behavior of HRS and the improvement

SUMMARY

Electric-pulse-induced reversible resistance switching effect has been observed in Ag/ $Pr_{1-x}Ca_xMnO_3$/Pt sandwiches. It is found that in such a structural system there exhibits memory effect and retention properties. The EPIR performance can be improved by post-annealing of the materials or by the local adjustment of the pulse applied mode. In addition, initialization of the perovskite film using voltage sweep or pulse sweep circles before application could optimize the EPIR devices.

ACKNOWLEDGEMENTS

This work was supported by the Natural "Science Foundation of China" under Grant No. 50202016 and 50325208.

REFERENCES

[1] S Q Liu, N J Wu, and Ignatiev A, "Electric-pulse-induced reversible resistance change effect in magnetoresistive films", *Appl Phys Lett.* , **76**, 2749-51(2000).

[2] S Q Liu, N J Wu, and Ignatiev A, US. Patent, 6204139B1, (2001)

[3] A Baikalov, Y Q Wang, B Shen, B Lorenz, S Tsui, Y Y Sun, Y Y Xue, and C W Chu, "Field-driven hysteretic and reversible resistive switch at the Ag–$Pr_{0.7}Ca_{0.3}MnO_3$ interface", *Appl Phys Lett.*, **83**, 957-59(2003).

[4] A Sawa, T Fujii, M Kawasaki, and Y Tokura, "Hysteretic current–voltage characteristics and resistance switching at a rectifying Ti/$Pr_{0.7}Ca_{0.3}MnO_3$ interface", *Appl Phys Lett.*, **85**: 4073-75(2004).

[5] L P Ma, Q F Xu , and Y Yang, "Organic nonvolatile memory by controlling the dynamic copper-ion concentration within organic layer", *Appl Phys Lett.*, **84**, 4908-10(2004).

[6] S Tsui, A Baikalov, J Cmaidalka, Y Y Sun, Y QWang and Y Y Xue , C W Chu, L Chen and A J Jacobson, "Field-induced resistive switching in metal-oxide interfaces", *Appl Phys Lett.*, **85**, 317-19(2004).

[7] J Sakai, and S Imai, "Room-temperature resistance switching and temperature hysteresis of $Pr_{0.7}Ca_{0.3}MnO_3$ junctions", *J Appl Phys.*, **97**, 10H709 (2005).

[8] J F Scott, et al, "Ferroelectric memories", *Science*, **246**, 1400-05(1989).

[9] D P Vijay, S B Desu, "Electrodes for PbZr/sub x/Ti/sub 1-x/O/sub 3/ ferroelectric thin films", *J Electrochem Soc.*,**140**, 2640-45(1993).

[10] R.Dong, Q. Wang, L.D.Chen, D.S.Shang, T.L.Chen, X.M.Li, and W.Q.Zhang, "Retention behavior of the electric-pulse-induced reversible resistance change effect in Ag-$La_{0.7}Ca_{0.3}MnO_3$-Pt sandwiches", *Appl. Phys. Lett.* **86**, 172107 (2005).

[11] J Q Guo, H Takeda, and N S Kazama, "Deposition conditions of magnetoresistance in $La_{0.67}Ca_{0.33}MnO_{3-\delta}$ thin film", *J Appl Phys.*, **81,** 7445-49(1997).

[12] H L Lu, J Gopalakrishnan, J L Peng, Q Li, G C Xiong, T Venkatesan, and R L Greene, "Dependence of giant magnetoresistance on oxygen stoichiometry and magnetization in polycrystalline $La_{0.67}Ba_{0.33}MnO_z$", *Phys Rev B.* **51**, 6143-46(1995).

[13] W Chang, J S Horwitz, A C Carter, J N Pond, S W Kirchoefer, C M Gilmore, and D B Chrisey, "The effect of annealing on the microwave properties of $Ba_{0.5}Sr_{0.5}TiO_3$ thin films", *Appl Phys Lett.*, **74,** 1033-35(1999).

[14] M J Rozenberg, I H Inoue, and M J Sánchez, "Nonvolatile Memory with Multilevel Switching: A Basic Model", *Phys Rev Lett.*, **92**, 178302-1(2004).

Electroceramic Materials and Applications
Edited by R. W. Schwartz

Author Index

Printed and bound by CPI Group (UK) Ltd, Croydon, CR0 4YY

16/09/2026

14963458-0002